Diagnostics in Plant Breeding

Thomas Lübberstedt • Rajeev K. Varshney
Editors

Diagnostics in Plant Breeding

Editors
Thomas Lübberstedt
Department of Agronomy
Iowa State University
Ames, Iowa
USA

Rajeev K. Varshney
ICRISAT
Centre of Excellence in Genomics (CEG)
Patancheru, Andra Pradesh
India

ISBN 978-94-007-5686-1   ISBN 978-94-007-5687-8 (eBook)
DOI 10.1007/978-94-007-5687-8
Springer Dordrecht Heidelberg New York London

Library of Congress Control Number: 2013933023

© Springer Science+Business Media Dordrecht 2013
This work is subject to copyright. All rights are reserved by the Publisher, whether the whole or part of the material is concerned, specifically the rights of translation, reprinting, reuse of illustrations, recitation, broadcasting, reproduction on microfilms or in any other physical way, and transmission or information storage and retrieval, electronic adaptation, computer software, or by similar or dissimilar methodology now known or hereafter developed. Exempted from this legal reservation are brief excerpts in connection with reviews or scholarly analysis or material supplied specifically for the purpose of being entered and executed on a computer system, for exclusive use by the purchaser of the work. Duplication of this publication or parts thereof is permitted only under the provisions of the Copyright Law of the Publisher's location, in its current version, and permission for use must always be obtained from Springer. Permissions for use may be obtained through RightsLink at the Copyright Clearance Center. Violations are liable to prosecution under the respective Copyright Law.
The use of general descriptive names, registered names, trademarks, service marks, etc. in this publication does not imply, even in the absence of a specific statement, that such names are exempt from the relevant protective laws and regulations and therefore free for general use.
While the advice and information in this book are believed to be true and accurate at the date of publication, neither the authors nor the editors nor the publisher can accept any legal responsibility for any errors or omissions that may be made. The publisher makes no warranty, express or implied, with respect to the material contained herein.

Printed on acid-free paper

Springer is part of Springer Science+Business Media (www.springer.com)

Preface

Recent advances in genomics including sequencing and genotyping technologies are revolutionizing plant breeding. DNA markers were a new concept about 30 years ago, but sequencing of the first complete plant genome was accomplished about 10 years ago. Meanwhile "third-generation" sequencing technologies have become available and started to enable sequencing of plant genomes within weeks rather than years. Low-cost and high-throughput genotyping technologies are enabling production of more than one million marker datapoints per day in leading breeding companies. Novel concepts such as the use of non-DNA-based biomarkers for prediction of complex characters as well as genomic selection for improving complex traits have emerged recently in both animal and plant breeding.

This book presents a collection of 19 cutting-edge research reviews on the development and application of molecular tools for the prediction of plant performance by the authorities in their respective fields. Given its significance for mankind and the available research resources, molecular diagnostics is the forefront area in the medical sciences. In plant sciences, DNA-based determination of yield potential and use of non-DNA biomarkers has just started. However, recent advances in genomics are expected to shift the focus of molecular breeding from "explanatory" to "predictive" in crop science. These issues have been presented by 46 authors from 13 countries in 19 chapters organized in 9 parts.

The first part, Introduction, includes two chapters that introduce DNA and non-DNA-based biomarkers. DNA-based markers enable reliable prediction of the yield or quality "potential" of plant genotypes. However, the ultimate yield or product quality substantially depends on environmental factors and genotype-by-environment interactions. Thus, compared to medical sciences, non-DNA biomarkers might for some applications enable a more accurate prediction of yield or quality traits in plants. Part II with four chapters deals with approaches towards identification of causative quantitative trait polymorphisms (QTPs). Forward genetics seeks to identify the genetic basis of a given phenotype. The identification of causative genetic variation is being facilitated by rapid progress in sequencing and genotyping technologies. Thus, marker density is no longer a bottleneck in the identification of genome regions affecting a given phenotype. Reverse genetics on the other hand seeks to identify which phenotype(s) result from a given sequence. This approach has been facilitated by the development of plant materials such as T-DNA mutant collections and induced TILLING populations and methods such as RNAi for silencing candidate genes. Recent advances in sequencing technologies are enabling targeted resequencing of candidate regions or even whole genomes. One challenge, however, will be efficient data handling for extracting biological meaning from the increasing amount of sequence information. Any type of genetic mapping is limited by the number of recombinations since the last common ancestor of the mapping population. Traditionally, genetic linkage mapping has been conducted within families derived from two parents after one or two generations. In contrast, association mapping, or linkage disequilibrium mapping (LD mapping), is conducted within populations in which relatedness is not established. Thus, the genetic resolutions, as defined by the number of recombinations since the last common ancestor, are magnitudes higher in 'population-mapping' as compared to 'family-mapping'. Recent efforts are seeking to combine population- and family-mapping methodology, thus creating unprecedented genetic resolution.

Two chapters included in Part III present validation of putative quantitative trait polymorphisms (QTPs). TILLING technology for allele mining can be used for this purpose. Recent years have seen a tremendous progress in this technology and platform (e.g. use of the next-generation sequencing technology). Conventional plant transformation suffers from the random manner in which transgenes are being incorporated in the target genome. This can both affect the expression of transgenes and disrupt genes in the target genome. In addition, allele substitutions are not possible by these conventional methods. Recently, however, reliable homologous recombination by zinc finger nucleases has been reported in maize and *Arabidopsis*. This technology allows targeted editing of plant genomes and will most likely be available for most major crop plants in the near future. Part IV dealing with conversion of QTPs into functional markers has three chapters that address technologies for detection of DNA markers. Single nucleotide polymorphism (SNP) markers have emerged in recent years as the markers of choice for plant genetics and breeding applications. One of the main reasons for this is the availability of high-throughput genotyping and sequencing technologies at decreasing costs.

Part V with three chapters presents methods for identification of methylation-based polymorphisms, RNA-based biomarkers, and metabolite-based biomarkers as well as their monitoring on a genome-wide scale, and examples of their application in plant breeding and plant genetics will be presented. It is anticipated that the coming years will see the generation of huge information on diagnostic markers in several plant species. For their successful utilization in plant breeding, it is essential to have this information in the databases so that the breeders can have the access to information in the user-friendly way. Although for some major crop species, some databases like GrainGenes (for small grain cereals), Gramene (for rice), Maize GDB (for maize), SoyBase (for soybean) exist, these databases need to be ready for storing the information on diagnostic markers. And there is also a challenge of storing such information for other plant species. Therefore, one chapter in Part VI summarizes databases and the above-mentioned issues. Similarly, the sole chapter in Part VII presents important challenges that restrict the use of molecular markers identified through linkage or association mapping studies for quantitative traits in plant breeding applications.

Finally, Parts VIII and IX include a set of chapters dealing with applications of DNA or non-DNA-based markers in plant breeding and crop- and trait-specific examples. One of the foremost applications of molecular markers has been the selection of the suitable parental lines to establish breeding populations or for exploiting heterosis. Registration of a new variety requires confirmation that the new variety is distinct from other varieties and is uniform and stable after repeated propagation. DNA-based functional markers might help to define and monitor criteria, when novel-derived varieties can be considered independent versus 'essentially derived'. A comprehensive overview has been provided on quantitative or qualitative trait polymorphisms for agronomic traits in maize and rice. Finally, an overview has been provided on quantitative or qualitative trait polymorphisms for selected traits across species. Information on the underlying genes and sequence motifs, genetic effects as well as derived markers assays are presented and thus made readily available to the plant scientific community.

In summary, this book is expected to provide an update as well as future plan on diagnostics in plant breeding that is cutting-edge and rapidly evolving. The editors are grateful to all the authors, who not only provided a timely review of the published research work in their area of expertise but also shared their unpublished results to offer an updated view. We also appreciate their cooperation in meeting the deadlines, revising the manuscripts and in checking the galley proofs. While editing this book, we received strong support from many reviewers, who willingly reviewed the manuscripts for their passion for the science of genomics and provided useful suggestions for improving the manuscripts. We would like to thank our colleagues, especially Bill Beavis at Iowa State University and Manish Roorkiwal, Reyazul Rouf Mir and Manjula Baddam at ICRISAT, for their help in various ways. Nevertheless, we take responsibility for any errors that might have crept in inadvertently during the editorial work.

We are thankful to Jacco Flipsen and Ineke Ravseloot of Springer during various stages of the development and completion of this project. The editors also recognize that the editorial work for this book took away precious time that they should have spent with their respective families. Rajeev K. Varshney is thankful to his wife Monika and children Prakhar and Preksha who allowed their time to be taken away to fulfil his editorial responsibilities in addition to research and other administrative duties at ICRISAT. Similarly, Thomas Lübberstedt promises to spend more time with his wife Uschi and children Paul and Arthur, rather than sitting absent-minded in front of his laptop.

We very much hope that our efforts will help those working in molecular breeding to better focus their research plans on crop improvement programmes. The book, in our opinion, should also help graduate students and teachers to develop a better understanding of this fundamental aspect of modern plant breeding.

Thomas Lübberstedt
Rajeev K. Varshney

Contents

Part I Introduction

1. **Diagnostics in Plant Breeding** 3
   Thomas Lübberstedt

2. **Non-DNA Biomarkers** 11
   K. Christin Falke and Gregory S. Mahone

Part II Identification of Quantitative Trait Polymorphisms (QTPs)

3. **Gene Identification: Forward Genetics** 41
   Qing Ji

4. **Gene Identification: Reverse Genetics** 61
   Erin Gilchrist and George Haughn

5. **Allele Re-sequencing Technologies** 91
   Stephen Byrne, Jacqueline D. Farrell, and Torben Asp

6. **Dissection of Agronomic Traits in Crops by Association Mapping** ... 119
   Yongsheng Chen

Part III Validation of QTPs

7. **TILLING and EcoTILLING** 145
   Gunter Backes

8. **Gene Replacement** 167
   Sylvia de Pater and Paul J.J. Hooykaas

Part IV Conversion of QTPs into Functional Markers

9 SNP Genotyping Technologies 187
Bruno Studer and Roland Kölliker

10 Insertion-Deletion Marker Targeting for Intron Polymorphisms 211
Ken-ichi Tamura, Jun-ichi Yonemaru, and Toshihiko Yamada

11 Evolving Molecular Marker Technologies in Plants: From RFLPs to GBS 229
Reyazul Rouf Mir, Pavana J. Hiremath, Oscar Riera-Lizarazu, and Rajeev K. Varshney

Part V Development of Non-DNA Biomarkers

12 Methylation-Based Markers 251
Emidio Albertini and Gianpiero Marconi

13 Transcriptome-Based Prediction of Heterosis and Hybrid Performance 265
Stefan Scholten and Alexander Thiemann

14 Metabolite-Based Biomarkers for Plant Genetics and Breeding 281
Olga A. Zabotina

Part VI Deposition of Diagnostic Marker Information

15 Plant Databases and Data Analysis Tools 313
Mary L. Schaeffer, Jack M. Gardiner, and Carolyn J. Lawrence

Part VII Statistical Considerations

16 Prospects and Limitations for Development and Application of Functional Markers in Plants 329
Everton A. Brenner, William D. Beavis, Jeppe R. Andersen, and Thomas Lübberstedt

Part VIII Applications in Plant Breeding

17 Parent Selection – Usefulness and Prediction of Hybrid Performance 349
Adel H. Abdel-Ghani and T. Lübberstedt

18 Variety Protection and Plant Breeders' Rights in the 'DNA Era' 369
Huw Jones, Carol Norris, James Cockram, and David Lee

Contents

Part IX Examples

19 **Qualitative and Quantitative Trait Polymorphisms in Maize** 405
   Qin Yang and Mingliang Xu

20 **Molecular Diagnostics in Rice (*Oryza sativa*)** 443
   Wenhao Yan, Zhongmin Han, and Yongzhong Xing

21 **Functional Marker Development Across Species in Selected Traits** 467
   Hélia Guerra Cardoso and Birgit Arnholdt-Schmitt

Index 517

About the Book

Diagnostics in Plant Breeding is systematically organizing cutting-edge research reviews on the development and application of molecular tools for the prediction of plant performance. Given its significance for mankind and the available research resources, medical sciences are leading the area of molecular diagnostics, where DNA-based risk assessments for various diseases and biomarkers to determine their onset become increasingly available. However, progress in plant genomics and in particular sequence technology in the past decade will shift the focus from "explanatory" to "predictive" in plant sciences. So far, most research in plant genomics has been directed towards understanding the molecular basis of biological processes or phenotypic traits. From a plant breeding perspective, however, the main interest is in predicting optimal genotypes based on molecular information for more time- and cost-efficient breeding schemes. In this book, we assemble chapters on all areas relevant to development and application of predictive molecular tools in plant breeding by leading authors in the respective areas.

About the Editors

Thomas Lübberstedt is Professor and K.J. Frey Chair in Agronomy at Iowa State University, USA. He is Director of the R.F. Baker Center for Plant Breeding and Founder of the Doubled Haploid Facility at Iowa State University, USA. His research revolves around application and development of tools and methods provided by genome analysis (especially DNA markers) to understand composition of complex traits and phenomena, to determine and exploit genetic diversity in elite and exotic germplasm and apply this knowledge to plant breeding. His crop focus is on maize and ryegrass; his traits focus on forage quality, vernalization response, resistance and heterosis. Most current activities are aiming at the development of gene-derived, "functional" markers and their implementation in plant breeding as well as on breeding bioenergy crops. He teaches Molecular Plant Breeding and is involved in developing a Distance MSc program in Plant Breeding.

Rajeev K. Varshney, with a basic background in molecular genetics, has over 10 years' research experience in international agriculture. Before joining the International Crops Research Institute for the Semi-Arid Tropics (ICRISAT) in 2005 and CGIAR Generation Challenge Program (GCP) in 2007, he worked at Leibniz Institute of Plant Genetics and Crop Plant Research (IPK), Germany, for 5 years. He is also at the Adjunct Professor position at The University of Western Australia. He has made significant contribution in the area of structural, comparative and applied genomics of cereal as well as legume species. Based on his contribution to crop genomics applied to breeding, he has won several awards including Elected Fellow of National Academy of Agricultural Sciences, India, and Young or Outstanding Scientist awards from other Indian science academies like Indian National Science Academy and National Academy of Sciences of India. He is serving several international journals as Associate or Subject Editor, and in the expert/review committee panels of several funding agencies in India, Germany, France, USA, EU and African/Arab countries. At present, he is an elected Chair for International Conference on Legume Genetics and Genomics-VI (ICLGG-VI, 2012).

Part I
Introduction

Chapter 1
Diagnostics in Plant Breeding

Thomas Lübberstedt

Introduction

Although the terms diagnosis and diagnostics are commonly associated with medicine (e.g., Zhou et al. 2008; http://www.openclinical.org/dss.html; http://www.roche.com/about_roche/business_fields/about-diagnostics.htm), both terms are broadly used. Important other areas include car diagnostics (http://www.automd.com/diagnose/), or more generally mechanics; weather and more generally climate (http://cires.colorado.edu/science/centers/cdc/); and computer and more specifically software applications (http://en.wikipedia.org/wiki/Diagnostic_program). In the broadest sense, diagnostics is about application of quantitative methods for interpretation of data (Fig. 1.1).

The terms diagnosis or diagnostics trace back to Greek and indo-European roots (http://www.collinsdictionary.com/dictionary/english). Dia means "apart", gno means to know or discern things. Diagnosis refers to discern or to distinguish something. In the medical area, the term diagnosis is used, to describe the process to identify and determine the nature and cause of symptoms through evaluation of pre-existing data (such as patient history), examination of patients by using conventional or laboratory methods to generate and ultimately interpret those different sources of information. In a biological sense, diagnosis deals with characterizing the distinguishing features of, e.g., an organism in taxonomic context. More generally, the definition of diagnosis refers to the critical analysis of the nature of something (http://www.collinsdictionary.com/dictionary/english). Diagnostics as opposed to diagnosis refers to a comprehensive set of procedures available for diagnoses in a particular area, such as plant breeding.

T. Lübberstedt (✉)
Department of Agronomy, Iowa State University, Ames, IA, USA
e-mail: thomasl@iastate.edu

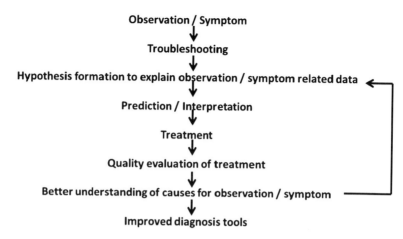

Fig. 1.1 Generic flow chart for any kind of diagnostics

In plant breeding, the major tasks are (1) generation of genetic variation as a source for (2) developing components of varieties, and (3) testing of experimental varieties (Becker 2010). All three of these key tasks can be performed intuitively based on experiences of plant breeders, but they increasingly benefit from diagnostic procedures. Central questions in plant breeding revolve around (i) identification of the best founder genotypes at the outset of breeding programs to generate genetic diversity, which relates to the usefulness concept in plant breeding, (ii) identification of the best variety components (such as inbred lines) or varieties, and (iii) evaluation of the performance of combinations of variety components such as experimental hybrids. Traditionally field trials (similar to clinical trials) are used, to address all three of those questions. Increasingly, DNA-based markers are used in marker-aided procedures to support or substitute field trial based evaluation. To a more limited extent compared to medicine, non-DNA based "biomarkers" are employed in plant breeding. However, in all cases, the purpose of using respective test procedures is to reliably predict optimal genotypes or genotype combinations. With technological progress in the area of genomics, the question becomes, whether novel procedures provide such predictions more reliably, in shorter time, and/or at lower costs compared to traditional procedures.

Classification of Diagnostic Technologies

There are different classifications of diagnostic tools (Table 1.1). Diagnostics can be based on phenotypic characters, or on molecular features. Phenotypic characterization can be based on destructive (after harvesting plant materials and any kind of treatment) or non-destructive methods (such as spectral characterization or seed color markers). Non-destructive methods have the advantage of not interfering

Table 1.1 Classification of diagnostic methods

Classification of diagnostic methods	Distinguishing features
Destructive versus non-destructive	Samples get destroyed with destructive methods, thus, non-destructive methods are preferable. A recent example is seed chipping, allowing characterization of seed fractions, without interfering with seed germination
Phenotypic versus molecular	Phenotypes can be strongly affected by non-inherited environmental factors. DNA-based methods exhibit much greater heritabilities, i.e., they are not as strongly influenced by environmental factors.
DNA- versus non-DNA biomarkers	DNA-markers report the potential or risk for target trait expression, whereas non-DNA biomarkers have the capability of reporting the onset or expression of a target trait (such as medical biomarkers for disease onset)
Functional versus random DNA-markers	Functional markers are derived from polymorphisms causally affecting target trait expression; in contrast, most random DNA-markers are effective by linkage with respective causal polymorphisms
Technical classification biomarkers	Depending on the molecular class: DNA, RNA, Proteins, metabolites
Technical classification DNA-markers	Can be depending on the underlying DNA polymorphism (SNP, INDEL, SSR) or detection technology

with normal growth and development of the organism. For example, seed can be classified and sorted into desirable and undesirable with regard to, e.g., oil content, before sowing. However, for several traits, such non-destructive methods are not available. An example might be inducible resistance in the absence of the pathogen.

A major reason for using molecular techniques is the ability to monitor or predict a trait of interest, before it becomes phenotypically visible. The best examples probably are related to human diseases. Based on molecular markers it is possible to predict the risk of individuals to suffer from a particular disease (based on DNA markers), but also to determine the onset of a disease such as cancer (based on non-DNA expression markers). Prediction of the onset of a disease might be crucial to determine the timing and mode of therapies. In plant breeding, seed chipping has been developed to allow selection prior to sowing of selected kernels based on DNA markers, which effectively reduces costs for cultivation and evaluation of undesirable genotypes.

For molecular markers it is practical to distinguish DNA-based and non-DNA based markers. Because DNA is present in each cell and not affected by environment, DNA-based information is consistent across plant organs, developmental stages, and environments or treatments. This can be an advantage in terms of robustness of information. However, the limitation of DNA-based markers is, that they do not provide information on changes in plant developmental or responses to environmental factors. Thus, DNA markers enable us to assess the potential of a particular genotype to develop a particular phenotype. However, they provide no information on actual metabolic processes that can be monitored by non-DNA molecular markers. Within both DNA and non-DNA markers, there are various technological and economic criteria for discrimination, which will be addressed in the following chapters of this book.

A final mode of discrimination of diagnostic procedures is based on the question, whether they report on causative factors resulting in phenotypic changes, or whether their predictive value is based on association. For DNA, so called "perfect", "ideal", or "functional" (Andersen and Lübberstedt 2003) markers have been described (FMs: will be used in the following for simplification). These FMs are derived from polymorphisms within genes, which cause trait variation. Thus, in case of presence of a particular allele at a polymorphic site within a resistance gene (as example), it can be predicted that the respective genotype will be resistant to a particular disease (isolate) (Ingvardsen et al. 2007). Once established, resistance assays on plants are no longer required for this particular disease. In contrast, if a DNA marker is linked to a resistance gene, its informativeness depends on the linkage disequilibrium present in the breeding population. "Blackbox" approaches based on random DNA markers are receiving increasing attention in plant breeding in relation to genomic selection strategies (Heffner et al. 2010). This is to a large extent driven by progress in sequencing and DNA marker technology, which allows genotyping of breeding populations with 1,000s of markers per genotype at low costs. Genomic selection has initially proven to be successful in animal breeding, and has more recently been employed in the plant breeding context (Asoro et al. 2011). With increasing information on genes affecting traits of interest and knowledge on causative polymorphisms, in the longer run combined approaches based on FMs and genomic selection for unexplained genetic variation will be developed.

Shift in Concepts of Using Diagnostic Tools

The initial boost in using diagnostics in plant breeding was stimulated by the advent of DNA marker technologies in the 1980s (Botstein et al. 1980). Compared to the earlier available isozymes, DNA markers such as RFLPs and later PCR based markers systems including SSRs, AFLPS, and RAPDs enabled generation of 100s and for some crops 1,000s of markers, and thus complete genome coverage of linkage mapping populations. However, generating marker data was still laborious and costly (Table 1.2). The main concept in the past decades for using markers in relation to mostly quantitative inherited agronomic traits was to first map quantitative trait loci (QTL), followed by either pyramiding QTL in the offspring of a given QTL mapping population, or by using QTL information with regard to effect and location in other populations based on marker-assisted selection (MAS) strategies (Lande and Thompson 1990). QTL mapping methodology has been refined over time from ANOVA over simple interval mapping to composite interval mapping to identify the position and effect of QTL more accurately (Tuberosa and Salvi 2006). QTL mapping is an effective approach for identifying genomic regions that segregate for traits with mono- or oligogenic inheritance. In this case,

Table 1.2 Optimal strategies for application of DNA markers for trait improvement, depending on the availability of DNA markers, and more specifically, functional markers

DNA markers		Application for trait improvement	
Availability	FMs	Mono- or oligogenic	Polygenic
Limited, expensive	None	MABC, pyramiding	QTL mapping, MAS
Unlimited	Limited	MABC, pyramiding	Genomic selection
Unlimited	Multiple	Genome engineering?	

use of respective diagnostic marker-based assays in backcross or gene pyramiding breeding schemes has been successful (Eathington et al. 2007). However, for quantitative traits successful strategies of QTL mapping followed by subsequent MAS have not been reported. Reasons include (i) that published QTL information are often not readily transferable to elite germplasm, if obtained in unrelated experimental populations; (ii) QTL effects are usually inflated due to limited sample sizes; and (iii) the majority of QTL are neglected because of the application of significance thresholds in the QTL mapping approach (Jannink 2010).

More recently, a paradigm shift occurred with the advent of large numbers of low cost markers by introduction of genomic selection procedures. Genomic selection was initially established in animal breeding (Meuwissen et al. 2001), but appears to also be a promising approach for marker-aided plant breeding (Bernardo and Yu 2007; Asoro et al. 2011). Instead of identifying QTL first before establishing subsequent MAS, the objective of genomic selection is to determine breeding values without need to identify QTL. The first step is to establish a training set of genotypes with both marker and phenotypic data, in order to predict breeding values based on estimated marker effects. This is followed by applying those markers to select among new breeding materials, which are not (necessarily) phenotyped, based on genomic estimated breeding values (GEBVs) derived from information obtained in the training set. Similar approaches based on estimation and prediction have been suggested to identify the most promising parental combinations to establish breeding populations (Zhong et al. 2009), and to use non-DNA markers to predict heterosis and hybrid performance (Frisch et al. 2010).

Although genomic selection appears to be successful in animal breeding, the number of respective experimental studies in plants is still limited. Important questions that will need to be addressed relate to (i) the marker density required for successful GS in plants, which will depend on the degree of linkage disequilibrium in the breeding populations under consideration, the haplotype diversity, and the genetic architecture of the traits of interest, (ii) the size of the training population and the impact of the degree genetic relationship to the breeding population on the success of GS, (iii) applicability across different testers, in polyploids, etc. Moreover, environment and thus genotype – by – environment interactions as well as phenotypic stability across environments play a much greater role in domesticated plants and will impact how well information gathered in training populations can be transferred to breeding populations.

Perspectives

Whereas genomic selection will likely become a major research area in plant breeding in the coming years, its objective is neither gene nor quantitative trait polymorphism (QTP) identification (Gianola et al. 2009; de los Campos et al. 2010). Nevertheless, progress in genetic studies of agronomic traits, driven by progress in sequencing technology, and based on genome-wide association studies, map-based gene isolation, among others, can be expected to lead to a dramatic increase in the number of genes and QTP identified with impact on agronomic traits in the next decades. The question then becomes in the longer run, whether more targeted approaches to select for optimal haplotypes and genotypes comparable to the "Breeding by Design" concept (Peleman and Rouppe van der Voort 2003) will be more effective than genomic selection, which might lead to fixation of unfavorable haplotypes (Hill and Robertson 1968). Another question is, how superior alleles, often rare, from novel germplasm such as from other geographic regions can be most effectively identified using diagnostics.

In medical sciences, non-DNA biomarkers play a much greater role than in plants. Whereas the risk as determined by DNA markers (equivalent to the term "potential" in plants) in medical sciences might be of some value for individuals, employers, insurances, it is more critical to know, whether a particular condition occurred, which requires a treatment. This is also true, because a genetic treatment by gene therapy is in most cases not available. Understanding the molecular mechanism(s) underlying a particular disease can be instrumental for developing a respective treatment. This concept might in the longer run also be of interest for crop sciences. If compounds would become available that help to counteract particular forms of stress, application of such compounds by spraying or seed coating might substitute or complement respective breeding efforts for improving agronomic performance.

Acknowledgements R.F. Baker Center for Plant Breeding, Iowa State University.

References

Andersen JR, Lübberstedt T (2003) Functional markers in plants. Trends Plant Sci 8:554–560
Asoro FG, Newell MA, Beavis WD, Scott MP, Jannink JL (2011) Accuracy and training population design for genomic selection on quantitative traits in elite North American oats. Plant Genome 4:132–144
Becker H (2010) Pflanzenzuechtung. Ulmer Verlag, Stuttgart, Germany
Bernardo R, Yu J (2007) Prospects for genome-wide selection for quantitative traits in maize. Crop Sci 47:1082–1090
Botstein D, White R, Skolnick M, Davis R (1980) Construction of a genetic linkage map in man using restriction fragment length polymorphisms. Am J Hum Genet 32:314–331
de los Campos G, Gianola D, Rosa GJM, Weigel KA, Crossa J (2010) Semi-parametric genomic-enabled prediction of genetic values using reproducing kernel Hilbert spaces methods. Genet Res 92:295–308

Eathington S et al (2007) Molecular markers in a commercial breeding program. Crop Sci 47 (Suppl 3): S154–S163

Frisch M, Thiemann A, Fu J, Schrag TA, Scholten S, Melchinger AE (2010) Transcriptome-based distance measures for grouping of germplasm and prediction of hybrid performance in maize. Theor Appl Genet 120:441–450

Gianola D, de los Campos G, Hill WG, Manfredi E, Fernando R (2009) Additive genetic variability and the Bayesian alphabet. Genetics 183:347–363

Heffner EL, Lorenz AJ, Jannink J, Sorrells ME (2010) Plant breeding with genomic selection: potential gain per unit time and cost. Crop Sci 50:1681–1690

Hill WG, Robertson A (1968) Linkage disequilibrium in finite populations. Theor Appl Genet 38:226–231

Ingvardsen C, Schejbel B, Lübberstedt T (2007) Functional markers for disease resistance in plants. Prog Bot 69:61–87

Jannink J-L (2010) Dynamics of long-term genomic selection. Genet Sel Evol 42:35

Lande R, Thompson R (1990) Efficiency of marker-assisted selection in the improvement of quantitative traits. Genetics 124:743–756

Meuwissen THE, Hayes BJ, Goddard ME (2001) Prediction of total genetic value using genome-wide dense marker maps. Genetics 157:1819–1829

Peleman JD, Rouppe van der Voort J (2003) Breeding by design. Trends Plant Sci 8:330–334

Tuberosa R, Salvi S (2006) Genomics approaches to improve drought tolerance in crops. Trends Plant Sci 11:405–412

Zhong S, Dekkers JCM, Fernando RL, Jannink J-L (2009) Factors affecting accuracy from genomic selection in populations derived from multiple inbred lines: a barley case study. Genetics 182:355–364

Zhou X-H, Obuchowski NA, McClish DK (2008) Statistical methods in diagnostic medicine. Wiley, New York

Chapter 2
Non-DNA Biomarkers

K. Christin Falke and Gregory S. Mahone

The central dogma of molecular biology (Crick 1970) describes the information flow from the level of DNA to phenotypic trait expression. DNA is transcribed to RNA, the RNA sequence is translated to amino acid sequences, the amino acid chains fold to proteins that act as enzymes and regulate physiological pathways, which are responsible for the concentration of metabolites (Fig. 2.1). The complex interactions between these levels are responsible for the phenotypic trait expression, in which plant breeders are interested. Classical genetics infers the presence of genes solely from the phenotypes of related individuals, while in genomics studies, the DNA itself is analyzed with molecular genetic methods. Recent advances in lab technology made it possible to assess with high throughput methods not only the DNA but also RNA, proteins, and metabolites. Analogous to the shift from genetics to genomics, these analysis possibilities led to the development of the so called "omics" technologies with the ambition to survey all mRNA, proteins, and metabolites – as opposed to a single molecule at a time. Transcriptomics analyzes the gene expression by evaluating mRNA, proteomics assesses proteins and metabolomics metabolites.

Since the beginnings of agriculture, crop improvement has been carried out by selecting plants with favorable phenotypes of desirable traits such as higher yield, better quality, higher resistance to disease, and better tolerance to abiotic stress. Selection of superior phenotypes led to superior genotypes and this phenotypic selection has been contributing substantially to the productivity increase in agronomy. A marker can be defined as an assay that is associated with a

---

K.C. Falke (✉)
Institute for Evolution and Biodiversity, Westfälische Wilhelms-University Münster, Münster, Germany
e-mail: k.christin.falke@uni-muenster.de

G.S. Mahone
Institute of Agronomy and Plant Breeding II, Justus-Liebig-University Giessen, Giessen, Germany

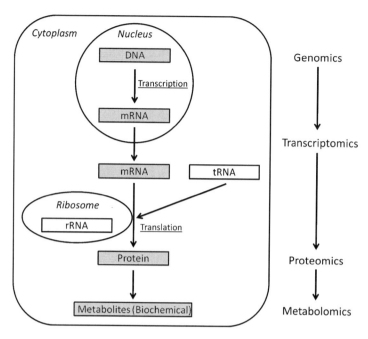

Fig. 2.1 General scheme of the transcription of DNA to RNA and translation to protein which forms the dogma of molecular biology. This schematic is simplistic and does not include all known aspects of the overall process as it is currently understood

measurable trait, the expression of which is highly correlated with the expression of a second trait, for which selection is carried out. The potential advantages of markers for selection have been known for more than 80 years by using morphological markers (Sax 1923) and the introduction of isoenzyme markers (Hunter and Markert 1957). These advantages were out of reach, however, until advances in molecular biology allowed researchers to apply DNA marker technologies. With the availability of endonucleases, restriction fragment length polymorphisms (RFLPs) were first introduced for tomato and maize in the mid 1980s (Botstein et al. 1980; Tanksley et al. 1989). A considerable advance for molecular marker techniques came through the application of the polymerase chain reaction method (Mullis and Faloona 1987). Since then, expectations increased that selection could shift gradually from phenotypes to genotypes (Walsh 2001). Currently, a number of DNA marker types exist which differ in technical requirements; the time, cost, and labor needed; density throughout the genome; and degree of polymorphism revealed. Expressed sequence tags (ESTs), genome sequencing experiments and in particular functionally characterized genes have enabled the development of DNA markers from transcribed regions of the genome. Simple sequence repeats (SSRs) and single-nucleotide polymorphisms (SNPs) are examples of DNA markers that can be derived from ESTs (Rafalski 2002; Varshney et al. 2005, 2006a).

DNA markers can decrease the effort and time requirements in breeding programs and are a valuable resource for characterizing breeding material. They have been used to (i) determine the genetic diversity in germplasm and its changes over time (Smith and Smith 1987, 1988; Tanksley and McCouch 1997; Dubreuil and Chacrosset 1998; Labate et al. 2003; Khlestkina et al. 2004), (ii) detect genetic relationships among germplasm in seed banks and applied breeding programs (Smith et al. 1990; Melchinger and Gumber 1998; Menkir et al. 2004; Geiger and Miedaner 2009), (iii) predict heterosis and hybrid performance (Melchinger 1999; Vuylsteke et al. 2000; Schrag et al. 2006, 2007, 2009a, b), (iv) map quantitative trait loci (QTL; Sax 1923; Thoday 1961; Geldermann 1975; Lander and Botstein 1989; Zeng 1993), (v) perform marker-assisted selection (MAS; Lande and Thompson 1990; Dudley 1993; Moreau et al. 2004; Collard and Mackill 2008), and (vi) perform genomic selection (Meuwissen et al. 2001; Bernardo and Yu 2007; Heffner et al. 2009; Piepho 2009; Jannink et al. 2010; Albrecht et al. 2011). The advent of DNA marker technologies has had a substantial impact on the methodology of plant breeding.

Classical plant breeding is based on population genetic and quantitative genetic concepts. Integrating knowledge gained from DNA markers in the theoretical framework of plant breeding is a very active research area. However, integrating data from transcriptomics, proteomics, and metabolomics into plant breeding is still in very early stages. The current article summarizes investigations on concepts where data from these technologies can serve as "non-DNA" markers, with the target to complement or replace the established DNA marker technology with these "non-DNA" markers.

Analysis of DNA Methylation

DNA methylation is a process in which cytosine in genomic DNA is converted to 5-methylcytosin. Methylation does not change the DNA sequence, but was shown to be associated with gene expression levels (Jones and Takai 2001). As such, DNA methylation is responsible for epigenetic changes and can be used to develop non-DNA markers. Changes in methylation patterns have proven to be heritable, causing changes not only to parent phenotypes but also to the offspring in some cases (Sano et al. 1990). Methylation has also been shown to react to environmental changes, particularly in regards to stress tolerance (Peng and Zhang 2009). As these changes may actually be heritable, methylation has the ability to induce a stress "memory" in offspring, leading to potentially important discoveries for specific stress tolerances breeding programs.

Multiple studies examined DNA methylation rates with regard to comparisons between parents and progeny. Early studies showed that methylation changes occurred in tissue culture (Kaeppler and Phillips 1993a; Phillips et al. 1994), but that the changes could show stability in subsequent selfing offspring in maize (Kaeppler and Phillips 1993b). In a study with potato, Nakamura and Hosaka (2010) showed that inbred lines, which displayed reduced performance due to inbreeding, had greater levels of DNA methylation than heterozygotes, which displayed heterosis. Similar studies have been performed in other crops, such as maize (Zhao et al. 2007) and rice (Xiong et al. 1999). These and other crop-specific studies have shown that methylation rates can be used to investigate heterosis. DNA methylation might therefore qualify as basis for non-DNA markers.

Genome-Wide Expression Profiling in Plant Breeding

In plant breeding experiments, differences in phenotypes among individuals can be observed due to both sequence polymorphisms resulting in changes to or absence of proteins, and/or, if there is no DNA sequence polymorphism, qualitative or quantitative differences in gene expression generating different amounts of protein in a cell (Druka et al. 2010). Comparative genomic analyses based on DNA are immune to environmental changes. In contrast, gene expression at the transcriptional level measures the dynamics of mRNA molecules quantitatively, and variation of gene expression can be determined at different states on a genome scale. Thus, gene expression studies can show if and to what extent genes are actively expressed at a specific location, time and/or developmental stage. A tool to measure the amount of mRNA molecules produced in a cell is genome-wide expression profiling (Druka et al. 2010; Zhang et al. 2010). In these analyses, the level of gene expression is mostly equated with the steady-state abundance of individual mRNA transcripts at a given time point. To determine the abundance information and to identify mRNAs differing in their expression status, a variety of techniques can be employed. These include quantitative reverse transcription polymerase chain reaction (RT-PCR) (Czechowski et al. 2004), DNA microarrays (Schena et al. 1995), massively parallel signature sequencing (MPSS; Brenner et al. 2000), next generation sequencing (NGS; Metzker 2010), serial analysis of gene expression (SAGE; Velculescu et al. 1995) or direct RNA sequencing (DRS; Ozsolak et al. 2009). Today, expression profiling has been applied to a wide range of plants including agronomically important crops like maize (Shi et al. 2006), rice (Ohdan et al. 2005), wheat (Jordan et al. 2007), and barley (Potokina et al. 2007). These studies have shown that the control of gene expression, while complex, can provide fundamental insight into the relation of gene expression and phenotype (Druka et al. 2008a). As the costs for gene expression profiling decrease, the number of studies using RNA has the potential to rapidly deepen our insights into how gene expression contributes to phenotype.

Assessing Germplasm Diversity

Genetic diversity of plants is not only the basis of survival and adaptation but is also a requirement for long-term selection response in breeding programs. It is increasingly necessary to monitor this genetic diversity since many breeding methods result in a reduction of genetic variation in breeding materials (Tanksley and McCouch 1997) as shown for major crops like wheat (Roussel et al. 2004; Fu and Somers 2009), barley (Russell et al. 2000), or maize (Labate et al. 2003). Decline of genetic diversity is often observed when the germplasm of entire breeding pools was derived solely from crosses of elite lines from the preceding breeding cycle (Bernardo 2002), which is common in several major crops (Yu and Bernardo 2004).

The application of DNA markers has provided a comparative basis for estimating genetic diversity in breeding programs. In maize breeding programs, Lu and Bernardo (2001) or Hagdorn et al. (2003) observed a slight reduction in genetic diversity after several cycles of selection by using SSR and RFLP markers, respectively. Similar results were obtained in breeding programs of other crops like oat (De Koeyer et al. 1999) and barley (Condón et al. 2008) with DNA markers.

Phenotypic variation or diversity is mainly the result of polymorphisms of the DNA sequence level. Availability of high-throughput gene expression profiling technologies offers the possibility for comparative transcriptome analyses of varieties, accessions, or genotypes in a breeding program. Differences at the transcript level, referred to as expression level polymorphisms (ELPs; Doerge 2002), have been shown to control natural phenotypic variation and associated with changes during maize domestication (Wang et al. 1999). Diversity in gene expression among 15 barley lines, which were obtained from several consecutive cycles of advanced cycle breeding for malting quality, was investigated in order to determine whether further selection response can be expected (Muñoz-Amatriaín et al. 2010). In accordance with studies in *Arabidopsis thaliana* (Kliebenstein et al. 2006a) and maize (Stupar et al. 2008), Muñoz-Amatriaín et al. (2010) found a significant correlation between DNA sequence polymorphisms and gene expression differences. However, they also assumed that their gene expression data are a better measure to determine the relationship between specific lines than DNA marker-based coefficients of parentages, indicating the usefulness of expression profiles as biomarkers. Moreover, an increasing genetic similarity and a decreasing number of differentially expressed genes was observed during the progress of the breeding program. This indicated a reduction of genetic and expression diversity through the breeding process. Comparing the results of Muñoz-Amatriaín et al. (2010) with other studies, the variation found in the transcriptome of the 15 barley lines was within the range of transcript level variation found between five maize inbred lines (Stupar et al. 2008) and within four barley varieties (Lapitan et al. 2009). Although expression of several genes was fixed by advanced cycle breeding, Muñoz-Amatriaín et al. (2010) identified differentially expressed genes between the lines of the breeding program associated with malting quality traits, in particular alpha-amylase activity and malt extract percentage, by using the Barley1 Gene chip and single-feature polymorphisms. These candidate genes have the potential for improving the malting quality within the breeding program by optimizing the selection for targets and are, therefore, useful as non-DNA biomarkers.

Grouping of Germplasm

Structuring of breeding germplasm is crucial for attaining sustainable long-term response to selection and, hence, success in the development of superior new varieties in case of hybrid breeding. The two most important issues in structuring germplasm are to establish heterotic groups, on which the success of hybrid breeding is based, and to assess the relatedness of breeding materials at the DNA level, which assures a broad genetic base for future breeding efforts. Up to a limit, increased genetic diversity results in increasing heterosis and hybrid performance. This indicates a link between genetic diversity and heterosis response (Melchinger 1999). Therefore, genetic distances based on DNA markers have been established as a valuable tool for these purposes in the last 30 years (Messmer et al. 1991; Melchinger et al. 1994; Melchinger 1999). Similar in concept to DNA marker-based genetic distances, Frisch et al. (2010) defined transcriptome-based distances.

A straightforward way for defining transcriptome-based distances is to regard the transcript abundance of a gene as one dimension in n_g (where n_g is the number of genes) dimensional space. The difference in gene expression between two genotypes is the distance with regard to that dimension, and the distances in each individual dimension can be summarized by the classical Euclidean distance. This yields the Euclidean transcriptome-based distance D_E between lines i and j on basis of n_g genes, calculated with the formula:

$$D_E(i,j) = \sqrt{\sum_{g=1}^{n_g} [l_g(i) - l_g(j)]^2}$$

where $l_g(i)$ and $l_g(j)$ are the transcript abundance of gene g in the inbred lines i and j. The authors used a logarithmic transformation for empirical reasons, but the concept can be employed without restrictions to untransformed values or to other transformations. An alternative definition of a transcriptome-based distance is the binary distance between lines i and j, calculated using the equation:

$$D_B(i,j) = \sqrt{\frac{1}{n_g} \sum_{g=1}^{n_g} [x_g(i) - x_g(j)]^2}$$

where $x_g(i)$ and $x_g(j)$ are indicator variables taking the values zero and one depending on differential gene expression of gene g in inbred line i and j. If gene g is differentially expressed in lines i and j, then:

$$x_g(i) = 1 \text{ and } x_g(j) = 0 \text{ for } l_g(i) > l_g(j), \text{ and}$$
$$x_g(i) = 0 \text{ and } x_g(j) = 1 \text{ for } l_g(i) \leq l_g(j).$$

The concept of transcriptome-based distances was employed on leaf material of seedlings for data of a maize factorial consisting of 7 flint and 14 dent lines

(Frisch et al. 2010). The 21 parental inbred lines were profiled with a 46-k oligonucleotide array and 98 hybrids were evaluated for grain yield and grain dry matter concentration in the field. For comparing genetic distances, the parental lines were also fingerprinted with ALFP markers.

With both the transcriptome data and the DNA marker data, cluster analysis and principal coordinate analysis were carried out (Fig. 2.2), which are the standard methods in grouping germplasm. The transcriptome-based distances separated the parental lines from the flint pool from those of the dent pool with the same accuracy as DNA marker data. From these results it can be concluded that the information content of transcriptome data is at least as high as that of DNA marker data.

Gene Expression and Heterosis

With the advent of high-throughput gene expression profiling technologies, several geneticists and plant breeders performed experiments in an attempt to gain a deeper knowledge on heterosis and hybrid performance. For example in maize, the experiments relied on hybrid plants and their parental lines, using different expression profiling platforms, experimental designs and tissues. Many of these studies examined gene expression and heterosis, such as the relative frequencies of additive and non-additive expression levels in the hybrid. Generally, the gene expression pattern should be additive in the hybrid compared to the expression in the parental lines (Birchler et al. 2010). Additive expression can be obtained if the hybrid expression level is equivalent to the mid-parent values (Stupar et al. 2008) and is expected if solely *cis*-regulatory differences are responsible for the expression regulation in the hybrid (Thiemann et al. 2010). However, in several studies it was observed that several genes displayed a non-additive expression pattern in the hybrid, meaning the hybrid expression level deviated from the mid-parent level (Swanson-Wagner et al. 2006; Meyer et al. 2007; Uzarowska et al. 2007; Guo et al. 2008; Hoecker et al. 2008; Stupar et al. 2008; Jahnke et al. 2010; Paschold et al. 2010; Riddle et al. 2010). Although a clear consensus about the differently expressed genes in the hybrids could not be observed over these studies (Birchler et al. 2010), a correlation between the number and fraction of genes showing non-additive expression patterns and the size of the heterosis response is inferred (Riddle et al. 2010). However, it has still not been conclusively determined whether varying regulation due to diverged alleles and regulatory elements at various loci contributes to heterosis (Birchler et al. 2010).

Prediction of Hybrid Performance

In hybrid breeding, two parental components from different genetic origins are crossed to generate hybrid seed. In developed countries, hybrids are the predominant type of variety for many crops, e.g., maize, sugar beet, rice, rye, sunflower, oil seed

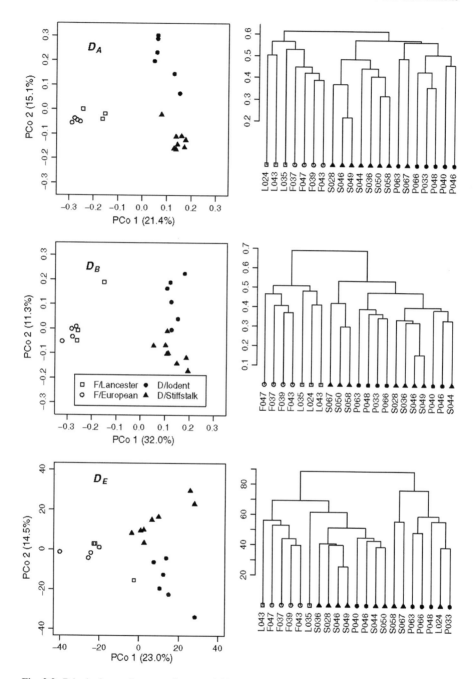

Fig. 2.2 Principal coordinate analyses and hierarchical cluster analyses based on the genetic distance D_A and the transcriptome-based distances D_B and D_E. The distances D_B and D_E were determined from the subset of genes S_P comprising 10,810 differently expressed genes (Frisch et al. 2010)

rape, and sorghum. In developing countries the number of hybrid varieties is still low but the interest is increasing considerably. In hybrid breeding, the identification of single-cross hybrids offering superior yield performance is one of the main challenges for plant breeders. Cross combinations are tested annually in extensive field trials, and the number of potential crosses increases rapidly with increasing numbers of inbred lines. However, the resources of breeding companies are limited and, therefore, only a small proportion of cross combinations can be evaluated in the field. Using the data available from related crosses to identify untested but promising hybrids is extremely important.

Prediction of hybrid performance with DNA markers is usually successful for hybrids belonging to one heterotic pool, so-called intra-pool crosses (Dhillon et al. 1990; Melchinger 1999). For hybrid seed production, however, only crosses between different pools, so-called inter-pool crosses, are important. A reliable prediction of hybrid yield is therefore not possible for this important area of application in applied breeding. Poor predictive power of genetic distances to identify the most promising inter-pool crosses is caused by linkage disequilibrium. A certain marker allele in one heterotic pool may be linked in coupling phase with a desirable QTL allele, while the same marker allele may be linked in repulsion phase with a desirable QTL allele in the opposite heterotic pool (Frisch et al. 2010). Thus, the inter-pool genetic distances at marker loci can provide only a poor estimate for the differences at functional genes between two lines belonging to different heterotic pools. For this reason, preferred approaches for marker-based prediction do not use genetic distances, but use instead linkage disequilibrium between marker and QTL (Vuylsteke et al. 2000; Schrag et al. 2006, 2007, 2009a, b, 2010).

For the prediction of hybrid performance and heterosis, transcriptome data have two advantages over DNA marker data: (i) they do not rely on linkage disequilibrium between marker alleles and QTL alleles, and (ii) they quantify directly the expression of genes, since this analysis not only determines if specific genes are present, but also the degree to which the genes are up or down-regulated. Consequently, it can be expected that transcriptome-based approaches are superior to DNA marker-based approaches.

For plant breeders, it may be more important to predict the value of heterosis of the hybrids, rather than to examine why hybrids typically show heterotic response and not randomly negative or positive effects. To predict heterosis and hybrid performance with transcriptome data, Birchler et al. (2003) suggested using the correlation of heterosis with the average expression of genes in the parental inbred lines. This approach was taken up by Fu et al. (2010) and Thiemann et al. (2010) for investigation of heterosis for grain yield and dry matter content in maize. Thiemann et al. (2010) designed a hybridization scheme for comparisons between parental inbred lines of inter-pool crosses with microarray analysis. They found functional groups and genes which were highly correlated to both heterosis and hybrid performance for the two investigated traits. The observed correlation offers first evidence towards prediction potential of the genes and their expression level. An alternative approach was reported by Frisch et al. (2010). Here, the authors disregarded the functional information of the analyzed genes and considered the transcript abundance level as quantitative variables characterizing a genotype. They then compared the efficiency of the prediction based on phenotypically estimated general combining ability (GCA) with the marker-based prediction and with transcriptome-based prediction at the same data set. Using leaves in the seedling stage, Frisch et al. (2010) showed that the correlation of observed with predicted values for heterosis and hybrid performance was higher with transcriptome-based distances (D_B and D_E) than for earlier prediction approaches based on phenotypic values and DNA markers. The best prediction accuracy was observed for the binary transcriptome distance (D_B) and 1,000–2,000 genes that were pre-selected on the basis of associations of differential gene expressions with hybrid performance and heterosis by using a binomial test.

The transcriptome-based prediction of heterosis and hybrid performance shows promise for practical application in breeding programs since the expression profiles of seedlings can be carried out directly after the production of the inbred lines. The performance of potential hybrids can be predicted by using transcriptome data, and promising hybrids can be subsequently evaluated in extensive field trials. Consequently, indirect pre-selection on the basis of expression profiles can efficiently increase the selection response of hybrid breeding programs. Due to recent advances in microarray technology, costs for gene expression profiles decrease and therefore transcriptome-based prediction of heterosis and hybrid performance will also allow a sizable decrease of the costs of hybrid breeding programs.

In conclusion, gene expression analyses are promising tools not only to establish heterotic groups but also to predict hybrid performance and heterosis which are the main tasks in hybrid breeding.

Application of Expressed Sequence tags

Expressed sequence tags are short cDNA sequences that enable the "tagging" of genes from which the translated mRNA originated. This allows researchers to determine if a gene was previously found in the species under study or in another organism, and in some cases infer the function of the gene. EST database creation lies at the intersection of transcriptomics and proteomics, since gene transcripts (mRNA) are sequenced to obtain the cDNA sequence. EST information exists in abundance for many crops. According to a recent review, rice, wheat, soybean, and cassava have over one million ESTs each currently, while maize has over two million (Langridge and Fleury 2011). EST resources can provide researchers with tools to analyze unsequenced genomes. Van der Hoeven et al. (2002) used EST data to investigate genome organization in tomato. They were able to predict that the tomato genome contains 35,000 genes, and that these genes were largely concentrated in euchromatic regions that comprise roughly one quarter of the genome. Large and complex genomes also benefit from EST-based analysis methods. Common wheat, for instance, is hexaploid with a large genome, many times larger than maize, which in turn is many times larger than rice. The size and redundancy of large genomes can pose problems to sequencing efforts. EST resources have been created (Lazo et al. 2004) and exploited to help organize and determine gene density in complex genome of hexaploid wheat (Qi et al. 2004). In this way, EST databases can be used to develop DNA markers as well as to determine gene hotspots to be given priority when sequencing.

Application of Expression QTL

Quantitatively inherited traits are affected by many genes throughout the genome with often individually small effects as well as environmental factors. In plant breeding research, generally segregating populations from a cross of two inbred lines with distinct quantitative variation are used as mapping populations to detect such complex traits. This analysis uses statistical methods to determine genomic regions, referred to as QTL, that display associations between genetic and phenotypic variation. Developments in molecular marker technologies have advanced QTL analysis which, through quick and cost-effective genotyping, led to the construction of detailed genetic linkage maps of both model and agronomically important species. Moreover, in the 1990s, new statistical tools were established (Lander and Botstein 1989; Haley and Knott 1992; Jansen and Stam 1994) and implemented in software packages (*e.g.*, Lincoln et al. 1993; Utz and Melchinger 1996) for the analysis of QTL mapping experiments. The most common method is "interval mapping". Here, every chromosome is divided into short intervals and each interval is treated separately for QTL detection and estimation by using the maximum likelihood method leading to LOD score statistics.

The detection of QTL has proven to be very useful in plant breeding. DNA markers associated with QTL involved in the inheritance of agronomically important traits have been detected in many species (Schön et al. 2004). The primary aim of QTL mapping in plant breeding experiments is to employ the identified marker-QTL associations for subsequent indirect selection in marker-assisted selection (MAS) programs, in particular for traits that show low heritability or that are difficult or expensive to assess phenotypically. However, for many traits this is still a huge challenge, since several investigations did not result in better varieties. The main shortcomings of the classical QTL detection are (1) the restricted genetic variation due to segregating populations obtained from biparental crosses, (2) a limited number of recombination events since the mapping populations consist of early-generation crosses (Ingvarsson and Street 2011), and (3) limited resolution due to insufficient sampling of the mapping population (Lande and Thompson 1990; Melchinger et al. 1998; Lübberstedt et al. 1998; Schön et al. 2004) or linked QTL in repulsion phase in the parents, cancelling out in the mapping population (Falke et al. 2007).

Classical QTL detection is usually performed to identify associations between DNA markers and phenotypic traits. However, similar to classical physiological traits, variation in gene expression can also have a quantitative distribution. Since

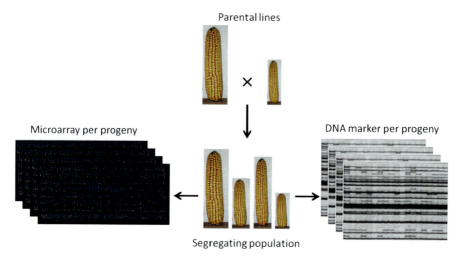

Fig. 2.3 eQTL mapping: Expression profiling in combination with DNA marker analysis of a segregating population enables the use of QTL analysis for detection of influential genes and gene products. Here, a segregating population was produced out of a cross between inbred maize lines. To analyze this population, each progeny undergoes microarray profiling and DNA marker analysis (Modified according to Jansen and Nap 2001)

advances in gene expression profiling and the decreasing cost of the respective high-throughput technologies have facilitated the analysis of whole genome transcript levels, variation in gene expression can be dissected by applying classical statistical tools and approaches for expression QTL detection. In this context Jansen and Nap (2001) introduced the concept of "genetical genomics", which combines expression profiles with DNA marker data of each genotype from a segregating population and detects genomic regions controlling the observed variation in expression by using levels of transcript abundance as a quantitative phenotype (Fig. 2.3). In general, the transcript abundance is measured for many genes simultaneously and transcript abundance profiles of single genes can each be thought of as individual traits, meaning that many such traits are recorded (Druka et al. 2010). This analysis, referred to as expression QTL (eQTL) mapping, attempts to attribute variation in the transcription level of a gene to genotypic differences (Doerge 2002). The significance of associations can be determined using LOD scores or Likelihood Ratio Statistic like in classical QTL mapping experiments, and the location in the genome can be revealed by plotting the LOD scores (Druka et al. 2010). A position in the genome showing a polymorphism associated with differential accumulation of a specific transcript is an eQTL (Jansen and Nap 2001; Kliebenstein 2009).

In general, eQTL can be divided into two major classes categorized as *cis* and *trans*-acting eQTL: *cis* eQTL represent a polymorphism physically located near the gene itself, whereas *trans* eQTL are located in the genome removed from the actual physical position of the gene whose transcript level has been measured (Gibson and Weir 2005; Hansen et al. 2008; Kliebenstein 2009). Thus, eQTL mapping provides the possibility to identify factors affecting the level of transcript expression (Varshney et al. 2005). In plants, the relationship between *cis* and *trans* eQTL has been investigated, *e.g.*, in *Arabidopsis thaliana* (Keurentjes et al. 2007; West et al. 2007) and barley (Potokina et al. 2008a). In these studies, the phenotypic effect of *cis*-acting eQTL was found to be greater than of *trans*-acting eQTLs.

The approach of eQTL mapping was first successfully applied in yeast (Brem et al. 2002). Recently, several studies have demonstrated the feasibility of this method in several plant species, mainly in the model species *Arabidopsis thaliana* (DeCook et al. 2006; Vuylsteke et al. 2006; Kliebenstein et al. 2006b; Keurentjes et al. 2007; West et al. 2007; Zhang et al. 2011), but also in trees species like poplar (Street et al. 2006; Drost et al. 2010) or eucalyptus (Kirst et al. 2004, 2005), and in some agronomically interesting crops including maize (Schadt et al. 2003; Shi et al. 2007; Swanson-Wagner et al. 2009), wheat (Jordan et al. 2007; Druka et al. 2008b) and barley (Potokina et al. 2008a, b; Chen et al. 2010). Moreover, the approach of eQTL mapping has been described and interpreted in several reviews (*e.g.* see Hansen et al. 2008; Kliebenstein 2009; Joosen et al. 2009; Druka et al. 2010; Holloway and Li 2010).

Today, modeling of integrated interaction networks underlying complex traits has been enabled by combining the power of large scale expression profiling platforms, full genome sequences or high throughput-genotyping technologies and precision phenotyping (Schadt et al. 2003; Druka et al. 2008b). The association of gene expression with genetic linkage maps can detect gene regulatory regions within the genome. Genes being controlled by these regions can become candidate genes for traits associated with the same region and can subsequently be used as sources for new genetic markers to increase the mapping efficiency of physiological QTL (Kliebenstein 2009) or to assist in map-based cloning of the gene of interest (Jordan et al. 2007; Li et al. 2010). Examples of phenotypic traits analyzed by combining expression profiling with classical genetic analyses include seed development in wheat (Jordan et al. 2007) and digestibility in maize (Shi et al. 2007). The eQTL mapping experiment of Shi et al. (2007) identified a hotspot of genes involved in cell-digestibility on chromosome 3, and in the same genomic region a QTL for lignin content was observed from Ralph et al. (2004). Thus, these results support that gene expression can possibly be used as biomarkers for classical phenotypes, and the gene underlying the eQTL can be a gene candidate.

In conclusion, the interest in eQTL mapping is increasing for several agronomically important species due to the prospect of reducing time and effort required to detect genes underlying quantitative traits (Kliebenstein 2009; Hansen et al. 2008; Druka et al. 2010). Moreover, expression profiling to identify eQTL can also be of great benefit for plant breeding not only as an important tool to analyze the relationship between genome and transcriptome but also to give information for new markers. Analogously, similar approaches can be used from the other 'omic' technologies to identify and use protein QTL (pQTL) and metabolite QTL (mQTL) (Joosen et al. 2009).

Analysis of Proteomic Data for Plant Breeding

While transcriptomics deals with the expression of genes, proteomics seeks to explore the translated products of gene expression. The original focus of proteomic analysis was on identification of all protein species in a specific cell or tissue, though the research applications have since diversified. Proteome studies involve analyses of structural and functional aspects of proteins (Park 2004). Proteome analyses have revealed that the number of expressed genes may underestimate the amount of produced proteins. It has been proposed that the 30,000 genes in humans translate to at least 90,000 proteins, which are achieved through alternative splicing and protein post-translational modification (Bolwell et al. 2004). Additionally, gene expression is also not always consistent with protein abundance (Ghazalpour et al. 2011; Gygi et al. 1999). The disparity between these two numbers highlights the importance of proteomic analyses. Originally, proteomics research, while quite advanced in humans and certain model organisms such as *E. coli* and yeast, lagged behind in plants (Chen and Harmon 2006). In plants, sequenced genomes and EST data have had an accelerating effect in advancing proteomic research. Improvements in the ability to retrieve, quantify, and study proteins from host plants have also made plant proteomics research more feasible and reproducible (Salekdeh and Komatsu 2007). Techniques such as two-dimensional gel electrophoresis (2DE) and mass spectrometry (MS) have been improved and expanded upon to aid in proteomic studies. Many reviews for these and other techniques and modifications are readily available, see reviews by Park (2004), Bolwell et al. (2004), Rossignol et al. (2006), and Jorrín-Novo et al. (2009).

Assessing Germplasm Diversity

Proteomic analyses may reveal variation in plants that can be used to assess diversity present in a population. As previously stated, variation at the protein level does not necessarily correspond to variation at the genomic and transcription levels. Proteins are closer to the phenotype of the plant, because they act directly on biochemical processes (Thiellement et al. 2002). This indicates that variation in the protein expression and abundance may be more useful than DNA markers to understanding phenotypes. Early work regarding proteomic diversity includes attempts to evaluate racial relationships in maize using isozymes (Stuber et al. 1977). Eivazi et al. (2008) calculated genetic distances in wheat using four different measures to compare diversity between genotypes. Quality trait data, AFLPs, SSRs, and proteomic markers were each used to calculate separate genetic distances and dendrograms. The study showed low levels of similarity between the resulting distance measures and dendrograms for each method. The abundance of studied proteins has shown a wide variation in maize (Jorgensen and Nguyen 1995; Riccardi et al. 2004), wheat (Zivy et al. 1984), and sugarcane (Ramagopal 1990). Furthermore, several studies have been published which use proteomic approaches to assess diversity in a range of crops such as lentil (Scippa et al. 2008), common beans (Mensack et al. 2010), and strawberries (Alm et al. 2007). The results of these studies indicate that proteomics research can be successfully used in diversity analysis. Presently, proteome diversity analysis is largely complementary to traditional diversity analysis methods employing DNA markers, but this may change as methods evolve and costs decrease.

Abiotic and Biotic Stresses

Plant breeding studies of crop stress response have utilized proteomic analysis methods to attempt to dissect this complex trait. Stress responses, both abiotic and biotic, are very important in plant breeding, because field conditions often include various stress factors, depending on the environment. Proteomics has been increasingly used to assess the response of stress-associated proteins to environmental stress (Vinocur and Altman 2005). Identifying proteins and their abundance in response to plant stress is potentially a better approach than using DNA markers, because, like the transcriptome, the proteome of plant tissues varies in different environments. Assessing the regulation of translational and post-translational products in response to stress has the potential to illuminate changes that cannot be detected with studies conducted at the DNA marker level. However, for this same reason, reproducibility is a concern and more care must therefore be taken when sampling, analyzing, and interpreting results. The first step in these studies involves comparing proteins in control and stressed plants to find stress response proteins. Identification of these stress response proteins may show that they have functions that directly relate to the trait of interest (Salekdeh and Komatsu 2007). In plants, several studies have focused particularly on proteomic analysis of abiotic stress response. One such study on field-grown plants under severe drought conditions was performed for sugar beet by Hajheidari et al. (2005). Protein differences under drought stress were observed in the two genotypes under consideration, with some differences occurring between genotypes as well.

Pathogen resistance is another important crop research area that has benefitted from proteomic analyses. Up-regulation of pathogenesis related (PR) proteins was found in several studies for plant tissues under biotic stress but also under abiotic stress. Wei et al. (2009) studied biotic stress caused by brown planthopper, a vascular feeder and rice pest. They observed that between lines resistant and susceptible to brown planthopper, 17 proteins were inversely regulated, leading to the possibility that these are PR proteins. The role of PR proteins, however, is not well understood, despite a strong correlation between PR protein expression and plant cell stress responses (Salekdeh and Komatsu 2007). Linking the results of these studies to other "omics" analyses might provide a better understanding of plant stress response. The presence or absence of PR proteins or related regulatory elements could be used as a non-DNA biomarker to differentiate resistant and susceptible plant materials in the future. Ideally, future research will enable measuring these proteins without testing the plant under full stress. Current research typically involves inducing stress in field or lab settings. By either inducing stress response directly on plant tissue through chemical or biological means, proteomic stress response research would see a large increase in efficiency.

Application of Protein Expression QTL Mapping

The previous transcriptomics section highlighted the use of eQTL analysis in order to associate DNA polymorphisms with changes in gene expression levels. While this technique shows promise, it may in some cases fail to find significant variation if the polymorphism doesn't affect expression itself, but instead affects post-translational behavior such as protein stability or enzymatic activity. Integrating both transcriptomic and proteomic analysis methods can, therefore, enhance detection of such variation (Stylianou et al. 2008). The earliest study attempting to locate QTL that affect protein expression was performed by Damerval et al. (1994). Using an F_2 maize population, 70 protein quantity loci, referred to as PQL, were detected for 42 of the 72 proteins analyzed. More than one PQL was detected for 20 of these 42 proteins. In a more recent study in barley, Witzel et al. (2011) mapped QTL that affected protein expression, referred to in the paper as pQTL. The pQTL detected in the study were further analyzed to reveal that many of the proteins were involved in metabolic or disease/defense-related processes. Proteomic methods such as the pQTL analysis can therefore provide breeders with new targets for agronomic improvement.

Analysis of Metabolic Networks for Plant Breeding

The analysis of the metabolome is another important area in the field of 'omic' technologies. Plant breeding research interest in this field is constantly increasing (for review see Fernie and Schauer 2009), since substantial variation exists for metabolite composition in plants (Keurentjes et al. 2006). The main goal of metabolic studies is to determine precisely the quantity and quality of all metabolites in a specific tissue sample (Kusano et al. 2007). Technological advances made several techniques like gas-chromatography-mass-spectrometry (GC-MS), liquid-chromatography-mass-spectrometry (LC-MC), capillary-electrophoresis-mass-spectrometry (CE-MS), fourier-transform-infrared-spectrometry (FT-IR), LC-photodiode-array (PDA), or nuclear magnetic resonance (NMR) available to assess variation in metabolite content present tissue under study (Mochida and Shinozaki 2010; Saito and Matsuda 2010), while MS and NMR dominate metabolite profiling. Currently, the limiting factor of most metabolomic tools is the high cost (Borrás and Slafer 2008). As costs decrease, metabolomic analysis will become an increasingly feasible and interesting selection tool for crop improvement. An important advantage of metabolomics is that it does not require the availability of genome sequences for the species of interest (Stitt and Fernie 2003). To date, researchers have mostly concentrated on single metabolic pathways like carotenoid biosynthesis of tomato (Liu et al. 2003) or simple metabolic processes like cold-sweetening in potato (Menéndez et al. 2002). However, since metabolomic approaches offer the opportunity to evaluate several metabolites simultaneously, they can be important for phenotyping and diagnostics in plants (Fernie and Schauer 2009).

Analysis of Heterosis and Hybrid Performance

The models of additive, (over-) dominant, and epistatic gene action of phenotypes have a counterpart at the molecular levels of the transcriptome, proteome, and metabolome. A first concept of how gene action can be modeled in metabolomic networks was proposed by Wright (1934), who investigated the activity of one enzyme in a linear metabolomic pathway and the outcome of the chain, i.e., the flux. The model was extended by Kacser and Burns (1981) who focused their study on chains of Michaelian reversible enzymes. They postulated that the difference caused by two different alleles of a gene is due to differences in catalytic activities. These catalytic activities of an enzyme are thought to be determined by its kinetic linkage to other enzymes via their substrates and products. The effect of such a system on the phenotype is modeled by the output, or flux, of an interacting system of enzymes. The models of Wright (1934) and Kacser and Burns (1981) were limited to linear chains of enzymes. Fievet et al. (2006) extended the theory to regulated enzymes and branched pathways.

The authors verified their model in vitro with the example of the first part of glycolysis. In a simulation study, the authors extended their approach and came to the conclusion that heterosis might be mainly caused by epistasis (Fievet et al. 2010). The above approaches attempt to model the metabolome on the level of one single enzyme chain or network. Modeling covers many details with respect to one single metabolite. However, with respect to application in plant breeding, two major tasks remain. The first is in vivo verification of the system dynamics. The new approach involving metabolite profiles and DNA markers as predictors for heterosis, could be the key for such experiments (Steinfath et al. 2010). The second task is to integrate the models in larger concepts. Therefore, systems biology approaches, which combine metabolic networks in order to use them for modeling traits of agronomic importance, seem suitable. While such functional models relating metabolome data to the phenotype are a target for long-term research, statistical models may be a quicker path to results that have practical applications in plant breeding. This was demonstrated with a statistical model ignoring the functional aspects, but combining metabolite abundance and SNP marker data. This model showed the ability to predict biomass yield in *Arabidopsis thaliana* (Steinfath et al. 2010).

Application of Metabolite QTL Mapping

Similar to the previous transcriptomic and proteomic approaches, researchers are attempting to use features of the plant metabolome in conjunction with QTL analyses. While many studies presently focus on *Arabidopsis thaliana*, QTL detection using metabolomic features as quantitative traits is branching out into crop species (Fernie and Schauer 2009). A study in tomato by Schauer et al. (2006) displayed the potential power of metabolomic methods to dissect traits for the benefit of crop breeding. Using introgression lines, they quantified 74 metabolites and proceeded to identify 889 single-trait QTL. Though most of the QTL were previously unknown, several had been previously reported providing a measure of validation. Use of metabolic profiling in agronomically important crops such as rice (Kusano et al. 2007) and maize (Harrigan et al. 2007) is increasing. Linking the plant metabolome to genetic polymorphisms will be the next step to dissect how DNA variation leads to metabolic variation in agronomic crop species.

Integrating Metabolomics and Breeding

Metabolomic analyses have only recently been incorporated into crop species. A recent study in maize examined possible relationships between physiological trait expression and molecular biomarkers to grain yield and its components (Cañas et al. 2011). This approach can be used to highlight breeding targets for crop improvement. Once certain aspects of a metabolic process are known, it may be possible to transfer entire biosynthetic pathways into agronomically important crops. Pfalz et al. (2011) engineered the indole glucosinolate biosynthesis pathway, which is found in *Arabidopsis thaliana* and mediates numerous biological interactions between the host plant and its natural enemies, into *Nicotiana benthamiana*. This study illuminated previously unknown aspects of this pathway, in addition to being an example of the ability to transfer metabolic pathway information between plants.

Outlook: Application of the 'Omics' Technologies as Non-DNA Markers in Plant Breeding

Most agronomically important traits are controlled by many genes with small effects and QTL analyses can only partially explain their genetic variance (Moreau et al. 2004), unless very large experiments are carried out (Schön et al. 2004). Therefore, the use of MAS in breeding programs for polygenic traits is still fragmentary (Varshney et al. 2006b). To overcome the limitations of the QTL mapping/MAS approach, Meuwissen et al. (2001) proposed 'genomic selection' (GS), which is being used in cattle breeding programs (Schaeffer 2006; Goddard and Hayes 2007, 2009). The major advantage of GS is its ability to provide improved prediction accuracy and to shorten the generation interval (Zhang et al. 2010). First results in maize and wheat (Piepho 2009; Crossa et al. 2010; Albrecht et al. 2011) indicate that GS can reduce the costs and simultaneously speed up selection gain also in plant breeding programs. While currently the main research focus lies on these DNA-based methods, non-DNA marker-based methods may complement the DNA-based methods for GS in the long-term.

The studies reviewed in this chapter have shown that non-DNA marker-based methods can provide more information than DNA-based methods alone, because they provide an insight into the physiological and biochemical processes and networks that relate the genotype to the phenotype. In principle, there are two approaches to make use of this information. Firstly, it can be employed to replace methods established already for DNA markers, such as transcriptome-based distances. It is still unknown whether advances in lab technology for non-DNA markers will be able to surpass the DNA-based methods with respect to technical feasibility, such that the additional information of non-DNA markers can be exploited with economic efficiency. The second approach is to establish more complex and realistic models for the relationship between genotype and phenotype using systems biology approaches. Combining information of different physiological levels (Fig. 2.1) with mathematical and statistical models may prove to be a tool that will allow breeders to incorporate new screening methods, such as techniques involving non-DNA biomarkers discussed in this chapter, into variety development, and will pave the way for future advances in plant breeding.

References

Albrecht T, Wimmer V, Auinger H-J, Erbe M, Knaak C, Ouzunova M, Simianer H, Schön CC (2011) Genome-based prediction of testcross values in maize. Theor Appl Genet 123:339–350

Alm R, Ekefjärd A, Krogh M, Häkkinen J, Emanuelsson C (2007) Proteomic variation is as large within as between strawberry varieties. J Proteome Res 6:3011–3020

Bernardo R (2002) Breeding for quantitative traits in plants. Stemma Press, Woodbury

Bernardo R, Yu J (2007) Prospects for genomewide selection for quantitative traits in maize. Crop Sci 47:1082–1090

Birchler J, Auger D, Riddle N (2003) In search of the molecular basis of heterosis. Plant Cell 15:2236–2239

Birchler JA, Yao H, Chudalayandi S, Vaiman D, Veitia RA (2010) Heterosis. Plant Cell 22:2105–2112

Bolwell GP, Slabas AR, Whitlegge JP (2004) Proteomics: empowering systems biology in plants. Phytochemistry 65:1665–1669

Borrás L, Slafer GA (2008) Agronomy and plant breeding are key to combating food crisis. Nature 453:1177

Botstein D, White RL, Skolnick M, Davis RW (1980) Construction of a genetic linkage map in man using restriction fragment length polymorphisms. Am J Hum Genet 32:314–331

Brem RB, Yvert G, Clinton R, Kruglyak L (2002) Genetic dissection of transcriptional regulation in budding yeast. Science 296:752–755

Brenner S, Johnson M, Bridgham J, Golda G, Lloyd DH, Johnson D, Luo S, McCurdy S, Foy M, Ewan M, Roth R, George D, Eletr S, Albrecht G, Vermaas E, Williams SR, Moon K,

Burcham T, Pallas M, DuBridge RB, Kirchner J, Fearon K, Mao J, Corcoran K (2000) Gene expression analysis by massively parallel signature sequencing (MPSS) on microbead arrays. Nat Biotechnol 18:630–634

Cañas RA, Amiour N, Quilleré I, Hirel B (2011) An integrated statistical analysis of the genetic variability of nitrogen metabolism in the ear of three maize inbred lines (*Zea mays* L.). J Exp Bot 62:2309–2318

Chen S, Harmon AC (2006) Advances in plant proteomics. Proteomics 6:5504–5516

Chen X, Hackett CA, Niks RE, Hedley PE, Booth C, Druka A, Marcel TC, Vels A, Bayer M, Milne I, Morris J, Ramsay L, Marshall D, Milne L, Waugh R (2010) An eQTL analysis of partial resistance to *Puccinia hordei* in barley. PLoS One 5:e8598

Collard BCY, Mackill DJ (2008) Marker-assisted selection: an approach for precision plant breeding in the twenty-first century. Philos Trans R Soc B 363:557–572

Condón F, Gustus C, Rasmusson DC, Smith KP (2008) Effect of advanced cycle breeding on genetic diversity in barley breeding germplasm. Crop Sci 48:1027–1036

Crick FH (1970) Central dogma of molecular biology. Nature 277:561–563

Crossa J, de los Campos G, Pérez P, Gianola D, Burgueño J, Araus JL, Makumbi D, Singh RP, Dreisigacker S, Yan J, Arief V, Bänziger M, Braun HJ (2010) Prediction of genetic values of quantitative traits in plant breeding using pedigree and molecular markers. Genetics 186: 713–724

Czechowski T, Bari RP, Stitt M, Scheible WR, Udvardi MK (2004) Real-time RT-PCR profiling of over 1400 Arabidopsis transcription factors: unprecedented sensitivity reveals novel root- and shoot-specific genes. Plant J 38:366–379

Damerval C, Maurice A, Josse JM, de Vienne D (1994) Quantitative trait loci underlying gene product variation: a novel perspective for analyzing regulation of genome expression. Genetics 137:289–301

De Koeyer DL, Phillips RL, Stuthman DD (1999) Changes in genetic diversity during seven cycles of recurrent selection for grain yield in oat, *Avena sativa* L. Plant Breed 118:37–43

DeCook R, Lall S, Nettleton D, Howell SH (2006) Genetic regulation of gene expression during shoot development in Arabidopsis. Genetics 172:1155–1164

Dhillon BS, Gurrath PA, Zimmer E, Wermke M, Pollmer WG, Klein D (1990) Analysis of diallel crosses of maize for variation and covariation in agronomic traits at silage and grain harvest. Maydica 35:297–302

Doerge RE (2002) Mapping and analysis of quantitative trait loci in experimental populations. Nat Rev Genet 3:43–52

Drost DR, Benedict CI, Berg A, Novaes E, Novaes CRDB, Yu Q, Dervinis C, Maia JM, Yap J, Miles B, Kirst M (2010) Diversification in the genetic architecture of gene expression and transcriptional networks in organ differentiation of Populus. Proc Natl Acad Sci USA 107:8492–8497

Druka A, Druka I, Centeno AG, Li H, Sun Z, Thomas WTB, Bonar N, Steffenson BJ, Ullrich SE, Kleinhofs A, Wise RP, Close TJ, Potokina E, Luo Z, Wagner C, Schweizer GF, Marshall DF, Kearsey MJ, Williams RW, Waugh R (2008a) Towards systems genetic analyses in barley: integration of phenotypic, expression and genotype data into GeneNetwork. BMC Genet 9:73

Druka A, Potokina E, Luo Z, Bonar N, Druka I, Zhang L, Marshall DF, Steffenson BJ, Close TJ, Wise RP, Kleinhofs A, Williams RW, Kearsey MJ, Waugh R (2008b) Exploiting regulatory variation to identify genes underlying quantitative resistance to the wheat stem rust pathogen *Puccinia graminis* f. sp. *tritici* in barley. Theor Appl Genet 117:261–272

Druka A, Potokina E, Luo Z, Jiang N, Chen X, Kearsey M, Waugh R (2010) Expression quantitative trait loci analysis in plants. Plant Biotechnol J 8:10–27

Dubreuil P, Chacrosset A (1998) Genetic diversity within and among maize populations: a comparison between isozyme and nuclear RFLP loci. Theor Appl Genet 96:577–587

Dudley JW (1993) Molecular markers in plant improvement: manipulation of genes affecting quantitative traits. Crop Sci 33:660–668

Eivazi AR, Naghavi MR, Hajheidari M, Pirseyedi SM, Ghaffari MR, Mohammadi SA, Majidi I, Salekdeh GH, Mardi M (2008) Assessing wheat (*Triticum aestivum* L.) genetic diversity using quality traits, amplified fragment length polymorphisms, simple sequence repeats and proteome analysis. Ann Appl Biol 152:81–91

Falke KC, Flachenecker C, Melchinger AE, Piepho H-P, Maurer HP, Frisch M (2007) Temporal changes in allele frequencies in two European F2 flint maize populations under modified recurrent full-sib selection. Theor Appl Genet 114:765–776

Fernie AR, Schauer N (2009) Metabolomics-assisted breeding: a viable option for crop improvement? Trends Genet 25:39–48

Fiévet JB, Dillmann C, Curien G, de Vienne D (2006) Simplified modeling of metabolic pathways for flux prediction and optimization: lessons from an in vitro reconstruction of the upper part of glycolysis. Biochem J 396:317–326

Fiévet JB, Dillmann C, de Vienne D (2010) Systemic properties of metabolic networks lead to an epistasis-based model for heterosis. Theor Appl Genet 120:463–473

Frisch M, Thiemann A, Fu J, Schrag TA, Scholten S, Melchinger AE (2010) Transcriptome-based distance measures for grouping of germplasm and prediction of hybrid performance in maize. Theor Appl Genet 120:441–450

Fu YB, Somers DJ (2009) Genome-wide reduction of genetic diversity in wheat breeding. Crop Sci 49:161–168

Fu J, Thiemann A, Schrag TA, Melchinger AE, Scholten S, Frisch M (2010) Dissecting grain yield pathways and their interactions with grain dry matter content by a two-step correlation approach with maize seedling transcriptome. BMC Plant Biol 10:63

Geiger HH, Miedaner T (2009) Rye breeding. In: Carena MJ (ed) Cereals, vol 3, Handbook of plant breeding. Springer, New York, pp 157–181

Geldermann H (1975) Investigations on inheritance of quantitative characters in animals by gene marker. I Methods. Theor Appl Genet 46:319–330

Ghazalpour A, Bennett B, Petyuk VA, Orozco L, Hagopian R, Mungrue IN, Farber CR, Sinsheimer J, Kang MH, Furlotte N, Park CC, Wen P-Z, Brewer H, Weitz K, Camp DG II, Pan C, Yordanova R, Neuhaus I, Tilford C, Siemers N, Gargalovic P, Eskin E, Kirchgessner T, Smith DJ, Smith RD, Lusis AJ (2011) Comparative analysis of proteome and transcriptome variation in mouse. PLoS Genet 7:e1001393

Gibson G, Weir B (2005) The quantitative genetics of transcription. Trends Genet 21:616–623

Goddard ME, Hayes BJ (2007) Genomic selection. J Anim Breed Genet 124:323–330

Goddard ME, Hayes BJ (2009) Mapping genes for complex traits in domestic animals and their use in breeding programmes. Nat Rev Genet 10:381–391

Guo M, Yang S, Rupe M, Hu B, Bickel DR, Arthur L, Smith O (2008) Genome-wide allele-specific expression analysis using Massively Parallel Signature Sequencing (MPSS) reveals *cis*- and *trans*-effects on gene expression in maize hybrid meristem tissue. Plant Mol Biol 66:551–563

Gygi SP, Rochon Y, Franza BR, Aebersold R (1999) Correlation between protein and mRNA abundance in yeast. Mol Cell Biol 19:1720–1730

Hagdorn S, Lamkey KR, Frisch M, Guimaraes PEO, Melchinger AE (2003) Molecular genetic diversity among progenitors and derived elite lines of BSSS and BSCB1 maize populations. Crop Sci 43:474–482

Hajheidari M, Abdollahian-Noghabi M, Askari H, Heidari M, Sadeghian SY, Ober ES, Salekdeh GH (2005) Proteome analysis of sugar beet leaves under drought stress. Proteomics 5:950–960

Haley CS, Knott SA (1992) A simple regression method for mapping quantitative trait loci in line crosses using flanking markers. Heredity 69:315–324

Hansen BG, Halkier BA, Kliebenstein DJ (2008) Identifying the molecular basis of QTLs: eQTLs add a new dimension. Trends Plant Sci 3:72–77

Harrigan GG, Stork LG, Riordan SG, Ridley WP, Macisaac S, Halls SC, Orth R, Rau D, Smith RG, Wen L, Brown WE, Riley R, Sun D, Modiano S, Pester T, Lund A, Nelson D (2007) Metabolite analyses of grain from maize hybrids grown in the United States under drought and watered conditions during the 2002 field season. J Agric Food Chem 55:6169–6176

Heffner EL, Sorrells ME, Jannink J-L (2009) Genomic selection for crop improvement. Crop Sci 49:1–12

Hoecker N, Keller B, Muthreich N, Chollet D, Descombe P, Piepho H-P, Hockholdinger F (2008) Comparison of maize (*Zea mays* L.) F1-hybrid and parental line primary root transcriptomes suggests organ-specific patterns of nonadditive gene expression and conserved expression trends. Genetics 179:1275–1283

Holloway B, Li B (2010) Expression QTLs: applications for crop improvement. Mol Breed 26:381–391

Hunter RL, Markert CL (1957) Histochemical demonstration of enzymes separated by zone electrophoresis in starch gels. Science 125:1294–1295

Ingvarsson PK, Street NR (2011) Association genetics of complex traits in plants. New Phytol 189:909–922

Jahnke S, Sarholz B, Thiemann A, Kühr V, Gutierrez-Marcos JF, Geiger HH, Piepho H-P, Scholten S (2010) Heterosis in early seed development: a comparative study of F1 embryo and endosperm tissues 6 days after fertilization. Theor Appl Genet 120:389–400

Jannink JL, Lorenz AJ, Iwata H (2010) Genomic selection in plant breeding: from theory to practice. Brief Funct Genomic Proteomic 9:166–177

Jansen RC, Nap JP (2001) Genetical genomics: the added value from segregation. Trends Genet 17:388–391

Jansen RC, Stam P (1994) High resolution of quantitative traits into multiple loci via interval mapping. Genetics 136:1447–1455

Jones PA, Takai D (2001) The role of DNA methylation in mammalian epigenetics. Science 293:1068–1070

Joosen RVL, Ligterink W, Hilhorst HWM, Keurentje JJB (2009) Advances in genetical genomics of plants. Curr Genomics 10:540–549

Jordan MC, Somers DJ, Banks TW (2007) Identifying regions of the wheat genome controlling seed development by mapping expression quantitative trait loci. Plant Biotechnol J 5:442–453

Jorgensen JA, Nguyen HT (1995) Genetic analysis of heat-shock proteins in maize. Theor Appl Genet 91:38–46

Jorrín-Novo JV, Maldonado AM, Echevarría-Zomeño S, Valledor L, Castillejo MA, Curto M, Valero J, Sghaier B, Donoso G, Redondo I (2009) Plant proteomics update (2007–2008): second-generation proteomic techniques, an appropriate experimental design, and data analysis to fulfill MIAPE standards, increase plant proteome coverage and expand biological knowledge. J Proteomics 72:285–314

Kacser H, Burns JA (1981) The molecular basis of dominance. Genetics 97:639–666

Kaeppler SM, Phillips RL (1993a) DNA methylation and tissue culture-induced variation in plants. In Vitro Cell Dev Biol 29:125–130

Kaeppler SM, Phillips RL (1993b) Tissue culture-induced DNA methylation variation in maize. Proc Natl Acad Sci 90:8773–8776

Keurentjes JJB, Fu J, de Vos RCH, Lommen A, Hall RD, Bino RJ, van der Plas LHW, Jansen RC, Vreugdenhil D, Koornneef M (2006) The genetics of plant metabolism. Nat Genet 38:842–849

Keurentjes JJB, Fu J, Terpstra IR, Garcia JM, van den Ackerveken G, Basten Snoek L, Peeters AJM, Vreugdenhil D, Koornneef M, Jansen RC (2007) Regulatory network construction in Arabidopsis by using genome-wide gene expression quantitative trait loci. Proc Natl Acad Sci USA 104:1708–1713

Khlestkina EK, Huang XQ, Quenum FJB, Chebotar S, Röder MS, Börner A (2004) Genetic diversity in cultivated plants – loss or stability? Theor Appl Genet 108:1466–1472

Kirst M, Myburg AA, De Leon JP, Kirst ME, Scott J, Sederoff R (2004) Coordinated genetic regulation of growth and lignin revealed by quantitative trait locus analysis of cDNA microarray data in an interspecific backcross of eucalyptus. Plant Physiol 135:2368–2378

Kirst M, Basten CJ, Myburg AA, Zeng ZB, Sederoff RR (2005) Genetic architecture of transcript-level variation in differentiating xylem of a eucalyptus hybrid. Genetics 169:2295–2303

Kliebenstein D (2009) Quantitative genomics: analyzing intraspecific variation using global gene expression polymorphisms or eQTLs. Annu Rev Plant Biol 60:93–114

Kliebenstein DJ, West MAL, van Leeuwen H, Kim K, Doerge RW, Michelmore RW, St. Clair DA (2006a) Genomic survey of gene expression diversity in *Arabidopsis thaliana*. Genetics 172:1179–1189

Kliebenstein DJ, West MAL, van Leeuwen H, Loudet O, Doerge RW, St Clair DA (2006b) Identification of QTLs controlling gene expression networks defined *a priori*. BMC Bioinformatics 7:308

Kusano M, Fukushima A, Kobayashi M, Hayashi N, Jonsson P, Moritz T, Ebana K, Saito K (2007) Application of a metabolomic method combining one-dimensional and two-dimensional gas chromatography-time-of-flight/mass spectrometry to metabolic phenotyping of natural variants in rice. J Chromatogr B 855:71–79

Labate JA, Lamkey KR, Mitchell SH, Kresovich S, Sullivan H, Smith JSC (2003) Molecular and historical aspects of Corn Belt dent diversity. Crop Sci 43:80–91

Lande R, Thompson R (1990) Efficiency of marker-assisted selection in the improvement of quantitative traits. Genetics 124:743–756

Lander ES, Botstein D (1989) Mapping Mendelian factors underlying quantitative traits using RFLP linkage maps. Genetics 121:185–199

Langridge P, Fleury D (2011) Making the most of 'omics' for crop breeding. Trends Biotechnol 29:33–40

Lapitan NLV, Hess A, Cooper B, Botha AM, Badillo D, Iyer H, Menert J, Close TJ, Wright L, Hanning G, Tahir M, Lawrence C (2009) Differentially expressed genes during malting and correlation with malting quality phenotypes in barley (*Hordeum vulgare* L.). Theor Appl Genet 118:937–952

Lazo GR, Chao S, Hummel DL, Edwards H, Crossman CC et al (2004) Development of an expressed sequence tag (EST) resource for wheat (*Triticum aestivum* L.): EST generation, unigene analysis, probe selection and bioinformatics for a 16,000-locus bin-delineated map. Genetics 168:585–593

Li Y, Huang Y, Bergelson J, Nordborg M, Borevitz JO (2010) Association mapping of local climate-sensitive quantitative trait loci in Arabidopsis thaliana. Proc Natl Acad Sci USA 107:21199–21204

Lincoln SE, Daly MJ, Lander ES (1993) Mapping genes controlling quantitative traits using MapMaker/QTL version 1.1. Whitehead Institute for Biomedical Research, Cambridge, MA

Liu YS, Gur A, Ronen G, Causse M, Damidaux R, Buret M, Hirschberg J, Zamir D (2003) There is more to tomato fruit colour than candidate genes. Plant Biotechnol J 1:195–207

Lu H, Bernardo R (2001) Molecular marker diversity among current and historical maize inbreds. Theor Appl Genet 103:613–617

Lübberstedt T, Melchinger AE, Fähr S, Klein D, Dally A, Westhoff P (1998) QTL mapping in testcrosses of flint lines of maize: III. Comparison across populations for forage traits. Crop Sci 38:1278–1289

Melchinger AE (1999) Genetic diversity and heterosis. In: Coors JG, Pandey S (eds) The genetics and exploitation of heterosis in crops. ASA-CSSA, Madison, pp 99–118

Melchinger AE, Gumber RK (1998) Overview of heterosis and heterotic groups in agronomic crops. In: Lamkey KR, Staub JE (eds) Concepts and breeding of heterosis in crop plants. CSSA, Madison, pp 29–44

Melchinger AE, Graner A, Singh M, Messmer MM (1994) Relationships among European barley germplasm: I. Genetic diversity among winter and spring cultivars revealed by RFLPs. Crop Sci 34:1191–1199

Melchinger AE, Utz HF, Schön CC (1998) Quantitative trait locus (QTL) mapping using different testers and independent population samples in maize reveals low power of QTL detection and large bias in estimates of QTL effects. Genetics 149:383–403

Menéndez CM, Ritter E, Schäfer-Pregl R, Walkemeier B, Kalde A, Salamini F, Gebhardt C (2002) Cold sweetening in diploid potato: mapping quantitative trait loci and candidate genes. Genetics 162:1423–1434

Menkir A, Melake-Berhan A, The C, Ingelbrecht I, Adepoju A (2004) Grouping of tropical mid-altitude maize inbred lines on the basis of yield data and molecular markers. Theor Appl Genet 108:1582–1590

Mensack MM, Fitzgerald VK, Ryan EP, Lewis MR, Thompson HJ, Brick MA (2010) Evaluation of diversity among common beans (*Phaseolus vulgaris* L.) from two centers of domestication using 'omics' technologies. BMC Genomics 11:686–696

Messmer MM, Melchinger AE, Lee M, Woodman WL, Lee EA, Lamkey KR (1991) Genetic diversity among progenitors and elite lines from the Iowa Stiff Stalk Synthetic (BSSS) maize population: comparison of allozyme and RFLP data. Theor Appl Genet 83:97–107

Metzker ML (2010) Sequencing technologies – the next generation. Nat Rev Genet 11:31–46

Meuwissen THE, Hayes BJ, Goddard ME (2001) Prediction of total genetic value using genome-wide dense marker maps. Genetics 157:1819–1829

Meyer S, Pospisil H, Scholten S (2007) Heterosis associated gene expression in maize embryos 6 days after fertilization exhibits additive, dominant and overdominant pattern. Plant Mol Biol 63:381–391

Mochida K, Shinozaki K (2010) Genomics and bioinformatics resources for crop improvement. Plant Cell Physiol 51:497–523

Moreau L, Charcosset A, Gallais A (2004) Experimental evaluation of several cycles of marker-assisted selection in maize. Euphytica 137:111–118

Mullis KB, Faloona FA (1987) Specific synthesis of DNA in vitro via a polymerase-catalyzed chain reaction. Methods Enzymol 155:335–350

Muñoz-Amatriaín M, Xiong Y, Schmitt MR, Bilgic H, Budde AD, Chao S, Smith KP, Muehlbauer GJ (2010) Transcriptome analysis of a barley breeding program examines gene expression diversity and reveals target genes for malting quality improvement. BMC Genomics 11:653

Nakamura S, Hosaka K (2010) DNA methylation in diploid inbred lines of potatoes and its possible role in the regulation of heterosis. Theor Appl Genet 120:205–214

Ohdan T, Francisco PB Jr, Sawada T, Hirose T, Terao T, Satoh H, Nakamura Y (2005) Expression profiling of genes involved in starch synthesis in sink and source organs of rice. J Exp Bot 56:3229–3244

Ozsolak F, Platt AR, Jones DR, Reifenberger JG, Sass LE, McInerney P, Thompson JF, Bowers J, Jarosz M, Milos PM (2009) Direct RNA sequencing. Nature 461:814–818

Park OK (2004) Proteomic studies in plants. J Biochem Mol Biol 37:133–138

Paschold A, Marcon C, Hoecker N, Hochholdinger F (2010) Molecular dissection of heterosis manifestation during early maize root development. Theor Appl Genet 120:441–450

Peng H, Zhang J (2009) Plant genomic DNA methylation in response to stresses: potential applications and challenges in plant breeding. Prog Nat Sci 19:1037–1045

Pfalz M, Mikkelsen MD, Bednarek P, Olsen CE, Halkier BA, Kroymann J (2011) Metabolic engineering in *Nicotiana benthamiana* reveals key enzyme functions in *Arabidopsis* indole glucosinolate modification. Plant Cell 23:716–729

Phillips RL, Kaeppler SM, Olhoft P (1994) Genetic instability of plant tissue cultures: breakdown of normal controls. Proc Natl Acad Sci USA 91:5222–5226

Piepho HP (2009) Ridge regression and extensions for genomewide selection in maize. Crop Sci 49:1165–1176

Potokina E, Druka A, Luo ZW, Wise R, Waugh R, Kearsey M (2007) eQTL analysis of 16,000 barley genes reveals a complex pattern of genome wide transcriptional regulation. Plant J 53:90–101

Potokina E, Druka A, Luo Z, Wise R, Waugh R, Kearsey M (2008a) Gene expression quantitative trait locus analysis of 16,000 barley genes reveals a complex pattern of genome-wide transcriptional regulation. Plant J 53:90–101

Potokina E, Druka A, Luo Z, Moscou M, Wise R, Waugh R, Kearsey M (2008b) Tissue-dependent limited pleiotropy affects gene expression in barley. Plant J 56:287–296

Qi LL, Echalier B, Chao S, Lazo GR, Butler GE, Anderson OD et al (2004) A chromosome bin map of 16,000 expressed sequence tag loci and distribution of genes among the three genomes of polyploidy wheat. Genetics 168:701–712

Rafalski A (2002) Applications of single nucleotide polymorphism in crop genetics. Curr Opin Plant Biol 5:94–100

Ralph J, Guillaumie S, Grabber JH, Lapierre C, Barriere Y (2004) Genetic and molecular basis of grass cell-wall biosynthesis and degradability. III. Towards a forage grass ideotype. C R Biol 327:467–479

Ramagopal S (1990) Protein polymorphism in sugarcane revealed by two-dimensional gel analysis. Theor Appl Genet 79:297–304

Riccardi F, Gazeau P, Jacquemot M-P, Vincent D, Zivy M (2004) Deciphering genetic variations of proteome responses to water deficit in maize leaves. Plant Physiol Biochem 42:1003–1011

Riddle NC, Jiang H, An L, Doerge RW, Birchler JA (2010) Gene expression analysis at the intersection of ploidy and hybridity in maize. Theor Appl Genet 120:341–353

Rossignol M, Peltier J-B, Mock H-P, Matros A, Maldonado AM, Jorrín JV (2006) Plant proteome analysis: a 2004–2006 update. Proteomics 6:5529–5548

Roussel V, Koenig J, Beckert M, Ballfourier F (2004) Molecular diversity in French bread wheat accessions related to temporal trends and breeding programmes. Theor Appl Genet 108:920–930

Russell JR, Ellis RP, Thomas WTB, Waugh R, Provan J, Booth A, Lawrence P, Young G, Powell W (2000) A retrospective analysis of spring barley germplasm development from 'foundation genotypes' to currently successful cultivars. Mol Breed 6:553–568

Saito K, Matsuda F (2010) Metabolomics for functional genomics, systems biology, and biotechnology. Annu Rev Plant Biol 61:463–489

Salekdeh GH, Komatsu S (2007) Crop proteomics: aim at sustainable agriculture of tomorrow. Proteomics 7:2976–2996

Sano H, Kamada I, Youssefian S, Katsumi M, Wabiko H (1990) A single treatment of rice seedlings with 5-azacytidine induces heritable dwarfism and undermethylation of genomic DNA. Mol Gen Genet 220:441–447

Sax K (1923) The association of size differences with seed coat pattern and pigmentation in *Phaseolus vulgaris*. Genetics 8:552–560

Schadt EE, Monks SA, Drake TA, Lusis AJ, Che N, Colinayo V, Ruff TG, Milligan SB, Lamb JR, Cavet G, Linsley PS, Mao M, Stoughton RB, Friend SH (2003) Genetics of gene expression surveyed in maize, mouse and man. Nature 422:297–302

Schaeffer LR (2006) Strategy for applying genome-wide selection in dairy cattle. J Anim Breed Genet 123:218–223

Schauer N, Semel Y, Roessner U, Gur A, Balbo I, Carrari F, Pleban T, Perez-Melis A, Bruedigam C, Kopka J, Willmitzer L, Zamir D, Fernie AR (2006) Comprehensive metabolic profiling and phenotyping of interspecific introgression lines for tomato improvement. Nat Biotechnol 24:447–454

Schena M, Shalon D, Davis RW, Brown PO (1995) Quantitative monitoring of gene expression patterns with a complementary DNA microarray. Science 270:467–470

Schön CC, Utz HF, Groh S, Truberg B, Openshaw S, Melchinger AE (2004) Quantitative trait locus mapping based on resampling in a vast maize testcross experiment and its relevance to quantitative genetics for complex traits. Genetics 167:485–498

Schrag TA, Melchinger AE, Sørensen AP, Frisch M (2006) Prediction of single-cross hybrid performance for grain yield and grain dry matter content in maize using AFLP markers associated with QTL. Theor Appl Genet 113:1037–1047

Schrag TA, Maurer HP, Melchinger AE, Piepho H-P, Peleman J, Frisch M (2007) Prediction of single-cross hybrid performance in maize using haplotype blocks associated with QTL for grain yield. Theor Appl Genet 114:1345–1355

Schrag TA, Möhring J, Maurer HP, Dhillon BS, Melchinger AE, Piepho H-P, Sørensen AP, Frisch M (2009a) Molecular marker-based prediction of hybrid performance in maize using unbalanced data from multiple experiments with factorial crosses. Theor Appl Genet 118:741–751

Schrag TA, Frisch M, Dhillon BS, Melchinger AE (2009b) Marker-based prediction of hybrid performance in maize single-crosses involving double haploids. Maydica 54:353–362

Schrag TA, Möhring J, Kusterer B, Dhillon BS, Melchinger AE, Piepho HP, Frisch M (2010) Prediction of hybrid performance in maize using molecular markers and joint analyses of hybrids and parental inbreds. Theor Appl Genet 120:451–461

Scippa GS, Trupiano D, Rocco M, Viscosi V, Di Michele M, D'Andrea A, Chiatante D (2008) An integrated approach to the characterization of two autochthonous lentil (*Lens culinaris*) landraces of Molise (south-central Italy). Heredity 101:136–144

Shi C, Koch G, Ouzunova M, Wenzel G, Zein I, Lübberstedt T (2006) Comparison of maize brown-midrib isogenic lines by cellular UV-microspectrophotometry and comparative transcript profiling. Plant Mol Biol 62:697–714

Shi C, Uzarowska A, Ouzunova M, Landbeck M, Wenzel G, Lübberstedt T (2007) Identification of candidate genes associated with cell wall digestibility and eQTL (expression quantitative trait loci) analysis in a Flint × Flint maize recombinant inbred line population. BMC Genomics 8:22

Smith JSC, Smith OS (1987) Associations among inbred lines of maize using electrophoretic, chromatographic, and pedigree data – 1. Multivariate and cluster analysis of data from 'Lancaster Sure Crop' derived lines. Theor Appl Genet 73:654–664

Smith JSC, Smith OS (1988) Associations among inbred lines of maize using electrophoretic, chromatographic, and pedigree data – 1. Multivariate and cluster analysis of data from Iowa stiff stalk synthetic derived lines. Theor Appl Genet 76:39–44

Smith OS, Smith JSC, Bowen SL, Tenborg RA, Wall SJ (1990) Similarities among a group of elite maize inbreds as measured by pedigree, F1 grain yield, grain yield, heterosis, and RFLPs. Theor Appl Genet 80:833–840

Steinfath M, Gärtner T, Lisec J, Meyer RC, Altmann T, Willmitzer L, Selbig J (2010) Prediction of hybrid biomass in Arabidopsis thaliana by selected parental SNP and metabolic markers. Theor Appl Genet 120:239–247

Stitt M, Fernie AR (2003) From measurements of metabolites to metabolomics: an 'on the fly' perspective illustrated by recent studies of carbon-nitrogen interactions. Curr Opin Biotechnol 14:136–144

Street NR, Skogstrom O, Sjodin A, Tucker J, Rodriguez-Acosta M, Nilsson P, Jansson S, Taylor G (2006) The genetics and genomics of the drought response in Populus. Plant J 48:321–341

Stuber CW, Goodman MM, Johnson FM (1977) Genetic control and racial variation of β-Glucosidase isozymes in maize. Biochem Genet 15:383–394

Stupar RM, Gardiner JM, Oldre AG, Haun WJ, Chandler V, Springer NM (2008) Gene expression analysis in maize hybrids and hybrids with varying levels of heterosis. BMC Plant Biol 8:33

Stylianou IM, Affourtit JP, Schockley KR, Wilpan RY, Abdi FA, Bhardwaj S, Rollins J, Churchill GA, Paigen B (2008) Applying gene expression, proteomics and single-nucleotide polymorphism analysis for complex trait gene identification. Genetics 178:1795–1805

Swanson-Wagner RA, Jia Y, DeCook R, Borsuk LA, Nettleton D, Schnable PS (2006) All possible modes of gene action observed in a global comparison of gene expression in a maize F1 hybrid and its inbred parents. Proc Natl Acad Sci USA 103:6805–6810

Swanson-Wagner RA, DeCook R, Jia Y, Bancroft T, Ji T, Zhao X, Nettleton D, Schnable PS (2009) Paternal dominance of *trans*-eQTL influences gene expression patterns in maize hybrids. Science 326:1118–1120

Tanksley SD, McCouch R (1997) Seed banks and molecular maps: unlocking genetic potential from the wild. Science 277:1063–1066

Tanksley SD, Young ND, Paterson AH, Bonierbale MW (1989) RFLP mapping in plant breeding: new tools for an old science. Nat Biotechnol 7:257–264

Thiellement H, Zivy M, Plomion C (2002) Combining proteomic and genetic studies in plants. J Chromatogr B 782:137–149

Thiemann A, Fu J, Schrag TA, Melchinger AE, Frisch M, Scholten S (2010) Correlation between parental transcriptome and field data for the characterization of heterosis in *Zea mays* L. Theor Appl Genet 120:401–413

Thoday JM (1961) Location of polygenes. Nature 191:368–370

Utz HF, Melchinger AE (1996) PLABQTL: a program for composite interval mapping of QTL. J Quant Trait Loci 2(1). Available at http://www.uni-hohenheim.de/âĹijipspwww/soft.html. Verified 10 Sept 2004

Uzarowska A, Keller B, Piepho H-P, Schwarz G, Ingvardsen C, Wenzel G, Lübberstedt T (2007) Comparative expression profiling in meristems of inbred-hybrid triplets of maize based on morphological investigations of heterosis for plant height. Plant Mol Biol 63:21–34

Van der Hoeven R, Ronning C, Giovannoni J, Martin G, Tanksley S (2002) Deductions about the number, organization, and evolution of genes in the tomato genome based on analysis of a large expressed sequence tag collection and selective genomic sequencing. Plant Cell 14:1441–1456

Varshney RK, Graner A, Sorrells ME (2005) Genomics-assisted breeding for crop improvement. Trends Plant Sci 10:621–630

Varshney RK, Graner A, Sorrells ME (2006a) Genic microsatellite markers in plants: features and applications. Trends Biotechnol 23:48–55

Varshney RK, Hoisington DA, Tyagi AK (2006b) Advances in cereal genomics and applications in crop breeding. Trends Biotechnol 24:490–499

Velculescu VE, Zhang L, Vogelstein B, Kinzler KW (1995) Serial analysis of gene expression. Science 270:484–487

Vinocur B, Altman A (2005) Recent advances in engineering plant tolerance to abiotic stress: achievements and limitations. Curr Opin Biotechnol 16:123–132

Vuylsteke M, Kuiper M, Stam P (2000) Chromosomal regions involved in hybrid performance and heterosis: their AFLP-based identification and practical use in prediction models. Heredity 85:208–218

Vuylsteke M, Van Den Daele H, Vercauteren A, Zabeau M, Kuiper M (2006) Genetic dissection of transcriptional regulation by cDNA-AFLP. Plant J 45:439–446

Walsh B (2001) Quantitative genetics in the age of genomics. Theor Popul Biol 59:175–184

Wang RL, Stec A, Hey J, Lukens L, Doebley J (1999) The limits of selection during maize domestication. Nature 398:236–239

Wei Z, Hu W, Lin Q, Cheng X, Tong M, Zhu L, Chen R, He G (2009) Understanding rice plant resistance to the brown planthopper (*Nilaparvata lugens*): a proteomic approach. Proteomics 9:2798–2808

West MAL, Kim K, Kliebenstein DJ, van Leeuwen H, Michelmore RW, Doerge RW, St Clair DA (2007) Global eQTL mapping reveals the complex genetic architecture of transcript level variation in Arabidopsis. Genetics 175:1441–1450

Witzel K, Pietsch C, Strickert M, Matros A, Röder MS, Weschke W, Wobus U, Mock H-P (2011) Mapping of quantitative trait loci associated with protein expression variation in barley grains. Mol Breed 27:301–314

Wright S (1934) Molecular and evolutionary theories of dominance. Am Nat 68:24–53

Xiong LZ, Xu CG, Saghai Maroof MA, Zhang Q (1999) Patterns of cytosine methylation in an elite rice hybrid and its parental lines, detected by a methylation-sensitive amplification polymorphism technique. Mol Gen Genet 261:439–446

Yu J, Bernardo R (2004) Changes in genetic variance during advanced cycle breeding in maize. Crop Sci 44:405–410

Zeng Z-B (1993) Theoretical basis of precision mapping of quantitative trait loci. Proc Natl Acad Sci USA 90:10972–10976

Zhang W, Li F, Nie L (2010) Integrating multiple 'omics' analysis for microbial biology: application and methodologies. Microbiology 156:287–301

Zhang X, Cal AJ, Borevitz JO (2011) Genetic architecture of regulatory variation in Arabidopsis thaliana. Genome Res 21:725–733

Zhao X, Chai Y, Liu B (2007) Epigenetic inheritance and variation of DNA methylation level and pattern in maize intra-specific hybrids. Plant Sci 172:930–938

Zivy M, Thiellement H, de Vienne D, Hofmann JP (1984) Study on nuclear and cytoplasmic genome expression in wheat by two-dimensional gel electrophoresis. 2. Genetic differences between two lines and two groups of cytoplasms at five developmental stages or organs. Theor Appl Genet 68:335–345

Part II
Identification of Quantitative Trait Polymorphisms (QTPs)

Chapter 3
Gene Identification: Forward Genetics

Qing Ji

Forward genetics aims to identify the sequence variation(s) responsible for a given phenotypic trait. Unlike reverse genetics, which seeks to find the mutated phenotype resulted from known sequence changes, gene identification via forward genetics has different process. It starts with the identification of the mutant phenotype (caused by either artificial mutagenesis or natural variation). Then a mapping population which segregates at the interested phenotype is produced. Together with high density molecular markers, the candidate genes are mapped and further narrowed down to a small chromosome region. Finally, genetic engineering (e.g. overexpression, gene knock down), complementation or reverse genetics methods are applied (Salvi and Tuberosa 2005) to identify the gene that is responsible for the phenotype. In this chapter, each step involved in gene identification by forward genetics (Fig. 3.1) will be reviewed in detail along with potential and limitations for each of them.

Generation and Identification of Mutants

This is the first step towards gene identification in the forward genetics approach. The mutants could be either naturally existed variations or artificially induced. Natural mutants often occur along with the evolution at very low rate of mutation, which tend to keep the adaptive traits, for example resistances to abiotic/biotic stresses, and discard the neutral or negative ones (Jiang and Ramachandran 2010). Therefore, it is usually hard to find the natural mutants for each trait of interest. Fortunately, artificial mutation provided ample resources for researchers. There are generally three types of artificial mutants: insertion mutation; physical and chemical mutation. T-DNA or transposons are used to create large number of

Q. Ji (✉)
Department of Agronomy, Iowa State University, Ames, IA 50011, USA
e-mail: qingji@iastate.edu

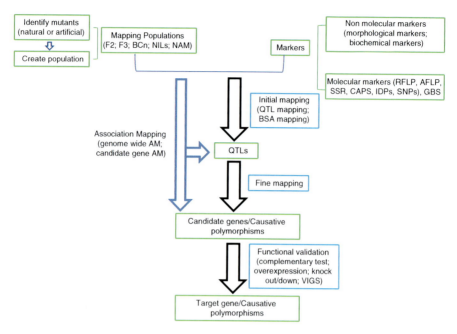

Fig. 3.1 Flowchart of gene identification by forward genetics approach

mutants in plants. If the sequences insert into a gene coding or regulating region, they will disrupt the gene function. Up to date, plenty of mutant databases for T-DNA or transposon insertions mutants have been available to public, mainly in Arabidopsis, rice and maize (Table 3.1). Although insertion mutagenesis is popular in the above mentioned experimental species, it is less feasible for other grasses, due to either immature transformation system or large size of genome (Kuromori et al. 2009). Comparatively, chemical mutagenesis by ethyl methanesulfonate (EMS), methylnitrosourea (MNU) or sodium azide, and physical mutagenesis by fast-neutrons, gamma rays or ion-beam irradiation have been more widely applied to generate mutants, contributed by the higher induction efficiency, broad induction spectrum and applicability to almost all plant species (Peters et al. 2003). The related databases are also available now, not only for rice and Arabidopsis, but also for sorghum, barley, tomato, soybean, *lotus japonicas* and so on (Table 3.1).

If the interested mutants are available from a mutant database, confirmation of the phenotype is desired before starting gene identification. However, if the mutants are generated without known phenotype, the phenotype of the potential mutants will be first identified by planting under different growth condition, enduring different growth stages and comparing with the wild type. Once the phenotype is decided, observe the phenotype in the F_1 and F_2 generations to determine if it is controlled by monogenic or multigenic factors. If monogenic, the dominance or recessiveness of the underlying gene can also be determined by analyzing the wild and mutant phenotype ratio in the F_2 progeny.

3 Gene Identification: Forward Genetics

Table 3.1 Artificial mutant databases

Species	Website	Type
Arabidopsis	http://www.arabidopsis.org/abrc/	T-DNA, transposon element, EMS
Rice	http://rmd.ncpgr.cn/index.cgi?nickname=	T-DNA insertion (enhancer trap)
	http://www.postech.ac.kr/life/pfg/risd/	T-DNA
	http://www.plantsignal.cn/zhcn/index.html	T-DNA
	http://trim.sinica.edu.tw/home	T-DNA
	http://tos.nias.affrc.go.jp/	tos17
	http://urgi.versailles.inra.fr/OryzaTagLine/	T-DNA
	http://tilling.ucdavis.edu/index.php/Main_Page	MNU
Maize	http://www.maizegdb.org/rescuemu-phenotype.php	RescueMu
	http://mtm.cshl.edu/	Mu
Tomato	http://tomatoma.nbrp.jp/	EMS; Gamma irradiation
	http://www.agrobios.it/tilling/index.html	EMS
	http://zamir.sgn.cornell.edu/mutants/links/abstract.html	EMS; fast-neutron
Barrel medic	http://bioinfo4.noble.org/mutant/	Fast Neutron; Tnt1 retrotransposon
Soybean	www.soybeantilling.org/index.jsp	EMS
	http://www.soybase.org/mutants/about.php	Fast neutron
Sorghum	http://www.lbk.ars.usda.gov/psgd/index-sorghum.aspx	EMS
Barley	http://bioinf.scri.ac.uk/barley/	EMS
Lotus japonicas	http://www.shigen.nig.ac.jp/bean/lotusjaponicus/top/top.jsp	EMS

Mapping Population Construction

Ideal mapping population is a key point for successful gene identification. To construct a mapping population, the first encountered question is how to choose the parents. Generally speaking, the selected parents should have sufficient genetic variation, especially for the initial/coarse mapping. Large phenotypic variation is another favorable factor to consider, but not a necessity. Although a large phenotypic variation is usually due to genetic difference, it can also be affected by the environment. Therefore, validation of the phenotypic variation at different environments is necessary to minimize environment effects. On the other hand, small phenotypic variances don't necessarily mean no genetic difference, since different loci could affect the same traits from different sides and thus result in similar traits. The third consideration is about the available resources. For example, to identify the maize brown midrib genes, the mapping population will be generated by crossing a wild type (green mid rib) inbred line with the mutant genotype. Two inbred lines, B73 and Z51 are candidates, B73 is preferred in this case because its whole genomics and EST/cDNA information and the BAC clones are publicly available now, which will provide massive information for marker development, candidate gene annotation and eventually validation. However, if the mutant is artificially generated, the background line (the wild type) is an ideal parent to cross with to produce the mapping population because it minimizes the background noise.

The second question is what kind of population needs to be generated for the mapping. Some phenotypes are simple Mendelian traits (qualitative trait), especially artificially mutagenesis traits. In those cases, a segregation ratio of 3:1 (dominant: recessive) is expected while observing the phenotype of the F_2 population. However, incomplete penetrance or environmental factors may sometimes skew the ratio, in which case, phenotyping the F_3 population to infer that of the F_2 individuals is necessary to get the accurate phenotypic data. For the mapping, a small number of F_2 or BC_1 population derived from the cross of the mutant and the wild type serves well to define a rough position of the responsible gene. Then, a large number of individuals which potentially include more recombinants are required for fine mapping. However, in most cases the target traits are contributed by multiple genetic factors (quantitative trait), for example, grain yield, flowering time, and plant height which are easily affected by the environments and often show epistatic interactions. The phenotypic data observed in corresponding F_2 population often exhibits continuous distribution. To fine map genes underlying a quantitative trait, an initial QTL (Quantitative Trait Loci) mapping with F_2 or backcross populations is usually required to identify the potential loci responsible for the interested trait. Based on the QTL mapping results, higher generations of backcross population need to be produced with the aid of adjacent markers to dissect the complex trait into single Mendelian trait for mapping and cloning each of the underlying genes respectively. For example, rice flowering time is regulated by multiple genetic factors. Yano et al. (1997) reported five potential QTL, *Hd1-Hd5*, controlling the heading date with an F_2 population derived from Nipponbare (*japonica*) and Kasalath (*indica*). Then three BC_3F_1/BC_3F_2 populations were generated from the same cross so that only the interested QTL (*Hd1*, *Hd2* and *Hd3*) will separate in each population to fine map each QTL (Yamamoto et al. 1998). Map-based cloning of *Hd1* was succeeded with a BC_3F_3 population consisting of over 9,000 of individuals (Yano et al. 2000). While *Hd3*, which was previously regarded as a single QTL for heading date, was later dissected into two closely linked gene, *Hd3a* and *Hd3b*, by analyzing advanced backcross progeny and nearly isogenic lines (NILs) (Monna et al. 2002). Further analysis of 2207 recombinant plants from the BC_3F_4 population narrowed down the *Hd3a* gene into a ~20 kb region (Kojima et al. 2002). The same strategy was used to obtain the mapping populations, that gave rise to the cloning of other genes involved in flowering time such as *Hd6* (Takahashi et al. 2001), *Ehd1* (Doi et al. 2004), *Ehd2* (Matsubara et al. 2008), *Ehd3* (Matsubara et al. 2011) and *DTH8* (Wei et al. 2010). However, only those major loci that explain large portion of phenotype variation can be realistically cloned. Up to date, all the successfully cloned QTL have more than 15% of contribution over the phenotype in the primary study.

Another very popular mapping population is nested association mapping (NAM) population, which contains 5,000 recombinant inbred lines (RILs) derived from the

F_2 crosses between B73 and 25 very diverse maize founder lines (Yu et al. 2008). The 5,000 RILs together with the IBM RIL population (intermated B73 × Mo17) comprise immortalized QTL mapping resource for researchers. The discovery of 1.6million SNPs from the 20% of maize low copy genome region with the panel of 27 NAM founders by Maize Hap project (Gore et al. 2009) has enabled the genome wide association study (GWAS) in maize. Furthermore, all the RILs have been genotyped using 1,106 SNPs with the coverage of the whole genome. Both the RILs and the genotype data are available to public now (http://www.panzea.org), therefore, no further genotyping is required for QTL analysis with NAM population. The only thing to do is to plant the seeds and phenotype the trait of interest. GWAS enables the identification of the QTL, candidate gene even causative SNPs underlying the phenotype of interest. NAM population has been successfully applied to dissect genetic factors that control leaf architecture, e.g. upper leaf angle, leaf width and leaf length in maize (Tian et al. 2011). Besides, resistance studies with NAM population were also reported. 32 QTL were identified conferring resistance to southern leaf blight disease (Kump et al. 2011) and 29 QTL were discovered conferring resistance to northern leaf blight (Poland et al. 2011). Multiple candidate genes and SNPs associated with resistance were also identified in those studies. Despite all the advantages of NAM population, the limitations were also reported, including the SNPs may not represent of all maize haploid map, sequence polymorphisms other than SNPs are not involved, and the number of population founders is limited which may lead to unable to dissect rare QTN (Tian et al. 2011).

Finally, one needs to consider how many progenies are required to ensure successful positional cloning of the causative gene. Dinka et al. (2007) proposed an equation to calculate it. To achieve over 95% of probability of success, the number of individuals $N = (100R \times \lambda_T)/\text{T-marker}$. Where R is the local recombination frequency and T-marker is distance between the closest two molecular markers, λ_T is the number of crossovers between the closest two molecular markers (≥ 2).

Marker System and New Marker Development

High density of polymorphic markers is another prerequisite for successful gene isolation. There are generally three types of markers based on their characteristics: morphological markers, e.g. the color of flowers or seeds; biochemical markers, e.g. isozymes; and molecular markers, which includes hybridization of based markers e.g. RFLP (restriction fragment length polymorphism) and PCR based markers e.g. AFLP (amplified fragment length polymorphism), RAPD (random amplified polymorphic DNA), SSR (simple sequence repeats), INDEL (insertion and deletion) and SNP (single nucleotide polymorphism). The criterion for the ideal marker includes: (1) highly polymorphic; (2) independent of environment or development stage; (3) codominant; (4) easy to detect and reproducible; (5) time and cost effective; (6) abundant in genomes. Since morphological and biochemical markers usually depend on environments and development stages, often very limited in numbers and detect methods, they are less used now. In contrast, molecular markers are independent of developmental stages or environments, and are abundant in plant genomes. With the advancement of automated marker detection technology and huge amount of sequence information available, molecular markers are widely used today. However, with so many kinds of molecular markers available, which differed on the amount of DNA required, degrees of polymorphisms, reproducibility and costs, the major question is how to decide which one serves best for a given mapping project.

RFLP markers are based on hybridization which is robust but very laborious and require using the radioactive materials (Williams et al. 1990). Comparatively, RAPD markers are much easier to use, however they have low reproducibility due to mismatch annealing which has limited its application (Neale and Harry 1994; Demeke et al. 1997; Karp et al. 1997). Garcia et al. (2004) reported that AFLP markers are fast and reliable and hence the best choice compared with RFLP and RAPD in evaluation maize diversity study. AFLP doesn't require any prior sequence information and therefore can be applied to almost any plant species, especially to those with very few sequence information available. Besides, AFLP is highly reproducible and informative. It is possible to obtain information for multiple loci (multiple polymorphic bands) based on one single core set of primers. AFLP markers can be detected either by polyacrylamide gel or an ABI DNA sequencer. Besides, the polymorphic AFLP markers are often amendable to be converted into PCR based markers, such as STS (sequence-tagged-site), CAPS (cleaved amplified polymorphic sequence) or SCAR (Sequence Characterized Amplified Regions) markers (Meksem et al. 2001; Xu et al. 2001; Dussle et al. 2002; Weerasena et al. 2004), which makes it easier to detect recombinants among large segregation population.

SSR marker was the marker of choice for most of the crops in the past decades (Bhattramakki et al. 2002; Jones et al. 2007). The frequency of SSR markers in maize is approximately one in every 8 kb (Wang et al. 1994). The biggest advantage of SSR is the amplification of multiple alleles at the same time, therefore it is more informative than SNP, which can have up to four alleles. Nonetheless, SNP are the most abundant DNA polymorphisms for almost every crops species, hence they are promising for constructing high density maps (Bhattramakki et al. 2002). It is estimated that the average frequency for one SNP is every 44 bp in the maize genome (Gore et al. 2009); 170 bp in rice (Yu et al. 2002); 130 bp in sugar beet (Schneider et al. 2001); 500 bp in cotton (Lu et al. 2005) and 3.3 and 6.6 kb in Arabidopsis (Jander et al. 2002). Many modern technologies have been applied to detect SNP in large scale, for example, the Infinium and GoldenGate system. The reproducibility of SNP varies from 98.1 to 99.3%, depending on the detection system, which is higher than SSR markers (91.7%) (Jones et al. 2007). The development of the next generation sequencing technologies has greatly decreased the cost for whole genome sequencing, which makes more and more sequence information publicly available. According to the report from NCBI (National Center for Biotechnology Information), there are 141 plant organisms so far have been sequenced and part of them have been completed and released to public (Table 3.2).

3 Gene Identification: Forward Genetics

Table 3.2 Sequenced plant genomes

Common name	Scientific name	Status	Link
Amborella	*Amborella trichopoda*	Incompleted	http://www.amborella.org/
Apple	*Malus × domestica*	Completed	http://genomics.research.iasma.it/gb2/gbrowse/apple/
Arabidopsis	*Arabidopsis thaliana*	Completed	http://arabidopsis.org/
Arabidopsis lyrata	*Arabidopsis lyrata*	Completed	http://genome.jgi-psf.org/Araly1/Araly1.download.html
Brachy	*Brachypodium distachyon*	Completed	http://www.brachypodium.org
Brassica rapa	*Brassica rapa*	Completed	http://www.brassica.info/
Cannabis	*Cannabis sativa*	Completed	http://genome.ccbr.utoronto.ca/cgi-bin/hgGateway
Capsella	*Capsella rubella*	Incompleted	http://www.phytozome.net/capsella.php
Cassava	*Manihot esculenta*	Incompleted	http://www.phytozome.net/cassava.php
Castor bean	*Ricinus communis*	Completed	http://castorbean.jcvi.org/index.php
Chocolate	*Theobroma cacao*	Completed	http://www.cacaogenomedb.org/main
Clementine orange	*Citrus clementina*	Incompleted	http://www.phytozome.net/clementine.php
Columbine	*Aquilegia sp.*	Unpublished	http://www.phytozome.net/aquilegia.php
Common bean	*Phaseolus vulgaris*	Incompleted	http://www.phytozome.org/commonbean.php
Cotton	*Gossypium raimonddi*	Incompleted	http://www.phytozome.net/cotton.php
Cucumber	*Cucumis sativus*	Completed	http://www.phytozome.net/cucumber.php
Date palm	*Phoenix dactylifera*	Completed	http://qatar-weill.cornell.edu/research/datepalmGenome/
Flax	*Linum usitatissimum*	Unpublished	http://www.phytozome.net/flax
Foxtail millet	*Setaria italica*	Unpublished	http://www.phytozome.net/foxtailmillet.php
Grape	*Vitis vinifera*	Completed	http://www.genoscope.cns.fr/externe/GenomeBrowser/Vitis/
Green agla	*Chlamydomonas reinhardtii*	Completed	http://genome.jgi-psf.org/Chlre3/Chlre3.home.html
Lotus	*Lotus japonicus*	Completed	http://www.kazusa.or.jp/lotus/index.html
Lycophyte	*Selaginella moellendorffii*	Completed	http://genome.jgi-psf.org/Selmo1/Selmo1.download.html

(continued)

Table 3.2 (continued)

Common name	Scientific name	Status	Link
Maize	*Zea mays*	Completed	http://www.maizesequence.org/index.html
Medicago	*Medicago truncatula*	Incompleted	http://www.medicagohapmap.org/?genome
Monkey flower	*Mimulus guttatus*	Incompleted	http://www.phytozome.net/mimulus
Moss	*Physcomitrella patens*	Completed	http://genome.jgi-psf.org/Phypa1_1/Phypa1_1.info.html
Papaya	*Carica papaya*	Completed	http://www.plantgdb.org/CpGDB/
Peach	*Prunus persica*	Unpublished	http://www.rosaceae.org/peach/genome
Pigeon pea	*Cajanus cajan*	Completed	http://dx.doi.org/10.1038/nbt.2022
Poplar	*Populus trichocarpa*	Completed	http://www.phytozome.net/poplar.php
Potato	*Solanum tuberosum*	Completed	http://potatogenomics.plantbiology.msu.edu/index.html
Rice	*Oryza sativa L. ssp. japonica*	Completed	http://rgp.dna.affrc.go.jp/E/IRGSP/rap-db1.html
Rice	*Oryza sativa L. ssp. indica*	Completed	http://rice.plantbiology.msu.edu/
Rose gum tree	*Eucalyptus grandis*	Unpublished	http://www.phytozome.net/eucalyptus.php
Salt cress	*Thellungiella parvula*	Pubmished	http://thellungiella.org/
Selaginella	*Selaginella moellendorffii*	Completed	http://genome.jgi-psf.org/Selmo1/Selmo1.download.html
Sorghum	*Sorghum bicolor*	Completed	http://www.phytozome.net/sorghum
Soybean	*Glycine max*	Completed	http://www.phytozome.net/soybean.php
Strawberry	*Fragaria vesca*	Completed	http://www.strawberrygenome.org/
Sweet orange	*Citrus sinensis*	Incompleted	http://www.phytozome.net/citrus.php
Thellungiella parvula	*Thellungiella parvula*	Completed	http://dx.doi.org/10.1038/ng.889
Tomato	*Solanum lycopersicum*	Incompleted	http://mips.helmholtz-muenchen.de/plant/tomato/about/releaseNotes.jsp

The sequence information has made the SNP marker development realistic in most of the plant species. Maize SNP database can be accessed from Panzea (http://www.panzea.org/index.html) and several maize SNP chips have been available as well. Rice SNP chips are also accessible through RiceSnp CONSORTIUM (http://

www.ricesnp.org/snpchips.aspx#). Now, genotyping by sequencing (GBS) marker system is gaining momentum by taking advantage of next generation sequencing technologies. GBS is quite straightforward in species with small genome size. However, when dealing with lager and more complicated genomes, for example, maize or barley, special care should be taken to reduce the redundancy and increase the efficiency. Introducing restriction digestions with a proper enzyme which rarely cut in the repetitive genome region can reduce the complexity and enrich the low copy region (usually the target region). Elshire et al. (2011) described a method for constructing a GBS library by digestion with a methylate sensitive enzyme *Ape*KI and application of barcode adaptors, which has greatly reduced genome complexity and enabled multiplex. For a 96-plex, the cost is reduced to $19/sample and 384-plex to $9/sample (http://www.maizegenetics.net/gbs-overview). Using this approach, maize IBM population which contains 276 RILs and barley DH populations were successfully genotyped with high quantity and quality (Elshire et al. 2011).

Initial Mapping

There are different mapping strategies depending on the goal of the research. QTL mapping is a general method to screen the whole genome for the putative loci controlling the interested trait. As a result, QTL mapping can provide a rough position of the candidate locus and a genetic map. However, most of the time, researchers are not interested in developing genetic maps, but in finding markers that are closely linked to genes coding for specific traits. In those cases, the Bulk Segregant Analysis (BSA) is a good choice. BSA was first proposed by Michelmore et al. (1991) for identifying closely linked markers to resistance gene against downy mildew in lettuce. Two bulks of DNA samples from the same segregating population are pooled based on their phenotype: highly resistant and extremely susceptible. The genetic compositions of the two bulks would be same except at the loci that are related to the phenotype. Molecular analysis of the two bulks together with the two parents revealed three RAPD markers that are closely linked to a resistance gene (Michelmore et al. 1991). Now BSA has been widely used in identification of closely linked markers to target loci, for example, loci related to leaf rust resistance in barley (Poulsen et al. 1995); scab resistance gene in apple (Yang et al. 1997); drought resistance in maize (Quarrie et al. 1999); heat tolerance in rice (Zhang et al. 2009) and so on. The bulked individuals are often from segregating F_2, double haploid or similar population. Currently more technologies are applied to increase the efficiency of BSA. For example, Borevitz et al. (2003) documented that application of the RNA expression array hybridization in combination with BSA to mapping dominant genes in Arabidopsis is feasible. Liu et al. (2010) also reported the successful mapping of mutants using high throughput SNP genotyping combined with BSA. Although BSA is a good choice to find closely linked markers, the disadvantages have to be considered as well. For quantitative traits controlled by more than two loci and each with a relatively small contribution to the phenotype, BSA mapping results may be confusing. Another pitfall for BSA analysis is the rare close recombinants are often been diluted, thus important information may be lost. Finally, BSA usually can't provide the distance between the detected marker and target trait, which is quite useful for later fine mapping.

Fine Mapping

Once the target locus was roughly mapped in a certain region of the genome, higher density of markers as well as plant materials with high frequency of recombination events are required for fine mapping. The rapid progress in sequencing and genotyping technologies make the development of markers no longer a limitation for most crops. Even for those less studied crops, sequence synteny or colinearity with the model species is often helpful for marker enrichment. Besides, if the BAC library is available and the mapped region is covered by a BAC contig, sequencing the BAC ends can also provide important information for marker development. As has been mentioned above, directly sequencing the recombinants is another choice, especially after the initial mapping which means only a small part of the genome needs to be analyzed. Also, bar code technology has allowed the sequencing of several individuals in one go (Binladen et al. 2007; Hoffmann et al. 2007; Parameswaran et al. 2007; Hamady et al. 2008).

With regard to the mapping materials, a proper segregating population with sufficient recombinants needs to be generated. Usually no problem is encountered for the Mendelian trait, a same F_2 or BC population used for the initial mapping with a larger number of individuals is sufficient to provide more recombinants. But for complex quantitative traits, only those loci with large effects can be further fine mapped and cloned. Mendelization by constructing a new experimental population, for example, the QTL-NILs (QTL nearly isogenic lines) population is indispensible to eliminate the effect of other loci and associate the phenotype with genotype. QTL-NILs are produced by several generations of backcrosses with the marker aided selection or from an introgression library, and a QTL-NIL population can be generated by crossing the QTL-NILs that differed only in the target QTL region.

Gene Validation

To validate the function of the candidate gene(s), gene overexpression, gene knockdown or knockout, gene complementation as well as other reverse genetics tools can be exploited separately or combined.

The full length sequence of the candidate gene driven by either its own promoter (complementation) or a stronger promoter like 35S cauliflower mosaic virus (CaMV) promoter (over expression) is inserted into an expression vector. The expression vector containing the candidate gene is then transformed into the receptor plants with homozygous recessive allele. The introduced genes will be (over) expressed in the transgenic plants and therefore complement the recessive alleles. Instead, in order to knockdown or knockout, antisense RNA or RNA interference construct targeting at specific region (usually the promoter or important domains) of the dominant allele will be transformed into the plants harboring dominant allele(s). The mRNA sequences transcribed from target dominant alleles will be degraded by the small RNA fragment introduced from the constructs, mediated by RISC (RNA-induced silencing complex), and therefore fail to translate into wildtype protein, which leads to the knock down or completely knock out of the target gene (Hannon 2002).

All the methods mentioned above demand mature transformation system. There are different transformation systems, for example, particle bombardment, electroporation, microinjection, silicon carbide, and chloroplast transformation available for transformation currently. And *Agrobacterium* mediated transformation remains the dominant system in plants (Barampuram and Zhang 2010). Although transformation mediated by *Agrobacterium* is already a mature system for some species, such as Arabidopsis, tobacco, switchgrass, soybean, sorghum, rice and maize, it remains to be established in many other crop species. Furthermore, transformation is usually genotype specific, mainly due to *in vitro* culture. For example, the hybrid maize line Hi-II, which is derived from the F_2 generation of A188 and B73, has been widely used as the receptor for maize transformation contributed by its high transformation efficiency (Vega et al. 2008). But the limitation lies on its genome heterozygosity. If A188 and B73 show different phenotypes in the target trait, which is controlled by more than one locus, it is usually difficult to decide the phenotype of the transgenic plants, therefore increase the difficulty in gene validation. Besides, transformation usually needs extra one or more generations of crosses to test the gene's function, which can be time consuming. Hence, finding alternative homozygous receptors with high efficiency and refining transformation systems are important for gene validation.

Virus Induced Gene Silence (VIGS) has provided another choice for gene identification. VIGS was previously used to describe the RNA-mediated antiviral defense mechanism in virus infected plants, which often get recovered from the infection (van Kammen 1997; Lu et al. 2003). The dsRNA was produced in infected cells and then processed into short interfering RNA (siRNA), which is part of the RISC to target homologous RNA for degradation (Bartel 2004). If the viral genome was modified by inserting part of host gene sequences, the silence will then target at the corresponding host gene and induce the gene silencing in host plant. The inserted host gene size usually ranges from 23 bp to 1.5 kb, however, 300–500 bp of DNA fragment with multiple stretches of over 23 bp can assure high efficient gene silencing (Burch-Smith et al. 2004). Different virus vector construct methods have facilitated various research purposes. The vectors integrated with Gateway (Invitrogen Crop. CA, USA) cloning site has made it easier to clone huge amount of cDNAs from EST library (Liu et al. 2002) and therefore allows high through-put of gene function analysis. Insertion of different gene fragments into a virus vector also allowed simultaneously silence of multiple distinct genes as have been demonstrated by Peele et al. (2001) and Turnage et al. (2002). If conserved sequences of a gene family are selected as the target site, it is possible to silence the whole gene family, thus avoiding redundancy.

In conclusion, VIGS has several advantages over other gene function analysis methods. First, it is a transient method and doesn't require transformation, which is much fast for traditional functional analysis. Second, VIGS is target specific and also capable of overcoming functional redundancy at the same time. Third, VIGS works in different genetic backgrounds and allows rapid comparisons of gene function between species. Last, VIGS is feasible for large scale of gene function identification contributed by the former advantages and the vector construction methods. Now VIGS has been widely used in gene function identification (Burch-Smith et al. 2004). Different plant viruses, including tobacco mosaic virus (TMV) (Lacomme et al. 2003), potato virus X (PVX) (Lu et al. 2003), barley stripe mosaic virus (BSMV) (Holzberg et al. 2002), Brome mosaic virus (BMV) (Scofield and Nelson 2009), Rice tungro bacilliform virus (RTBV) (Purkayastha et al. 2010) and so on, have been modified for VIGS in different species.

However, the disadvantages VIGS in gene identification also include several aspects as have been commented by Lu et al. (2003). Firstly, VIGS may fail to produce the phenotype due to the incomplete gene knock out. The residual expression may suffice the wild phenotype. Secondly, the similarity of gene sequences may interfere with the result analysis. But this can be addressed by designing the target sequence on both gene specific region and gene family conserved region. However, it remains as a problem in those species with few sequence information available. Thirdly, pleiotropy needs to be considered when interpreting the results, especially in plant resistance gene analysis. Besides, Burch-Smith et al. (2004) also mentioned the non-uniform silencing of the gene within single plant and between different plants and environments. Using an internal positive control that can mark the silenced region with visible phenotypes could be a solution to this problem. Finally, current reliable VIGS vectors usually have limited host ranges. Therefore, development of stable and efficient virus vector with wider host ranges is necessary for functional analysis in different species.

Reverse genetics tools, for example T-DNA insertion or TILLING (Targeting induced local lesions in genomes) populations can also be used for gene function validation (McCallum et al. 2000) if available. More information regarding reverse genetics methods can be found in this book Chap. 4.

Association Mapping

Traditional mapping method (linkage mapping, LM) requires high density markers and segregating populations. Marker density is no longer a bottleneck in gene identification, thanks to the modern sequencing technology, and the ever increasing genomic sequence resources available for more and more plant species also facilitate identification of candidate genes. However, creating a proper mapping population with enough recombination events remains a limiting factor for LM. Fortunately, another method called association mapping (AM) is now available for researchers. AM also known as linkage disequilibrium (LD) mapping, is a mapping method which takes advantage of historically accumulated linkage disequilibrium to link phenotypes with genotypes. The details for AM can be found in this book Chap. 6.

Compared with LM, AM has three main advantages: higher mapping resolution, higher number of evaluated alleles, broader reference population and less research time (Flint-Garcia et al. 2003). To detect LD between molecular markers and functional loci, LM exploits a large experimental population, which is often costly and time-consuming to generate and accompanied with limited number of recombination events, therefore is less powerful in detection. While for AM, a collected population from different germplasm sources including landraces, cultivars, lines or varieties from regional breeding programs can be used to analyze specific trait underlying genes. Based on the genetics context, four types of populations are used for AM: (1) natural population; (2) germplasm bank collections; (3) synthetic populations and (4) elite breeding materials (Stich and Melchinger 2010). AM also has potentially high genetic resolution contributed by many meiotic events accumulated in history. Another advantage AM has over LM is that AM can estimate the effects of multiple loci responsible for a complex trait at the same time, while LM often aim at isolating one major effect gene. The above advantages have made AM an increasingly popular mapping method in recent years. So far, it has been successfully applied into various plant species, including many important crop species, e.g. rice, wheat, sorghum (*Sorghum bicolor* L.), barley (*Hordeum vulgare* L.), soybean (*Glycine max* L.), potato, sugarcane, and some trees, grasses as well as Arabidopsis (Zhu et al. 2008).

However, AM has its own disadvantages. For example, given the balanced design, LD can only be affected by recombination for LM. While for AM, LD can be affected not only by recombination, but also by various factors such as population structure, mating type, selection, mutation, genetic drift and ascertainment bias (Clark et al. 2005; Stich et al. 2005; Yu et al. 2006; Stich and Melchinger 2010). Therefore, AM usually needs to deal with population structure and familial relatedness. Different degrees of population structure and familial relatedness ask for the corresponding methods for association study. But if the interested trait is strongly associated with population structure, the AM usually has less power to detect association, for example, the flowering time in maize (Thornsberry et al. 2001). Besides, AM is limited in identifying the phenotypic effects of those alleles with relatively high frequency in the investigated population (Sorkheh et al. 2008). Rare alleles usually cannot be evaluated because there are not enough individuals carrying this allele available. In the two cases mentioned above, LM is more efficient than AM. In conclusion, LM evaluates two alleles with low resolution by segregation population while AM evaluates multiple alleles with high resolution. Stich and Melchinger (2010) proposed that LM and AM are complement to each other and combining them together enables the successfully dissection of quantitative trait down to a single gene level.

Case Study for Gene Identification

So far many genes have been successfully identified in crops. Among them, the wide compatibility gene *S5n* in rice is of special interest due to both its importance in rice hybrid breeding and the complexity of the trait, which demanded extra efforts to determine the phenotype. Up to date, more than 50 loci have been reported to be involved in rice hybrid sterility (Ouyang et al. 2009), which is a major obstacle in exploiting heterosis of *indica/japonica* hybrid rice. The *S5* is one of the major loci responsible for the embryo sterility and the allele from wide compatibility varieties (WCV) is *S5n*, so called wide compatibility gene (WCG). The WCV can produce fully fertile offsprings when crossed with either *indica* or *japonica*. A prevalent genetic model, named 'one-locus sporo-gametophytic interaction' has been proposed by Ikehashi and Araki (1986) to explain this phenomenon. According to this model, there are three alleles at the S5 locus, *S5i*, *S5j*, and *S5n*, present in *indica*, *japonica*, and WCVs, respectively. The heterozygote of *indica* and *japonica* (*S5iS5j*) produces semi-sterile panicles, resulting from partial abortion of female gametes carrying *S5j* (Ikehashi and Araki 1986). However, this fertility barrier can be overcome by the *S5n* allele, resulting in normal fertile panicles in either *indica* and WCV (*S5iS5n*) or *japonica* and WCV (*S5j S5n*) hybrids.

The effort of mapping and identifying this gene dates back to the 1980s. Since the hybrid sterility between inter-subspecies is a quantitative trait and controlled by multiple genetic factors, a QTL mapping will be the start point to find out where the *S5* located in the rice genome. The wide compatibility trait of the plant is tested by investigating the fertility of its offspring derived from the crossing with corresponding test lines, for example, the typical *indica* line 'Nanjing11' or the *japonica* line 'Balilla'. The special phenotyping method decided that F_2 or backcross population doesn't work conveniently for *S5* QTL mapping. Instead the three-way crossing populations were produced, with one parent from the cross of *japonica* (or *indica*) and wide compatibility variety, and the other one from the test line, *indica* (or *japonica*) parent. The phenotype of the individuals derived from the three-way cross can be easily determined by examining the seed setting rates of the spikelets of each individual, with no extra cross required. Based on the previous study, the *S5* was located on Chromosome 6 between the markers R2349 and RZ450 (Liu et al. 1997). However, three-way crossing population has the genetic components from three parents. The complex genetic background may lead to unreliable association between the phenotype and genotype, which is obviously unfavorable for fine mapping. To minimize the background noise, different *S5n* near isogenic lines (NIL) with either *indica* or *japonica* background were generated for the fine mapping. Ji et al. (2005) generated the NILs by introducing the *S5n* from '02428' and 'Dular' (both are WCVs), to 'Nanjing11' and 'Balilla', respectively. The backcross was conducted for 10–13 generations followed by 2–3 generations of selfings. For each generation, the WC trait was assayed by testcrossing to ensure the presence of *S5n*. A total of 549 isogenic lines were obtained for the fine mapping. The mapping result delimited the *S5n* into a 50 kb region. The fine mapping work was also reported by another research group, which exploited a NIL three-way crossing population (Qiu et al. 2005). The *S5j* from 'Balilla' and the *S5n* from '02428' were introduced into 'Nanjing11' through three and six generations of backcross followed by one generation of selfing, respectively. For each generation, the molecular markers are used to ensure the presence of introduced alleles. Then the *S5nS5i* individuals were selected for further crossing with the *S5jS5j* genotypes to make a new three-way cross, *S5n/S5i//S5j* with the unique background from 'Nanjing11'. A total of 8,000 individuals from this cross were selected for fine mapping. Qiu et al. (2005) finally narrowed down the *S5* into a 40 kb region, with a 20 kb overlap compared to the result of Ji et al. (2005). Further mapping work of the *S5n* was carried out by Ji et al. (2010) using 11 NILs derived three way cross families. The results pinpointed the *S5* to a single gene encoding aspartic protease (*Asp*). And interestingly, intragenic recombination within this gene was also detected in a number of recombinants.

To validate the function of this gene, both over expression and complementation experiment were conducted by introducing the *S5n* allele from '02428' into *japonica* rice 'Balilla'. However, the test results indicated that *S5n* allele failed to enhance fertilities of the *indica/japonica* hybrids (Ji et al. 2010). But transform the *S5i* from *indica* 'Nanjing11' into *japonica* significantly reduced the fertility, which proved the association of *Asp* gene with the incompatibility of *indica/japonica* hybrids (Chen et al. 2008; Ji et al. 2010).

Sequence analysis of a collection of 36 randomly selected rice varieties revealed that, compared with non-WCVs, all the *S5n* had a 136 bp deletion in the 5′ end and another 1 bp of SNP near the 3′ end (1233:A/C), which led to the missing of the signal peptide and the change of the amino acid in the encoded protein respectively. Furthermore, study has also been conducted to find out the causative site within the *Asp* gene in relation to the WC trait by RNAi and site directed mutation. Interestingly, although the 136 bp deletion caused the miss of the signal peptide and therefore led to the mislocation of the encoded protein (Chen et al. 2008), the 1 bp SNP seemed to have direct relation to the *indica/japonica* hybrid sterility (Ji et al. 2012). The *S5n* allele was believed to be a loss of function allele, while the *S5i* and *S5j* are functional by incompatible due to the SNP (1233: A/C) (Chen et al. 2008; Ji et al. 2010, 2012). Based on the sequence difference, the functional markers AD1 and AC2 have been developed to help select the WCVs, which will greatly facilitate the rice breeding program (Ji et al. 2010).

Conclusions and Perspectives

Gene Identification in crops is of great interest for both theoretical study and practical breeding program. Both forward genetics and reverse genetics are powerful tools in gene identification for crops. Reverse genetics starts from the gene of interest to its function. The biggest challenge often lies in finding the gene-indexed mutants, which includes constructing a mutant library and selection of targeted gene mutants. Although the process may take several years (Alonso and Ecker 2006), once the gene indexed mutants database is established, high throughput of gene identification is possible. Forward genetics is a 'phenotype-centric' process and provides reliable means in gene identification. The most attractive point is unbiased and no preconceived ideas regarding the nature of the gene to be identified are involved (Alonso and Ecker 2006). But the cost in constructing mapping population and typically working with one gene per time has hindered the high throughput of gene identification. The development of molecular platform, advances in bioinformatics and availability of new tools for candidate genes validation have been and will keep facilitating the developments of both forward and reverse genetics. Using reverse genetic tools to accelerate gene identification in forward genetics is an exciting perspective in the future. For example the completion of gene-indexed Arabidopsis mutant collections will enable the whole genome forward genetics selection and systemic study of given gene. Once a gene is identified, reverse genetic tools, such as, gene fragment replacement, site directed mutation and so on can be exploited to study the causative site within the identified gene. Based on the causative sites, functional markers will be developed to assist selection in breeding program.

References

Alonso JM, Ecker JR (2006) Moving forward in reverse: genetic technologies to enable genome-wide phenomic screens in Arabidopsis. Nat Rev Genet 7:524–536

Barampuram S, Zhang ZJ (2010) Recent advances in plant transformation. In: Birchler J A (ed) Plant chromosome engineering: methods and protocols, methods in molecular biology, vol 701. DOI:10.1007/978-1-61737-957-4_1

Bartel DP (2004) MicroRNAs: genomics, biogenesis, mechanism, and function. Cell 116:281–297

Bhattramakki D, Dolan M, Hanafey M, Wineland R, Vaske D, Register JC 3rd, Tingey SV, Rafalski A (2002) Insertion-deletion polymorphisms in 3′ regions of maize genes occur frequently and can be used as highly informative genetic markers. Plant Mol Biol 48:539–547

Binladen J, Gilbert MT, Bollback JP, Panitz F, Bendixen C, Nielsen R, Willerslev E (2007) The use of coded PCR primers enables high-throughput sequencing of multiple homolog amplification products by 454 parallel sequencing. PLoS One 2:e197

Borevitz JO, Liang D, Plouffe D, Chang HS, Zhu T, Weigel D, Berry CC, Winzeler E, Chory J (2003) Large-scale identification of single-feature polymorphisms in complex genomes. Genome Res 13:513–523

Burch-Smith TM, Anderson JC, Martin GB, Dinesh-Kumar SP (2004) Applications and advantages of virus-induced gene silencing for gene function studies in plants. Plant J 39:734–746

Chen JJ, Ding JH, Ouyang YD, Du HY, Yang JY, Cheng K, Zhao J, Qiu SQ, Zhang XL, Yao JL, Liu KD, Wang L, Xu CG, Li XH, Xue YB, Xia M, Ji Q, Lu JF, Xu ML, Zhang Q (2008) A triallelic system of S5 is a major regulator of the reproductive barrier and compatibility of indica–japonica hybrids in rice. Proc Natl Acad Sci USA 105:11436–11441

Clark AG, Hubisz MJ, Bustamante CD, Williamson SH, Nielsen R (2005) Ascertainment bias in studies of human genomewide polymorphism. Genome Res 15:1496–1502

Demeke T, Sasikumar B, Hucl P, Chibbar RN (1997) Random Amplified Polymorphic DNA (RAPD) in cereal improvement. Maydica 42:133–142

Dinka SJ, Campbell MA, Demers T, Raizada MN (2007) Predicting the size of the progeny mapping population required to positionally clone a gene. Genetics 176:2035–2054

Doi K, Izawa T, Fuse T, Yamanouchi U, Kubo T, Shimatani Z, Yano M, Yoshimura A (2004) Ehd1, a B-type response regulator in rice, confers short-day promotion of flowering and controls FT-like gene expression independently of Hd1. Genes Dev 18:926–936

Dussle CM, Quint M, Xu ML, Melchinger AE, Lübberstedt T (2002) Conversion of AFLP fragments tightly linked to SCMV resistance genes Scmv1 and Scmv2 into simple PCR-based markers. Theor Appl Genet 105:1190–1195

Elshire RJ, Glaubitz JC, Sun Q, Poland JA, Kawamoto K, Buckler ES, Mitchell SE (2011) A robust, simple genotyping-by-sequencing (GBS) approach for high diversity species. PLoS One 6:e19379

Flint-Garcia SA, Thornsberry JM, Buckler ES (2003) Structure of linkage disequilibrium in plants. Annu Rev Plant Biol 54:357–374

Garcia AAF, Benchimol LL, Barbosa AMM, Geraldi IO, Souza CL Jr, de Souza AP (2004) Comparison of RAPD, RFLP, AFLP and SSR markers for diversity studies in tropical maize inbred lines. Genet Mol Biol 27:579–588

Gore MA, Chia RJ, Elshire Sun Q, Ersoz ES, Hurwitz BL, Peiffer JA, McMullen MD, Grills GS, Ross-Ibarra J (2009) A first generation haplotype map of maize. Science 326:1115–1117

Hamady M, Walker JJ, Harris JK, Gold NJ, Knight R (2008) Error-correcting barcoded primers for pyrosequencing hundreds of samples in multiplex. Nat Methods 5:235–237

Hannon GJ (2002) RNA interference. Nature 418:244–251

Hoffmann C, Minkah N, Leipzig J, Wang G, Arens MQ, Tebas P, Bushman FD (2007) DNA bar coding and pyrosequencing to identify rare HIV drug resistance mutations. Nucleic Acids Res 35:e91

Holzberg S, Brosio P, Gross C, Pogue GP (2002) Barley stripe mosaic virus-induced gene silencing in a monocot plant. Plant J 30:315–327

Ikehashi H, Araki H (1986) Genetics of F1 sterility in remote crosses in rice. Rice Genetics. In: IRRI (ed) Rice genetics proceedings of the first rice genetics symposium. Manila: IRRI, pp 119–130

Jander G, Norris SR, Rounsley SD, Bus DF, Levin IM, Last RL (2002) *Arabidopsis* map-based cloning in the postgenome era. Plant Physiol 129:440–450

Ji Q, Lu JF, Chao Q, Gu MH, Xu ML (2005) Delimiting a rice wide-compatibility gene S5n to a 50 kb region. Theor Appl Genet 111:1495–1503

Ji Q, Lu JF, Chao Q, Zhang Y, Zhang MJ, Gu MH, Xu ML (2010) Two sequence alterations, a 136 bp InDel and an A/C polymorphic site, in the *S5* locus are associated with spikelet fertility of *indica-japonica* hybrid in rice. J Genet Genomics 37:57–68

Ji Q, Zhang MJ, Lu JF, Wang HM, Lin B, Liu QQ, Chao Q, Zhang Y, Liu CX, Gu MH, Xu ML (2012) Molecular basis underlying the S5-dependent reproductive isolation and compatibility of indica/japonica rice hybrids. Plant Physiol 158:1319–1328.

Jiang SY, Ramachandran S (2010) Natural and artificial mutants as valuable resources for functional genomics and molecular breeding. Int J Biol Sci 6:228–251

Jones ES, Sullivan H, Bhattramakki D, Smith JS (2007) A comparison of simple sequence repeat and single nucleotide polymorphism marker technologies for the genotypic analysis of maize (*Zea mays* L.). Theor Appl Genet 115:361–371

Karp A, Edwards K, Bruford M, Vosman B, Morgante M, Seberg O, Kremer A, Boursot P, Arctander P, Tautz D, Hewitt G (1997) Newer molecular technologies for biodiversity evaluation: opportunities and challenges. Nat Biotechnol 15:625–628

Kojima S, Takahashi Y, Kobayashi Y, Monna L, Sasaki T, Araki T, Yano M (2002) Hd3a, a rice ortholog of the Arabidopsis FT gene, promotes transition to flowering downstream of Hd1 under short-day conditions. Plant Cell Physiol 43:1096–1105

Kump KL, Bradbury PJ, Wisser RJ, Buckler ES, Belcher AR, Oropeza-Rosas MA, Zwonitzer JC, Kresovich S, McMullen MD, Ware D, Balint-Kurti PJ, Holland JB (2011) Genome-wide association study of quantitative resistance to southern leaf blight in the maize nested association mapping population. Nat Genet 43:163–168

Kuromori T, Takahashi S, Kondou Y, Shinozaki K, Matsui M (2009) Phenome analysis in plant species using loss-of-function and gain-of-function mutants. Plant Cell Physiol 50:1215–1231

Lacomme C, Hrubikova K, Hein I (2003) Enhancement of virus-induced gene silencing through viral-based production of inverted-repeats. Plant J 34:543–553

Liu KD, Wang J, Li HB, Xu CG, Liu AM, Li XH, Zhang Q (1997) A genome-wide analysis of wide compatibility in rice and the precise location of the S5 locus in the molecular map. Theor Appl Genet 95:809–814

Liu Y, Schiff M, Dinesh-Kumar SP (2002) Virus-induced gene silencing in tomato. Plant J 31:777–786

Liu S, Chen HD, Makarevitch I, Shirmer R, Emrich SJ, Dietrich CR, Barbazuk WB, Springer NM, Schnable PS (2010) High-throughput genetic mapping of mutants via quantitative single nucleotide polymorphism typing. Genetics 184:19–26

Lu R, Martin-Hernandez AM, Peart JR, Malcuit I, Baulcombe DC (2003) Virus-induced gene silencing in plants. Methods 30:296–303

Lu Y, Curtiss J, Zhang J, Percy RG, Cantrell RG (2005) Discovery of single nucleotide polymorphisms in selected fiber genes in cultivated tetraploid cotton. In: Beltwide cotton conference, National Cotton Council of America, Memphis, TN

Matsubara K, Yamanouchi U, Wang ZX, Minobe Y, Izawa T, Yano M (2008) Ehd2, a rice ortholog of the maize INDETERMINATE1 gene, promotes flowering by up-regulating Ehd1. Plant Physiol 148:1425–1435

Matsubara K, Yamanouchi U, Nonoue Y, Sugimoto K, Wang ZX, Minobe Y, Yano M (2011) Ehd3, encoding a PHD finger-containing protein, is a critical promoter of rice flowering. Plant J 66:603–612

McCallum CM, Comai L, Greene EA, Henikoff S (2000) Targeted screening for induced mutations. Nat Biotechnol 18:455–457

Meksem K, Ruben E, Hyten D, Triwitayakorn K, Lightfoot DA (2001) Conversion of AFLP bands into high-throughput DNA markers. Mol Genet Genomics 265:207–214

Michelmore RW, Paran I, Kesseli RV (1991) Identification of markers linked to disease-resistance genes by bulked segregant analysis: a rapid method to detect markers in specific genomic regions by using segregating populations. Proc Natl Acad Sci USA 88:9828–9832

Monna L, Lin HX, Kojima S, Sasaki T, Yano M (2002) Genetic dissection of a genomic region for quantitative trait locus, Hd3, into two loci, Hd3a and Hd3b, controlling heading date in rice. Theor Appl Genet 104:772–778

Neale DB, Harry DE (1994) Genetic mapping in forest trees: RFLPs, RAPDs and beyond. Ag Biotechnol News Info 6:107N–114N

Ouyang YD, Chen JJ, Ding JH et al (2009) Advances in the understanding of inter-subspecific hybrid sterility and wide-compatibility in rice. Chin Sci Bull 54:2332–2341

Parameswaran P, Jalili R, Tao L, Shokralla S, Gharizadeh B, Ronaghi M, Fire AZ (2007) A pyrosequencing-tailored nucleotide barcode design unveils opportunities for large-scale sample multiplexing. Nucleic Acids Res 35:e130

Peele C, Jordan CV, Muangsan N, Turnage M, Egelkrout E, Eagle P, Hanley-Bowdoin L, Robertson D (2001) Silencing of a meristematic gene using geminivirus-derived vectors. Plant J 27:357–366ee

Peters JL, Cnudde F, Gerats T (2003) Forward genetics and map-based cloning approaches. Trends Plant Sci 8:484–491

Poland JA, Bradbury PJ, Buckler ES, Nelson RJ (2011) Genome-wide nested association mapping of quantitative resistance to northern leaf blight in Maize. Proc Natl Acad Sci USA 108:6893–6898

Poulsen DME, Henry RJ, Johnston RP, Irwin JAG, Rees RG (1995) The use of bulk segregant analysis to identify a RAPD marker linked to leaf rust resistance in barley. Theor Appl Genet 91:270–273

Purkayastha A, Mathur S, Verma V, Sharma S, Dasgupta I (2010) Virus-induced gene silencing in rice using a vector derived from a DNA virus. Planta 232:1531–1540

Qiu SQ, Liu KD, Jiang JX, Song X, Xu CG, Li XH, Zhang Q (2005) Delimitation of the rice wide compatibility gene S5n to a 40-kbDNA fragment. Theor Appl Genet 111:1080–1086

Quarrie SA, Lazić-Jančić V, Kovačević D, Steed A, Pekić S (1999) Bulk segregant analysis with molecular markers and its use for improving drought resistance in maize. J Exp Bot 50: 1299–1306

Salvi S, Tuberosa R (2005) To clone or not to clone plant QTLs: present and future challenges. Trends Plant Sci 10:297–304

Schneider K, Weisshaar B, Borchardt DC, Salamini F (2001) SNPs frequency and allelic haplotypes structure of *Beta vulgaris* expressed genes. Mol Breed 8:63–74

Scofield SR, Nelson RS (2009) Resources for virus-induced gene silencing in the grasses. Plant Physiol 149:152–157

Sorkheh K, Malysheva-Otto LV, Wirthensohn MG, Tarkesh-Esfahani S, Martínez-Gómez P (2008) Linkage disequilibrium, genetic association mapping and gene localization in crop plants. Genet Mol Biol 31:805–814

Stich B, Melchinger AE (2010) An introduction to association mapping in plants. CAB Rev Perspect Agric Vet Sci Nutr Nat Resour 5:1–9

Stich B, Melchinger AE, Frisch M, Maurer HP, Heckenberger M, Reif JC (2005) Linkage disequilibrium in European elite maize germplasm investigated with SSRs. Theor Appl Genet 111:723–730

Takahashi Y, Shomura A, Sasaki T, Yano M (2001) Hd6, a rice quantitative trait locus involved in photoperiod sensitivity, encodes the α subunit of protein kinase CK2. Proc Natl Acad Sci USA 98:7922–7927

Thornsberry JM, Goodman MM, Doebley J, Kresovich S, Nielsen D, Buckler ES (2001) Dwarf8 polymorphisms associate with variation in flowering time. Nat Genet 28:286–289

Tian F, Bradbury PJ, Brown PJ, Hung H, Sun Q, Flint-Garcia S, Rocheford TR, McMullen MD, Holland JB, Buckler ES (2011) Genome-wide association study of leaf architecture in the maize nested association mapping population. Nat Genet 43:159–162

Turnage MA, Muangsan N, Peele CG, Robertson D (2002) Geminivirus-based vectors for gene silencing in Arabidopsis. Plant J 30:107–117

van Kammen A (1997) Virus-induced gene silencing in infected and transgenic plants. Trends Plant Sci 2:409–411

Vega JM, Yu W, Kennon AR, Chen X, Zhang ZJ (2008) Improvement of Agrobacterium-mediated transformation in Hi-II maize (*Zea mays*) using standard binary vectors. Plant Cell Rep 27: 297–305

Wang Z, Weber JL, Zhong G, Tanksley SD (1994) Survey of plant short tandem DNA repeats. Theor Appl Genet 88:1–6

Weerasena JS, Steffenson BJ, Falk AB (2004) Conversion of an amplified fragment length polymorphism marker into a co-dominant marker in the mapping of the Rph15 gene conferring resistance to barley leaf rust, Puccinia hordei Otth. Theor Appl Genet 108:712–719

Wei X, Xu J, Guo H, Jiang L, Chen S, Yu C, Zhou Z, Hu P, Zhai H, Wan J (2010) DTH8 suppresses flowering in rice, influencing plant height and yield potential simultaneously. Plant Physiol 153:1747–1758

Williams JG, Kubelik AR, Livak KJ, Rafalski JA, Tingey SV (1990) DNA polymorphisms amplified by arbitrary primers are useful as genetic markers. Nucleic Acids Res 18:6531–6535

Xu M, Huaracha E, Korban SS (2001) Development of sequence-characterized amplified regions (SCARs) from amplified fragment length polymorphism (AFLP) markers tightly linked to the Vf gene in apple. Genome 44:63–70

Yamamoto T, Kuboki Y, Lin SY, Sasaki T, Yano M (1998) Fine mapping of quantitative trait loci Hd-1, Hd-2 and Hd-3, controlling heading date of rice, as single Mendelian factors. Theor Appl Genet 97:37–44

Yang HY, Korban SS, Krüger J, Schmidt H (1997) The use of a modified bulk segregant analysis to identify a molecular marker linked to a scab resistance gene in apple. Euphytica 94:175–182

Yano M, Harushima Y, Nagamura Y, Kurata N, Minobe Y, Sasaki T (1997) Identification of quantitative trait loci controlling heading date in rice using a high-density linkage map. Theor Appl Genet 95:1025–1032

Yano M, Katayose Y, Ashikari M, Yamanouchi U, Monna L, Fuse T, Baba T, Yamamoto K, Umehara Y, Nagamura Y, Sasaki T (2000) Hd1, a major photoperiod sensitivity quantitative trait locus in rice, is closely related to the Arabidopsis flowering time gene CONSTANS. Plant Cell 12:2299–2301

Yu J, Holland JB, McMullen MD, Buckler ES (2008) Genetic design and statistical power of nested association mapping in maize. Genet 178:539–551

Yu J, Hu S, Wang J, Wong GK, Li S, Liu B, DengY DL, Zhou Y, Zhang X (2002) A draft sequence of the rice genome (*Oryza sativa* L. ssp. *indica*). Science 296:79–92

Yu J, Pressoir G, Briggs WH et al (2006) A unified mixed-model method for association mapping that accounts for multiple levels of relatedness. Nat Genet 38:203–208

Zhang GL, Chen LY, Xiao GY, Xiao YH, Chen XB, Zhang ST (2009) Bulked segregant analysis to detect QTL related to heat tolerance in rice (*Oryza sativa* L.) using SSR markers. Agric Sci China 8:482–487

Zhu C, Gore M, Buckler ED, Yu J (2008) Status and prospects of association mapping in plants. Plant Genome 1:5–20

Chapter 4
Gene Identification: Reverse Genetics

Erin Gilchrist and George Haughn

Introduction

The number of sequenced genes whose function remains unknown continues to climb with the continuing decrease in the cost of genome sequencing. Comparative genetics and bioinformatics have been invaluable in investigating the function of the genes that have been sequenced, but the elucidation of gene function *in planta* remains a huge challenge. Many gene functions have been defined through the use of forward genetics, where a phenotype is identified and used to clone the gene responsible. However, in most instances, genes of known sequence are not associated with a phenotype. This is particularly true in non-model species where forward genetics can be more challenging due to genetic redundancy. Reverse genetics is a powerful tool that can be used to identify the phenotype that results from disruption of a specific sequenced gene, even with no prior knowledge of its function. Several approaches have been developed in plants that have led to the production of resources including collections of T-DNA insertion mutants, RNAi-generated mutants, and populations carrying point mutations that can be detected by TILLING, direct sequencing or high resolution melting analysis (Table 4.1). These reverse genetics resources allow for the identification of mutations in candidate genes and subsequent phenotypic analysis of these mutants. In addition, new advances in technology and reduction in technical costs may soon make it practical to use whole genome sequencing or gene targeting on a routine basis to identify or generate mutations in specific genes in a variety of different plant species. This chapter will present the current status and promising prospects for the future of reverse genetics in plants.

E. Gilchrist (✉) • G. Haughn
Department of Botany, University of British Columbia, Vancouver, BC, Canada
e-mail: erin.gilchrist@ubc.ca; george.haughn@ubc.ca

Table 4.1 Advantages and disadvantages of different reverse genetics techniques in plants

Technique	Advantages	Disadvantages	Species	References
Chemical mutagenesis and TILLING	Loss-of-function, reduction-of-function, and gain-of-function phenotypes	Requires the construction of a mutagenised population	Arabidopsis	Till et al. (2003), Martin et al. (2009)
			Barley	Caldwell et al. (2004), Talame et al. (2008)
			Brassica species	Wang et al. (2008), Himelblau et al. (2009), and Stephenson et al. (2010)
	Not labelled as genetic engineering	Few cost-effective ways of screening for individuals from the mutagenised population	Legumes	Perry et al. (2003), Dalmais et al. (2008), and Calderini et al. (2011)
			Melon	Dahmani-Mardas et al. (2010)
			Oats	Chawade et al. (2010)
			Potato	Elias et al. (2009)
		Position of mutation is random	Rice	Wu et al. (2005), Till et al. (2007)
			Sorghum	Xin et al. (2008)
			Soybean	Cooper et al. (2008)
			Tomato	Minoia et al. (2010), Piron et al. (2010)
			Wheat	Slade et al. (2005), Uauy et al. (2009)
T-DNA mutagenesis	Individuals carrying an insertion can be identified using PCR	Very large populations must be screened to achieve genome saturation	Apple	Smolka et al. (2010)
			Arabidopsis	Clough and Bent (1998)
			Banana	Sun et al. (2011)

4 Gene Identification: Reverse Genetics

Can be used for eliminating (knock-out) or enhancing (activation) of gene function	Position of mutation is random	Birch tree	Zeng et al. (2010)
	Requires transformation	Blueberry	Song and Sink (2004)
	Labelled as genetically engineered	Brachypodium	Thole et al. (2010)
		Brassica rapa	Lee et al. (2004)
		Carrot	Chen and Punja (2002)
		Cassava	Taylor et al. (2004)
		Chickpea	Indurker et al. (2010)
		Cucumber	Unni and Soniya (2010)
		Eucalyptus	Chen et al. (2007)
		Grape	Bouquet et al. (2007)
		Jute	Chattopadhyay et al. (2011)
		Lettuce	Michelmore et al. (1987)
		Lotus	Imaizumi et al. (2005)
		Medicago	Tadege et al. (2005)
		Peanut	Anuradha et al. (2006)
		Pear	Sun et al. (2011)
		Perilla	Ghimire et al. (2011)
		Pigeonpea	Krishna et al. (2010)
		Pine	Grant et al. (2004)
		Poplar	Busov et al. (2005)
		Potato	Barrell and Conner (2011)
		Rice	Jeon et al. (2000), Wan et al. (2009)
		Safflower	Belide et al. (2011)

(continued)

Table 4.1 (continued)

Technique	Advantages	Disadvantages	Species	References
			Sorghum	Kumar et al. (2011)
			Soybean	Widholm et al. (2010)
			Strawberry	Oosumi et al. (2010)
			Sugarcane	Arencibia and Carmona (2007)
			Switchgrass	Li and Qu (2011)
			Tomato	Mathews et al. (2003)
Transposon mutagenesis	Individuals carrying an insertion can be identified using PCR	Very large populations must be screened to achieve genome saturation	Arabidopsis	D'Erfurth et al. (2003), Marsch-Martínez (2011), and Nishal et al. (2005)
			Aspen	Kumar and Fladung (2003)
	Can be used for eliminating (knock-out) or enhancing (activation) of gene function	Position of mutation is random	Barley	Ayliffe and Pryor (2011)
			Beet	Kishchenko et al. (2010)
		Can be epigenetic effects	Legumes	D'Erfurth et al. (2003)
			Maize	May and Martienssen (2003)
	Produce many unique insertion lines from a few initial plant lines		Rice	Upadhyaya et al. (2011), Zhu et al. (2007)
			Tobacco	D'Erfurth et al. (2003)
Radiation or fast-neutron mutagenesis	Completely eliminates gene function	Very large number of plants must be screened	Arabidopsis	Li and Zhang (2002)
			Clementine	Rios et al. (2008)
			Legumes	Rios et al. (2008), Rogers et al. (2009)

4 Gene Identification: Reverse Genetics

		Size limitation of deletions detected	*Noccaeacaerulescens*	Lochlainn et al. (2011)
			Rice	Bruce et al. (2009)
			Soybean	Bolon et al. (2011)
			Tomato	Dor et al. (2010)
Virus-induced gene silencing (VIGS)	Relatively inexpensive	Not heritable	Apple	Sasaki et al. (2011)
			Arabidopsis	Burch-Smith et al. (2006)
			Barley	Scofield and Nelson (2009)
	Homologous genes may be affected with a single construct	Homologous genes may be affected with a single construct	Brachypodium	Demircan and Akkaya (2010)
			Brassica nigra	Zheng et al. (2010)
			California poppy	Wege et al. (2007)
	Phenotype is transient	Phenotype is transient	Cassava	Fofana et al. (2004)
			Chilli pepper	Chung et al. (2004)
	Does not require transformation	Side-effects of the infection may interfere with phenotype	Columbine	Gould and Kramer (2007)
			Cotton	Tuttle et al. (2008)
	Delivers rapid results	Level of silencing of target gene is variable	Cucurbit species	Igarashi et al. (2009)
			Ginger	Renner et al. (2009)
			Haynaldia	Wang et al. (2010)
			Jatropha	Ye et al. (2009)
			Legumes	Igarashi et al. (2009)
			Maize	Ding et al. (2006)
			Ornamental plants	Jiang et al. (2011)

(continued)

Table 4.1 (continued)

Technique	Advantages	Disadvantages	Species	References
			Pea	Constantin et al. (2004)
			Pear	Sasaki et al. (2011)
			Potato	Brigneti et al. (2004)
			Rice	Ding et al. (2006)
			Soybean	Zhang and Ghabrial (2006), Yamagishi and Yoshikawa (2009)
			Tomato	Fu et al. (2005)
			Wheat	Scofield and Nelson (2009)
RNA interference (RNAi)	Heritable	Some genes are resistant to silencing	Arabidopsis	Ossowski et al. (2008)
			Artemesinin	Zhang et al. (2009)
	Partial loss of function can be achieved	Expression is rarely completely silenced	Banana	Angaji et al. (2010)
			Barley	Angaji et al. (2010)
			Brassica species	Wood et al. (2011)
	Silencing is directed against a specific gene(s)	Long-term expression levels are variable	Coffee	Angaji et al. (2010)
			Cotton	Angaji et al. (2010)
	Transcripts of multiple genes can be silenced by a single construct	Silencing level may vary	Tomato	Fernandez et al. (2009)
			Wheat	Fu et al. (2007)
		'Off-target' silencing		
	Induced phenotypes are dominant			

4 Gene Identification: Reverse Genetics

New generation sequencing (NGS)	Vast amount of information obtained directly	High cost of data analysis and storage	Oat	Oliver et al. (2011)
			Tomato	Rigola et al. (2009), Tsai et al. (2011)
High resolution melting curve analysis (HRM)	Simple technology	Only small PCR fragments can be screened in one reaction	Almond	Wu et al. (2008)
			Barley	Hofinger et al. (2009)
	Inexpensive		Chilli pepper	Park et al. (2009)
			Maize	Li et al. (2010)
			Oat	Oliver et al. (2011)
			Olive	Muleo et al. (2009)
			Peach	Chen and Wilde (2011)
			Potato	De Koeyer et al. (2010)
			Ryegrass	Studer et al. (2009)
			Tomato	Gady et al. (2009)
			Wheat	Dong et al. (2009)
Gene targeting	Mutations in single, targeted genes	Difficulty and high cost of designing target zinc finger motifs	Arabidopsis	Zhang and Voytas (2011)
			Maize	Shukla et al. (2009)
			Soybean	Curtin et al. (2011)
			Tobacco	Townsend et al. (2009)

Established Techniques

Chemical Mutagenesis

Chemical mutagenesis was used to generate populations of mutants for forward genetics long before the advent of DNA sequencing and reverse genetics. Point mutations are, generally, less deleterious than large rearrangements and so a high degree of saturation can be achieved in a mutant population using chemicals that generate single base pair changes or small insertions and deletions. This approach is, therefore, useful for the examination of gene function using genome-wide approaches. Two chemicals, in particular, are known to cause primarily single base pair mutations in DNA in all organisms in which they have been tested: ethylmethane sulphonate (EMS) and ethyl nitrosourea (ENU).

While many reverse genetics techniques provide only loss-of-function alleles, chemical mutagenesis can result in either loss-of-function, reduction-of-function, or gain-of-function phenotypes. In fact, the frequency of induced missense alleles is, on average, three times higher than that of nonsense alleles. Many missense alleles will not have an effect on gene function since they may not alter the gene product(s) significantly, but examples of dominant point mutations caused by missense alleles have been well documented, including ones that affect plant hormone responses (Wang et al. 2006; Biswas et al. 2009), leaf polarity (Juarez et al. 2004; Byrne 2006), and host-pathogen defence (Eckardt 2007). The difficulty with using point mutations for reverse genetics screens is that there are few cost-effective ways of screening the mutagenised population for individuals that carry mutations in specific genes. The advent of TILLING, New Generation Sequencing (NGS), and High Resolution Melting (HRM) analysis, however, have made possible the screening of large populations, at a reasonable cost, within an acceptable time frame (Fig. 4.1). TILLING operations use a variety of techniques for creating mutant populations and screening them, including that described by (Colbert et al. 2001) which employs a mismatch-specific endonuclease for identifying point mutations in the target gene of interest. Generally, in this procedure the mutagenised generation (M_1) is grown up and then the progeny of these plants (the M_2 generation) are used for screening. This ensures that the mutations that are identified in this process are heritable and eliminates the background somatic mutations that may be present in the M_1 generation. After collecting seeds and DNA from the M_2 plants, the DNA from several mutagenised individuals is pooled, and then the polymerase chain reaction (PCR) is used to amplify a target gene of interest. In conventional TILLING, the PCR products (amplicons) are denatured and allowed to randomly re-anneal before being digested with a celery juice extract (CJE) (Till et al. 2003). Mismatches in the amplicons occur when mutant and wild-type strands of DNA are re-annealed together to form a heteroduplex. This heteroduplex then becomes a target for the mismatch-specific enzyme. Only the samples carrying a mismatch are cleaved, and these novel fragments can be detected using DNA separation technology such as the LI-COR DNA Analyser (LI-COR Biosciences, Lincoln, NE, USA), or AdvanCE

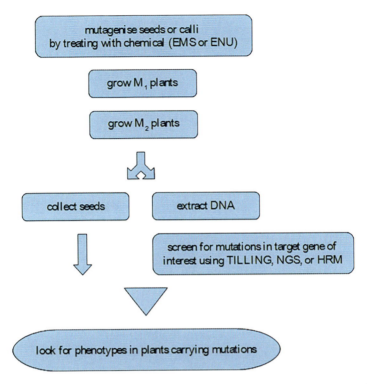

Fig. 4.1 Chemical mutagenesis for reverse genetics. Flow chart describing the procedure for TILLING, high-throughput new generation sequencing (NGS) or high resolution melting (HRM) analysis of a mutagenised population. M_1 refers to the mutagenized generation; M_2 refers to the progeny of the mutagenized generation

F96 (Advanced Analytical Technologies, Inc., Ames, IA, USA). The drawbacks of TILLING are that it requires the construction of a mutagenised population, and for many species the development of such a population is challenging. Further, the technique itself is labour-intensive, relatively expensive, and requires a high rate of mutagenesis to make the effort cost-effective. Nonetheless, TILLING has worked well in a wide variety of model and non-model plants as listed in Table 4.1.

Insertional Mutagenesis

One of the most established methods for reverse genetics is the production of populations of individuals that have insertions that disrupt gene function at unique sites in their genomes (Fig. 4.2). The advantage of insertional mutagenesis is that individuals carrying an insertion of known sequence in a specific gene can be identified in a population using PCR, a simple and relatively inexpensive technique.

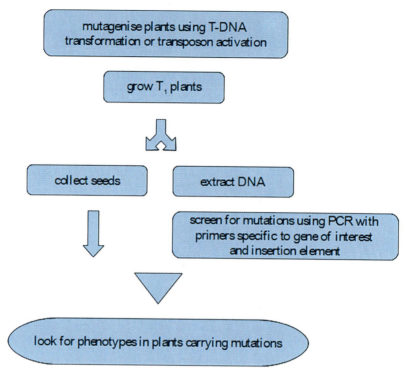

Fig. 4.2 Insertional mutagenesis for reverse genetics. Flow chart describing the procedure for insertional mutagenesis using either T-DNA transformation or transposon activation. T_1 refers to the first generation after transformation of the T-DNA or transposon

PCR amplification is performed with one host gene-specific primer and one vector-based primer from the insertion element. Thus, an amplification product will only be observed, when the insertion is present in close proximity to the target gene (from which the host primer was designed). Gene disruption using this technique typically results in a total loss of gene function.

Insertional mutagenesis can also be used for activation tagging. Activation tagging is a method of causing over-expression or ectopic expression of a gene of interest. A construct is engineered such that it carries a strong promoter or enhancer element which, when introduced into the genome, can insert at random positions. Some of these insertion sites will be upstream of the target gene of interest where they can enhance transcription of that gene. The position of the insertion is determined using PCR as described above. Such enhanced expression can create a phenotype even in cases where loss-of-function mutations are not able to do so because of redundancy or lethality.

For either activation or disruption of gene function, the insertion can be detected in the first generation following transformation (T_1), and can be easily followed in a population of plants where the element will segregate in Mendelian fashion.

4 Gene Identification: Reverse Genetics

Insertional mutagenesis is generally generated either by transformation using an *Agrobacterium*-derived T-DNA construct or by transposon activation. Each of these options is discussed below.

Transfer-DNA (T-DNA) Mutagenesis

There are a number of different transformation techniques that can be used in plants, but by far the most established is *Agrobacterium*-mediated gene transfer using some form of T-DNA construct. In this process, the T-DNA segment of the tumour-inducing (T_i) plasmid from an *Agrobacterium* species integrates randomly into the plant genome and causes disruption or activation of the gene of interest depending on the construct used (Hellens et al. 2000). It is technically difficult to clone a gene directly into the T-DNA region of the Ti plasmid because the plasmid is large, making it challenging to isolate directly from *Agrobacterium*. Therefore, a binary vector system is typically employed (Lee and Gelvin 2008). This technique involves the use of two separate plasmids, one carrying the insert DNA flanked by the left and right border sequences of the T-DNA, and the other carrying the virulence genes from the Ti plasmid needed for infection and transfer of the of the T-DNA into the host. Using this system, the first vector can be constructed and grown in *E. coli* before transformation into an *Agrobacterium* strain that has been engineered to transfer the cloned DNA fragment into the plant without causing the typical symptoms of *Agrobacterium* infection.

There are several transformation techniques that can be used depending on the host plant (for review see Meyers et al. (2010)). The simplest of these is the floral dip method that involves simply dipping developing flowers into media containing the transgenic *Agrobacterium* and then planting the seeds from these plants on selective media so that only transgenic plants can germinate. This is the technique most commonly used in the model *Arabidopsis* (Clough and Bent 1998). For most plants, leaf-disc inoculation is used instead. This technique involves soaking the leaf discs in the *Agrobacterium* solution and then placing them on callus-inductive media containing the herbicide against which one of the transgenes on the T-DNA confers resistance (Barampuram and Zhang 2011). For plants that are resistant to *Agrobacterium*, electroporation or biolistic transformation of plant protoplasts is sometimes used, where the transforming DNA is introduced using an electrical pulse or bombardment with particles to which the transforming DNA constructs are attached respectively (Meyers et al. 2010).

One of the disadvantages of using T-DNA vectors to create insertional libraries is that very large populations must be screened to achieve genome saturation (a mutation in every gene). In addition, insertion is generally random so that activation of the introduced DNA may or may not be successful depending on the site of integration. For some species, large insertion libraries have been generated allowing researchers to access mutations in almost any gene of interest through comprehensive databases that have been set up for this purpose (for examples see Alonso et al. 2003; Krishnan et al. 2009). T-DNA transformation strategies have been used successfully in many plants for both applied and basic research purposes. Aside from the model *Arabidopsis*, some plant species where this technology has been successful are listed in Table 4.1.

Transposon Mutagenesis

Transposon mutagenesis has been used for over half a century to create mutations that were originally detected using forward genetic screens. For the past three decades it has been used in reverse genetic screens that identify disruptions in target genes of interest (May and Martienssen 2003). Transposon-based reverse genetics usually involves two components: an autonomous element that includes the transposase gene, and one or more non-autonomous elements that are only active when the transposase produced by the autonomous element is active.

The first gene to be cloned using transposon tagging employed the *Activator/Dissociation* (*Ac/Ds*) transposon system from maize (reviewed in May and Martienssen (2003)). *Ac* is a member of the *hAT* cut-and-paste family of transposable elements, some of which have been shown to be controlled by environmental factors. Another *hAT* element, *Tam3*, has been extensively used in *Antirrhinum* because of its unique temperature-controlled characteristic activation at 15°C but not at temperatures above 25°C (Schwarz-Sommer et al. 2003). While, originally, transposon mutagenesis was only possible in plants like maize and *Antirrhinum* which had active and well-understood transposon systems, technological and intellectual advances in the understanding of transposition have made it possible to use some elements heterologously. The *Ac/Ds* system, along with the maize *Suppressor-mutator*(*Spm*) has been shown to work in many species other than maize, (for review see Candela and Hake (2008)). Systems in which *Ac/Ds* or *Spm* transposon-tagging has been effective include aspen trees (Kumar and Fladung 2003), barley and other cereals (Ayliffe and Pryor 2011), beet (Kishchenko et al. 2010), rice (Upadhyaya et al. 2011), and *Arabidopsis* (Marsch-Martínez 2011) among others.

Another transposon family, *Mu* (and *Mu-like* elements), includes the most widely spread and most mutagenic transposons found in plants. This transposon system is commonly used for reverse genetics in maize (Lisch and Jiang 2009), but the high activity level of the transposon can lead to deleterious somatic mutations and so *Mu* has been difficult to use in some heterologous systems, including rice, because epigenetic silencing occurs within a few generations (Diao and Lisch 2006).

The *Tos* retrotransposons were the first endogenous transposons demonstrated to be active in rice and remain the most commonly used in this species for a number of reasons, not the least of which is that because *Tos17* is derive from rice, affected lines can be grown and used without the regulatory problems associated with genetically modified organisms (GMOs) (Miyao et al. 2007). Several other transposons have also been used to create tagged populations in rice (Zhu et al. 2007).

Other transposons such as the Tobacco *Tnt1* element have also been used for transposon-tagging in systems such as in tobacco itself (Grandbastien et al. 1989), and in *Arabidopsis* (Courtial et al. 2001) as well as in the legume *Medicago truncatula* (D'Erfurth et al. 2003) and lettuce (Mazier et al. 2007).

While most approaches to transposon mutagenesis result in random insertion of elements throughout the genome, transposition from a T-DNA construct carrying both the transposase and the non-autonomous element is effective for generating multiple insertion events within one region of the genome. In this system if transposition is inducible, for example through the use of a heat shock promoter, then induction of the transposase can result in transposition of the non-autonomous element from the T-DNA into flanking genomic regions, generating new mutant lines that have insertions in close proximity to the site of insertion of the T-DNA construct. Subsequent heat shock treatments can generate novel mutations by causing reactivation of the transposase, and the cycle can be repeated as many times as necessary to achieve saturation of mutations in this region (Nishal et al. 2005).

While T-DNA insertion systems are more popular than transposons, efficient transformation systems are still lacking in many monocot crop species so that transposon-tagging continues to hold a useful position in the arsenal of reverse genetics techniques (Ayliffe and Pryor 2011). In addition, transposon-generated populations have the advantage of being able to produce many unique insertion lines from a few initial plant lines and lack epigenetic changes associated with T-DNA-based insertions (Upadhyaya et al. 2011).

Fast-Neutron Mutagenesis

Another form of mutagenesis that causes physical disruption in genes is radiation or fast neutron bombardment (Li and Zhang 2002). In this technique, seeds are irradiated using fast neutrons and deletions are identified using PCR primers that flank the gene of interest (Fig. 4.3). Amplification time is restricted so as to preferentially allow amplification of the mutant (deleted) DNA where a smaller PCR product is synthesised. One advantage of using this technique is that the deletions produced via physical mutagenesis will almost certainly completely eliminate any gene function. The most useful benefit of this technology, however, may be the fact that tandemly linked gene duplications may be deleted in the same line. Mutation of tandemly-linked genes in the same line is difficult to achieve with other reverse genetics technologies commonly used in plants. The limitations of this technique include the fact that a very large number of plants must be laboriously screened, and that there are constraints on the size of deletions that can be recovered. Nonetheless, this technology has been effective in creating mutant populations in *Arabidopsis* (Li and Zhang 2002), legumes (Rios et al. 2008; Rogers et al. 2009), rice (Bruce et al. 2009), soybean (Bolon et al. 2011), tomato (Dor et al. 2010), the citrus clementine (Rios et al. 2008), and in the metal-tolerant plant species *Noccaeacae rulescens* (Ó Lochlainn et al. 2011).

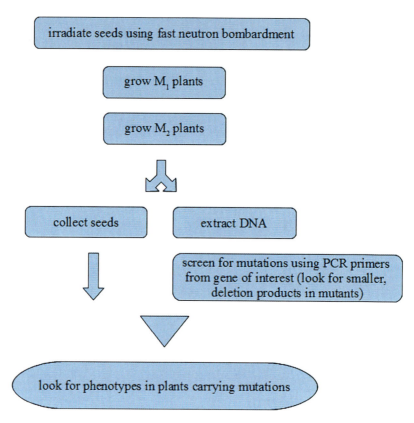

Fig. 4.3 Fast neutron mutagenesis. Flow chart describing the procedure for fast neutron mutagenesis. M_1 refers to the mutagenized generation; M_2 refers to the progeny of the mutagenized generation

Virus-Induced Gene Silencing (VIGS)

This technology began to be used extensively in the 1990s and is based on post-transcriptional gene silencing (PTGS) (Burch-Smith et al. 2004). The term VIGS was created by van Kammen (1997) to describe the development of a plant's resistance to virus infection after introduction of a viral transgene. VIGS is a very adaptable technique and has been used in many species. It has some advantages over other reverse techniques, such as the fact that it is relatively inexpensive, delivers rapid results and does not require transformation. In addition, because the phenotype is transient, deleterious effects of loss of gene function may be observed without causing lethality or infertility. The drawbacks include the fact that transient effects cannot be following using classical genetic studies, and that the vectors exhibit some host and/or tissue specificity. There can also be side-effects of the infection that may interfere with the silencing phenotype. In addition, the function

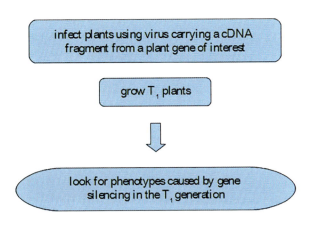

Fig. 4.4 Virus-induced gene silencing (*VIGS*). Flow chart describing the procedure for using virus-induced gene silencing to create transient loss-of-function mutations in specific genes. The cDNA fragment is part of the coding region of the gene. T_1 refers to the first generation after transformation of the cloned fragment into the plant

of several homologous genes may be affected with a single construct, complicating the interpretation of observed phenotypes. Finally, the level of silencing of the target gene is variable depending on the construct and the growth conditions used, and it is rare that genes will be completely silenced.

The protocol for VIGS involves cloning a 200–1,300 bp cDNA fragment from a plant gene of interest into a DNA copy of the genome of a plant virus (usually a RNA virus) and transfecting the plant with this construct (Hayward et al. 2011) (Fig. 4.4). Double-stranded RNA from the viral genome, including the sequence from the gene of interest, is formed during viral replication. The double-stranded RNA molecules are degraded into small interfering RNA (siRNA) molecules by the plant Dicer-like enzymes, thus activating the siRNA silencing pathway (for review see Chen (2009)) resulting in the degradation of the target gene transcript leading to a knockout or knockdown phenotype for the gene of interest.

The earliest vectors used for VIGS in plants included the *Tobacco mosaic virus*, *Potato virus X*, and *Tomato golden mosaic virus* but these had disadvantages such as infection symptoms that interfered with or complicated a mutant phenotype, or lack of infection in certain tissues (Ratcliff et al. 2001). Currently the most widely-used VIGS vector for dicotyledonous species is the *Tobacco rattle virus* (TRV) which has a broad plant host range, infects many different tissue types, and produces relatively mild disease symptoms in most plants (Hayward et al. 2011). The TRV vector has been used successfully for VIGS in the model species *Arabidopsis* (Burch-Smith et al. 2006), as well as a number of crop species including tomato (Fu et al. 2005), potato (Brigneti et al. 2004), *Jatropha* (Ye et al. 2009), chilli pepper (Chung et al. 2004), and *Brassica nigra* (Zheng et al. 2010). This vector has also been used for VIGS in a number of ornamental plants such as petunia, *Impatiens* and chrysanthemum (Jiang et al. 2011), California poppy (Wege et al. 2007), and columbine (Gould and Kramer 2007) to name a few.

In spite of it's broad host-range, however, some dicotyledonous and all monocotyledonous plants are resistant to infection by TRV. The main virus vector used for VIGS in monocots has been the *Barley stripe mosaic virus* (Holzberg et al. 2002).

This is currently the vector most commonly used in barley and wheat (reviewed in Scofield and Nelson 2009) and *Brachypodium* (Demircan and Akkaya 2010), and it has also been shown to be effective in less well-studied monocots such as the wheat-relative *Haynaldia* (Wang et al. 2010) and culinary ginger (Renner et al. 2009). More recently, the *Brome mosaic virus* has been used efficiently for VIGS in both rice and maize (Ding et al. 2006) and continued improvements to both of these vectors show promise for future studies using VIGS in monocot species.

There are also several other new virus vectors that are being used or studied for VIGS in plants. The *Apple latent spherical virus* is reported to be even broader in its host range than TRV, and to have minimal side effects. It has been used in both model and non-model dicot plants including legumes and cucurbit species (Igarashi et al. 2009), soybean (Yamagishi and Yoshikawa 2009) and fruit trees (Sasaki et al. 2011) as well as many species that are also susceptible to TRV.

In addition to the established VIGS vectors that can be used in many different species, new species-specific vectors are being developed for a diverse range of plants that will allow this technique to be used to study gene function in an even wider range of species. These include the *Pea early browning virus* for *Pisum sativum* (Constantin et al. 2004), the *African cassava mosaic virus* for cassava (Fofana et al. 2004), *Bean pod mottle virus* for soybean and other P*haseolus* species (Zhang and Ghabrial 2006) and the *Cotton leaf crumple virus* for cotton (Tuttle et al. 2008).

RNA Interference or Artificial MicroRNA Gene Silencing

RNA interference (RNAi), or RNA-induced gene silencing is similar to VIGS mechanistically, but the former is heritable in nature and so offers a different scope for investigation (McGinnis 2010). For the RNAi technique, a construct that produces double-stranded RNA complementary to the gene of interest is introduced into a cell where it activates the RNA silencing pathway and degrades some or all of the transcripts from the gene(s) of interest (for review see Chen 2009; Huntzinger and Izaurralde 2011) (Fig. 4.5). There are several techniques commonly used to activate the RNAi pathway in plants but the most popular strategies involve transformation with a construct encoding a hairpin RNA structure (hpRNA) or the use of an artificial microRNA (amiRNA) targeting the gene of interest (Eamens and Waterhouse 2011). For the hpRNA technique, reverse transcriptase PCR (RT-PCR) is used to amplify a region in the gene of interest, which is then cloned into a vector that creates inverted repeats of this region. The vector will also, typically, carry a promoter that will allow expression of the transgene at the time and in the tissue desired, along with a selectable marker for detection (Doran and Helliwell 2009). When this region is transcribed, the products act as dsRNA targets for the small RNA silencing pathway genes that normally target endogenous transcript(s) for degradation.

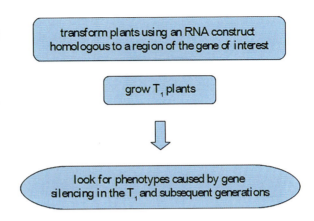

Fig. 4.5 RNA interference (*RNAi*). Flow chart describing the procedure for using RNA interference to create (usually) heritable loss-of-function mutations in specific plant genes. T_1 refers to the first generation after transformation of the RNAi construct into the plant

For amiRNA gene silencing, either ectopic or constitutive expression of an endogenous miRNAs is used to silence a target gene of interest (Alvarez et al. 2006), or an artificially constructed microRNA gene carrying a 21 bp insert complementary to the target gene is transformed into the plant where it acts in the endogenous miRNA silencing pathway (Ossowski et al. 2008). In addition, modern RNAi techniques may involve the use of promoters that are temporally or spatially specific, or that are inducible by some chemical or abiotic factor (Masclaux and Galaud 2011).

The advantages of using RNAi and amiRNA-based technology for reverse genetics in plants include the fact that a partial loss of function can be achieved when complete loss of function might be lethal, and that silencing is directed against a specific gene(s) so the screening of large populations is not required. In addition, transcripts of multiple genes from the same family can be silenced by a single construct (Alvarez et al. 2006; Schwab et al. 2006). This latter advantage is especially useful in plants since many plants have undergone partial or complete polyploidisation at some stage during their evolution, and a public website has been created to assist in the design of amiRNAs in more than 90 different plant species: http://wmd3.weigelworld.org (Ossowski Stephan, Fitz Joffrey, Schwab Rebecca, Riester Markus and Weigel Detlef, pers. comm.). Other advantages of this technique are that the induced phenotypes are dominant, they can be observed in the T_1 generation, and that stable inheritance of the transgenic RNAi gene makes the technique suitable for genetic engineering of traits into crop species in a manner that can be propagated from generation to generation.

Some disadvantages of the RNAi technology include the fact that some genes are resistant to silencing, possibly because of sequence or structural features of these genes. In addition, transcripts of genes that are similar in sequence to the target locus may be concomitantly down-regulated as well as the transcripts from the actual target gene. Although this is less of a problem in plants than in animals, 'off-target' silencing must be considered when planning experiments (Senthil-Kumar and Mysore 2011). The silencing level also may vary depending on the construct and the species, and gene expression is rarely completely silenced. The long-term effects of RNAi are also variable and expression of the transgenic RNAi constructs is often less effective in succeeding generations of transgenic lines.

In spite of these disadvantages, however, RNAi has been used in many plant species for both experimental and applied purposes such as nutritional improvement, pest resistance, reduction of toxins and improved response to abiotic stresses (Jagtap et al. 2011). In addition, RNAi constructs that target pathogenesis genes in insects, nematodes, or fungal parasites has been very successful at creating crops and other plants that are resistant to infection by these pathogens (Niu et al. 2010). This technique has been used successfully to improve a number of crops including several Brassica species (reviewed in Wood et al. (2011)), banana, cotton, barley, and coffee (Angaji et al. 2010), wheat (Fu et al. 2007), tomato (Fernandez et al. 2009), and the anti-malarial *Artemesinin* (Zhang et al. 2009).

Emerging Techniques

Promising Technologies for Screening Mutagenised Populations

New Generation Sequencing (NGS)

Direct sequencing would be the simplest method for screening mutagenised populations, and this possibility may become a reality in the near future as sequencing costs continue to decline spurred by the astounding advances in NGS techniques (Niedringhaus et al. 2011). Two types of sequencing technologies have now been tested on mutant TILLING populations and both groups report success using this strategy on tomato (Rigola et al. 2009; Tsai et al. 2011). Most recently, sequencing of whole genomes using this new technology has also proven that single-nucleotide polymorphism analysis is possible using sequencing, even in complex genomes such as oat where there is no previous reference sequence available (Oliver et al. 2011).

High Resolution Melting Curve Analysis (HRM)

Melting curve analysis has been used to identify DNA variants since the late 1990s (Wittwer et al. 1997), but was of limited use because of the technical limitations imposed by instrumentation and dye technologies. With the development of more sensitive double-stranded DNA (dsDNA) dyes and improvements in instrumentation that allow more accurate measurements of amplicon melting behaviour (Vossen et al. 2009) it is now possible to use HRM analysis for genomic-scale screening such as is required for TILLING or other SNP-detection or genotyping projects. The process is based on the fact that a dsDNA binding dye is intercalated between each base pair of a double-stranded PCR amplicon. When the DNA is heated it starts to denature, thus releasing the encaptured dye which then no longer fluoresces. This

4 Gene Identification: Reverse Genetics

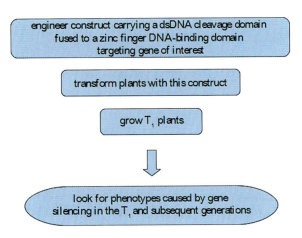

Fig. 4.6 Gene targeting using zinc-finger nucleases (*ZFNs*). Flow chart describing the procedure for using zinc-finger nuclease technology to create heritable loss-of-function mutations in specific genes. T_1 refers to the first generation after transformation of the ZFN construct into the plant

decrease in fluorescence is recorded by a camera and visualized on the screen. The rate of fluorescence decay is dependent on the sequence of the DNA, but also on the fidelity of the match of the two strands. Thus, any amplicon containing a mutation will produce a mismatch when paired with a wild-type amplicon, and this can be detected by a more rapid decrease in fluorescence than with homozygous wild-type amplicons. Only PCR fragments of a few hundred base pairs can be screened in one reaction, but this technology has been used for many medical applications and has now been tested in several plant species for detecting SNPs in both mutagenized and natural populations. This technology appears to be very versatile, inexpensive and has been successful in most systems in which it has been tested including almond (Wu et al. 2008), tomato (Gady et al. 2009), wheat (Dong et al. 2009), barley (Hofinger et al. 2009), ryegrass (Studer et al. 2009), olive (Muleo et al. 2009), chilli pepper (Park et al. 2009), maize (Li et al. 2010), potato (De Koeyer et al. 2010), peach (Chen and Wilde 2011), and oat (Oliver et al. 2011).

Gene Targeting

Random mutagenesis often results in mutations in single loci that do not have an effect on phenotype, and knock-down strategies rarely silence a gene or gene family entirely. Thus, in plants where gene redundancy is high and where polyploidy is often the rule rather than the exception, gene targeting allows for the isolation of plants carrying mutations in single, defined genes or multiple genes of a gene family within the same plant. Gene targeting involves the integration or removal of a piece of DNA from a specific target sequence in the host plant (Fig. 4.6). In theory, this enables the generation of specific alleles of any gene in the plant. It has been successfully used in fungi for many years, but remained elusive in plants until the recent improvements in synthetic Zinc finger nucleases (ZFNs) that were first created in the late 1990s (Chandrasegaran and Smith 1999). Modern ZFN's are engineered by combining two zinc finger proteins that recognise a specific DNA sequence, with an endonuclease that causes non-specific double-stranded breaks in DNA. They were first used in plants in 2005 where they were shown to cause mutations at site-specific locations at a rate of approximately 20% (Lloyd et al. 2005). Most mutations created by this approach are small deletions or insertions of a few base pairs that can be attributed to the nonhomologous end joining (NHEJ) DNA repair mechanism found in all species (reviewed in Mladenov and Iliakis (2011)). Recently, this system has been improved so that targeting of specific ZFNs to plant genes can cause mutation rates of from 30 to 70% at these sites when the toxicity of the construct is controlled by making them heat or hormone-inducible (Zhang et al. 2010). One of the drawbacks of this technology has been the difficulty and high cost of designing appropriate zinc finger motifs to target the selected region in the genome, but this has been simplified by the creation of the publicly available OPEN (Oligomerized Pool Engineering) platform for engineering zinc-finger constructs (Maeder et al. 2008). A new development in ZFN technology is the context-dependent assembly (CoDA) platform recently published by the Zinc Finger Consortium (Sander et al. 2011). This strategy uses archived information from several hundred existing zinc finger arrays to automatically design new ZFN constructs that have different sequence specificity without requiring technical expertise beyond standard cloning techniques. CoDA appears to be as specific and less labour-intensive than OPEN and holds much promise for gene targeting in non-model, polyploid species (Curtin et al. 2011). ZFN technology has, to date, been successfully employed in a number of animal and plant species with equal success. Plants in which it has demonstrated utility include *Arabidopsis* (Zhang and Voytas 2011), soybean (Curtin et al. 2011), maize (Shukla et al. 2009) and tobacco (Townsend et al. 2009). Transcription activator-like (TAL) DNA-binding proteins have been developed for targeted gene modulation in plants as an alternative to ZFN technology. TAL proteins have been shown to be useful as a tool to study for gene activation (for review see Bogdanove et al. 2010). More recently, constructs known as TALENs have been successfully used for sequence-specific gene disruption by combining the catalytic domain of the FokI nuclease with specific TAL effector constructs (Cermak et al. 2011).

Conclusion

There are more and more resources available for reverse-genetic studies in plants. Each has its advantages and disadvantages depending on the species being targeted and the questions addressed (Table 4.1). Both VIGS and RNAi remain attractive, in part, because of their low cost. They are very useful for studying genes of unknown function in species for which these techniques have been developed. The availability of T-DNA and transposon insertion lines that are accessible to the public makes those resources attractive as well. It is more expensive to perform physical mutagenesis using, for example, radiation, followed by reverse genetic analysis of the mutagenised population, but this approach is, nonetheless, useful in cases where other techniques have not been successful. Chemical mutagenesis and TILLING provide more varied types of mutations than other techniques, but can be more time-consuming and costly. One of the newest and most promising techniques is the fine-tuning of the zinc-finger nuclease and TALEN techniques that, for the first time, allow targeted mutagenesis in plants at an acceptable cost and in a reasonable amount of time using technology that is available in most laboratories. Finally, with the continuous development of new technologies, the most efficient technique for examining gene function in plants in the future may involve direct sequencing of part or complete genomes of individual plants, or some completely novel technology that has yet to be developed.

References

Alonso JM, Stepanova AN, Leisse TJ, Kim CJ, Chen H, Shinn P, Stevenson DK, Zimmerman J, Barajas P, Cheuk R, Gadrinab C, Heller C, Jeske A, Koesema E, Meyers CC, Parker H, Prednis L, Ansari Y, Choy N, Deen H, Geralt M, Hazari N, Hom E, Karnes M, Mulholland C, Ndubaku R, Schmidt I, Guzman P, Aguilar-Henonin L, Schmid M, Weigel D, Carter DE, Marchand T, Risseeuw E, Brogden D, Zeko A, Crosby WL, Berry CC, Ecker JR (2003) Genome-wide insertional mutagenesis of *Arabidopsis thaliana*. Science 301:653–657

Alvarez JP, Pekker I, Goldshmidt A, Blum E, Amsellem Z, Eshed Y (2006) Endogenous and synthetic microRNAs stimulate simultaneous, efficient, and localized regulation of multiple targets in diverse species. Plant Cell 18:1134–1151

Angaji S, Hedayati SS, Hosein R, Samad S, Shiravi S, Madani S (2010) Application of RNA interference in plants. Plant Omics 3:77–84

Anuradha T, Jami S, Datla R, Kirti P (2006) Genetic transformation of peanut (*Arachis hypogaea* L.) using cotyledonary node as explant and a promoterless gus::npt II fusion gene based vector. J Biosci 31:235–246

Arencibia AD, Carmona ER (2007) Sugarcane (Saccharum spp.). In: Wang K (ed) Agrobacterium Protocols, vol 2. Humana Press, pp 227–235

Ayliffe MA, Pryor AJ (2011) Activation tagging and insertional mutagenesis in barley. In: Pereira A (ed) Plant reverse genetics: methods and protocols, methods in molecular biology, vol 678. Humana Press, New York, pp 107–128

Barampuram S, Zhang ZJ (2011) Recent advances in plant transformation. In: Birchler JA (ed) Plant chromosome engineering: methods and protocols, methods in molecular biology, vol 701. Humana Press, New York, pp 1–35

Barrell P, Conner A (2011) Facilitating the recovery of phenotypically normal transgenic lines in clonal crops: a new strategy illustrated in potato. Theor Appl Genet 122:1171–1177

Belide S, Hac L, Singh S, Green A, Wood C (2011) Agrobacterium-mediated transformation of safflower and the efficient recovery of transgenic plants via grafting. Plant Methods 7:12

Biswas B, Pick Kuen C, Gresshoff PM (2009) A novel ABA insensitive mutant of lotus japonicus with a wilty phenotype displays unaltered nodulation regulation. Mol Plant 2:487–499

Bogdanove AJ, Schornack S, Lahaye T (2010) TAL effectors: finding plant genes for disease and defense. Curr Opin Plant Biol 13:394–401

Bolon Y-T, Haun WJ, Xu WW, Grant D, Stacey MG, Nelson RT, Gerhardt DJ, Jeddeloh JA, Stacey G, Muehlbauer GJ, Orf JH, Naeve SL, Stupar RM, Vance CP (2011) Phenotypic and genomic analyses of a fast neutron mutant population resource in soybean. Plant Physiol 156:240–253

Bouquet A, Torregrosa L, Iocco P, Thomas MR (2007) Grapevine (*Vitis vinifera* L.). In: Wang K (ed) Agrobacterium protocols, vol 2. Humana Press, Totowa, pp 273–285

Brigneti G, Martín-Hernández AM, Jin H, Chen J, Baulcombe DC, Baker B, Jones JDG (2004) Virus-induced gene silencing in Solanum species. Plant J 39:264–272

Bruce M, Hess A, Bai J, Mauleon R, Diaz MG, Sugiyama N, Bordeos A, Wang G-L, Leung H, Leach J (2009) Detection of genomic deletions in rice using oligonucleotide microarrays. BMC Genomics 10:129

Burch-Smith TM, Anderson JC, Martin GB, Dinesh-Kumar SP (2004) Applications and advantages of virus-induced gene silencing for gene function studies in plants. Plant J 39:734–746

Burch-Smith TM, Schiff M, Liu Y, Dinesh-Kumar SP (2006) Efficient virus-induced gene silencing in Arabidopsis. Plant Physiol 142:21–27

Busov VB, Brunner AM, Meilan R, Filichkin S, Ganio L, Gandhi S, Strauss SH (2005) Genetic transformation: a powerful tool for dissection of adaptive traits in trees. New Phytologist 167:9–18

Byrne ME (2006) Shoot meristem function and leaf polarity: the role of class III HD–ZIP genes. PLoS Genet 2:e89

Calderini O, Carelli M, Panara F, Biazzi E, Scotti C, Tava A, Porceddu A, Arcioni S (2011) Collection of mutants for functional genomics in the legume *Medicago truncatula*. Plant Genet Resour 9:174–176

Caldwell DG, McCallum N, Shaw P, Muehlbauer GJ, Marshall DF, Waugh R (2004) A structured mutant population for forward and reverse genetics in Barley (*Hordeum vulgare* L.). Plant J 40:143–150

Candela H, Hake S (2008) The art and design of genetic screens: maize. Nat Rev Genet 9:192–203

Cermak T, Doyle EL, Christian M, Wang L, Zhang Y, Schmidt C, Baller JA, Somia NV, Bogdanove AJ, Voytas DF (2011) Efficient design and assembly of custom TALEN and other TAL effector-based constructs for DNA targeting. Nucleic Acids Res 39:e82

Chandrasegaran S, Smith J (1999) Chimeric restriction enzymes: what is next? Biol Chem 380:841–848

Chattopadhyay T, Roy S, Mitra A, Maiti MK (2011) Development of a transgenic hairy root system in jute (*Corchorus capsularis* L.) with gusA reporter gene through Agrobacterium rhizogenes mediated co-transformation. Plant Cell Rep 30:485–493

Chawade A, Sikora P, Brautigam M, Larsson M, Vivekanand V, Nakash M, Chen T, Olsson O (2010) Development and characterization of an oat TILLING-population and identification of mutations in lignin and beta-glucan biosynthesis genes. BMC Plant Biol 10:86

Chen W, Punja Z (2002) Transgenic herbicide- and disease-tolerant carrot (*Daucus carota* L.) plants obtained through *Agrobacterium*-mediated transformation. Plant Cell Rep 20:929–935

Chen X (2009) Small RNAs and their roles in plant development. Annu Rev Cell Dev Biol 25:21–44

Chen Y, Wilde HD (2011) Mutation scanning of peach floral genes. BMC Plant Biol 11:96

Chen Z-Z, Ho C-K, Ahn I-S, Chiang VL (2007) Eucalyptus. In: Wang K (ed) Agrobacterium protocols, vol 2. Humana Press, Totowa, pp 125–134

Chung E, Seong E, Kim YC, Chung EJ, Oh SK, Lee S, Park JM, Joung YH, Choi D (2004) A method of high frequency virus-induced gene silencing in chili pepper (Capsicum annuum L. cv. Bukang). Mol Cells 17:377–380

Clough SJ, Bent AF (1998) Floral dip: a simplified method for Agrobacterium-mediated transformation of Arabidopsis thaliana. Plant J 16:735–743

Colbert T, Till BJ, Tompa R, Reynolds S, Steine MN, Yeung AT, McCallum CM, Comai L, Henikoff S (2001) High-throughput screening for induced point mutations. Plant Physiol 126:480–484

Constantin GD, Krath BN, MacFarlane SA, Nicolaisen M, Elisabeth Johansen I, Lund OS (2004) Virus-induced gene silencing as a tool for functional genomics in a legume species. Plant J 40:622–631

Cooper J, Till B, Laport R, Darlow M, Kleffner J, Jamai A, El-Mellouki T, Liu S, Ritchie R, Nielsen N, Bilyeu K, Meksem K, Comai L, Henikoff S (2008) TILLING to detect induced mutations in soybean. BMC Plant Biol 8:9

Courtial B, Feuerbach F, Eberhard S, Rohmer L, Chiapello H, Camilleri C, Lucas H (2001) Tnt transposition events are induced by in vitro transformation Arabidopsis thaliana and transposed copies integrate into genes. Mol Genet Genomics 265:32–42

Curtin SJ, Zhang F, Sander JD, Haun WJ, Starker C, Baltes NJ, Reyon D, Dahlborg EJ, Goodwin MJ, Coffman AP, Dobbs D, Joung JK, Voytas DF, Stupar RM (2011) Targeted mutagenesis of duplicated genes in soybean with zinc-finger nucleases. Plant Physiol 156:466–473

Dahmani-Mardas F, Troadec C, Boualem A, Lévêque S, Alsadon AA, Aldoss AA, Dogimont C, Bendahmane A (2010) Engineering melon plants with improved fruit shelf life using the TILLING approach. PLoS ONE 5:e15776

Dalmais M, Schmidt J, Le Signor C, Moussy F, Burstin J, Savois V, Aubert G, Brunaud V, de Oliveira Y, Guichard C, Thompson R, Bendahmane A (2008) UTILLdb, a *Pisum sativum* in silico forward and reverse genetics tool. Genome Biol 9

De Koeyer D, Douglass K, Murphy A, Whitney S, Nolan L, Song Y, De Jong W (2010) Application of high-resolution DNA melting for genotyping and variant scanning of diploid and autotetraploid potato. Mol Breed 25:67–90

Demircan T, Akkaya M (2010) Virus induced gene silencing in *Brachypodium distachyon*, a model organism for cereals. Plant Cell Tissue Organ Cult 100:91–96

D'Erfurth I, Cosson V, Eschstruth A, Lucas H, Kondorosi A, Ratet P (2003) Efficient transposition of the Tnt1 tobacco retrotransposon in the model legume *Medicago truncatula*. Plant J 34:95–106

Diao X-M, Lisch D (2006) Mutator transposon in maize and MULEs in the plant genome. Acta Genet Sin 33:477–487

Ding XS, Schneider WL, Chaluvadi SR, Mian MAR, Nelson RS (2006) Characterization of a brome mosaic virus strain and its use as a vector for gene silencing in monocotyledonous hosts. Mol Plant Microbe Interact 19:1229–1239

Dong C, Vincent K, Sharp P (2009) Simultaneous mutation detection of three homoeologous genes in wheat by High Resolution Melting analysis and Mutation Surveyor(R). BMC Plant Biol 9:143

Dor E, Alperin B, Wininger S, Ben-Dor B, Somvanshi V, Koltai H, Kapulnik Y, Hershenhorn J (2010) Characterization of a novel tomato mutant resistant to the weedy parasites Orobanche and Phelipanche spp. Euphytica 171:371–380

Doran T, Helliwell C (2009) RNA interference: methods for plants and animals, xii, 257 p edn. Cabi, Wallingford/Cambridge, MA

Eamens AL, Waterhouse PM (2011) Vectors and methods for hairpin RNA and artificial microRNA-mediated gene silencing in plants. In: Birchler JA (ed) Plant chromosome engineering: methods and protocols, methods in molecular biology, vol 701. Humana Press, New York, pp 179–197

Eckardt NA (2007) Positive and negative feedback coordinate regulation of disease resistance gene expression. Plant Cell 19:2700–2702

Elias R, Till B, Mba C, Al-Safadi B (2009) Optimizing TILLING and Ecotilling techniques for potato (*Solanum tuberosum* L). BMC Res Notes 2:141

Fernandez AI, Viron N, Alhagdow M, Karimi M, Jones M, Amsellem Z, Sicard A, Czerednik A, Angenent G, Grierson D, May S, Seymour G, Eshed Y, Lemaire-Chamley M, Rothan C, Hilson P (2009) Flexible tools for gene expression and silencing in tomato. Plant Physiol 151:1729–1740

Fofana IBF, Sangaré A, Collier R, Taylor C, Fauquet CM (2004) A geminivirus-induced gene silencing system for gene function validation in cassava. Plant Mol Biol 56:613–624

Fu D-Q, Zhu B-Z, Zhu H-L, Jiang W-B, Luo Y-B (2005) Virus-induced gene silencing in tomato fruit. Plant J 43:299–308

Fu D, Uauy C, Blechl A, Dubcovsky J (2007) RNA interference for wheat functional gene analysis. Transgenic Res 16:689–701

Gady A, Hermans F, Van de Wal M, van Loo E, Visser R, Bachem C (2009) Implementation of two high through-put techniques in a novel application: detecting point mutations in large EMS mutated plant populations. Plant Methods 5:13

Ghimire BK, Seong ES, Lee CO, Lim JD, Lee JG, Yoo JH, Chung I-M, Kim NY, Yu CY (2011) Enhancement of alpha-tocopherol content in transgenic *Perilla frutescens* containing the gamma-TMT gene. Afr J Biotechnol 10:2430–2439

Gould B, Kramer E (2007) Virus-induced gene silencing as a tool for functional analyses in the emerging model plant Aquilegia (columbine, Ranunculaceae). Plant Methods 3:6

Grandbastien M-A, Spielmann A, Caboche M (1989) Tnt1, a mobile retroviral-like transposable element of tobacco isolated by plant cell genetics. Nature 337:376–380

Grant JE, Cooper PA, Dale TM (2004) Transgenic *Pinus radiata* from *Agrobacterium tumefaciens*-mediated transformation of cotyledons. Plant Cell Rep 22:894–902

Hayward A, Padmanabhan M, Dinesh-Kumar SP (2011) Virus-induced gene silencing in nicotiana benthamiana and other plant species. In: Pereira A (ed) Plant reverse genetics: methods and protocols, methods in molecular biology, vol 678. Humana Press, New York, pp 55–63

Hellens R, Mullineaux P, Klee H (2000) Technical focus: a guide to agrobacterium binary Ti vectors. Trends Plant Sci 5:446–451

Himelblau E, Gilchrist E, Buono K, Bizzell C, Mentzer L, Vogelzang R, Osborn T, Amasino R, Parkin I, Haughn G (2009) Forward and reverse genetics of rapid-cycling *Brassica oleracea*. Theor Appl Genet 118:953–961

Hofinger BJ, Jing H-C, Hammond-Kosack KE, Kanyuka K (2009) High-resolution melting analysis of cDNA-derived PCR amplicons for rapid and cost-effective identification of novel alleles in barley. TAG Theor Appl Genet 119:851–865

Holzberg S, Brosio P, Gross C, Pogue GP (2002) Barley stripe mosaic virus-induced gene silencing in a monocot plant. Plant J 30:315–327

Huntzinger E, Izaurralde E (2011) Gene silencing by microRNAs: contributions of translational repression and mRNA decay. Nat Rev Genet 12:99–110

Igarashi A, Yamagata K, Sugai T, Takahashi Y, Sugawara E, Tamura A, Yaegashi H, Yamagishi N, Takahashi T, Isogai M, Takahashi H, Yoshikawa N (2009) Apple latent spherical virus vectors for reliable and effective virus-induced gene silencing among a broad range of plants including tobacco, tomato, *Arabidopsis thaliana*, cucurbits, and legumes. Virology 386:407–416

Imaizumi R, Sato S, Kameya N, Nakamura I, Nakamura Y, Tabata S, Ayabe S-I, Aoki T (2005) Activation tagging approach in a model legume, *Lotus japonicus*. J Plant Res 118:391–399

Indurker S, Misra H, Eapen S (2010) Agrobacterium-mediated transformation in chickpea (*Cicer arietinum* L.) with an insecticidal protein gene: optimisation of different factors. Physiol Mol Biol Plants 16:273–284

Jagtap U, Gurav R, Bapat V (2011) Role of RNA interference in plant improvement. Naturwissenschaften 98:473–492

Jeon J-S, Lee S, Jung K-H, Jun S-H, Jeong D-H, Lee J, Kim C, Jang S, Lee S, Yang K, Nam J, An K, Han M-J, Sung R-J, Choi H-S, Yu J-H, Choi J-H, Cho S-Y, Cha S-S, Kim S-I, An G (2000) T-DNA insertional mutagenesis for functional genomics in rice. Plant J 22:561–570

Jiang C-Z, Chen J-C, Reid M (2011) Virus-induced gene silencing in ornamental plants. In: Kodama H, Komamine A (eds) Methods in molecular biology, vol 744. Humana Press, New York, pp 81–96

Juarez MT, Twigg RW, Timmermans MCP (2004) Specification of adaxial cell fate during maize leaf development. Development 131:4533–4544

Kishchenko E, Komarnitskii I, Kuchuk N (2010) Transposition of the maize transposable element Spm in transgenic sugar beets. Cytol Genet 44:200–205

Krishna G, Reddy P, Ramteke P, Bhattacharya P (2010) Progress of tissue culture and genetic transformation research in pigeon pea [*Cajanus cajan* (L.) Millsp.]. Plant Cell Rep 29:1079–1095

Krishnan A, Guiderdoni E, An G, Hsing TC, Han C, Lee MC, Yu S-M, Upadhyaya N, Ramachandran S, Zhang Q, Sundaresan V, Hirochika H, Leung H, and Pereira A (2009) Mutant resources in rice for functional genomics of the grasses. Plant Physiol 149:165–170

Kumar S, Fladung M (2003) Somatic mobility of the maize element Ac and its utility for gene tagging in aspen. Plant Mol Biol 51:643–650

Kumar V, Campbell L, Rathore K (2011) Rapid recovery- and characterization of transformants following *Agrobacterium*-mediated T-DNA transfer to sorghum. Plant Cell Tissue Organ Cult 104:137–146

Lee L-Y, Gelvin SB (2008) T-DNA binary vectors and systems. Plant Physiol 146:325–332

Lee M-K, Kim H-S, Kim J-S, Kim S-H, Park Y-D (2004) Agrobacterium-mediated transformation system for large-scale producion of transgenic chinese cabbage (*Brassica rapa* L. ssp. *pekinensis*) plants for insertional mutagenesis. J Plant Biol 47:300–306

Li R, Qu R (2011) High throughput Agrobacterium-mediated switchgrass transformation. Biomass Bioenergy 35:1046–1054

Li X, Zhang Y (2002) Reverse genetics by fast neutron mutagenesis in higher plants. Funct Integr Genomics 2:254–258

Li Y-D, Chu Z-Z, Liu X-G, Jing H-C, Liu Y-G, Hao D-Y (2010) A cost-effective high-resolution melting approach using the EvaGreen dye for DNA polymorphism detection and genotyping in plants. J Integr Plant Biol 52:1036–1042

Lisch D, Jiang N (2009) Mutator and MULE transposons. In: Bennetzen JL, Hake S (eds) Handbook of maize. Springer, New York, pp 277–306

Lloyd A, Plaisier CL, Carroll D, Drews GN (2005) Targeted mutagenesis using zinc-finger nucleases in Arabidopsis. Proc Natl Acad Sci USA 102:2232–2237

Lochlainn SÓ, Fray RG, Hammond JP, King GJ, White PJ, Young SD, Broadley MR (2011) Generation of nonvernal-obligate, faster-cycling *Noccaea caerulescens* lines through fast neutron mutagenesis. New Phytol 189:409–414

Maeder ML, Thibodeau-Beganny S, Osiak A, Wright DA, Anthony RM, Eichtinger M, Jiang T, Foley JE, Winfrey RJ, Townsend JA, Unger-Wallace E, Sander JD, Müller-Lerch F, Fu F, Pearlberg J, Göbel C, Dassie Justin P, Pruett-Miller SM, Porteus MH, Sgroi DC, Iafrate AJ, Dobbs D, McCray PB Jr, Cathomen T, Voytas DF, Joung JK (2008) Rapid "open-source" engineering of customized zinc-finger nucleases for highly efficient gene modification. Mol Cell 31:294–301

Marsch-Martínez N (2011) A transposon-based activation tagging system for gene function discovery in arabidopsis. In: Yuan L, Perry SE (eds) Plant transcription factors, methods in molecular biology, vol 754. Humana Press, New York, pp 67–83

Martin B, Ramiro M, Martinez-Zapater J, Alonso-Blanco C (2009) A high-density collection of EMS-induced mutations for TILLING in *Landsberg erecta* genetic background of *Arabidopsis*. BMC Plant Biol 9:147

Masclaux F, Galaud J-P (2011) Heat-inducible RNAi for gene functional analysis in plants. In: Kodama H, Komamine A (eds) RNAi and plant gene function analysis, methods in molecular biology, vol 744. Humana Press, New York, pp 37–55

Mathews H, Clendennen SK, Caldwell CG, Liu XL, Connors K, Matheis N, Schuster DK, Menasco DJ, Wagoner W, Lightner J, Wagner DR (2003) Activation tagging in tomato identifies a transcriptional regulator of anthocyanin biosynthesis, modification, and transport. Plant Cell Online 15:1689–1703

May BP, Martienssen RA (2003) Transposon mutagenesis in the study of plant development. Crit Rev Plant Sci 22:1

Mazier M, Botton E, Flamain F, Bouchet J-P, Courtial B, Chupeau M-C, Chupeau Y, Maisonneuve B, Lucas H (2007) Successful gene tagging in lettuce using the Tnt1 retrotransposon from tobacco. Plant Physiol 144:18–31

McGinnis KM (2010) RNAi for functional genomics in plants. Brief Funct Genomics 9:111–117

Meyers B, Zaltsman A, Lacroix B, Kozlovsky SV, Krichevsky A (2010) Nuclear and plastid genetic engineering of plants: comparison of opportunities and challenges. Biotechnol Adv 28:747–756

Michelmore R, Marsh E, Seely S, Landry B (1987) Transformation of lettuce (*Lactuca sativa*) mediated by *Agrobacterium tumefaciens*. Plant Cell Rep 6:439–442

Minoia S, Petrozza A, D'Onofrio O, Piron F, Mosca G, Sozio G, Cellini F, Bendahmane A, Carriero F (2010) A new mutant genetic resource for tomato crop improvement by TILLING technology. BMC Res Notes 3:69

Miyao A, Iwasaki Y, Kitano H, Itoh J-I, Maekawa M, Murata K, Yatou O, Nagato Y, Hirochika H (2007) A large-scale collection of phenotypic data describing an insertional mutant population to facilitate functional analysis of rice genes. Plant Mol Biol 63:625–635

Mladenov E, Iliakis G (2011) Induction and repair of DNA double strand breaks: the increasing spectrum of non-homologous end joining pathways. Mutat Res/Fundam Mol Mech Mutagen 711:61–72

Muleo R, Colao MC, Miano D, Cirilli M, Intrieri MC, Baldoni L, Rugini E (2009) Mutation scanning and genotyping by high-resolution DNA melting analysis in olive germplasm. Genome/Natl Res Counc Can 52:252–260

Niedringhaus TP, Milanova D, Kerby MB, Snyder MP, Barron AE (2011) Landscape of next-generation sequencing technologies. Anal Chem 83:4327–4341

Nishal B, Tantikanjana T, Sundaresan V (2005) An inducible targeted tagging system for localized saturation mutagenesis in Arabidopsis. Plant Physiol 137:3–12

Niu JH, Jian H, Xu JM, Guo YD, Liu QA (2010) RNAi technology extends its reach: engineering plant resistance against harmful eukaryotes. Afr J Biotechnol 9:7573–7582

Oliver R, Lazo G, Lutz J, Rubenfield M, Tinker N, Anderson J, Wisniewski Morehead N, Adhikary D, Jellen E, Maughan PJ, Brown Guedira G, Chao S, Beattie A, Carson M, Rines H, Obert D, Bonman JM, Jackson E (2011) Model SNP development for complex genomes based on hexaploid oat using high-throughput 454 sequencing technology. BMC Genomics 12:77

Oosumi T, Ruiz-Rojas JJ, Veilleux RE, Dickerman A, Shulaev V (2010) Implementing reverse genetics in *Rosaceae*: analysis of T-DNA flanking sequences of insertional mutant lines in the diploid strawberry, *Fragaria vesca*. Physiologia Plantarum 140:1–9

Ossowski S, Schwab R, Weigel D (2008) Gene silencing in plants using artificial microRNAs and other small RNAs. Plant J 53:674–690

Park S-W, An S-J, Yang H-B, Kwon J-K, Kang B-C (2009) Optimization of high resolution melting analysis and discovery of single nucleotide polymorphism in capsicum. Hortic Environ Biotech 50:31–39

Perry JA, Wang TL, Welham TJ, Gardner S, Pike JM, Yoshida S, Parniske M (2003) A TILLING reverse genetics tool and a web-accessible collection of mutants of the legume Lotus japonicus. Plant Physiol 131:866–871

Piron F, Nicolaï M, Minoïa S, Piednoir E, Moretti A, Salgues A, Zamir D, Caranta C, Bendahmane A (2010) An induced mutation in tomato eIF4E leads to immunity to two Potyviruses. PLoS ONE 5:e11313

Ratcliff F, Martin-Hernandez AM, Baulcombe DC (2001) Technical advance: tobacco rattle virus as a vector for analysis of gene function by silencing. Plant J 25:237–245

Renner T, Bragg J, Driscoll HE, Cho J, Jackson AO, Specht CD (2009) Virus-induced gene silencing in the culinary ginger (*zingiber officinale*): an effective mechanism for down-regulating gene expression in tropical monocots. Mol Plant 2:1084–1094

Rigola D, van Oeveren J, Janssen A, Bonné A, Schneiders H, van der Poel HJA, van Orsouw NJ, Hogers RCJ, de Both MTJ, van Eijk MJT (2009) High-throughput detection of induced mutations and natural variation using KeyPoint™ technology. PLoS One 4:e4761

Rios G, Naranjo M, Iglesias D, Ruiz-Rivero O, Geraud M, Usach A, Talon M (2008) Characterization of hemizygous deletions in Citrus using array-Comparative Genomic Hybridization and microsynteny comparisons with the poplar genome. BMC Genomics 9:381

Rogers C, Wen J, Chen R, Oldroyd G (2009) Deletion based reverse genetics in *Medicago truncatula*. Plant Physiol 151(3):1077–1086

Sander JD, Dahlborg EJ, Goodwin MJ, Cade L, Zhang F, Cifuentes D, Curtin SJ, Blackburn JS, Thibodeau-Beganny S, Qi Y, Pierick CJ, Hoffman E, Maeder ML, Khayter C, Reyon D, Dobbs D, Langenau DM, Stupar RM, Giraldez AJ, Voytas DF, Peterson RT, Yeh J-RJ, Joung JK (2011) Selection-free zinc-finger-nuclease engineering by context-dependent assembly (CoDA). Nat Methods 8:67–69

Sasaki S, Yamagishi N, Yoshikawa N (2011) Efficient virus-induced gene silencing in apple, pear and Japanese pear using Apple latent spherical virus vectors. Plant Methods 7:15

Schwab R, Ossowski S, Riester M, Warthmann N, Weigel D (2006) Highly specific gene silencing by artificial microRNAs in Arabidopsis. Plant Cell Online 18:1121–1133

Schwarz-Sommer Z, Davies B, Hudson A (2003) An everlasting pioneer: the story of Antirrhinum. Nat Reverse Genet 4:655–664

Scofield SR, Nelson RS (2009) Resources for virus-induced gene silencing in the grasses. Plant Physiol 149:152–157

Senthil-Kumar M, Mysore KS (2011) Caveat of RNAi in plants: the off-target effect. In: Kodama H, Komamine A (eds) RNAi and plant gene function analysis, methods in molecular biology, vol 744. Humana Press, New York, pp 13–25

Shukla VK, Doyon Y, Miller JC, DeKelver RC, Moehle EA, Worden SE, Mitchell JC, Arnold NL, Gopalan S, Meng X, Choi VM, Rock JM, Wu Y-Y, Katibah GE, Zhifang G, McCaskill D, Simpson MA, Blakeslee B, Greenwalt SA, Butler HJ, Hinkley SJ, Zhang L, Rebar EJ, Gregory PD, Urnov FD (2009) Precise genome modification in the crop species *Zea mays* using zinc-finger nucleases. Nature 459:437–441

Slade AJ, Fuerstenberg SI, Loeffler D, Steine MN, Facciotti D (2005) A reverse genetic, nontransgenic approach to wheat crop improvement by TILLING. Nat Biotechnol 23:75–81

Smolka A, Li X-Y, Heikelt C, Welander M, Zhu L-H (2010) Effects of transgenic rootstocks on growth and development of non-transgenic scion cultivars in apple. Transgenic Res 19:933–948

Song G-Q, Sink KC (2004) Agrobacterium tumefaciens-mediated transformation of blueberry (*Vaccinium corymbosum* L.). Plant Cell Rep 23:475–484

Stephenson P, Baker D, Girin T, Perez A, Amoah S, King G, Ostergaard L (2010) A rich TILLING resource for studying gene function in *Brassica rapa*. BMC Plant Biol 10:62

Studer B, Jensen L, Fiil A, Asp T (2009) "Blind" mapping of genic DNA sequence polymorphisms in *Lolium perenne* L. by high resolution melting curve analysis. Mol Breed 24:191–199

Sun Q, Zhao Y, Sun H, Hammond R, Davis R, Xin L (2011) High-efficiency and stable genetic transformation of pear (*Pyrus communis* L.) leaf segments and regeneration of transgenic plants. Acta Physiologiae Plantarum 33:383–390

Tadege M, Ratet P, Mysore KS (2005) Insertional mutagenesis: a Swiss army knife for functional genomics of *Medicago truncatula*. Trends Plant Sci 10:229–235

Talame V, Bovina R, Sanguineti MC, Tuberosa R, Lundqvist U, Salvi S (2008) TILLMore, a resource for the discovery of chemically induced mutants in barley. Plant Biotechnol J 6:477–485

Taylor N, Chavarriaga P, Raemakers K, Siritunga D, Zhang P (2004) Development and application of transgenic technologies in cassava. Plant Mol Biol 56:671–688

Thole V, Worland B, Wright J, Bevan MW, Vain P (2010) Distribution and characterization of more than 1000 T-DNA tags in the genome of *Brachypodium distachyon* community standard line Bd21. Plant Biotechnol J 8:734–747

Till BJ, Colbert T, Tompa R, Enns LC, Codomo CA, Johnson JE, Reynolds SH, Henikoff JG, Greene EA, Steine MN, Comai L, Henikoff S (2003) High-throughput TILLING for functional genomics. Methods Mol Biol (Clifton, NJ) 236:205–220

Till BJ, Cooper J, Tai TH, Colowit P, Greene E, Henikoff S, Comai L (2007) Discovery of chemically induced mutations in rice by TILLING. BMC Plant Biol 7:19

Townsend JA, Wright DA, Winfrey RJ, Fu F, Maeder ML, Joung JK, Voytas DF (2009) High-frequency modification of plant genes using engineered zinc-finger nucleases. Nature 459: 442–445

Tsai H, Howell T, Nitcher R, Missirian V, Watson B, Ngo KJ, Lieberman M, Fass J, Uauy C, Tran RK, Khan AA, Filkov V, Tai TH, Dubcovsky J, Comai L (2011) Discovery of rare mutations in populations: TILLING by sequencing. Plant Physiol 156:1257–1268

Tuttle JR, Idris AM, Brown JK, Haigler CH, Robertson D (2008) Geminivirus-mediated gene silencing from cotton leaf crumple virus is enhanced by low temperature in cotton. Plant Physiol 148:41–50

Uauy C, Paraiso F, Colasuonno P, Tran R, Tsai H, Berardi S, Comai L, Dubcovsky J (2009) A modified TILLING approach to detect induced mutations in tetraploid and hexaploid wheat. BMC Plant Biol 9:115

Unni S, Soniya E (2010) Transgenic *Cucumis sativus* expressing the hepatitis B surface antigen. Plant Mol Biol Rep 28:627–634

Upadhyaya NM, Zhu Q-H, Bhat RS (2011) Transposon insertional mutagenesis in rice. In: Pereira A (ed) Plant reverse genetics: methods and protocols, methods in molecular biology, vol 678. Humana Press, New York, pp 147–177

van Kammen A (1997) Virus-induced gene silencing in infected and transgenic plants. Trends Plant Sci 2:409–411

Vossen RHAM, Aten E, Roos A, den Dunnen JT (2009) High-Resolution Melting Analysis (HRMA)—more than just sequence variant screening. Hum Mutat 30:860–866

Wan S, Wu J, Zhang Z, Sun X, Lv Y, Gao C, Ning Y, Ma J, Guo Y, Zhang Q, Zheng X, Zhang C, Ma Z, Lu T (2009) Activation tagging, an efficient tool for functional analysis of the rice genome. Plant Mol Biol 69:69–80

Wang N, Wang Y, Tian F, King GJ, Zhang C, Long Y, Shi L, Meng J (2008) A functional genomics resource for *Brassica napus*: development of an EMS mutagenized population and discovery of FAE1 point mutations by TILLING. New Phytologist 180:751–765

Wang W, Esch JJ, Shiu S-H, Agula H, Binder BM, Chang C, Patterson SE, Bleecker AB (2006) Identification of important regions for ethylene binding and signaling in the transmembrane domain of the ETR1 ethylene receptor of Arabidopsis. Plant Cell Online 18:3429–3442

Wang X, Cao A, Yu C, Wang D, Wang X, Chen P (2010) Establishment of an effective virus induced gene silencing system with BSMV in *Haynaldia villosa*. Mol Biol Rep 37:967–972

Wege S, Scholz A, Gleissberg S, Becker A (2007) Highly efficient Virus-Induced Gene Silencing (VIGS) in California poppy (*Eschscholzia californica*): an evaluation of VIGS as a strategy to obtain functional data from non-model plants. Ann Bot 100:641–649

Widholm JM, Finer JJ, Vodkin LO, Trick HN, LaFayette P, Li J, Parrott W (2010) Soybean. In: Kempken F, Lorz H, Nagata T (eds) Genetic modification of plants: agriculture, horticulture and forestry. Springer, Berlin, pp 473–498

Wittwer CT, Ririe KM, Andrew RV, David DA, Gundry RA, Balis UJ (1997) The LightCycler™ a microvolume multisample fluorimeter with rapid temperature control. Biotechniques 22:176–181

Wood T, Stephenson P, Østergaard L (2011) Resources for reverse genetics approaches in Brassica species. In: Schmidt R, Bancroft I (eds) Genetics and genomics of the Brassicaceae. Springer, New York, pp 561–583

Wu JL, Wu CJ, Lei CL, Baraoidan M, Bordeos A, Madamba MRS, Ramos-Pamplona M, Mauleon R, Portugal A, Ulat VJ, Bruskiewich R, Wang GL, Leach J, Khush G, Leung H (2005) Chemical- and irradiation-induced mutants of indica rice IR64 for forward and reverse genetics. Plant Mol Biol 59:85–97

Wu S-B, Wirthensohn M, Hunt P, Gibson J, Sedgley M (2008) High resolution melting analysis of almond SNPs derived from ESTs. TAG Theor Appl Genet 118:1–14

Xin Z, Li Wang M, Barkley N, Burow G, Franks C, Pederson G, Burke J (2008) Applying genotyping (TILLING) and phenotyping analyses to elucidate gene function in a chemically induced sorghum mutant population. BMC Plant Biol 8:103

Yamagishi N, Yoshikawa N (2009) Virus-induced gene silencing in soybean seeds and the emergence stage of soybean plants with Apple latent spherical virus vectors. Plant Mol Biol 71:15–24

Ye J, Qu J, Bui HTN, Chua N-H (2009) Rapid analysis of *Jatropha curcas* gene functions by virus-induced gene silencing. Plant Biotechnol J 7:964–976

Zeng F-S, Zhan Y-G, Zhao H-C, Xin Y, Qi F-H, Yang C-P (2010) Molecular characterization of T-DNA integration sites in transgenic birch. Trees Struct Funct 24:753–762

Zhang C, Ghabrial SA (2006) Development of bean pod mottle virus-based vectors for stable protein expression and sequence-specific virus-induced gene silencing in soybean. Virology 344:401–411

Zhang F, Voytas DF (2011) Targeted mutagenesis in Arabidopsis using zinc-finger nucleases. In: Birchler JA (ed) Plant chromosome engineering: methods and protocols, methods in molecular biology, vol 701. Humana Press, New York, pp 167–177

Zhang L, Jing F, Li F, Li M, Wang Y, Wang G, Sun X, Tang K (2009) Development of transgenic *Artemisia annua* (Chinese wormwood) plants with an enhanced content of artemisinin, an effective anti-malarial drug, by hairpin-RNA-mediated gene silencing. Biotechnol Appl Biochem 52:199–207

Zhang F, Maeder ML, Unger-Wallace E, Hoshaw JP, Reyon D, Christian M, Li X, Pierick CJ, Dobbs D, Peterson T, Joung JK, Voytas DF (2010) High frequency targeted mutagenesis in *Arabidopsis thaliana* using zinc finger nucleases. Proc Natl Acad Sci 107:12028–12033

Zheng S-J, Snoeren TAL, Hogewoning SW, van Loon JJA, Dicke M (2010) Disruption of plant carotenoid biosynthesis through virus-induced gene silencing affects oviposition behaviour of the butterfly *Pieris rapae*. New Phytol 186:733–745

Zhu Q-H, Eun M, C-d H, Kumar C, Pereira A, Ramachandran S, Sundaresan V, Eamens A, Upadhyaya N, Wu R (2007) Transposon insertional mutants: a resource for rice functional genomics. In: Upadhyaya NM (ed) Rice functional genomics. Springer, New York, pp 223–271

Chapter 5
Allele Re-sequencing Technologies

Stephen Byrne, Jacqueline D. Farrell, and Torben Asp

Introduction

The first plant to have its genome sequenced was *Arabidopsis thaliana* in the year 2000 (Arabidopsis Genome 2000). The 125Mb genome was sequenced using a BAC-by-BAC approach and Sanger based sequencing technology, and was completed at an estimated cost of US$70 million (Feuillet et al. 2011). The impact of the sequenced genome on our understanding of plant biology in the intervening years has been immense, recently the subject of a special issue of *The Plant Journal* (McCourt and Benning 2010). The introduction of next-generation sequencing technologies has led to a rapid increase in the number of plant genomes being sequenced (Feuillet et al. 2011), which is largely being driven by reducing sequencing costs and increasing throughput. We have already begun to see the release of sequenced genomes for a number of important crop plants, largely or exclusively generated using next-generation sequencing technologies (Xu et al. 2011; Argout et al. 2011; Shulaev et al. 2011; Wang et al. 2011; Al-Dous et al. 2011). Once a genome is established *de novo*, efforts will turn to characterizing genetic variation across the genome. The opportunity this offers to plant breeders is enormous as they begin to exploit the potentials of genomic selection for their breeding programs.

Twelve years on from the release of the first plant genome an ambitious project is underway to completely re-sequence 1001 *Arabidopsis* genotypes (http://1001genomes.org/). As more and more plant genomes are sequenced *de novo*, we will see an increasing move towards larger and larger re-sequencing projects to capture genetic variation. This is therefore a timely chapter outlining re-sequencing

S. Byrne • J.D. Farrell • T. Asp (✉)
Department of Molecular Biology and Genetics, Faculty of Science and Technology,
Research Centre Flakkebjerg, Aarhus University, Forsøgsvej 1, 4200 Slagelse, Denmark
e-mail: Torben.Asp@agrsci.dk

technologies, beginning with an introduction to the available sequencing platforms and then moving to the application of these technologies for target allele re-sequencing through to whole genome re-sequencing.

Sequencing Technology Overview

The most widely used sequencing platforms at present are the 454 GS FLX from Roche, the HiSeq and Genome Analyzer systems from Illumina, the SOLiD and Ion Torrent systems from Life Technologies, the HeliScope system from Helicos Biosciences and PacBio RS from Pacific Biosciences. An eagerly anticipated high throughput (HTP) sequencing platform from Oxford Nanopore Technologies has been flagged for commercial availability at the end of 2012 (http://www.nanoporetech.com/news/press-releases/view/39). In the following sections we will provide an overview of these sequencing technologies, starting with a description of the underlying chemistries that these technologies are built on.

Template Preparation and Sequencing Chemistries

Currently, the majority of the HTP sequencing platforms require a template DNA library for amplification. During this process the DNA or cDNA is fragmented by random shearing, and size selected (Zhou et al. 2010). A single stranded template library is introduced into the instrument and is immobilized on beads or on a solid surface depending on the instrument specifications. The process in which the template library is amplified varies by the sequencing platform. The first commercially available HTP sequencer used emulsion PCR to amplify the template library (Dressman et al. 2003). As technology advanced, other means of amplification were developed, including bridge PCR (Adessi et al. 2000), and *in situ* polymerase colonies (Mitra and Church 1999; Shendure et al. 2005).

In 2005, the first commercially available HTP sequencing platform was the 454, which was later purchased by Roche (www.my454.com). The amplification process used by Roche 454 is emulsion PCR (Margulies et al. 2005), where the amplification process begins with a single template strand attached to a bead. The bead and attached template strand are enclosed in oil with aqueous solution containing PCR reagents. Thousands of these beads are used during the process and are added to the PicoTiterPlate™ (Huse et al. 2007). These wells are specifically designed so that only one bead with the attached template strand can enter the well. During amplification the template strand will multiply to hundreds of cloned strands, after which the process of sequencing will begin. The Roche 454 sequencing platform use a chemical sequencing by synthesis procedure called pyrosequencing (Margulies et al. 2005).

Sequencing by synthesis (SBS) pyrosequencing was first demonstrated by Ronaghi et al. (1996), by using a pyrophosphate, ATP sulfurylase, and luciferase-catalyzed reaction to document the extension of the DNA chain. During pyrosequencing, enzymes are used to start a biochemical chain reaction that creates a biochemiluminescent light after a nucleotide is added (Ronaghi et al. 1996). The amount of light generated after the incorporation of nucleotides to the DNA chain is proportional to the number of nucleotides added (Fuller et al. 2009).

After the introduction of the Roche 454 sequencing platform, the company Solexa, now part of Illumina (www.illumina.com), introduced a short read HTP sequencing platform in 2006. The Illumina platform implements bridge PCR as the amplification method and a SBS method for sequencing.

Illumina sequencing platform uses SBS with reversible terminator chemistry. A reagent mixture containing all four fluorescently labelled nucleotides with reversible terminator and DNA polymerase are added to the flow cell (Ju et al. 2006). The fluorescently labeled nucleotide with a reversible terminator is incorporated in the DNA chain, the flow cell is scanned documenting the fluorescent attached to the nucleotide and position, the terminator is cleaved, the flow is washed and the cycle repeats (Ju et al. 2006; Turcatti et al. 2008).

Applied Biosystems (www.appliedbiosystems.com), now part of Life Technologies, commercially released a new HTP sequencing platform; the Sequencing by Oligonucleotide Ligation Detection system (SOLiD) that uses a sequencing by ligation (SBL) approach (Shendure et al. 2005; Mardis 2008a, b). Similar to the Roche 454 amplification method the single strand templates are amplified by emulsion PCR using paramagnetic beads. After which the beads with the amplified template library are fixed to a glass substrate (Metzker 2010; Zhou et al. 2010). The SOLiD system uses a two nucleotide sequencing system in which two nucleotides are associated with a particular dye. The SBL process begins with an universal primer annealing to the SOLiD specific adapters (Mardis 2008a, b), then an eight-nucleotide primer is ligated to the template, scanned and three bases and fluorescent labels are cleaved off (Shendure et al. 2005). This process is repeated ten times generating ten base calls in five nucleotide intervals. The process repeats again with one nucleotide shorter primer and repeating the SBL process (Shendure et al. 2005; Mardis 2008a, b). The fluorescent color images generated during each cycle are ordered in a linear sequence and aligned to a reference sequence (Metzker 2010).

The Ion Torrent (www.iontorrent.com) sequencing platform uses emulsion PCR for template strand amplification and then measures H^+ ions to determine the template strand sequence. During the sequencing process there are no lasers, optics or fluorescent labeling, instead the Ion Torrent sequencing platform uses Ion-Sensitive Field Effect Transistors (ISFET) positioned under the microwells. When the polymerase incorporates the nucleotide in the extending strand, a H^+ ion is generated and the ISFETs document the pH change as a recordable voltage change. If homopolymer regions are encountered during the sequencing process the ISFET will record a voltage reading proportional to the number of nucleotides incorporated (Rothberg et al. 2011).

The first demonstration of single molecule sequencing (SMS) was published by Braslavsky et al. (2003). The absence of a template strand amplification step bypasses extensive lab work and biological bias (Pushkarev et al. 2009). Next generation sequencing platforms that implement SMS have been deemed as the third-generation sequencers because of the technological advancement (Ozsolak 2012).

The first commercially available sequencer using the SMS platform was the Heliscope Single Molecule Sequencer (www.helicosbio.com) (Pushkarev et al. 2009). Randomly fragmented RNA strands with a poly-A tail are added to the flow cell surface and hybridize to the poly-T oligomers on the flow cell (Treffer and Deckert 2010; Su et al. 2011). Double stranded DNA can also be used in the Helicos platform with the extra step of denaturing the DNA into single strands and the addition of a 3' poly-A tail (Treffer and Deckert 2010). During the sequencing process one of the four fluorescently labeled nucleotides and DNA polymerase are added to the flow cell, and if the complement nucleotide is available it is added to the extending DNA chain. The flow cells are scanned with a sensitive optic system to document the nucleotide addition, then the fluorescent label is cleaved and the flow cell is washed for the next nucleotide cycle (Treffer and Deckert 2010; Zhou et al. 2010; Su et al. 2011).

The PacBio sequencing platform developed by Pacific Biosciences (www.pacificbiosciences.com) also use SMS, however, this instrument has the individual DNA polymerases attached to the slide (Eid et al. 2009). The PacBio sequencing platform use Single Molecule Real Time (SMRT) real time amplification, using an array of nano well structures called Zero Mode Waveguides (ZMWs) (Levene et al. 2003). The polymerase attached to the bottom of the ZMW well (McCarthy 2010) adds a fluorescent-labeled nucleotide to the complementary DNA strand (Flusberg et al. 2010). When the nucleotide is added to the extending DNA chain, a fluorescent pulse is detected and identifies the nucleotide. The ZMWs reduces the interference of back ground fluorescence (Levene et al. 2003); created by other biological materials used during sequencing (McCarthy 2010). The fluorescent-label is cleaved by the polymerase before moving to the next base. When the polymerase encounters regions of methylated nucleotides, the enzyme kinetics change (McCarthy 2010). This kinetic information is collected and used to identify methylation in the samples (Flusberg et al. 2010). The PacBio sequencing platform is able to generate long reads and has the potential for both strobe sequencing and mapping methylation patterns (Flusberg et al. 2010; McCarthy 2010).

Current State of Sequencing Technologies

Since the release of the first sequenced human genome the price of sequencing has dropped considerably. Currently, the average cost for sequencing is dropping by 50% every 5 months (Pennisi 2011), and the plethora of service providers now

5 Allele Re-sequencing Technologies

Table 5.1 Information regarding the various HTP sequencing platforms including sequence length, sequences per run and estimated error rate

Company	Platform	Read length (bp)	Reads per run (estimated)	Estimated error rate (%)
Roche 454	GS FLX Titanium XL+	400–1,000	1,000,000	0.4–1.5
	GS Junior			
Illumina	HiSeq 2000	2 × 36–151	3.4 million–6 billion	0.5–2
	HiSeq 1000			
	Genome Analyzer IIX			
	Genome Analyzer IIE			
	iScanSQ			
Life Sciences	5500 SOLiD	2 × 35–75	700–1,000 million	0.06–0.2
	5500xl SOLiD			
Ion Torrent	Ion Torrent 314	200	800 million	1–3
	Ion Torrent 318			
	Ion Proton			
	Ion Proton II			
Helicos Biosciences	HeliScope	55	100,000–8 million	3–5
Pacific Biosciences	PacBio RS	700–6,000	10,000	13–15

Sources: (Glenn 2011; Ozsolak 2012), www.illumina.com, Ion Torrent Application Note Spring 2011, www.appliedbiosystems.com, www.my454.com, www.helicobios.com, www.pacificbiosciences.com

offering NGS services has resulted in the complete suite of technologies being accessible to all in the scientific community. Each platform can generate a wide range of read lengths and throughput, both of which will come into consideration when determining which type of platform is best suited for the task in hand.

The Roche 454 System

The Roche 454 was the first sequencing platform commercially available and as the technology has advanced so has the sequence length. Currently, Roche 454 can generate an average read length of 600–800 bp (Table 5.1) (Shokralla et al. 2012; www.my454.com), which is the main advantage of this platform when compared to others (Zhou et al. 2010). The long reads are beneficial when constructing a genome, as they have the ability to sequence through short repetitive regions.

However, the Roche 454 has a high reagent and sequencing cost per base pair (Glenn 2011; Zhang et al. 2011). The emulsion PCR requires training, proper template quantification, and expensive reagents, increasing the cost of sequencing on a Roche 454 platform (Zhou et al. 2010; Glenn 2011). The process of pyrosequencing can encounter problems when sequencing a homopolymer region (e.g. GGGG). When a homopolymer region is sequenced the optical strength of the signal determines the length, but this process is prone to errors (Zhou et al. 2010).

The most common errors that occur are deletions and insertions and this platform has an estimated raw base accuracy rate of 98.5–99.6% (Table 5.1) (Zhang et al. 2011; Ozsolak 2012).

The Illumina GA and HiSeq Systems

When the Illumina HTP sequencing platform was introduced it could produce read lengths up to 25 bp, over the years the technology has advanced and the Illumina Genome Analyser can now generate reads up to 150 bp in length (Table 5.1). Many labs and researchers prefer the Illumina sequencing platform because of its broad utility and its low cost (Glenn 2011). The Illumina platform also supports the use of paired ends and mate pairs and has an estimated error rate of 0.1–2% (Ozsolak 2012).

The Illumina sequencing platform uses the process of bridge PCR for library template DNA amplification; this process can be efficient when no problems occur. If one or more lanes on the flow cell have poor bridge PCR amplification, the sample must be redone which increases the cost for sequencing (Glenn 2011). During the process of strand extension it is important that all the template strands elongate in unison. If the reversible terminator is not removed the strands within the cluster are extended become out of synch or could fail to extend (Zhou et al. 2010).

The SOLiD System

The SOLiD sequencing platform, as with the Ion Torrent and the Roche 454, use emulsion PCR for library template amplification. However, the SOLiD sequencing platform uses magnetic beads and a system called EZ-bead system, which make the process of emulsion PCR less time consuming and tedious (Glenn 2011; Zhang et al. 2011). The SOLiD sequencing platform has a read length of approximately 75 bp (Table 5.1) and has the added feature of processing two slides at a time (Zhang et al. 2011). The SOLiD platform also supports the use of paired ends and mate pairs. The SOLiD sequencing platform has a high raw error rate but since each base is sequenced independently multiple times the final error rate is 0.06–0.2% (Zhou et al. 2010; Glenn 2011; Ozsolak 2012).

The Ion Torrent System

The Ion Torrent platform uses emulsion PCR for amplification, the reagents are expensive, and a high amount of training is required for this method (Glenn 2011). The Ion Torrent platform has an estimated 98.7% accuracy rate and with a 5-nucleotide homopolymer length the platform has a 97.32% accuracy rate (Rothberg et al. 2011; Ozsolak 2012).

In 2010, the Ion Torrent sequencing platform could generate from 10 Mb up to 1 Gb of data per run (Pennisi 2010), with advancements in semi-conductor technology the amount of data this platform can generate has increased to 10 Gb (Shokralla et al. 2012). In 2011, Life Technologies released an application note for the Ion Torrent platform with a novel protocol to generate 100 bp pair-end reads; with the anticipation that improvements can generate 200 bp and possibly 400 bp reads (Ion Torrent Application Note: Paired-End Sequencing 2011).

The Helicos System

The Helicos Heliscope was the first SMS sequencing platforms on the market and uses real time sequencing that detects fluorescence labeled nucleotides. The Helicos Heliscope generates up to 55 bp reads, (Table 5.1) (Ozsolak 2012) and the most common errors that occur in the read are deletion and insertions (Pushkarev et al. 2009; Zhang et al. 2011) with an error rate of 3–5% (Ozsolak 2012). The Helicos Heliscope slows down the rate of enzyme kinetics during sequence extension which reduces the chance of homopolymer sequencing errors (Zhou et al. 2010). The lack of template library preparations and low input amounts (400–1,000 attomoles) makes the Helicos platform advantageous with working with limited nucleic acid quantities (Ozsolak 2012).

The PacBio System

The PacBio sequencing platform has a promising potential for low cost and fast rate sequencing with a long read length as well as strobe read sequencing. The median read length of this platform is 2,000 bp with a current maximum read length of 6,000 bp (Table 5.1); there is potential for both of these values to increase (Ozsolak 2012). An advantage of the PacBio platform is that no other sequencing platform commercially available conducts both SMS and documents methylation status (Flusberg et al. 2010). The PacBio sequencing platform has been documented to sequence 1–3 bases per second (Eid et al. 2009). The fast sequencing rate combined with not as fast imaging technology is thought to be a potential reason for the high error rate of 13–15% (Table 5.1) (Ozsolak 2012). The current imaging technologies limits the number of ZMWs that can be scanned per run, however, the company states that with improvements in imaging technology the number of ZMWs per run will increase (Ozsolak 2012).

Future Sequencing Platforms

Recent technology advancements have brought about another generation of sequencing platforms, which show great promise. The development of these HTP sequencing platforms has been driven by the human medical field, which has set the goal to generate a human genome for 1,000 USD (Service 2006). The main advancement of these new sequencing platforms is SMS; this technological advancement has overcome the bias caused by template amplification while reducing tedious lab work. Other advancements include; faster sequencing time, higher throughput, longer reads with higher accuracy, and lower amounts of starting material (Pareek et al. 2011; Ozsolak 2012).

Future HTP sequencing platforms are moving towards SMS, and many different companies are in the process of developing new sequencing platforms. VisiGen Biotechnologies (which is now part of Life Technologies) is developing a HTP sequencing platform in which a DNA polymerase is fixed on a substrate surface while the DNA is extended untethered (Hardin et al. 2008; Zhou et al. 2010). The sequencing process is done in real time using a Fluorescence Resonance Energy Transfer (FRET), which can generate longer sequences (Zhou et al. 2010; Zhang et al. 2011). When the polymerase incorporates a nucleotide a fluorescent light is created, the light is analyzed and the cycle continues until the polymerase can no longer continue. There is no removal of fluorescent label or blocking group, and sequence analysis is done in real time (Zhou et al. 2010).

Non-optical Microscopic Imaging

In the microscopy community proof-of-concept work is being conducted to determine if it is possible to use scanning tunneling microscopy or atomic force microscopy to visually distinguish the four nucleotides along the DNA, allowing the DNA chain to be visually read. Tanaka and Kawai (2009) has been able to distinguish guanine from the other three nucleotides using a scanning tunneling microscope. Companies including ZS Genetics (www.zsgenetics.com) and Halcyon Molecular (www.halcyonmolecular.com) are working on establishing this form of sequencing platform, which could potentially create very long reads during a short amount of time (Ozsolak 2012).

Nanopore

Many companies including Agilent (www.agilent.com), DNA Electronics (http://www.dnae.co.uk), IBM (www-03.ibm.com), NabSys (www.nabsys.com), and Oxford Nanopore Technologies (www.nanoporetech.com) are exploring the idea of using nanopore technology (Niedringhaus et al. 2011; Ozsolak 2012). There are two main challenges that must be addressed in any nanopore sequencing; (i) how to distinguish the nucleotides as the strand passes through the nanopore, and (ii) how to control the speed of the DNA strand as it passes through the nanopore (Branton et al. 2008). Even with these unique challenges the nanopore sequencing platform is thought to be simple and straightforward because theoretically very long reads can be generated from a low quantity of nucleic acid (Anselmetti 2012; Ozsolak 2012).

The concept of this technology is that by using a membrane with perforated nanopores, the DNA stand will be electrophysically driven through the pores, and as the DNA strand passes through the pores, the DNA will be read by electrophysical means (Kasianowicz et al. 1996). It has been shown that individual nucleotides can be identified while a DNA strand is moving through a nanopore. However, the speed at which the DNA strand travels through the nanopore is too fast for accurate identification (Clarke et al. 2009). In 2012, Cherf et al. and Manrao et al. independently published data showing that they were able to slow the speed of the DNA translocation through a protein nanopore so that it would be possible to sequence the nucleotides.

IBM and Roche together are developing a new sequencing technology described as "DNA transistor" which could potentially record the nucleotide sequence as the template is pulled through the nanopore sensor (Zhang et al. 2011; Ozsolak 2012).

Through genetic engineering, University of Oxford in collaboration with Oxford Nanopore Technologies has been able to generate a biochemical nanopore. They have recently demonstrated that engineered biochemical nanopores were able to reduce the speed at which the DNA strand passes through the nanopore (Clarke et al. 2009). In early 2012 the company Oxford Nanopore Technologies demonstrated their HTP sequencing platform and stated that the platform would be commercially available late 2012. Currently there is no peer-reviewed data from the Oxford Nanopore platform; however, the company has made many statements that have excited the NGS community. Oxford Nanopore Technologies has stated that it was able to sequence 10 kb of a single sense and anti-sense DNA strand (http://www.nanoporetech.com/news/press-releases/view/39). Oxford Nanopore Technologies has not released an in depth explanation of their platform but some basic information is available. The protein nanopore and enzyme were designed to control a single strand of DNA, and as the DNA goes through the nanopore a direct electronic analysis is conducted. The protein nanopore is inserted in a polymer bilayer membrane across the top of a microwell. Each microwell has a sensor chip that measures the ionic current as the single molecule passes through the nanopore (www.nanoporetech.com).

Target-Enrichment Strategies for Re-sequencing

Re-sequencing of candidate genes or other genomic regions of interest in plants is a key step in detection of genetic variations associated with various traits. Re-sequencing techniques can be divided into those which test for known polymorphisms (genotyping) and those which scan for polymorphisms such as substitutions, insertions and deletions in a given target region (variation analysis). The focus of this section will be on target-enrichment strategies for variation analysis.

The development of next-generation sequencing technologies has made sequencing an affordable approach to study genetic variation. However, the cost of whole genome re-sequencing still remains too high to be feasible for many plant species with large and complex genomes.

Target-enrichment re-sequencing refers to sequencing a targeted region of a species' genome from multiple individuals. Recent developments in target-enrichment strategies allows for enrichment for regions of interest at a scale that is matched to the throughput of next-generation sequencing platforms, and has emerged as a promising alternative to whole genome re-sequencing (Turner et al. 2009b). Target-enrichment enables scientists to investigate variations of selected genomic regions or genes with high coverage and lower costs (Harismendy and Frazer 2009). Thus, targeted re-sequencing allows for a larger numbers of samples to be investigated compared to whole genome sequencing.

Genomic target-enrichment can target specific areas of the genome, including genes of interest and linkage regions, however, this target-enrichment strategy is limited to what is already known (Teer and Mullikin 2010). In contrast, transcriptome re-sequencing, partial genome re-sequencing and whole-genome re-sequencing allows for an unbiased investigation of both known and unknown regions in the genome.

For comparison and evaluation of the performance of different target-enrichment strategies parameters such as multiplexity, specificity, uniformity, input requirements, probe design, cost, and overall workflows with implications for automation must be taken into consideration. In this section we review the recent developments in target-enrichment strategies suitable for genomic and exome capture, and discuss how these technologies can be integrated with next-generation sequencing technologies. The focus will be on technological and conceptual differences rather than absolute performance parameters, since experiments for direct comparisons are not available so far in plants.

Polymerase-Mediated Target-Enrichment

Polymerase Chain Reaction (PCR) has been extensively used to amplify genomic regions with the use of specific primer pairs. However, it becomes laborious and expensive to sequence thousands of genes by PCR. To overcome this problem, multiplex PCR has been an effective method to prepare samples for sequencing. This has recently been further developed by using microfluidics. Instead of using plates with hundreds of wells, aqueous microdroplets can segregate thousands of individual reactions in the same tube, allowing for large-scale multiplexing PCR (Tewhey et al. 2009). The technology is based on picoliter-volume droplets, where each droplet is the functional equivalent of an individual test tube and contains all the components for a single PCR reaction. The droplets are processed on a disposable chip and are suitable for large-scale multiplexing and automatized workflows. This technology is commercially available from RainDance Technologies (www.raindancetechnologies.com).

The RainDance technology has recently been used for microdroplet PCR target-enrichment in humans with 1.5 million parallel amplifications targeting 435 exons of 47 genes. 84% of the uniquely mapping reads fell within the targeted sequences and the coverage was approximately 90% of the targeted bases, demonstrating that the RainDance microdroplet technology is well suited for parallel target-enrichment for re-sequencing (Tewhey et al. 2009).

Solid- Versus Solid-Phase Target-Enrichment

There are several target-enrichment strategies, each with unique advantages and disadvantages, but common to them all is that they rely on sequence-specific nucleic acid hybridization (Albert et al. 2007; Bau et al. 2009; Summerer et al. 2009; Dahl et al. 2007; Gnirke et al. 2009; Hodges et al. 2007; Okou et al. 2007; Porreca et al. 2007; Turner et al. 2009a). The target-enrichment strategies can be divided into two groups; solid-phase- and solution-phase hybridization.

Solid-phase hybridization relies on hybridization between oligonucleotide probes and the target DNA on a solid-phase support such as a microarray surface, where a probe covalently bound to a microarray surface will hybridize to its specific target. The non-hybridized genomic material is removed by washing, and the target sequences are eluted and used for sequencing. Solution-based strategies for target enrichment apply nested multiplex PCR or oligonucleotide constructs that conjugate designated primer pairs or select genomic DNA by circularization. Subsequently, target-enrichment is performed by amplification of the genomic target DNA. An alternative method is hybrid-phase target-enrichment, which utilize beads and hybridization to capture target sequences in a solution.

Solid-Phase Target-Enrichment

The principle of solid-phase target enrichment is well-established. Genomic DNA is fragmented, linkers are added to both ends, and hybridized to microarrays containing customizable sets of single stranded probes (50–200 bp). After stringent washing, the genomic fraction bound to the array is enriched for regions targeted by the probes on the microarray. An elution step is performed to release the selected DNA fragments from the microarray, and the eluted DNA fragments are sequenced. Solid-phase target enrichment platforms include the Sequence Capture Arrays from Roche NimbleGen (www.Nimblegen.com) and the SureSelect DNA Capture Array from Agilent Technologies (www.Agilent.com).

The NimbleGen Sequence Capture Array platform utilizes high-density oligonucleotide microarrays for target-enrichment of genome regions for sequencing using 454 GS FLX sequencing. The NimbleGen Sequence Capture Array can capture up to 50 Mb total regions on a single 2.1M array and up to 5 Mb on a single 385K array.

The SureSelect DNA Capture Array from Agilent Technologies can include up to 244,000 60-mer probes. Designed for smaller studies, the SureSelect DNA

Capture Arrays complement Agilent Technologies' in-solution SureSelect Target Enrichment System, which is designed for medium to large-scale studies of tens through thousands of samples. Microarray-based target-enrichment has been extensively applied to mammalian genomes for re-sequencing of exons, large genomic loci and candidate gene sets (Albert et al. 2007; Hodges et al. 2007; Okou et al. 2007), whereas there is only one published study in plants. In maize, the Roche NimbleGen Sequence Capture Array has been used to re-sequence non-repetitive portions of an approximately 2.2 Mb chromosomal interval and a set of 43 genes dispersed in the maize genome. Sequencing of the target-enriched regions showed approximately 1,800–3,000-fold enrichment and 80–98% coverage of targeted bases. In the targeted regions more than 2,500 SNPs were identified and it was possible to recover novel sequences from non-reference alleles (Fu et al. 2010), demonstrating that solid-phase target-enrichment is a promising technology for targeted re-sequencing of complex plant genomes.

There are a number of advantages and disadvantages of using custom designed microarray chips for target-enrichment. The custom design option has the advantage of a high flexibility where both long contiguous regions and many short discontiguous regions can be targeted on the same technology platform. Another advantage is that the size of the target region(s) of custom designed microarrays can be varied from kilobases to megabases. In addition, solid-phase target-enrichment has been observed to perform better with respect to target-enrichment uniformity compared to solution-phase target-enrichment. The disadvantages are the high costs for custom microarray design, and a lower scalability for hundreds and thousands of samples compared to solution-phase target-enrichment, (Turner et al. 2009b). Finally, to have enough DNA for a solid-phase target-enrichment a relatively large amount of DNA of 10–15 μg is required (Mamanova et al. 2010).

In-Solution Target-Enrichment

To overcome some of the limitations of solid-phase target-enrichment, solution-based target-enrichment strategies have been developed. The general principle of hybridization of specific probes designed to target regions is similar to solid-phase target-enrichment, but in contrast to solid-phase target-enrichment which has a high excess of DNA over probes, solution-based target-enrichment has an excess of probes over DNA, requiring a smaller quantity of DNA (Gnirke et al. 2009).

Two approaches have been developed for efficient single-tube multiplexed amplification of genomic loci. These approaches are Molecular Inversion Probes (MIP) probes (Dahl et al. 2007; Porreca et al. 2007; Turner et al. 2009a) and Selector probes (Dahl et al. 2005, 2007). Both methods are based on ligase-assisted DNA-circularization reactions that, similar to PCR, rely on enzymatic specificity and dual hybridization recognition.

For MIP probes the primary capture event involves circularization of targets. MIP probes have two terminal target recognition sequences that are connected by a common linker (Dahl et al. 2007; Porreca et al. 2007; Turner et al. 2009a). Probes are added to the genomic DNA sample. After a denaturation followed by an annealing step, the target-complementary ends of the probes are hybridized to the target DNA loci to form a gap of ~60–190 bp that is subsequently filled in by a DNA polymerase. The non-circularized probes are removed by an exonuclease digest, resulting in a circular library of target loci. The circular DNA targets are amplified, providing an enriched amplification product ready for sequencing.

MIP probes have recently been used in combination with Illumina next-generation sequencing in two studies with 13,000 and 55,000 exon targets, respectively. A total of 91–98% of the targets were successfully captured with no allelic bias (Porreca et al. 2007; Turner et al. 2009a). Modified MIP probes have also recently been used for multiplexed capture and quantitation of individual splice events in human tissues targeting up to 20,000 splice junctions in a single reaction (Lin et al. 2010).

An alternative in-solution target-enrichment approach is selective circularization using Selector probes (Dahl et al. 2005, 2007), where circularization is performed by ligation of targeted restriction fragments from genomic DNA. The DNA sample is fragmented by a restriction enzyme and denatured. The Selector probe library is added and the probes hybridize to the targeted fragments. Each Selector probe is an oligonucleotide designed to hybridize to both ends of a target DNA restriction fragment, guiding the targeted fragments to form circular DNA molecules. The Selector probes are biotinylated and the targeted fragments can therefore be retrieved with paramagnetic streptavidin beads. The circular molecules are closed by ligation and amplified by PCR. Only circular DNA targets are amplified, providing an enriched amplification product ready for sequencing (Porreca et al. 2007). Selector probes have been commercialized by companies such as Halogenomics (www.Halogenomics.com).

In humans, Selector probes have recently been used to capture and re-sequence 501 exons from 28 genes involved in cancer, where a specificity of 94% and a coverage of 98% of the targeted regions was achieved (Johansson et al. 2010).

Advantages of MIP and Selector probe approaches includes: (i) the approaches are highly specific, (ii) high levels of multiplexing can be achieved, and (iii) captured amplicons can be directly sequenced. Major disadvantages of MIP and Selector probes for target-enrichment is the lower capture uniformity as well as the high cost associated with covering a large number of target loci (Kahvejian et al. 2008).

In-Solution Hybrid-Selection Target-Enrichment

In-solution hybrid-selection target-enrichment strategies combine the advantages of the multiplexing capabilities of the hybridization-based approach and the flexibility provided by the solution-based strategies for target-enrichment.

The solution hybrid-selection system use either biotin-labeled RNA- or DNA baits. Agilent Technologies' (www.Agilent.com) RNA-driven DNA capture

SureSelect Target Enrichment System use biotin labeled RNA baits for capture of target-regions. RNA bait-DNA hybrids are then retrieved from the complex mixture by streptavidin labeled paramagnetic beads. The system is designed for large-scale multiplexing and next-generation sequencing. This system has been used in humans where 15,000 exons and four genomic regions were targeted using 22,000 baits (Gnirke et al. 2009).

A similar target-enrichment system, the DNA-driven DNA capture SeqCap EZ has been developed by Roche NimbleGen and use DNA baits for capture of target regions. DNA bait-DNA hybrids are then retrieved from the complex mixture by streptavidin labeled paramagnetic beads. SeqCap EZ Library is a solution-based capture method that enables enrichment of the whole exome or custom target regions of interest in a single test tube. SeqCap EZ Choice Libraries enable target-enrichment of genome regions of interest and are offered in two configurations: SeqCap EZ Choice Library can capture up to 7 Mb and the SeqCap EZ Choice XL Library can capture up to 50 Mb custom regions.

Partial Genome Re-sequencing

In many cases there will be a need to sequence much more of the genome than has been described thus far. This may include re-sequencing of whole chromosomes or whole transcriptomes, or the re-sequencing of genome fractions after sequence capture or complexity reduction with restriction enzymes. It follows that the larger the fraction of the genome to be sequenced the higher the cost, therefore having the ability to precisely select the fraction of the genome necessary to achieve study objectives is attractive. The following sections give an overview of some approaches for partial genome re-sequencing.

Transcriptome Re-sequencing

Sequencing the transcriptome is one approach to focus on a restricted portion of the genome. For the purposes of this discussion we are not interested in quantification of transcript levels, but rather on uncovering genetic variation in gene coding regions. For this reason it may be advantageous to normalize the transcriptome prior to sequencing. Fortunately, simple strategies for transcriptome normalization are available (Zhulidov et al. 2004) and effective normalization combined with sample multiplexing may make this a cost effective strategy to characterize variation in gene rich regions of the genome. Difficulties in SNP calling may arise due to paralogous genes and alternatively spliced transcripts, especially when a reference genome is not available. It may be possible, however, to overcome some of these difficulties through *de novo* assembly of the transcriptome from the generated read data using various computational tools (Grabherr et al. 2011; Robertson et al. 2010).

Transcriptome re-sequencing has already been exploited to discover SNPs in maize (Barbazuk et al. 2007), bovine (Canovas et al. 2010), black cottonwood (Geraldes et al. 2011), potato (Hamilton et al. 2011) and rye (Haseneyer et al. 2011). Many of these studies focused on identifying a SNP panel suitable for the development of genotyping platforms, e.g. SNP arrays. An alternative would be to re-sequence the transcriptomes of all plants to be genotyped and make genotype calls directly from the sequence data. This has particular advantages for polyploidy species where hybridization based SNP arrays may struggle to accurately determine allele dosage states. This approach was recently used to generate genetic linkage maps in the polyploid crop oilseed rape (Bancroft et al. 2011).

Exome Re-sequencing

An alternative approach to transcriptome re-sequencing in order to target the genic regions of the genome is through the use of exome re-sequencing technologies. A number of different technologies, including Agilent SureSelect Target Enrichment, Roche NimbleGen, RainDance Technologies, and Illumina TrueSeq Target Enrichment Kits have been developed recently that enable the re-sequencing of the entire exomes of specific organisms. These technologies have already been discussed in this chapter in relation to targeted allele re-sequencing applications. In addition to being able to target specific genomic fragments of interest, they can be used to capture large numbers of target genes, gene families or entire exomes. Examples of this are the targeting of the entire exome of the human X chromosome for sequencing using RainDance Technologies (Mondal et al. 2011), and the development of a custom array for exome sequence capture in soybean using Roche NimbleGen technology (Haun et al. 2011). At present the technologies have predominantly been applied to the enrichment of human and mouse exomes and commercial kits are available. However, the development of kits to enrich the exomes of major crop species may follow as high quality reference genomes become available.

Re-sequencing After Genome Complexity Reduction

Genome complexity reduction involves reducing the portion of the genome to be sequenced with the aid of restriction endonucleases. Initial studies focused on the creation of Reduced Representation Libraries (RRLs), which involved the use of restriction enzymes, followed by size selection and sequencing. It was commonly carried out on a pool of genotypes to allow the identification of a panel of SNPs to be subsequently used in the development of genotyping platforms (Van Tassell et al. 2008; Wu et al. 2010; Myles et al. 2010). The reducing costs and increasing throughput of sequencing machines is now making it feasible to directly re-sequence a portion of the genome in all genotypes under study, thus bypassing the development of genotyping platforms. The goal is to sequence the same fraction of the genome in all samples under study, and thus be able to directly compare allele diversity. This has been elegantly demonstrated in a maize study that used restriction enzymes to reduce the genome complexity down to ~20%, and sequenced this fraction in the 27 founder lines of the maize nested association mapping population (Gore et al. 2009). This enabled the construction of a haplotype map in maize and ultimately has led to the dissection of many traits in maize by genome wide association studies (Poland et al. 2011; Tian et al. 2011; Kump et al. 2011).

Different approaches for achieving genome complexity reduction for subsequent sequencing have been developed, such as Restriction site Associated DNA sequencing (RAD-seq) (Baird et al. 2008), and Genotyping by Sequencing (GBS) (Elshire et al. 2011). The use of a bar-coding system allows the pooling of multiple samples, with the capacity for multiplexing ever increasing as sequence throughput increases. The fraction of the genome re-sequenced in both approaches is ultimately controlled by the choice of restriction enzyme. Frequent cutting enzymes will sample a larger fraction of the genome but require higher sequencing throughput to cover all sites in all samples, whereas less frequent enzymes will sample a smaller portion of the genome but require less sequencing throughput. Methylation sensitive enzymes can also be employed to avoid cutting in repetitive portions of the genome (Elshire et al. 2011), which is particularly problematic for highly repetitive plant genomes. To date the RAD-seq and GBS approaches have been utilized in plants mainly for the construction of genetic linkage maps (Pfender et al. 2011; Chutimanitsakun et al. 2011; Elshire et al. 2011) but in future we will see them being more and more applied to genome wide association and genomic selection studies.

Whole Chromosome Re-sequencing

The ability to separate chromosomes through flow cytometry enables the sequencing of individual chromosomes or chromosome arms. The process of separating chromosomes via flow cytometry involves passing them in single file through a light beam, and measuring any fluorescence and scattering of light. The scattered light and fluorescence will vary according to chromosome size, enabling chromosomes of a particular size to be separated and collected for sequencing. Our capacity to successfully separate chromosomes will ultimately depend on our ability to sufficiently discriminate between them. One powerful use of flow sorting chromosomes is in *de novo* sequencing projects, particularly for very large repetitive genomes like those of the cereals (Dolezel et al. 2007). In addition to its use for the establishment of chromosome specific BAC libraries, sequencing can be directly carried out on individual chromosomes. This serves to help reduce the complexity of assembling short read sequence data by allowing the sequencing and assembly of individual chromosomes one at a time. Examples of this approach can already be found for barley and wheat. In barley, flow sorted chromosome 1H was sequenced by 454 GS FLX pyrosequencing to only a 1.3X coverage, and with the aid of the rice and sorghum genomes it enabled a virtual gene order for the entire chromosome to be proposed (Mayer et al. 2009). In wheat, chromosome arm 7DS was sequenced to approximately 34X coverage, allowing the assembly of low copy and genic regions (Berkman et al. 2011). This approach is also currently being taken by the International Wheat Genome Sequencing Consortium (IWGSC) to initially provide a low coverage sequencing survey of all 21 wheat chromosomes (http://www.wheatgenome.org/). The relevance of chromosome sorting should go beyond *de novo* sequencing projects. Even with a high quality assembly, it may still be advantageous in some of the cereal genomes to flow sort chromosomes prior to whole genome re-sequencing projects. In many cases linkage mapping and association studies have narrowed down the location of loci controlling various traits to specific chromosomes or chromosome arms. Re-sequencing of these chromosome arms in large panels of diverse genotypes will make the pin pointing of causative loci and the identification of agriculturally valuable alleles possible.

Whole Genome Re-sequencing

The ability to re-sequence the complete genomes of a large collection of plants would significantly enhance our ability to associate genetic variation with phenotypic variation. In such a scenario we are not reliant on the presence of LD between our marker and QTL, since the causative polymorphism will be sequenced. Furthermore, re-sequencing complete genomes is also very motivating in terms of truly understanding the genome structure of that species, especially as we begin to learn more and more that genome structure is not fixed within species. However, a prerequisite for re-sequencing of complete genomes at low coverage will be the availability of a high quality reference genome.

Plant species previously deemed unsuitable for *de novo* genome sequencing due to financial constraints are now becoming more amenable with the advent of next-generation sequencing technologies. This has led to an upsurge in the number of plant genomes published that have been predominantly sequenced using next generation machines (Al-Dous et al. 2011; Argout et al. 2011; Shulaev et al. 2011; Wang et al. 2011; Xu et al. 2011) , and it is likely many more are under way. The quality of assemblies varies between species and a set of qualifiers was recently introduced to help describe the status of a genome assembly (Chain et al. 2009). At present it is unfeasible to produce finished genomes for all plant species, and to date only *Arabidopsis* and rice are regarded as having achieved the gold standard (Feuillet et al. 2011). However, for re-sequencing purposes an '*Improved High-Quality Draft*' may enable the goals of many studies to be achieved. This implies that additional work has been done beyond initial shotgun sequencing and that contigs or scaffolds have been ordered and orientated.

De novo genome sequencing based on short read data involves sequencing to a very high coverage, and due to the repetitive nature of genomes, it also requires the generation of a number of libraries with varying insert sizes to enable resolution of repeat regions. As more and more genomes are being sequenced we are constantly accumulating knowledge on how best to use our resources for *de novo* genome sequencing (Schatz et al. 2010). The same will be true for re-sequencing and the most cost effective approaches will depend on the size and complexity of the genome, with key factors for consideration being coverage and library insert size. Large scale re-sequencing of multiple genotypes is in its very early days but results from a limited number of studies have begun to emerge. A study in soybean re-sequenced 31 genomes in order to look at genome wide genetic variation in wild and cultivated soybean genomes (Lam et al. 2011). The re-sequenced genomes were aligned back to the Williams 82 reference, which is considered an '*Improved high-quality draft genome*' by Feuillet et al. (2011). Sequence depths ranged from 1.6X to 7.9X, and encouragingly this corresponded to a coverage of the reference genome of between 74% and 96%. A larger scale re-sequencing study is being carried out in *Arabidopsis*, called the 1001 genomes project (http://1001genomes.org/). The project aims to re-sequence 1,001 geographically diverse genomes by 2012 and as of October 2011, the sequence of 465 genomes have been released. The first phase of the project involved sequencing of 80 strains to an average coverage of 17X with 200 bp insert libraries and read lengths of between 42 and 64 bp. After aligning reads against the reference genome, 67.3% of the genome was covered in at least 75 of the re-sequenced strains and 43.6% in all 80 strains (Cao et al. 2011). One of the largest scale re-sequencing projects being undertaken is the 1000 genomes project in humans (http://www.1000genomes.org/). It aims to re-sequence the genomes of 2,500 people representing various world populations (Altshuler et al. 2010), and is expected to be completed in 2012. In order to generate the complete sequence of each individual it was estimated to require coverage of 28X. However, taking advantage of the complete dataset and imputation approaches, the project team believes it will be possible to identify variants and genotypes for each sample with coverage of 4X. The completion of both these ambitious projects will represent an important milestone towards our understanding of the impact of genetic variation on phenotypic variation in these species. They should also offer valuable lessons on how future large scale re-sequencing projects should proceed.

Informatics Challenges

We have already discussed the various next-generation sequencing platforms currently available, and applications from targeting specific genes to complete genome re-sequencing. One thing that has not been mentioned so far is how to handle the data from re-sequencing projects. The storage and analysis of this deluge of data is non-trivial and requires significant investment in informatics resources. Current challenges for genome informatics stem from the fact that sequencers are doubling output in a shorter time span than compute power and hard disk capacity can be doubled (for review see Kahn 2011; Stein 2010). There are also significant financial challenges due to the fact that the cost of sequencing a base is dropping at a faster rate than the cost of storing a byte of data (Baker 2010).

The level of investment in informatics required will ultimately depend on the ambitions of the research facility or breeding company. Assembly of large plant genomes for example requires significant computational power and enough RAM to store large de-bruijn graphs in memory. The RedHat Linux system at The Genome Analysis Centre (TGAC) in Norwich, reported to be one of the largest on the planet, consists of 6 terabytes of RAM and 600 terabytes of fast disk storage (http://www.tgac.ac.uk/news/6/15/Record-breaking-data-centre-for-genome-sequencing-opened-in-Norwich/). It is hoped that this facility will enable scientists to assemble very large and complex genomes such as bread wheat.

Ultimately, any informatics resources will need to be scalable to cope with both the ever increasing throughput of sequencing machines and increasingly ambitious re-sequencing projects. A case in point is the 1000 Genomes Project, which has generated more sequence data in its first 6 months than GenBank has collect in 21 years of operation (Pennisi 2011). In terms of plants, large scale re-sequencing projects will become more and more common place in breeding companies as they look to emulate the achievements made in genomic selection for animal breeding. Re-sequencing of material from large breeding programs, either by partial or whole genome re-sequencing, will invariable generate large amounts of data for storage and backup. The speedy analysis of this data will also require the ability to farm jobs out across many processors. Fortunately, many high performance computer clusters allow for additional RAM and disk storage to be added, which should allow the scaling of resources as more and more breeding material is sequenced. A final, but important challenge is the recruitment of skilled IT professionals and bioinformaticians to establish and maintain the necessary resources, and ultimately convert data into knowledge.

Cloud Computing

Not every research group has access to, or the resources to develop and maintain, the kind of high performance computing clusters discussed above. Moreover, investment in such resources may not be prudent for groups only requiring sporadic access to such resources. Cloud computing may be a viable alternative in these circumstances. We are already beginning to see the impact cloud computing is having on how society uses computers. More and more, personal computer users are turning to the cloud to store their digital content; examples include the Amazon Cloud Drive and Apple's iCloud. This is being driven by a consumers desire to have access to all their digital content on all their devices, many of which have a limited storage capacity. In addition to its obvious role in data storage, cloud computing can also enable access to powerful machines to analyze large data sets via the web. Such a 'computer on demand' service negates the need for investment in costly computer resources.

It is not hard to imagine situations where cloud computing would be of benefit to smaller plant breeding companies without access to high performance computer clusters. Breeders engaged in genomic selection may wish to re-sequence material from their breeding program intermittently, with the goal of mapping to a reference genome and calling SNPs. The effective use of cloud-computing for such a task has already been demonstrated by the cloud-computing software tool, Crossbow (Langmead et al. 2009). Crossbow's utility was demonstrated by aligning a 38X coverage of the human genome onto the reference, and calling SNPs in less than 3 h using rented computer resources from Amazon.

The above example demonstrates the power of cloud computing for a specific purpose. Ideally, researchers would be able to access a plethora of sequence analysis tools through a single user-friendly interface. Fortunately, such a tool, Galaxy, has been in existence for a number of years (Giardine et al. 2005) and is being continually developed. Galaxy enables users to upload their data to a free public server and perform analysis using a multitude of sequence analysis tools. As it is run on a free public server there is limits to the size of data sets that can be uploaded and stored, and more computationally demanding tools are unavailable. However, Galaxy CloudMan (Afgan et al. 2010) was recently developed to enable the private installation of Galaxy on a cloud computer resource, such as Amazon's EC2 cloud. This provides a ready to use environment to harness the power of cloud computing for sequence analysis. Recently BGI announced the launch of a cloud computing service (Callaway 2011) to add to their already impressive capabilities in DNA-sequencing (Cyranoski 2010). Ultimately, these resources will assist biologists to make sense of DNA sequence data without high capital investments in computer resources.

Summary

There are many platforms available for generating an ever increasing amount of sequence data, and more technologies are sure to be introduced. This chapter has described a number of technologies that can be exploited in order to harness the power of next-generation machines for allele re-sequencing. These ranged from the targeting of very specific portions of the genome, e.g. kb region underlying a QTL, right up to whole genome re-sequencing to characterize genome wide variation. As more and more plants have their genomes de-coded, the focus will increasingly turn to large scale re-sequencing projects to capture genetic variation and relate this to our understanding of agriculturally important traits. One thing is certain, it is an exciting time for plant scientists and the deluge of re-sequencing data will be sure to keep us busy for some time to come.

References

Adessi C, Matton G, Ayala G, Turcatti G, Mermod JJ, Mayer P, Kawashima E (2000) Solid phase DNA amplification: characterisation of primer attachment and amplification mechanisms. Nucleic Acids Res 28(20):E87. doi:10.1093/nar/28.20.e87

Afgan E, Baker D, Coraor N, Chapman B, Nekrutenko A, Taylor J (2010) Galaxy CloudMan: delivering cloud compute clusters. BMC Bioinformatics 11. doi:10.1186/1471-2105-11-s12-s4

Albert TJ, Molla MN, Muzny DM, Nazareth L, Wheeler D, Song X, Richmond TA, Middle CM, Rodesch MJ, Packard CJ, Weinstock GM, Gibbs RA (2007) Direct selection of human genomic loci by microarray hybridization. Nat Methods 4(11):903–905. doi:http://www.nature.com/nmeth/journal/v4/n11/suppinfo/nmeth1111_S1.html

Al-Dous EK, George B, Al-Mahmoud ME, Al-Jaber MY, Wang H, Salameh YM, Al-Azwani EK, Chaluvadi S, Pontaroli AC, DeBarry J, Arondel V, Ohlrogge J, Saie IJ, Suliman-Elmeer KM, Bennetzen JL, Kruegger RR, Malek JA (2011) De novo genome sequencing and comparative genomics of date palm (Phoenix dactylifera). Nat Biotechnol 29(6):521–U584. doi:10.1038/nbt.1860

Altshuler DL, Durbin RM, Abecasis GR, Bentley DR, Chakravarti A, Clark AG, Collins FS, De la Vega FM, Donnelly P, Egholm M, Flicek P, Gabriel SB, Gibbs RA, Knoppers BM, Lander ES, Lehrach H, Mardis ER, McVean GA, Nickerson D, Peltonen L, Schafer AJ, Sherry ST, Wang J, Wilson RK, Deiros D, Metzker M, Muzny D, Reid J, Wheeler D, Li JX, Jian M, Li G, Li RQ, Liang HQ, Tian G, Wang B, Wang W, Yang HM, Zhang XQ, Zheng HS, Ambrogio L, Bloom T, Cibulskis K, Fennell TJ, Jaffe DB, Shefler E, Sougnez CL, Gormley N, Humphray S, Kingsbury Z, Koko-Gonzales P, Stone J, McKernan KJ, Costa GL, Ichikawa JK, Lee CC, Sudbrak R, Borodina TA, Dahl A, Davydov AN, Marquardt P, Mertes F, Nietfeld W, Rosenstiel P, Schreiber S, Soldatov AV, Timmermann B, Tolzmann M, Affourtit J, Ashworth D, Attiya S, Bachorski M, Buglione E, Burke A, Caprio A, Celone C, Clark S, Conners D, Desany B, Gu L, Guccione L, Kao K, Kebbel A, Knowlton J, Labrecque M, McDade L, Mealmaker C, Minderman M, Nawrocki A, Niazi F, Pareja K, Ramenani R, Riches D, Song W, Turcotte C, Wang S, Dooling D, Fulton L, Fulton R, Weinstock G, Burton J, Carter DM, Churcher C, Coffey A, Cox A, Palotie A, Quail M, Skelly T, Stalker J, Swerdlow HP, Turner D, De Witte A, Giles S, Bainbridge M, Challis D, Sabo A, Yu F, Yu J, Fang XD, Guo XS, Li YR, Luo RB, Tai S, Wu HL, Zheng HC, Zheng XL, Zhou Y, Marth GT, Garrison EP, Huang W, Indap A, Kural D, Lee WP, Leong WF, Huang WC, Quinlan AR, Stewart C, Stromberg MP, Ward AN, Wu JT, Lee C, Mills RE, Shi XH, Daly MJ, DePristo MA, Ball AD, Banks E, Browning BL, Garimella KV, Grossman SR, Handsaker RE, Hanna M, Hartl C, Kernytsky AM, Korn JM, Li H, Maguire JR, McCarroll SA, McKenna A, Nemesh JC, Philippakis AA, Poplin RE, Price A, Rivas MA, Sabeti PC, Schaffner SF, Shlyakhter IA, Cooper DN, Ball EV, Mort M, Phillips AD, Stenson PD, Sebat J, Makarov V, Ye K, Yoon SC, Bustamante CD, Boyko A, Degenhardt J, Gravel S, Gutenkunst RN, Kaganovich M, Keinan A, Lacroute P, Ma X, Reynolds A, Clarke L, Cunningham F, Herrero J, Keenen S, Kulesha E, Leinonen R, McLaren W, Radhakrishnan R, Smith RE, Zalunin V, Zheng-Bradley XQ, Korbel JO, Stutz AM, Bauer M, Cheetham RK, Cox T, Eberle M, James T, Kahn S, Murray L, Fu YT, Hyland FCL, Manning JM, McLaughlin SF, Peckham HE, Sakarya O, Sun YA, Tsung EF, Batzer MA, Konkel MK, Walker JA, Albrecht MW, Amstislavskiy VS, Herwig R, Parkhomchuk DV, Agarwala R, Khouri H, Morgulis AO, Paschall JE, Phan LD, Rotmistrovsky KE, Sanders RD, Shumway MF, Xiao CL, Auton A, Iqbal Z, Lunter G, Marchini JL, Moutsianas L, Myers S, Tumian A, Knight J, Winer R, Craig DW, Beckstrom-Sternberg SM, Christoforides A, Kurdoglu AA, Pearson J, Sinari SA, Tembe WD, Haussler D, Hinrichs AS, Katzman SJ, Kern A, Kuhn RM, Przeworski M, Hernandez RD, Howie B, Kelley JL, Melton SC, Li Y, Anderson P, Blackwell T, Chen W, Cookson WO, Ding J, Kang HM, Lathrop M, Liang LM, Moffatt MF, Scheet P, Sidore C, Snyder M, Zhan XW, Zollner S, Awadalla P, Casals F, Idaghdour Y, Keebler J, Stone EA, Zilversmit M, Jorde L, Xing JC, Eichler EE, Aksay G, Alkan C, Hajirasouliha I, Hormozdiari F, Kidd JM, Sahinalp SC, Sudmant PH, Chen K, Chinwalla A, Ding L, Koboldt DC, McLellan MD, Wallis

JW, Wendl MC, Zhang QY, Albers CA, Ayub Q, Balasubramaniam S, Barrett JC, Chen YA, Conrad DF, Danecek P, Dermitzakis ET, Hu M, Huang N, Hurles ME, Jin HJ, Jostins L, Keane TM, Le SQ, Lindsay S, Long QA, MacArthur DG, Montgomery SB, Parts L, Tyler-Smith C, Walter K, Zhang YJ, Gerstein MB, Abyzov A, Balasubramanian S, Bjornson R, Du JA, Grubert F, Habegger L, Haraksingh R, Jee J, Khurana E, Lam HYK, Leng J, Mu XJ, Urban AE, Zhang ZD, Coafra C, Dinh H, Kovar C, Lee S, Nazareth L, Yu FL, Wilkinson J, Khouri HM, Scott C, Gharani N, Kaye JS, Kent A, Li T, McGuire AL, Ossorio PN, Rotimi CN, Su YY, Toji LH, Brooks LD, Felsenfeld AL, McEwen JE, Abdallah A, Christopher R, Clemm NC, Duncanson A, Green ED, Guyer MS, Peterson JL, Genomes Project C (2010) A map of human genome variation from population-scale sequencing. Nature 467(7319):1061–1073. doi:10.1038/nature09534

Anselmetti D (2012) NANOPORES tiny holes with great promise. Nat Nanotechnol 7:81–82

Arabidopsis Genome I (2000) Analysis of the genome sequence of the flowering plant *Arabidopsis thaliana*. Nature 408(6814):796–815

Argout X, Salse J, Aury JM, Guiltinan MJ, Droc G, Gouzy J, Allegre M, Chaparro C, Legavre T, Maximova SN, Abrouk M, Murat F, Fouet O, Poulain J, Ruiz M, Roguet Y, Rodier-Goud M, Barbosa-Neto JF, Sabot F, Kudrna D, Ammiraju JSS, Schuster SC, Carlson JE, Sallet E, Schiex T, Dievart A, Kramer M, Gelley L, Shi Z, Berard A, Viot C, Boccara M, Risterucci AM, Guignon V, Sabau X, Axtell MJ, Ma ZR, Zhang YF, Brown S, Bourge M, Golser W, Song XA, Clement D, Rivallan R, Tahi M, Akaza JM, Pitollat B, Gramacho K, D'Hont A, Brunel D, Infante D, Kebe I, Costet P, Wing R, McCombie WR, Guiderdoni E, Quetier F, Panaud O, Wincker P, Bocs S, Lanaud C (2011) The genome of *Theobroma cacao*. Nat Genet 43(2):101–108. doi:10.1038/ng.736

Baird NA, Etter PD, Atwood TS, Currey MC, Shiver AL, Lewis ZA, Selker EU, Cresko WA, Johnson EA (2008) Rapid SNP discovery and genetic mapping using sequenced RAD markers. PLoS One 3(10). doi:doi:10.1371/journal.pone.0003376

Baker M (2010) Next-generation sequencing: adjusting to data overload. Nat Methods 7(7):495–499

Bancroft I, Morgan C, Fraser F, Higgins J, Wells R, Clissold L, Baker D, Long Y, Meng JL, Wang XW, Liu SY, Trick M (2011) Dissecting the genome of the polyploid crop oilseed rape by transcriptome sequencing. Nat Biotechnol 29(8):762–766. doi:10.1038/nbt.1926

Barbazuk WB, Emrich SJ, Chen HD, Li L, Schnable PS (2007) SNP discovery via 454 transcriptome sequencing. Plant J 51(5):910–918. doi:10.1111/j.1365-313X.2007.03193.x

Bau S, Schracke N, Kränzle M, Wu H, Stähler P, Hoheisel J, Beier M, Summerer D (2009) Targeted next-generation sequencing by specific capture of multiple genomic loci using low-volume microfluidic DNA arrays. Anal Bioanal Chem 393(1):171–175. doi:10.1007/s00216-008-2460-7

Berkman PJ, Skarshewski A, Lorenc MT, Lai KT, Duran C, Ling EYS, Stiller J, Smits L, Imelfort M, Manoli S, McKenzie M, Kubalakova M, Simkova H, Batley J, Fleury D, Dolezel J, Edwards D (2011) Sequencing and assembly of low copy and genic regions of isolated *Triticum aestivum* chromosome arm 7DS. Plant Biotechnol J 9(7):768–775. doi:10.1111/j.1467-7652.2010.00587.x

Branton D, Deamer DW, Marziali A, Bayley H, Benner SA, Butler T, Di Ventra M, Garaj S, Hibbs A, Huang X, Jovanovich SB, Krstic PS, Lindsay S, Ling XS, Mastrangelo CH, Meller A, Oliver JS, Pershin YV, Ramsey JM, Riehn R, Soni GV, Tabard-Cossa V, Wanunu M, Wiggin M, Schloss JA (2008) The potential and challenges of nanopore sequencing. Nat Biotechnol 26(10):1146–1153. doi:10.1038/nbt.1495

Braslavsky I, Hebert B, Kartalov E, Quake SR (2003) Sequence information can be obtained from single DNA molecules. Proc Natl Acad Sci USA 100(7):3960–3964. doi:10.1073/pnas.0230489100

Callaway E (2011) Genome giant offers data service. Nature 475(7357):435–437

Canovas A, Rincon G, Islas-Trejo A, Wickramasinghe S, Medrano JF (2010) SNP discovery in the bovine milk transcriptome using RNA-Seq technology. Mamm Genome 21(11–12):592–598. doi:10.1007/s00335-010-9297-z

Cao J, Schneeberger K, Ossowski S, Gunther T, Bender S, Fitz J, Koenig D, Lanz C, Stegle O, Lippert C, Wang X, Ott F, Muller J, Alonso-Blanco C, Borgwardt K, Schmid KJ, Weigel D (2011) Whole-genome sequencing of multiple *Arabidopsis thaliana* populations. Nat Genet 43(10):956–U960. doi:10.1038/ng.911

Chain PSG, Grafham DV, Fulton RS, FitzGerald MG, Hostetler J, Muzny D, Ali J, Birren B, Bruce DC, Buhay C, Cole JR, Ding Y, Dugan S, Field D, Garrity GM, Gibbs R, Graves T, Han CS, Harrison SH, Highlander S, Hugenholtz P, Khouri HM, Kodira CD, Kolker E, Kyrpides NC, Lang D, Lapidus A, Malfatti SA, Markowitz V, Metha T, Nelson KE, Parkhill J, Pitluck S, Qin X, Read TD, Schmutz J, Sozhamannan S, Sterk P, Strausberg RL, Sutton G, Thomson NR, Tiedje JM, Weinstock G, Wollam A, Detter JC, Genomic Stand C, Human Microbiome Project J (2009) Genome project standards in a new era of sequencing. Science 326(5950):236–237. doi:10.1126/science.1180614

Chutimanitsakun Y, Nipper RW, Cuesta-Marcos A, Cistue L, Corey A, Filichkina T, Johnson EA, Hayes PM (2011) Construction and application for QTL analysis of a Restriction Site Associated DNA (RAD) linkage map in barley. BMC Genomics 12. doi:10.1186/1471-2164-12-4

Clarke J, Wu H-C, Jayasinghe L, Patel A, Reid S, Bayley H (2009) Continuous base identification for single-molecule nanopore DNA sequencing. Nat Nanotechnol 4(4):265–270. doi:10.1038/nnano.2009.12

Cyranoski D (2010) The sequence factory. Nature 464(7285):22–24. doi:10.1038/464022a

Dahl F, Gullberg M, Stenberg J, Landegren U, Nilsson M (2005) Multiplex amplification enabled by selective circularization of large sets of genomic DNA fragments. Nucleic Acids Res 33(8):e71. doi:10.1093/nar/gni070

Dahl F, Stenberg J, Fredriksson S, Welch K, Zhang M, Nilsson M, Bicknell D, Bodmer WF, Davis RW, Ji H (2007) Multigene amplification and massively parallel sequencing for cancer mutation discovery. Proc Natl Acad Sci USA 104(22):9387–9392. doi:10.1073/pnas.0702165104

Dolezel J, Kubalakova M, Paux E, Bartos J, Feuillet C (2007) Chromosome-based genomics in the cereals. Chromosome Res 15(1):51–66. doi:10.1007/s10577-006-1106-x

Dressman D, Yan H, Traverso G, Kinzler KW, Vogelstein B (2003) Transforming single DNA molecules into fluorescent magnetic particles for detection and enumeration of genetic variations. Proc Natl Acad Sci USA 100(15):8817–8822. doi:10.1073/pnas.1133470100

Eid J, Fehr A, Gray J, Luong K, Lyle J, Otto G, Peluso P, Rank D, Baybayan P, Bettman B, Bibillo A, Bjornson K, Chaudhuri B, Christians F, Cicero R, Clark S, Dalal R, deWinter A, Dixon J, Foquet M, Gaertner A, Hardenbol P, Heiner C, Hester K, Holden D, Kearns G, Kong X, Kuse R, Lacroix Y, Lin S, Lundquist P, Ma C, Marks P, Maxham M, Murphy D, Park I, Pham T, Phillips M, Roy J, Sebra R, Shen G, Sorenson J, Tomaney A, Travers K, Trulson M, Vieceli J, Wegener J, Wu D, Yang A, Zaccarin D, Zhao P, Zhong F, Korlach J, Turner S (2009) Real-time DNA sequencing from single polymerase molecules. Science 323(5910):133–138. doi:10.1126/science.1162986

Elshire RJ, Glaubitz JC, Sun Q, Poland JA, Kawamoto K, Buckler ES, Mitchell SE (2011) A robust, simple Genotyping-by-Sequencing (GBS) approach for high diversity species. PLoS One 6(5). doi:doi:10.1371/journal.pone.0019379

Feuillet C, Leach JE, Rogers J, Schnable PS, Eversole K (2011) Crop genome sequencing: lessons and rationales. Trends Plant Sci 16(2):77–88. doi:10.1016/j.tplants.2010.10.005

Flusberg BA, Webster DR, Lee JH, Travers KJ, Olivares EC, Clark TA, Korlach J, Turner SW (2010) Direct detection of DNA methylation during single-molecule, real-time sequencing. Nat Methods 7(6):461–U472. doi:10.1038/nmeth.1459

Fu Y, Springer NM, Gerhardt DJ, Ying K, Yeh C-T, Wu W, Swanson-Wagner R, D'Ascenzo M, Millard T, Freeberg L, Aoyama N, Kitzman J, Burgess D, Richmond T, Albert TJ, Barbazuk WB, Jeddeloh JA, Schnable PS (2010) Repeat subtraction-mediated sequence capture from a complex genome. Plant J 62(5):898–909. doi:10.1111/j.1365-313X.2010.04196.x

Fuller CW, Middendorf LR, Benner SA, Church GM, Harris T, Huang XH, Jovanovich SB, Nelson JR, Schloss JA, Schwartz DC, Vezenov DV (2009) The challenges of sequencing by synthesis. Nat Biotechnol 27(11):1013–1023. doi:10.1038/nbt.1585

Geraldes A, Pang J, Thiessen N, Cezard T, Moore R, Zhao YJ, Tam A, Wang SC, Friedmann M, Birol I, Jones SJM, Cronk QCB, Douglas CJ (2011) SNP discovery in black cottonwood (*Populus trichocarpa*) by population transcriptome resequencing. Mol Ecol Resour 11:81–92. doi:DOI 10.1111/j.1755-0998.2010.02960.x

Giardine B, Riemer C, Hardison RC, Burhans R, Elnitski L, Shah P, Zhang Y, Blankenberg D, Albert I, Taylor J, Miller W, Kent WJ, Nekrutenko A (2005) Galaxy: a platform for interactive large-scale genome analysis. Genome Res 15(10):1451–1455. doi:10.1101/gr.4086505

Glenn TC (2011) Field guide to next-generation DNA sequencers. Mol Ecol Resour 11(5): 759–769. doi:10.1111/j.1755-0998.2011.03024.x

Gnirke A, Melnikov A, Maguire J, Rogov P, LeProust EM, Brockman W, Fennell T, Giannoukos G, Fisher S, Russ C, Gabriel S, Jaffe DB, Lander ES, Nusbaum C (2009) Solution hybrid selection with ultra-long oligonucleotides for massively parallel targeted sequencing. Nat Biotechnol 27(2):182–189. doi:http://www.nature.com/nbt/journal/v27/n2/suppinfo/nbt.1523_S1.html

Gore MA, Chia JM, Elshire RJ, Sun Q, Ersoz ES, Hurwitz BL, Peiffer JA, McMullen MD, Grills GS, Ross-Ibarra J, Ware DH, Buckler ES (2009) A first-generation haplotype map of maize. Science 326(5956):1115–1117. doi:10.1126/science.1177837

Grabherr MG, Haas BJ, Yassour M, Levin JZ, Thompson DA, Amit I, Adiconis X, Fan L, Raychowdhury R, Zeng QD, Chen ZH, Mauceli E, Hacohen N, Gnirke A, Rhind N, di Palma F, Birren BW, Nusbaum C, Lindblad-Toh K, Friedman N, Regev A (2011) Full-length transcriptome assembly from RNA-Seq data without a reference genome. Nat Biotechnol 29(7):644–652. doi:10.1038/nbt.1883

Hamilton JP, Hansey CN, Whitty BR, Stoffel K, Massa AN, Van Deynze A, De Jong WS, Douches DS, Buell CR (2011) Single nucleotide polymorphism discovery in elite north american potato germplasm. BMC Genomics 12. doi:10.1186/1471-2164-12-302

Hardin S, Gao X, Briggs J, Willson R, Tu S-C (2008) Methods for real-time single molecule sequence determination. Official Gazette of the United States Patent and Trademark Office Patents

Harismendy O, Frazer KA (2009) Method for improving sequence coverage uniformity of targeted genomic intervals amplified by LR-PCR using Illumina GA sequencing-by-synthesis technology. Biotechniques 46(3):229–231

Haseneyer G, Schmutzer T, Seidel M, Zhou RN, Mascher M, Schon CC, Taudien S, Scholz U, Stein N, Mayer KFX, Bauer E (2011) From RNA-seq to large-scale genotyping – genomics resources for rye (*Secale cereale* L.). BMC Plant Biol 11:131. doi:10.1186/1471-2229-11-131

Haun WJ, Hyten DL, Xu WW, Gerhardt DJ, Albert TJ, Richmond T, Jeddeloh JA, Jia GF, Springer NM, Vance CP, Stupar RM (2011) The composition and origins of genomic variation among individuals of the soybean reference cultivar Williams 82. Plant Physiol 155(2):645–655. doi:10.1104/pp. 110.166736

Hodges E, Xuan Z, Balija V, Kramer M, Molla MN, Smith SW, Middle CM, Rodesch MJ, Albert TJ, Hannon GJ, McCombie WR (2007) Genome-wide in situ exon capture for selective resequencing. Nat Genet 39(12):1522–1527. doi:http://www.nature.com/ng/journal/v39/n12/suppinfo/ng.2007.42_S1.html

Huse SM, Huber JA, Morrison HG, Sogin ML, Mark Welch D (2007) Accuracy and quality of massively parallel DNA pyrosequencing. Genome Biol 8(7):R143. doi:doi:10.1186/gb-2007-8-7-r143

Information Technolgies Pvt Ltd (2011) Application note: performance, Spring 2011 [Online]. Available: http://www.iontorrent.com/lib/images/PDFs/performance_overview_application_note_041211.pdf

Johansson H, Isaksson M, Falk Sörqvist E, Roos F, Stenberg J, Sjöblom T, Botling J, Micke P, Edlund K, Fredriksson S, Göransson Kultima H, Ericsson O, Nilsson M (2010) Targeted resequencing of candidate genes using selector probes. Nucleic Acids Res. doi:10.1093/nar/gkq1005

Ju J, Kim DH, Bi L, Meng Q, Bai X, Li Z, Li X, Marma MS, Shi S, Wu J, Edwards JR, Romu A, Turro NJ (2006) Four-color DNA sequencing by synthesis using cleavable fluorescent nucleotide reversible terminators. Proc Natl Acad Sci USA 103(52):19635–19640. doi:10.1073/pnas.0609513103

Kahn SD (2011) On the future of genomic data. Science 331(6018):728–729. doi:10.1126/science.1197891

Kahvejian A, Quackenbush J, Thompson JF (2008) What would you do if you could sequence everything? Nat Biotechnol 26(10):1125–1133

Kasianowicz JJ, Brandin E, Branton D, Deamer DW (1996) Characterization of individual polynucleotide molecules using a membrane channel. Proc Natl Acad Sci USA 93(24):13770–13773. doi:10.1073/pnas.93.24.13770

Kump KL, Bradbury PJ, Wisser RJ, Buckler ES, Belcher AR, Oropeza-Rosas MA, Zwonitzer JC, Kresovich S, McMullen MD, Ware D, Balint-Kurti PJ, Holland JB (2011) Genome-wide association study of quantitative resistance to southern leaf blight in the maize nested association mapping population. Nat Genet 43(2):163–168. doi:10.1038/ng.747

Lam HM, Xu X, Liu X, Chen WB, Yang GH, Wong FL, Li MW, He WM, Qin N, Wang B, Li J, Jian M, Wang JA, Shao GH, Wang J, Sun SSM, Zhang GY (2011) Resequencing of 31 wild and cultivated soybean genomes identifies patterns of genetic diversity and selection (vol 42, pg 1053, 2010). Nat Genet 43(4):387–387. doi:10.1038/ng0411-387

Langmead B, Schatz MC, Lin J, Pop M, Salzberg SL (2009) Searching for SNPs with cloud computing. Genome Biol 10(11). doi:doi:10.1186/gb-2009-10-11-r134

Levene MJ, Korlach J, Turner SW, Foquet M, Craighead HG, Webb WW (2003) Zero-mode waveguides for single-molecule analysis at high concentrations. Science 299(5607):682–686. doi:10.1126/science.1079700

Lin S, Wang W, Palm C, Davis R, Juneau K (2010) A molecular inversion probe assay for detecting alternative splicing. BMC Genomics 11(1):712

Mamanova L, Coffey AJ, Scott CE, Kozarewa I, Turner EH, Kumar A, Howard E, Shendure J, Turner DJ (2010) Target-enrichment strategies for next-generation sequencing. Nat Methods 7(2):111–118. doi:http://www.nature.com/nmeth/journal/v7/n2/suppinfo/nmeth.1419_S1.html

Mardis ER (2008a) The impact of next-generation sequencing technology on genetics. Trends Genet 24:133–141

Mardis ER (2008b) Next-generation DNA sequencing methods. Annu Rev Gen Hum Genet 9:387–402

Margulies M, Egholm M, Altman WE, Attiya S, Bader JS, Bemben LA, Berka J, Braverman MS, Chen YJ, Chen ZT, Dewell SB, Du L, Fierro JM, Gomes XV, Godwin BC, He W, Helgesen S, Ho CH, Irzyk GP, Jando SC, Alenquer MLI, Jarvie TP, Jirage KB, Kim JB, Knight JR, Lanza JR, Leamon JH, Lefkowitz SM, Lei M, Li J, Lohman KL, Lu H, Makhijani VB, McDade KE, McKenna MP, Myers EW, Nickerson E, Nobile JR, Plant R, Puc BP, Ronan MT, Roth GT, Sarkis GJ, Simons JF, Simpson JW, Srinivasan M, Tartaro KR, Tomasz A, Vogt KA, Volkmer GA, Wang SH, Wang Y, Weiner MP, Yu PG, Begley RF, Rothberg JM (2005) Genome sequencing in microfabricated high-density picolitre reactors. Nature 437(7057): 376–380. doi:10.1038/nature03959

Mayer KFX, Taudien S, Martis M, Simkova H, Suchankova P, Gundlach H, Wicker T, Petzold A, Felder M, Steuernagel B, Scholz U, Graner A, Platzer M, Dolezel J, Stein N (2009) Gene content and virtual gene order of barley chromosome 1H. Plant Physiol 151(2):496–505. doi:10.1104/pp. 109.142612

McCarthy A (2010) Third generation DNA sequencing: Pacific biosciences' single molecule real time technology. Chem Biol 17:675–676

McCourt P, Benning C (2010) Arabidopsis: a rich harvest 10 years after completion of the genome sequence. Plant J 61(6):905–908. doi:10.1111/j.1365-313X.2010.04176.x

Metzker ML (2010) Applications of next-generation sequencing. Sequencing technologies – the next generation. Nat Rev Genet 11(1):31–46. doi:10.1038/nrg2626

Mitra RD, Church GM (1999) In situ localized amplification and contact replication of many individual DNA molecules. Nucleic Acids Res 27(24):e34. doi:doi:10.1093/nar/27.24.e34

Mondal K, Shetty AC, Patel V, Cutler DJ, Zwick ME (2011) Targeted sequencing of the human X chromosome exome. Genomics 98(4):260–265. doi:10.1016/j.ygeno.2011.04.004

Myles S, Chia JM, Hurwitz B, Simon C, Zhong GY, Buckler E, Ware D (2010) Rapid genomic characterization of the genus vitis. PLoS One 5(1). doi:doi:10.1371/journal.pone.0008219

Niedringhaus TP, Milanova D, Kerby MB, Snyder MP, Barron AE (2011) Landscape of next-generation sequencing technologies. Anal Chem 83(12):4327–4341. doi:10.1021/ac2010857

Okou DT, Steinberg KM, Middle C, Cutler DJ, Albert TJ, Zwick ME (2007) Microarray-based genomic selection for high-throughput resequencing. Nat Methods 4(11):907–909. doi:http://www.nature.com/nmeth/journal/v4/n11/suppinfo/nmeth1109_S1.html

Ozsolak F (2012) Third-generation sequencing techniques and applications to drug discovery. Expert Opin Drug Dis 7:231–243

Pareek CS, Smoczynski R, Tretyn A (2011) Sequencing technologies and genome sequencing. J Appl Genet 52(4):413–435. doi:10.1007/s13353-011-0057-x

Pennisi E (2010) GENOMICS semiconductors inspire new sequencing technologies. Science 327(5970):1190

Pennisi E (2011) Will computers crash genomics? Science 331(6018):666–668. doi:10.1126/science.331.6018.666

Pfender WF, Saha MC, Johnson EA, Slabaugh MB (2011) Mapping with RAD (restriction-site associated DNA) markers to rapidly identify QTL for stem rust resistance in *Lolium perenne*. Theor Appl Genet 122(8):1467–1480. doi:10.1007/s00122-011-1546-3

Poland JA, Bradbury PJ, Buckler ES, Nelson RJ (2011) Genome-wide nested association mapping of quantitative resistance to northern leaf blight in maize. Proc Natl Acad Sci USA 108(17):6893–6898. doi:10.1073/pnas.1010894108

Porreca GJ, Zhang K, Li JB, Xie B, Austin D, Vassallo SL, LeProust EM, Peck BJ, Emig CJ, Dahl F, Gao Y, Church GM, Shendure J (2007) Multiplex amplification of large sets of human exons. Nat Methods 4(11):931–936. doi:http://www.nature.com/nmeth/journal/v4/n11/suppinfo/nmeth1110_S1.html

Pushkarev D, Neff NF, Quake SR (2009) Single-molecule sequencing of an individual human genome. Nat Biotechnol 27(9):847–850. doi:10.1038/nbt.1561

Robertson G, Schein J, Chiu R, Corbett R, Field M, Jackman SD, Mungall K, Lee S, Okada HM, Qian JQ, Griffith M, Raymond A, Thiessen N, Cezard T, Butterfield YS, Newsome R, Chan SK, She R, Varhol R, Kamoh B, Prabhu AL, Tam A, Zhao YJ, Moore RA, Hirst M, Marra MA, Jones SJM, Hoodless PA, Birol I (2010) De novo assembly and analysis of RNA-seq data. Nat Methods 7(11):909–U962. doi:10.1038/nmeth.1517

Ronaghi M, Karamohamed S, Pettersson B, Uhlen M, Nyren P (1996) Real-time DNA sequencing using detection of pyrophosphate release. Anal Biochem 242(1):84–89. doi:10.1006/abio.1996.0432

Rothberg JM, Hinz W, Rearick TM, Schultz J, Mileski W, Davey M, Leamon JH, Johnson K, Milgrew MJ, Edwards M, Hoon J, Simons JF, Marran D, Myers JW, Davidson JF, Branting A, Nobile JR, Puc BP, Light D, Clark TA, Huber M, Branciforte JT, Stoner IB, Cawley SE, Lyons M, Fu YT, Homer N, Sedova M, Miao X, Reed B, Sabina J, Feierstein E, Schorn M, Alanjary M, Dimalanta E, Dressman D, Kasinskas R, Sokolsky T, Fidanza JA, Namsaraev E, McKernan KJ, Williams A, Roth GT, Bustillo J (2011) An integrated semiconductor device enabling non-optical genome sequencing. Nature 475(7356):348–352. doi:10.1038/nature10242

Schatz MC, Delcher AL, Salzberg SL (2010) Assembly of large genomes using second-generation sequencing. Genome Res 20(9):1165–1173. doi:10.1101/gr.101360.109

Service RF (2006) Gene sequencing - the race for the $1000 genome. Science 311:1544–1546

Shendure J, Porreca GJ, Reppas NB, Lin XX, McCutcheon JP, Rosenbaum AM, Wang MD, Zhang K, Mitra RD, Church GM (2005) Accurate multiplex polony sequencing of an evolved bacterial genome. Science 309(5741):1728–1732. doi:10.1126/science.1117389

Shokralla S, Spall JL, Gibson JF, Hajibabaei M (2012) Next-generation sequencing technologies for environmental DNA research. Mol Ecol 21:1794–1805

Shulaev V, Sargent DJ, Crowhurst RN, Mockler TC, Folkerts O, Delcher AL, Jaiswal P, Mockaitis K, Liston A, Mane SP, Burns P, Davis TM, Slovin JP, Bassil N, Hellens RP, Evans C, Harkins T, Kodira C, Desany B, Crasta OR, Jensen RV, Allan AC, Michael TP, Setubal JC, Celton JM, Rees DJG, Williams KP, Holt SH, Rojas JJR, Chatterjee M, Liu B, Silva H, Meisel L, Adato A, Filichkin SA, Troggio M, Viola R, Ashman TL, Wang H, Dharmawardhana P, Elser J, Raja R, Priest HD, Bryant DW, Fox SE, Givan SA, Wilhelm LJ, Naithani S, Christoffels A, Salama DY,

Carter J, Girona EL, Zdepski A, Wang WQ, Kerstetter RA, Schwab W, Korban SS, Davik J, Monfort A, Denoyes-Rothan B, Arus P, Mittler R, Flinn B, Aharoni A, Bennetzen JL, Salzberg SL, Dickerman AW, Velasco R, Borodovsky M, Veilleux RE, Folta KM (2011) The genome of woodland strawberry (*Fragaria vesca*). Nat Genet 43(2):109–U151. doi:10.1038/ng.740

Stein LD (2010) The case for cloud computing in genome informatics. Genome Biol 11(5). doi:doi:10.1186/gb-2010-11-5-207

Su Z, Ning B, Fang H, Hong H, Perkins R, Tong W, Shi L (2011) Next-generation sequencing and its applications in molecular diagnostics. Expert Rev Mol Diagn 11(3):333–343. doi:10.1586/erm.11.3

Summerer D, Wu H, Haase B, Cheng Y, Schracke N, Stähler CF, Chee MS, Stähler PF, Beier M (2009) Microarray-based multicycle-enrichment of genomic subsets for targeted next-generation sequencing. Genome Res 19(9):1616–1621. doi:10.1101/gr.091942.109

Tanaka H, Kawai T (2009) Partial sequencing of a single DNA molecule with a scanning tunnelling microscope. Nat Nanotechnol 4:518–522

Teer JK, Mullikin JC (2010) Exome sequencing: the sweet spot before whole genomes. Hum Mol Genet 19(R2):R145–R151. doi:10.1093/hmg/ddq333

Tewhey R, Warner JB, Nakano M, Libby B, Medkova M, David PH, Kotsopoulos SK, Samuels ML, Hutchison JB, Larson JW, Topol EJ, Weiner MP, Harismendy O, Olson J, Link DR, Frazer KA (2009) Microdroplet-based PCR enrichment for large-scale targeted sequencing. Nat Biotechnol 27(11):1025–1031. doi:http://www.nature.com/nbt/journal/v27/n11/suppinfo/nbt.1583_S1.html

Tian F, Bradbury PJ, Brown PJ, Hung H, Sun Q, Flint-Garcia S, Rocheford TR, McMullen MD, Holland JB, Buckler ES (2011) Genome-wide association study of leaf architecture in the maize nested association mapping population. Nat Genet 43(2):159–162. doi:10.1038/ng.746

Treffer R, Deckert V (2010) Recent advances in single-molecule sequencing. Curr Opin Biotechnol 21(1):4–11. doi:10.1016/j.copbio.2010.02.009

Turcatti G, Romieu A, Fedurco M, Tairi A-P (2008) A new class of cleavable fluorescent nucleotides: synthesis and optimization as reversible terminators for DNA sequencing by synthesis. Nucleic Acids Res 36(4):e25. doi:doi:10.1093/nar/gkn021

Turner EH, Lee C, Ng SB, Nickerson DA, Shendure J (2009a) Massively parallel exon capture and library-free resequencing across 16 genomes. Nat Methods 6(5):315–316. doi:http://www.nature.com/nmeth/journal/v6/n5/suppinfo/nmeth.f.248_S1.html

Turner EH, Ng SB, Nickerson DA, Shendure J (2009b) Methods for genomic partitioning. Annu Rev Genomics Hum Genet 10(1):263–284. doi:doi:10.1146/annurev-genom-082908-150112

Van Tassell CP, Smith TPL, Matukumalli LK, Taylor JF, Schnabel RD, Lawley CT, Haudenschild CD, Moore SS, Warren WC, Sonstegard TS (2008) SNP discovery and allele frequency estimation by deep sequencing of reduced representation libraries. Nat Methods 5(3):247–252. doi:10.1038/nmeth.1185

Wang XW, Wang HZ, Wang J, Sun RF, Wu J, Liu SY, Bai YQ, Mun JH, Bancroft I, Cheng F, Huang SW, Li XX, Hua W, Wang JY, Wang XY, Freeling M, Pires JC, Paterson AH, Chalhoub B, Wang B, Hayward A, Sharpe AG, Park BS, Weisshaar B, Liu BH, Li B, Liu B, Tong CB, Song C, Duran C, Peng CF, Geng CY, Koh CS, Lin CY, Edwards D, Mu DS, Shen D, Soumpourou E, Li F, Fraser F, Conant G, Lassalle G, King GJ, Bonnema G, Tang HB, Wang HP, Belcram H, Zhou HL, Hirakawa H, Abe H, Guo H, Wang H, Jin HZ, Parkin IAP, Batley J, Kim JS, Just J, Li JW, Xu JH, Deng J, Kim JA, Li JP, Yu JY, Meng JL, Wang JP, Min JM, Poulain J, Hatakeyama K, Wu K, Wang L, Fang L, Trick M, Links MG, Zhao MX, Jin MN, Ramchiary N, Drou N, Berkman PJ, Cai QL, Huang QF, Li RQ, Tabata S, Cheng SF, Zhang S, Zhang SJ, Huang SM, Sato S, Sun SL, Kwon SJ, Choi SR, Lee TH, Fan W, Zhao X, Tan X, Xu X, Wang Y, Qiu Y, Yin Y, Li YR, Du YC, Liao YC, Lim Y, Narusaka Y, Wang YP, Wang ZY, Li ZY, Wang ZW, Xiong ZY, Zhang ZH (2011) The genome of the mesopolyploid crop species *Brassica rapa*. Nat Genet 43(10):1035–U1157. doi:10.1038/ng.919

Wu XL, Ren CW, Joshi T, Vuong T, Xu D, Nguyen HT (2010) SNP discovery by high-throughput sequencing in soybean. BMC Genomics 11. doi:10.1186/1471-2164-11-469

Xu X, Pan SK, Cheng SF, Zhang B, Mu DS, Ni PX, Zhang GY, Yang S, Li RQ, Wang J, Orjeda G, Guzman F, Torres M, Lozano R, Ponce O, Martinez D, De la Cruz G, Chakrabarti SK, Patil VU, Skryabin KG, Kuznetsov BB, Ravin NV, Kolganova TV, Beletsky AV, Mardanov AV, Di Genova A, Bolser DM, Martin DMA, Li GC, Yang Y, Kuang HH, Hu Q, Xiong XY, Bishop GJ, Sagredo B, Mejia N, Zagorski W, Gromadka R, Gawor J, Szczesny P, Huang SW, Zhang ZH, Liang CB, He J, Li Y, He Y, Xu JF, Zhang YJ, Xie BY, Du YC, Qu DY, Bonierbale M, Ghislain M, Herrera MD, Giuliano G, Pietrella M, Perrotta G, Facella P, O'Brien K, Feingold SE, Barreiro LE, Massa GA, Diambra L, Whitty BR, Vaillancourt B, Lin HN, Massa A, Geoffroy M, Lundback S, DellaPenna D, Buell CR, Sharma SK, Marshall DF, Waugh R, Bryan GJ, Destefanis M, Nagy I, Milbourne D, Thomson SJ, Fiers M, Jacobs JME, Nielsen KL, Sonderkaer M, Iovene M, Torres GA, Jiang JM, Veilleux RE, Bachem CWB, de Boer J, Borm T, Kloosterman B, van Eck H, Datema E, Hekkert BTL, Goverse A, van Ham R, Visser RGF, Potato Genome Sequencing C (2011) Genome sequence and analysis of the tuber crop potato. Nature 475(7355):189–U194. doi:10.1038/nature10158

Zhang J, Chiodini R, Badr A, Zhang GF (2011) The impact of next-generation sequencing on genomics. J Genet Genomics 38(3):95–109. doi:10.1016/j.jgg.2011.02.003

Zhou X, Ren L, Li Y, Zhang M, Yu Y, Yu J (2010) The next-generation sequencing technology: a technology review and future perspective. Sci China Life Sci 53(1):44–57. doi:10.1007/s11427-010-0023-6

Zhulidov PA, Bogdanova EA, Shcheglov AS, Vagner LL, Khaspekov GL, Kozhemyako VB, Matz MV, Meleshkevitch E, Moroz LL, Lukyanov SA, Shagin DA (2004) Simple cDNA normalization using kamchatka crab duplex-specific nuclease. Nucleic Acids Res 32(3). doi:doi:10.1093/nar/gnh031

Chapter 6
Dissection of Agronomic Traits in Crops by Association Mapping

Yongsheng Chen

Introduction

Association mapping, also known as linkage disequilibrium mapping, has been widely used to dissect complex traits in plants after the pioneer candidate gene association in structured population in maize by Thornsberry et al. (2001). Its principle, similar as linkage analysis, relies on the linkage disequilibrium between a marker and functional alleles. When two groups of individuals with contrasting phenotypes have different functional alleles at a locus contributing to the trait, a marker LD with the causal allele will also show phenotypic difference between its allele classes, no matter segregation of other genes responsible for the same trait (Rafalski and Ananiev 2009). As association mapping is applied to populations which accumulated recombinants in long history, it has higher resolution to detect QTL than conventional linkage analysis, and even can find candidate causal DNA polymorphism(s). In addition, it samples and estimates effects of multiple alleles on a trait simultaneously, therefore it has potential to explore more useful alleles than conventional bi-parents linkage mapping, and therefore can be directly applied in breeding populations. Due to these two advantages, we have been seeing its flourish in methods development, improvement, and applications. To date, this technique has been successfully applied into various plant species, including many important crop species for agronomically important traits, *e.g.* rice (*Oryza sativa* L.), maize (*Zea mays*, L.), wheat (*Triticum aestivum* L.), sorghum (*Sorghum bicolor* L.), barley (*Hordeum vulgare* L.), soybean (*Glycine max* (L.) Merr.), potato (*Solanum tuberosum* L.) (reviewed by Zhu et al. (2008), and Abdurakhmonov and Abdukarimov (2008)). These association mapping were done at different genomic levels. In this review, we described how (and why) association mapping being used

Y. Chen (✉)
Department of Agronomy, Iowa State University, Ames, IA 50011, USA
e-mail: cdjys2007@gmail.com

in different genomic levels and reviewed different methods used in association studies with some examples. We also summarized some genetic resources for association mapping generated in research community. Finally we discussed the prospects of association mapping in dissecting agronomic important traits and in application in plant breeding. As the design of association mapping, factors affecting association mapping were discussed in many reviews, they might be mentioned but not the focus of this review.

Association Mapping at Different Genomic Levels

Association mapping can be used as either forward or reverse tool for different purposes which correspond to association mapping at four different genomic scales (Balding 2006), that means at QTL level, candidate gene level, candidate polymorphism level, and whole genome level (Fig. 6.1). Prior knowledge is required for the first three level association studies, but not necessary for the whole genome level. However, prior candidate gene and/or pathway information, if available, could help to interpret the whole genome association results (Atwell et al. 2010). Some samples done at these four different levels were summarized in Table 6.1.

Association Study at QTL Level

It is used to confirm a linkage mapping result, and/or fine mapping a QTL region, or even search for a candidate gene within a QTL confidence interval. This is done by testing the relationship between a few markers across and/or within the QTL region which usually spanned a few cM and the trait(s) in question. The markers could be selected from those nearby QTL regions from literatures or public resources, for example "MaizeGDB" for maize, "SoyBase and the Soybean Breeder's Toolbox" for soybean. Alternatively, several to dozens of diverse germplasm lines (accessions) which represent the most genetic diversity of association panel are selected for sequencing. Then markers within different LD blocks and/or located within biologically important domain or motifs are selected to assay the whole association panel. With the development of next generation sequencing technologies, in near future it is possible to identify all the polymorphisms within the target region by re-sequencing.

A good example to illustrate this approach for confirming QTL and fine mapping is the study reported by Stich et al. (2008). In their study, several SSR markers around the QTLs were selected to confirm and fine map the QTLs. Finally, a QTL region around the marker M18 on linkage group E was further defined into an interval less than 1 cM.

Linkage QTL mapping and followed QTL cloning have been used for understanding the genetic basis of agronomically important traits. QTL cloning is

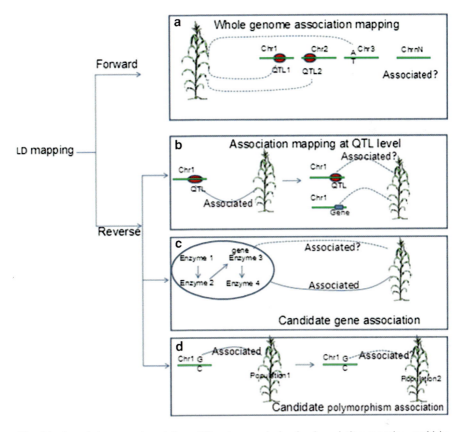

Fig. 6.1 Association mapping at four different genomic levels. Association mapping could be conducted at four different genomic level: (**a**) whole genome association mapping to identify genetic factors across the whole genome contributing to the trait in question; (**b**) association mapping at QTL level, which can be employed to confirm a previous identified QTL in a different (larger) germplasm, or to fine map a QTL; (**c**) candidate gene association mapping, which takes advantage of prior (inferred) functional information of candidate genes; (**d**) candidate polymorphism association mapping, which can be employed to develop functional markers. The whole genome association mapping is forward genetic approach, while the other three are reverse genetic approach

feasible, but with only a few successful examples, as positional cloning is very time, labor, and money consuming. To facilitate map based cloning, candidate gene strategy is combined with linkage mapping. However, candidate gene is not always obvious in QTL region. Association mapping could be employed to fine mapping the QTL region and even for candidate gene prediction (at least to reduce the candidate genes numbers, which depends on the LD across the QTL region), to facilitate the complementation experiment. Sugarcane mosaic virus (SCMV) is a serious disease in maize, to which two major QTLs conferring resistance have been mapped in chromosome 3 and 6, respectively (Xia et al. 1999; Xu et al. 1999). Although,

Table 6.1 Examples of association at different genomic levels with different statistic methods

Crop	Phenotype	Marker	Association level	Association panel	Association methods	Ref.
Rice	Rice strip virus	QTL flanking markers	QTL	148 landrace varieties	Student's T test	Zhang et al. (2011)
Sugarcane	Pachymetra root rot	QTL linked markers	QTL	154 diverse cultivars	ANOVA	McIntyre et al. (2005)
	Brown rust	QTL linked markers	QTL			
Potato	Late blight disease	QTL flanking/inside markers	QTL	415 diverse cultivars	Non-parametric Wilcoxon rank-sum test	Gebhardt et al. (2004)
Sugar beet	Sugar content	QTL flanking/inside markers	QTL	111 diploid inbreds	Multivariate association	Stich et al. (2008)
Maize	SCMV	QTL flanking/inside markers	QTL	192 diverse inbred lines	Mixed model	Lübberstedt and Xu unpublished
Maize	Flowering time	Nine markers in Dwarf8 gene	Candidate gene	92 inbred lines	Logistic regression	Thornsberry et al. (2001)
Maize	Provitamin A compounds	Markers in lcyE gene	Candidate gene	282 lines	Mixed model	Harjes et al. (2008)
Maize	Vitamin components	Markers in crtRB1 gene	Candidate gene	Three different sets of Inbred lines(281, 245, and 155 diverse lines)	Mixed model	Yan et al. (2010)
Maize	Cell wall digestibility	Markers in 10 monolignol genes	Candidate gene	40 inbred lines with contrasting CWD	Logistic, GLM, MLM	Brenner et al. (2010)

Maize	Biomass yield traits	Markers in 10 monolignol genes	Candidate gene	40 inbred lines with contrasting CWD	GLM, MLM	Chen et al. (2010)
Maize	Kernel color	Phytoene synthase gene Y1	Candidate gene	41 yellow vs 34 white inbred lines	Case control	Palaisa et al. (2003)
Maize	Flowering time	One indel marker in Dwarf8	Candidate polymorphism	375 inbred lines and 275 landraces	Linear regression and logistic regression	Camus-Kulandaivelu et al. (2006)
Maize	Flowering time	Nine markers in Dwarf8 gene	Candidate polymorphism	71 inbred lines	Logistic regression and linear regression	Andersen et al. (2005)
Rice	Culm traits, flowering, leaf length, panicle length, grain width	63 RAPD markers	Less strictly speaking GWAS	48 accessions selected from 200 to maximum representing diversity	Multiple regression	Virk et al. (1996)
Oat	13 quantitative traits	157 AFLP bands	Less strictly speaking GWAS	64 cultivars and landraces	Student's T test	Beer et al. (1997)
Maize	Oleic acid content	8,590 SNP	Strictly speaking GWAS	553 inbred lines	Kolmogorov-Smirnov test	Belo et al. (2008)
Arabidopsis	107 phenotypes	250,000 SNP	Strictly speaking GWAS	191 lines	Non-parametric Wilcoxon rank-sum test for qualitative traits; MLM for quantitative traits Fisher's exact test for categorical traits	Atwell et al. (2010)

(continued)

Table 6.1 (continued)

Crop	Phenotype	Marker	Association level	Association panel	Association methods	Ref.
Maize	Leaf length, width, and upper leaf angle	1.6 million SNP	Strictly speaking GWAS	5,000 RILs nested in 25 populations	Subsampling-based multiple SNP model with forward regression	Tian et al. (2011)
Maize	Southern leaf blight disease	1.6 million SNP	Strictly speaking GWAS	5,000 RILs nested in 25 populations	Subsampling-based multiple SNP model with forward regression	Kump et al. (2011)
Maize	Northern leaf blight disease	1.6 million SNP	Strictly speaking GWAS	5,000 RILs nested in 25 populations	Subsampling-based multiple SNP model with forward regression	Poland et al. (2011)
Rice	14 agronomic traits		Strictly speaking GWAS	373 indica rice	MLM	Huang et al. (2010)
Rice	34 agronomic traits	44,100 SNP	Strictly speaking GWAS	413 domesticated Asian rice varieties	MLM, simple linear regression, and logistic regression	Zhao et al. (2011)

GLM general linear model which considers the effect of population structure in the association model, *MLM* mixed model which considers both effect of population structure and internal relatedness within sub-population, *GWAS* genome wide association analysis fine mapping by screening very large bi-parental segregating populations helped to narrow the QTL interval, there are still several genes within the target region. Unfortunately, no obvious genes involved in disease resistance in these two regions exist. In order to reduce the candidate gene numbers to minimize the complementary tests by transformation, several SSR and gene derived markers within the two target regions were used to test the LD around the QTL regions, and to test the association between markers and phenotypes in a 192 inbred lines (Lübberstedt and Xu unpublished data).

Association Study at Candidate Gene Level

Association mapping is often used to validate a candidate gene and then search for causal polymorphisms within the interested candidate gene. In this case, it is called candidate gene association mapping. The candidate gene can be sequenced for the whole association panel for molecular marker identification, which is the best for searching causal polymorphisms. Alternatively, molecular marker selection could be done as mentioned above by sequencing a small set of germplasm accessions and identified markers then are used for whole population genotyping.

As prior knowledge about the candidate gene is required, several ways help to select candidate genes. Positional candidate genes (Pflieger et al. 2001) were selected based on results of linkage mapping, small genomic (QTL) region association, genome wide association studies (GWAS), comparison of isogenic lines, and comparisons of genome segment substitute lines. Another type was functional candidate genes (Pflieger et al. 2001), which belong to a biochemical/regulatory pathway involved in expression of the target trait, or their orthologs in other species being proved to express the interested trait. In addition, genes could also be candidates for association if their transcripts are associated with phenotype (Passador-Gurgel et al. 2007). Finally, different types of mutations can provide candidate genes, especially, TILLING (McCallum et al. 2000) for forward genetics (Rakshit et al. 2010) is able to provide genes with no obvious function clue but causing phenotypic variation. It should be noted that above methods, however, could be complemented to each other for better prediction and selection of candidate genes.

The most famous candidate gene association study in plants is the association between *Dwarf8* gene and flowering time and plant height in maize (Thornsberry et al. 2001). After the first time of successfully controlling the population structure inferred by randomly distributed SSR markers, nine polymorphisms across this gene were found to be associated with flowering time, but no significant association for plant height.

Association Study at Candidate DNA Polymorphism Level

In some cases, previous studies have revealed some potential causal polymorphisms within a candidate gene associated with the target trait, and following researchers want to validate their results in other (even larger) germplams, or in other words to test the transferability of marker-trait associations, for breeding purpose. After Thornsberry et al. (2001) reported nine polymorphisms associated with flowering time in maize, Andersen et al. (2005) tested the nine marker-trait associations in a different set of maize inbred lines. Their results showed that six of the nine marker-trait associations were also significant in their materials. Later, one of the nine markers reported by Thornsberry et al. (2001) which is near the SH2 like domain, was validated to be related with flowering time in a larger population consisting of more than 300 maize inbred lines (Camus-Kulandaivelu et al. 2006).

Association Study at Whole Genome Level

As many agronomic important traits are quantitative traits, which are determined by numerous genetic factors with small effects, interactions between them, and with environmental factors. What we got from the association mapping at candidate gene and QTL level provided only the tip of the iceberg of the genetic architecture of quantitative traits. To capture as much as possible of genetic architecture of quantitative traits, GWAS is being proving to be a very useful tool. When using association mapping as forward approach, essentially it can be grouped into two categories: less stringent speaking GWAS and stringent speaking GWAS. In less stringent speaking association mapping, a small or medium size of molecular markers randomly spread across genome was employed for searching QTLs for traits in question. This was done much often than the strict speaking GWAS in the past, partly because it is much more challenging to do the latter due to the limited molecular markers, high costs in both money and labor, which imposed great burden for small individual labs. Besides its feasibility for small individual labs, the less stringent speaking GWAS design has a few other merits (Virk et al. 1996). Although only a limit number of markers across the whole genome are used, it can provide information for selecting parents for bi-parent QTL mapping, wherein the accessions with extreme phenotypes and with polymorphisms around the QTL could be selected for crossing to maximize the power of bi-parent QTL mapping. Another advantage of this design is nearly no limitation on the number of phenotypes, because the population is diverse and phenotypes could be analyzed in the same investigation. Consequently, a small lab can even do QTL searching for multiple traits. Finally, this could be used to predict the performance of genotypes prior to phenotypic evaluation, which could save resources in plant breeding. However, nowadays, (close to) stringent speaking GWAS has begun to emerge in maize, rice, barley, and the model specials Arabidopsis, thanks to the large amount of molecular markers and multiple molecular marker assay techniques. In maize, this is achieved by doing association in nested association mapping (NAM) population (McMullen et al. 2009).

Strictly speaking, whole genome association needs tens of hundreds of thousands of molecular markers to cover the whole genome for large number of individuals. The marker numbers depends on the extent of LD within target species. Some studies on whole genome LD are available for rice (Oryza sativa) (Zhao et al. 2011; Huang et al. 2010), maize (Tenaillon et al. 2001; Ching et al. 2002), barley (Rostoks et al. 2006), and Oat (Newell et al. 2011). Although the LD decay varies across inter, intra-chromosomes, and species, self pollination species generally have higher LD than out-crossing species (Abdurakhmonov and Abdukarimov 2008). Higher LD indicates less molecular markers required in GWAS, lower resolution, and higher false positives due to LD. In contrast, lower LD suggests more markers required for GWAS, and higher resolution, lower false positives. Ideally, every gene has at least one marker being tested in whole genome association. Owing to the next generation sequence technology, large amount of DNA markers could be obtained to satisfy the strictly speaking GWAS.

The pioneer LD mapping in plant was less strictly speaking GWAS, which was done in rice (Virk et al. 1996). In this study, 48 rice accessions were selected to represent phenotype extremes of 200 accessions. Subsequently, these 48 samples were genotyped by 63 RAPD markers, which were randomly spread in chromosomes. 12–32 markers were showed association with six traits before multiple adjustments. Even after Bonferoni multiple adjustments, a few markers remained significance, which explained 5.78–49.84% of phenotypic variation. In addition, they demonstrated that with a few markers, it was possible to predict phenotypic variation to a good extent. The first more strictly speaking whole genome association mapping was done in maize with 8,590 SNPS (Beló et al. 2008). Their study confirmed the feasibility of whole genome association study and showed its resolution could reach at gene level. In their study, a SNP marker with highest p-value located within 1.7 kb of fatty acid desaturase 2 (fad2) gene, which is responsible for converting oleic acid (18:1) to linoliec acid (18:2), was associated with oleic acid content. Recently, with good experiment design and next generation sequencing, a few whole genome association studies with millions of molecular markers were realized in maize (Tian et al. 2011; Kump et al. 2011; Poland et al. 2011) and rice (Huang et al. 2010).

Methodologies for Association Mapping

The association mapping methods ranged from the simplest student's test to mix model which considers population structure as well as relatedness between association members. Student's t test was one of the earliest statistical methods used in both candidate gene and GWAS and still being employed so far. It compares the phenotypic difference between two genotypic classes, and significant difference indicates that loci might LD with loci (or itself) contributing to the trait in question. Examples employing this statistic method include the reports by Beer et al. (1997) and Zhang et al. (2011) (Table 6.1). The student's t test requires normal distribution of phenotype. If the phenotypic record is not normal distributed or data is ordinal scale, non-parametric Mann-Whitney U test (or called Wilcoxon rank-sum test) can be employed to substitute the student's t test. This statistic approach tests the difference of population mean ranks between two samples, and has been used for both GWAS (Atwell et al. 2010) and QTL level associations (Gebhardt et al. 2004) (Table 6.1).

Some agronomic traits are recorded as categorical data. For example, disease phenotype is quite often recorded by scales (*e.g.* scale 1–9), or even recorded as either resistant or susceptible type. For this type of traits, Fisher's exact test or case control method are options to study the underlying genetic basis. In the GWAS in Arabidopsis, Atwell et al. (2010) used Fisher exact test for the categorical traits (Table 6.1). Case control was first used in human disease study (Herbst et al. 1971) and then widely applied to LD association to search for human disease genes or polymorphisms. A couple hundreds of reports employed this method to detect association between genes or polymorphisms and diseases (verified in May of 2011 by ISI web of Science by Keywords of case control study and gene). Case control is to test genotype (allele) frequency difference for a given phenotype. To employ this method, susceptible (case) and normal (control) group should be carefully selected first. Significant difference of allele frequency at a locus between affected (case) and unaffected (control) groups indicate this locus is LD with the causative allele. To detect the association between a locus and the target trait, contingency table is the common method for this analysis. 2×2 contingency table could be used to detect the relationship of two alleles and the target trait (actually where Chi-square or Fisher exact test could be used), while 2 by multiple contingency tables could detect the relationship between multiple alleles and the target trait. Although this method could be applied into any molecular markers, Lander and Schork (1994) suggested that it is much more meaningful when it is applied to candidate genes. Crucial issues related to the case control study design were discussed in the review of Zondervan and Cardon (2007), including control selection for candidate gene association or GWAS, sample size requirements, *etc*. In case control study, genomic control could be employed to reduce the bias of population structures. A set of random markers across the genome was used to infer the system inflation effect, which was used as a denominator for P value of each candidate polymorphism (Devlin et al. 2001). The limitation of genomic control was the large variation of inflation factor and loss of power with increasing population stratification level (Köhler and Bickeböller 2006). This method is mostly used in the simplest single-SNP analyses in human (Balding 2006), and was seldom used in plant. The only example in plant based on this method was reported by Palaisa et al. (2003). They compared the allele frequency between yellow and white color types of inbred lines, and found Y1 (Phytoene synthase 1) rather than PSY2 (Phytoene synthase 2) is under selection for kernel color.

Other than case control method, which could use genomic control to correct population structure, the student's t test, non-parametric Mann-Whitney-U test, non-parametric rank-sum test, and Fisher's exact test do not consider the population structure which usually leads to high frequency of false positives. Thus, it is suggested to define the population structure first and then conduct association in sub-populations to avoid the noise from population stratification, if possible. For instance, in the report of Beló et al. (2008), they employed Kolmogorov-Smirnov test in sub-populations. This non-parametric method tests whether the distance between two distributions is significant, or in other words, whether the two distributions are equal. However, lower statistic power and higher false positives are expected when association tests is individually done in sub-populations as sample size is reduced.

As population structure is common in plant due to selection and genetic drift (Gupta et al. 2005), structured association methods are more popular than above mentioned methodologies. The pioneer structured association in plant employed modified logistic regression-logistic regression ratio test (Thornsberry et al. 2001). They calculated the likelihoods of association between candidate polymorphisms with population structure, and likelihoods of association between candidate polymorphisms with both phenotypic variation and population structure. The statistic test of the ratio of the two likelihoods indicates whether a polymorphism was associated with the target trait after correcting the population stratification. Although this method initiated the structured association mapping in plant and was also used in some studies (Andersen et al. 2007; Brenner et al. 2010), it has been substituted by general linear model (GLM) and mixed model recently, probably due to its inability to correct relatedness and that the dependent variable is binary in logistic regression, but most agronomic traits are quantitative traits. GLM also takes the population structure into the statistical model, testing marker trait relationships by least squares solution according to the method of Searle (1987). Its principle is to test the residual association between markers and the interested phenotype after excluding the proportion of variation attributed to the population membership (Mackay and Powell 2007). Besides population membership, mixed model takes internal relatedness within sub-population into account (Yu and Buckler 2006). Therefore this model which is called as an unified MLM (Yu and Buckler 2006) or standard MLM (Zhang et al. 2010) is expected to have less false positives than GLM (Yu and Buckler 2006).

To solve MLM in GWAS with huge amount of molecular data and large same size, computation time is unbelievable tremendous. To reduce the computation time, one approach is to reduce the number of random effects. Zhang et al. (2010) proposed a compressed MLM base on this idea. By grouping the n individuals (the sample size) to k (k <= s) group based on the kinship, the compressed MLM reduces the calculation time by a factor of $c = n/k$ (c is called compression level). In addition, the authors also proposed 'population parameters previously determined' P3D to avoid the re-computing variance components in estimating population parameters. Combining the compressed MLM with P3D, the new methodology substantially reduces the computation time. For example, standard MLM consisting of 1,315 genotypes and one millions markers requires 26 years to solve the model at a Dell computer (Optiplex 755) with two physical CPUs (E6850 @ 3.00 GHz) and 3.25 GB RAM operated under Windows XP, while the compression MLM with P3D takes only 2.7 days on the same computer (Zhang et al. 2010). Moreover, the compression MLM does not sacrifice the statistic power compared with standard MLM, even perform better if appropriate compressed level is defined (Zhang et al. 2010). Combining compressed MLM with P3D will be a good choice for strictly speaking GWAS in future.

Multiple linear regressions were also employed in both candidate gene and whole genome association analyses for complex traits. Recently, this method was also used in GWAS in NAM populations. The population confounding effect is greatly reduced because of controlled crossing in NMA population (Tian et al. 2011). To operate GWAS in NAM, the association was combined with joint linkage mapping. The residues of each chromosome were calculated first by joint linkage and then used as phenotype in association, which substantially reduced the noise of genetic background on the association. In the association analyses, the forward regression was adopted to detect the relationship between SNPs and a phenotype. Moreover, a sub-sampling strategy was employed to test the robustness of SNP association by sampling 80% of the recombinant inbred lines to construct 100 subsample data sets to measure the association reliability (Tian et al. 2011; Kump et al. 2011).

To connect real breeding practice, transmission disequilibrium test (TDT) association mapping could be employed in breeding materials, which may include inbred lines, landraces, and collected samples from natural populations (Mackay and Powell 2007). TDT tests the ratio of transmission of a certain allele from a heterozygous parent to progeny with extreme phenotype to its non-transmission across the whole selected families (Mackay and Powell 2007). TDT is proposed for qualitative traits (Sun et al. 2000), while quantitative transmission disequilibrium test (QTDT) is developed for quantitative traits for family based samples (Abecasis et al. 2000). However, TDT and QTDT were seldom used in plant. Stich et al. (2006) developed a family-based association mapping called quantitative inbred pedigree disequilibrium test for plant breeding and was proved to be effective for inbred lines with pedigree information. But the power of their approach was not as good as the logistic regression test. However, the family based method is not sensitive to linkage disequilibrium (Spielman et al. 1993) and could be a time and money saving method compared with GLM and MLM when it is used in candidate gene association, because it is not required to run a certain number of markers across the genome for population structure inference (Stich et al. 2006).

Above mentioned methods could be executed in SAS Institute (1999) or R (Ihaka and Gentleman 1996) software. The association method employed in NAM requires professional programming ability. For the complex models, they have been implemented in user-friendly software. For example, the TASSEEL software (Bradbury et al. 2007) implements logistic regression (or called SA association), general linear model (GLM), standard mixed model (Yu and Buckler 2006), and

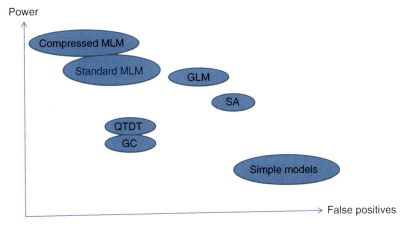

Fig. 6.2 General trend of different methods of association mapping in terms of statistical power and false positives. *MLM* mixed linear model, *GLM* general linear model, *GC* genomic control, *SA* structural association (for example, modified logistic regression), *QTDT* quantitative trait disequilibrium test; simple models include student's T test, fisher exact test, Wilcoxon rank-sum test, Kolmogorov-Smirnov test, and simple regression (without permutation or sub-sampling strategy like in the report of Tian et al. 2011). It should be noted the robustness of each method will be population, sample, and trait dependent

compression MLM (Zhang et al. 2010). Mixed model was also built into EMMA (Kang et al. 2008). For population structure inference, model based methods were often employed, which was implemented in the software called Structure (Pritchard et al. 2000; Falush et al. 2003). Principle component analysis could also infer population structure (Price et al. 2006). For internal membership (kinship matrix) calculation, either SPAGeDi (Hardy and Vekemans 2002) or Tassel (Bradbury et al. 2007) is capable.

Comparison the results from different methods could indicate the impact of the population structure and/or relatedness on association mapping, and help researchers to select the appropriate method in their association study. Generally, compressed MLM will have less false positives than standard MLM without sacrificing statistical power, or even can improve statistical power if appropriate compressed level is chosen (Zhang et al. 2010). Other methods, for example, GLM, GC, QTDT, SA, will have lower statistical power and more false positives as they do not control relatedness (Yu and Buckler 2006). But GC and QTDT will be usually better and inferior than GLM and SA in terms of detection power, respectively (Yu and Buckler 2006). For other simple models, like simple regression, student's T test, they will be inferior to the complex models because they cannot control both population and relatedness. However, the robustness of each method will be population, sample, and trait dependent. The general trend regarding power and false positives of each method is depicted in Fig. 6.2.

Genetic Resources for Association Mapping

Association mapping could be applied in both segregating breeding populations and fixed non-segregating populations, *e.g.* elite inbred lines, recombination inbred lines, double haploid (DH) lines. Although applying association mapping in segregating breeding populations could directly combine breeding actives with discovering useful alleles or lines with beneficial alleles, the effectiveness would be lower than it is applied in fixed non-segregating populations, where genotypic information and structure information could be repeatedly used for other labs and agronomic, physiological, and biochemical traits. In addition, the non-segregating lines can be grown in different environments with replicates, which leads to increased statistic power as replicates could help reduce the environmental noise (replicates could increase heritability). Moreover any trait can be theoretically measured in the same population. Association with multiple traits in the same population might enable dissection of the trait correlations into different genomic levels (Chen et al. 2010; Chen and Lübberstedt 2010). Selfing can occur in crops either by natural or artificial manipulation (except for some with selfing incompatibility, if there is), fixed (inbred) lines are thus possible. Therefore, current focus on creating and evaluating association panels are almost conducted with elite inbred lines, RIL, or DH lines (Table 6.2).

In maize, the NAM population, which was created by crossing B73 with 25 genetic diverse founders, will become a good genetic resource for GWAS in maize (McMullen et al. 2009). In order to utilize heterosis in maize, thousands of diverse elite inbred lines have been developed, all of which are potential association

Table 6.2 Examples of genetic resources for association mapping

Crop	Sample size	Markers	Ref.
Maize	2,500 RILS	1.6 million SNP	Gore et al. (2009), McMullen et al. (2009)
Maize	537 (tropical, subtropical, temperate) inbred lines	1,536 SNP	Yang et al. (2011)
Maize	359 inbred lines collected for stress tolerance study	1,260 SNP	Wen et al. (2011)
Maize	50 BC1DH lines from GEM (more will come soon)	235 SNP	Brenner et al. (2012)
Barley	500 cultivars register in UK over 20 years	1,536 SNP	Cockram et al. (2010)
Oat	1,205 lines from US, Canada, and Europe	402 DArt	Newell et al. (2011)
Rice	413 diverse Asian O. Sativa accessions	44,100 SNP	Zhao et al. (2011)
Barley	192 cultivated lines	3,060 SNP	Waugh et al. (2010)
Arabidopsis	191 lines	250,000 SNP	Atwell et al. (2010)
Arabidopsis	1,001 lines	Re-sequencing	Weigel and Mott (2009)

resources. By assessing over 1,000 maize elite inbred lines from temperate, tropic, and subtropical regions by 1,536 SNPs as well as phenotypic variation of 12 traits, Yang et al. (2011) selected 527 inbred lines to represent the global maize diversity. And their "concept proven" marker trait association suggested this panel is suitable for association mapping. In addition, trait specific association panel, for example, stress tolerance association panel including 359 advanced elite inbred lines were collected for trait dissection of drought, low nitrogen, soil acidity, pest, disease resistance (Wen et al. 2011). To broaden the germplasm base of commercial hybrids in USA, a program called Germplasm Enhancement of Maize (GEM) was set up by collaborations between several public universities and private companies to collect exotic maize races across the world, and to introduce them into two expired Plant Variety Protection (PVP) elite lines (Pollak 2003). In the GEM, higher genetic diversity and much more useful alleles for breeding are expected than in inbred lines, which are reflected in the report of Chen (2011). Therefore, the GEM derived materials will be good resources for genetics as well as breeding. Brenner et al. (2012) developed 50 BC_1DH derived from GEM with donor (exotic) fragments cover $\sim$93% of the genome. They found wide genetic variation of cell wall digestibility (CWD) as well as other agronomic traits for these 50 DH lines. Although, the sample size of this population was very small, they still detected SNPs associated with CWD, flowering time, and lodging. Later, more BC_1DH derived from the GEM will be available. Based on the results of Brenner et al. (2012), the GEM derived DH lines will be great resources for association studies. In barley, Cockram et al. (2011) collected $\sim$500 cultivars and genotyped with 1,536 SNPs. They found high level of LD within and between chromosomes, which suggests low marker density requirement for GWAS in this species. In oat, Newell et al. (2011) genotyped 1,205 lines by 402 diversity array technology markers. Surprisingly, weak population structure was observed in their wide collection, although barley is known with strong population structure (Hamblin et al. 2010). In addition, the LD was relatively consistent across the majority clusters of the collections (Newell et al. 2011). So they suggested that their wide collections are suitable for GWAS. In rice, Zhao et al. (2011) genotyped the 413 diverse pure *O. Sativa* accessions, and made the genotypic information and their 34 phenotype traits open to the public domain. Arabidopsis is not crop, but it reproduces by selfing, which results in fixed lines. To be a classic model species, it has the largest research community in plant kingdom. The information obtained from this species could be transferred or used for references for other crops. Atwell et al. (2010) already showed the feasibility to dissect the genetic basis into gene level with small sample size ($\sim$96 or 192 lines) by GWAS. The 1,001 project, which is planning to re-sequence 1,001 lines (Weigel and Mott 2009), will definitely increase the power for association. In future, besides creating new association panel for each crop, increasing the marker densities for association panels, and making them available to public research community will greatly facilitate the genetic dissection of agronomic traits and thus enhance the genetic improvement in breeding.

Prospects of Association Mapping in Dissecting Traits and Application in Breeding

Currently, breeding has been switched from phenotypic evaluation based selection to modern breeding, wherein genomic tools are involved to assist phenotypic evaluation. In the past decades, large number of conventional linkage mapping studies reported numerous QTLs for agronomic traits, most of which were not applied to breeding application (Bernardo 2008), probably due to the reported QTLs either being false positives, or only specific for the mapping populations, or the confidence interval being too large for efficient marker assisted selection (MAS). Association mapping can be employed at QTL level in other and larger germplasm to confirm the QTLs reported by linkage mapping, and narrow down its confidence interval. In addition, association mapping can explore multiple alleles across wide germplasm, therefore it has potential to search for favorable allele(s) around the reported QTL region in large germplasm or directly in the breeding populations (Hamblin et al. 2011). Accordingly, future association mapping could greatly take advantage of the QTLs in library shelves, to execute QTL level association mapping for confirmation of previously reported QTLs, and precisely located the causative genes, or at least to identify tightly linked markers to improve the efficiency of MAS.

Genetic dissection of agronomical important traits is finally intended to identify causative gene(s) and/or polymorphism(s), which in turn can be employed in MAS to enhance breeding efficiency. However, if the significant markers in association tests are not causative polymorphisms, the linkage phase between markers and causative polymorphisms can change during the breeding reproductive procedures, and even might be different across germplasm populations. Because of the different nature of agronomical important traits, different strategies will be employed in MAS. When traits are determined by major effect genes, it is better to search for the functional markers (FM) (Andersen and Lübberstedt 2003) underlying the major effect gene for the target trait. When molecular markers in the causative genes are not causative sites, they still do not guarantee the accuracy of MAS, as low LD or high recombination within gene could break the linkage phase between the markers and functional sites. For example, the LD within *COMT* gene could only persist about 100 bp with $r^2 > 0.1$ in exotic populations, while persist about 500 bp in elite lines (Chen 2011). Development of FMs could avoid the loss of favorable sites by MAS, and avoid the repeated mapping or the LD test around the target locus across other germplasm. In crops, some traits are qualitative traits, for example, purple leaf color, blast resistance in rice, anthocyanin pigmentation, opaque endosperm in maize, and semidwarfism in rice (reviewed by Badu et al. 2004). It is possible to search FM markers for these traits. In addition, major effect loci exist for simple biochemical or developmental traits (Hamblin et al. 2011), for example, the lignin content/cell wall digestibility, caretonoid composition and content, and starch composition and content. All these traits are deserved to identify causative polymorphisms for effective marker assisted/based breeding through either candidate gene or candidate polymorphism association mapping.

In self-crossing species, most likely major effect genes also exist for complex traits, even for grain yield. For instance, *Ghd7* gene in rice has strong effect on the number of grains per panicle (Xue et al. 2008), which is an important component of grain yield. In another study, Huang et al. (2010) found a few loci (no more than 12) could explain 20–60% of phenotypic variation for the 14 agronomic traits except Hull color. These results suggested the feasibility of development functional markers for the major effect loci in self-crossing species by two association steps: first, to identify major effective QTL by either conventional QTL mapping or GWAS, second, to search for functional cites by employing candidate gene/polymorphism association mapping.

For complex traits, a large number of loci with minor effects are expected to control the expression of traits. In this case, it is not feasible and worthy to resolve each QTL into functional alleles because there is larger number of effective loci and each one has a small effect, which prevents searching of functional alleles. Unless the complex traits are dissected into component traits, each might own major effect genes. However, some QTL for complex traits might also deserve to search for FM. An example is combing a few QTLs for heterosis into hybrids by MAS could increase the heterosis by 12–15% (Stuber et al. 1992, 1999). Alternatively, genomic selection (GS) of individuals with higher estimated breeding values is much more promising for complex traits. Although, randomly markers across the genome are used for GS, some proposed to combine both FM markers and random markers to improve the efficiency of GS (Brenner 2011). In the report of Kump et al. (2011), 51 SNPs identified by GWAS could explain 74% genetic variation of NCLB. This indicates that GWAS could help to predict the phenotypic performance. In addition, if these markers identified by GWAS are included in future mapping studies for the same trait as covariate, it will increase the power to discover other QTLs by reducing genetic background noise.

Unlike in human association study, GWAS in plants is much more successful with fewer resources to dissect complex traits (Hamblin et al. 2011). In addition, plants can be subjected to artificial crossing for specific experiment design, for example NAM in maize (McMullen et al. 2009), which could increase the power of association mapping. Currently, only a few of strictly speaking GWAS are reported in plants, mostly likely due to lack of high density of molecular markers. Thanks to next generation of sequencing (NGS) techniques, it is expected an explosion of molecular markers by re-sequencing and *de novo* sequencing. For example, re-sequencing of six inbred lines (Lai et al. 2010) and 517 rice landraces (Huang et al. 2010) obtained a huge amount of new SNPs. In addition, NGS does genotyping by sequencing. It provides polymorphisms immediately after re-sequencing or *de novo* sequencing, and thus no marker assay development is required (only bioinformatics is required). Besides discovering SNP polymorphisms, NGS are capable to discover structural variants (SVs), such as presence-absence variants (PAVs), copy number variants (CNVs), deletions, translocations, inversions etc. (Hall and Quinlan 2012). SVs are abundant in at least some plant specials. For example, PNVs and CNV are abundant in maize (Springer et al. 2009). In rice, there are 5% of genes showing presence-absence when comparing *japonica* and *indica* species (Ding et al. 2007).

As whole genome sequences are only available for a few species, whether this rule applies to other species remain to be uncovered. SVs might be involved in phenotypic variation. In the report of Springer et al. (2009), the identified PAVs include hundreds of single copy and expressed genes, which they thought might be involved in heterosis. Cannon et al. (2004) also reported that tandem duplications may play import roles in evolution of traits in Arabidopsis. With NGS, it is feasible to discover and include structural variants in future GWAS. Including the SVs in the GWAS will help to find part of the missing heritability (Manolio et al. 2009). Moreover, it is also possible to reveal the whole genome of each individual in the association panel by NGS. The genome structure information will help to interpret the association result either from the genome itself or through comparative genomics within the association panel and across species. Furthermore, the capability to assay the transcripts of association panel by NGS will make WGAS to map eQTLs for transcripts related to target traits possible, which would help to discover *cis* and *trans*-elements for agronomic important traits, providing a novel type of markers connecting transcripts with phenotypes, and leading more insights into the formation of traits. Thereby, NGS will be an impetus to wide application of WGAS in crops. However, current NGS identified only SNP polymorphisms for whole genome association study. The depth of sequencing is not high enough to give a good sequencing quality, which might result in the false positive polymorphisms. In addition, there are still great challenges in identifying SNPs or even obtaining whole genome sequence for polyploidy crops by current NGS (Varshney et al. 2009; Feuillet et al. 2011), which represents an obstacle of whole genome association in polyploidy crops.

For the GWAS populations, once their genotype and population structure are once obtained, they can be repeatedly used for any biological, evolution, biochemical, and agronomic traits. Undoubtedly, GWAS will become more popular as forward genetic tools to dissect the genetic architecture of agronomically important traits. The focus of future GWAS will shift into accurate and efficient phenotyping (Myles et al. 2009). As numerous traits could be measured in the same association panel, GWAS will play an important role in understanding the molecular basis of trait correlations at polymorphism level. In terms of complex traits, its genetic basis is supposed to be complicated. Decomposing the complex traits into (relative) simple component traits will help us to finally understand the genetic basis of complex traits. For example, drought tolerance trait is suggested to be dissected to secondary traits (Campos et al. 2004), *e.g.* stress for flowering (measured by anther silking interval) and rooting systems (depth and intensity of rooting).

Another focus in future of association mapping would be exploration of the wild genetic resources for basic research and breeding application. Because after long breeding history, favorable alleles for major QTL, if there are, have been fixed (Doebley 2006). New "big" genes need to be mined out from wild resources (Hospital 2009) for genetic improvement. In human genetics, the lost of heritability is a severe problem (Eichler et al. 2010). Although very few GWAS demonstrated the loss of heritability is not so serious in plants as in human (Hamblin et al. 2011), it is still a problem. And whether the lost of heritability is serious for other traits need to be revealed. Therefore, exploiting wild germplasm collections by association mapping will help to find the lost heritability and dig novel "big" genes (alleles) for further genetic improvement.

Remarks

Association studies could be used at four different genomic levels for different purposes. QTL level region could take advantage of large amount reported QTLs to precisely localize the underlying gene and to provide tightly linked markers for MAS. Candidate gene and polymorphism association mapping could be employed for FM development. With advancement of NGS, genotyping technology, statistic model improvement, the GWAS will become a popular forward tool to dissect the genetic nature of complex traits, and can help predict the agronomic performance before selection. In addition, as multiple traits could be measured for the association panel, association mapping will be a good choice for dissection of trait correlations into polymorphism level. In future, besides accurate genotyping and phenotyping, enabling the genetic association resources open to the whole world research community could improve the efficiency of association mapping and facilitate the understanding of genetics of complex traits. Finally, association mapping will also takes on the responsibility of mining "big" novel genes or alleles for genetic improvement in breeding and searching for the lost of heritability.

References

Abdurakhmonov IY, Abdukarimov A (2008) Application of association mapping to understanding the genetic diversity of plant germplasm resources. Int J Plant Genomics 2008:574927

Abecasis GR, Cardon LR, Cookson WO (2000) A general test of association for quantitative traits in nuclear families. Am J Hum Genet 66:279–292

Andersen JR, Lübberstedt T (2003) Functional markers in plants. Trends Plant Sci 8:554–560

Andersen JR, Schrag T, Melchinger AE, Zein I, Lübberstedt T (2005) Validation of Dwarf8 polymorphisms associated with flowering time in elite European inbred lines of maize (Zea mays L.). Theor Appl Genet 111:206–217

Andersen JR, Zein I, Wenzel G, Krützfeldt B, Eder J, Ouzunova M, Lübberstedt T (2007) High levels of linkage disequilibrium and associations with forage quality at a phenylalanine ammonia-lyase locus in European maize (Zea mays L.) inbreds. Theor Appl Genet 114: 307–319

Atwell S, Huang YS, Vilhjálmsson BJ, Willems G, Horton M, Li Y, Meng D, Platt A, Tarone AM, Hu TT, Jiang R, Muliyati NW, Zhang X, Amer MA, Baxter I, Brachi B, Chory J, Dean C, Debieu M, de Meaux J, Ecker JR, Faure N, Kniskern JM, Jones JD, Michael T, Nemri A, Roux F, Salt DE, Tang C, Todesco M, Traw MB, Weigel D, Marjoram P, Borevitz JO, Bergelson J, Nordborg M (2010) Genome-wide association study of 107 phenotypes in Arabidopsis thaliana inbred lines. Nature 465:627–631

Badu R, Nair SK, Prasanna BM, Gupta HS (2004) Integrating marker-assisted selection in crop breeding-prospects and challenges. Curr Sci 87:607–619

Balding DJ (2006) A tutorial on statistical methods for population studies. Nat Rev Genet 7:781–791

Beer SC, Siripoonwiwat W, O'donoughue LS, Souza E, Matthews D, Sorrels ME (1997) Associations between molecular markers and quantitative traits in an oat germplasm pool: can we infer linkages. J Agric Genom 3:197

Beló A, Zheng P, Luck S, Shen B, Meyer DJ, Li B, Tingey S, Rafalski A (2008) Whole genome scan detects an allelic variant of fad2 associated with increased oleic acid levels in maize. Mol Genet Genomics 279:1–10

Bernardo R (2008) Molecular markers and selection for complex traits in plants: learning from the last 20 years. Crop Sci 48:1649–1664

Bradbury PJ, Zhang Z, Kroon DE, Casstevens TM, Ramdoss Y, Buckler ES (2007) TASSEL: software for association mapping of complex traits in diverse samples. Bioinformatics 23:2633–2635

Brenner EA (2011) Diagnostic DNA markers for lignocellulosic conversion of maize stover into biofuels. Ph.D. thesis, Iowa State University, Ames

Brenner EA, Zein I, Chen Y, Andersen JR, Wenzel G, Ouzunova M, Eder J, Darnhofer B, Frei U, Barrière Y, Lübberstedt T (2010) Polymorphisms in O-methyltransferase genes are associated with stover cell wall digestibility in European maize (Zea mays L.). BMC Plant Biol 10:27

Brenner EA, Blanco M, Gardner C, Lübberstedt T (2012) Genotypic and phenotypic characterization of isogenic doubled haploid exotic introgression lines in maize. Mol Breed. doi:1.1007/s11032-001-9684-5

Campos H, Cooper M, Habben JE, Edmeades GO, Schussler JR (2004) Improving drought tolerance in maize: a view from industry. Field crops Res 90:19–34

Camus-Kulandaivelu L, Veyrieras JB, Madur D, Combes V, Fourmann M, Barraud S, Dubreuil P, Gouesnard B, Manicacci D, Charcosset A (2006) Maize adaptation to temperate climate: relationship between population structure and polymorphism in the Dwarf8 gene. Genetics 172:2449–2463

Cannon SB, Mitra A, Baumgarten A, Young ND, May G (2004) The roles of segmental and tandem gene duplication in the evolution of large gene families in *Arabidopsis thaliana*. BMC Plant Biol 4:10

Chen Y (2011) Pleiotropic effects of genes involved in cell wall lignification on agronomic characters. Ph.D. thesis, Iowa State University, Ames

Chen Y, Lübberstedt T (2010) Molecular basis of trait correlations. Trends Plant Sci 15:454–461

Chen Y, Zein I, Brenner EA, Andersen JR, Landbeck M, Ouzunova M, Lubberstedt T (2010) Polymorphisms in monolignol biosynthetic genes are associated with biomass yield and agronomic traits in European maize (Zea mays L.). BMC Plant Biol 10:12

Ching A, Caldwell KS, Jung M, Dolan M, Smith OS, Tingey S, Morgante M, Rafalski AJ (2002) SNP frequency, haplotype structure and linkage disequilibrium in elite maize inbred lines. BMC Genet 3:19

Cockram J, White J, Zuluaga DL, Smith D, Comadran J, Macaulay M, Luo Z, Kearsey MJ, Werner P, Harrap D, Tapsell C, Liu H, Hedley PE, Stein N, Schulte D, Sterernagel B, Marshall DF, Thomas WTB, Ramsay L, Mackay I, Balding DJ, AGOUEB Consortium, Waugh R, O'Sullivan DM (2010) Genome-wide association mapping to candidate polymorphism resolution in the un-sequenced barley genome. Proc Natl Acad Sci USA 107:21611–21616

Devlin B, Roeder K, Wasserman L (2001) Genomic control, a new approach to genetic-based association studies. Thero Popul Biol 60:155–166

Ding J, Araki H, Wang Q, Zhang P, Yang S, Chen JQ, Tian D (2007) Highly asymmetric rice genomes. BMC Genomics 8:154

Doebley J (2006) Unfallen grains: how ancient farmers turned weeds into crops. Science 312:1318–1319

Eichler EE, Flint J, Gibson G, Kong A, Leal SM, Moore JH, Nadeau JH (2010) Missing heritability and strategies for finding the underlying causes of complex disease. Nat Rev Genet 11:446–450

Falush D, Stephens M, Pritchard JK (2003) Inference of population structure using multilocus genotype data: linked loci and correlated allele frequencies. Genetics 164:1567–1587

Feuillet C, Leach JE, Rogers J, Schnable PS, Eversole K (2011) Crop genome sequencing: lessons and rationales. Trends Plant Sci 16:77–89

Gebhardt C, Ballvora A, Walkemeier B, Oberhagemann P, Schüler K (2004) Assessing genetic potential in germplasm collections of crop plants by marker-trait association: a case study for potatoes with quantitative variation of resistance to late blight and maturity type. Mol Breed 13:93–102

Gore MA, Wright MH, Ersoz ES, Bouffard P, Szekeres ES, Jarvie TP, Hurwitz BL, Narechania A, Harkins TT, Grills GS, Ware DH, Buckler ES (2009) Large-scale discovery of gene-enriched SNPs. Plant Genome 2:121–133

Gupta P, Rustgi S, Kulwal PL (2005) Linkage disequilibrium and association studies in higher plants: present status and future prospects. Plant Mol Biol 57:461–485

Hall IM, Quinlan AR (2012) Detection and interpretation of genomic structural variation in mammals. Methods Mol Biol 838:225–248

Hamblin MT, Close TJ, Bhat PR, Chao S, King JG, Abraham KJ, Blake T, Brooks WS, Cooper B, Griffey CA, Hayes PM, Hole DJ, Horsley RD, Obert DE, Smith KP, Ullrich SE, Muehlbauer GJ, Jannink J (2010) Population structure and linkage disequilibrium in US barley germplasm: implications for association mapping. Crop Sci 50:556–566

Hamblin MT, Buckler ES, Jannink JL (2011) Population genetics of genomics-based crop improvement methods. Trends Genet 27:98–106

Hardy OJ, Vekemans X (2002) spagedi: a versatile computer program to analyse spatial genetic structure at the individual or population levels. Mol Ecol Notes 2:618–620

Harjes CE, Rocheford TR, Bai L, Brutnell TP, Kandianis CB, Sowinski SG, Stapleton AE, Vallabhaneni R, Williams M, Wurtzel ET, Yan J, Buckler ES (2008) Natural genetic variation in lycopene epsilon cyclase tapped for maize biofortification. Science 319:330–333

Herbst AL, Ulfelder H, Poskanzer DC (1971) Adenocarcinoma of the vagina: association of maternal stilbestrol therapy with tumor appearance in young women. N Engl J Med 284:878–881

Hospital F (2009) Challenges for effective marker-assisted selection in plants. Genetica 136:303–310

Huang X, Wei X, Sang T, Zhao Q, Feng Q, Zhao Y, Li C, Zhu C, Lu T, Zhang Z, Li M, Fan D, Guo Y, Wang A, Wang L, Deng L, Li W, Lu Y, Weng Q, Liu K, Huang T, Zhou T, Jing Y, Li W, Lin Z, Buckler ES, Qian Q, Zhang QF, Li J, Han B (2010) Genome-wide association studies of 14 agronomic traits in rice landraces. Nat Genet 42:961–967

Ihaka R, Gentleman R (1996) R: a language for data analysis and graphics. J Comput Graph Stat 5:299–314

Kang HM, Zaitlen NA, Wade CM, Kirby A, Heckerman D, Daly MJ, Eskin E (2008) Efficient control of population structure in model organism association mapping. Genetics 178:1709–1723

Köhler K, Bickeböller H (2006) Case-control association tests correcting for population stratification. Ann Hum Genet 70:89–115

Kump KL, Bradbury PJ, Buckler ES, Belcher AR, Oropeza-Rosas M, Wisser RJ, Zwonitzer JC, Kresovich S, McMullen MD, Ware D, Balint-Kurti PJ, Holland JB (2011) Genome-wide association study of quantitative resistance to southern leaf blight in the maize nested association mapping population. Nat Genet 43:163–168

Lai J, Li R, Xu X, Jin W, Xu M, Zhao H, Xiang Z, Song W, Ying K, Zhang M, Jiao Y, Ni P, Zhang J, Li D, Guo X, Ye K, Jian M, Wang B, Zheng H, Liang H, Zhang X, Wang S, Chen S, Li J, Fu Y, Springer NM, Yang H, Wang J, Dai J, Schnable PS, Wang J (2010) Genome-wide patterns of genetic variation among elite maize inbred lines. Nat Genet 42:1027–1030

Lander ES, Schork N (1994) Genetic dissection of complex traits. Science 265:2037–2048

Mackay I, Powell W (2007) Methods for linkage disequilibrium mapping in crops. Trends Plant Sci 12:57–63

Manolio TA, Collins FS, Cox NJ, Goldstein DB, Hindorff LA, Hunter DJ, McCarthy MI, Ramos EM, Cardon LR, Chakravarti A, Cho JH, Guttmacher AE, Kong A, Kruglyak L, Mardis E, Rotimi CN, Slatkin M, Valle D, Whittemore AS, Boehnke M, Clark AG, Eichler EE, Gibson G,

Haines JL, Mackay TF, McCarroll SA, Visscher PM (2009) Finding the missing heritability of complex diseases. Nature 461:747–753

McCallum CM, Comai L, Greene EA, Henikoff S (2000) Targeted screening for induced mutations. Nat Biotechnol 18:455–457

McIntyre CL, Whan VA, Croft B, Magarey R, Smith GR (2005) Identification and validation of molecular markers associated with Pachymetra root rot and brown rust resistance in sugarcane using map- and association-based approaches. Mol Breed 16:151–161

McMullen MD, Kresovich S, Villeda HS, Bradbury PJ, Li H, Sun Q, Flint-Garcia S, Thornsberry J, Acharya C, Bottoms C, Brown P, Browne C, Eller M, Guill K, Harjes C, Kroon D, Lepak N, Mitchell SE, Peterson B, Pressoir G, Romero S, Oropeza Rosas M, Salvo S, Yates H, Hanson M, Jones E, Smith S, Glaubitz JC, Goodman M, Ware D, Holland JB, Buckler ES (2009) Genetic properties of the maize nested association mapping population. Science 325:737–740

Myles S, Peiffer J, Brown PJ, Ersoz ES, Zhang Z, Costich DE, Buckler ES (2009) Association mapping: critical considerations shift from genotyping to experimental design. Plant Cell 21:2194–2202

Newell MA, Cook D, Tinker NA, Jannink JL (2011) Population structure and linkage disequilibrium in oat (*Avena Sativa* L.): implications for genome-wide association studies. Theor Appl Genet 122:623–632

Palaisa KA, Morgante M, Williams M, Rafalski A (2003) Contrasting effects of selection on sequence diversity and linkage disequilibrium at two phytoene synthase loci. Plant Cell 15:1795–1806

Passador-Gurgel G, Hsieh WP, Hunt P, Deighton N, Gibson G (2007) Quantitative trait transcripts for nicotine resistance in *Drosophila melanogaster*. Nat Genet 39:264–268

Pflieger S, Lefebvre V, Causse M (2001) The candidate gene approach in plant genetics: a review. Mol Breed 7:275–291

Poland JA, Bradbury PJ, Buckler ES, Nelson RJ (2011) Genome-wide nested association mapping of quantitative resistance to northern leaf blight in maize. Proc Natl Acad Sci 108:6893–6898

Pollak LM (2003) The history and success of the public-private project on germplasm enhancement of maize (GEM). Adv Agron 78:45–87

Price AL, Patterson NJ, Plenge RM, Weinblatt ME, Shadick NA, Reich D (2006) Principal components analysis corrects for stratification in genome-wide association studies. Nat Genet 38:904–909

Pritchard JK, Stephens M, Donnelly P (2000) Inference of population structure using multilocus genotype data. Genetics 155:945–959

Rafalski A, Ananiev E (2009) Genetic diversity, linkage disequilibrium and association mapping. In: Bennetzen JL, Hake S (eds) Handbook of maize: genetics and genomics. Springer, New York, pp 201–220

Rakshit S, Kanzaki H, Matsumura H, Rakshit A, Fujibe T, Okuyama Y, Yoshida K, Tamiru MO, Shenton M, Utsushi H, Mitsuoka C, Abe A, Kiuchi Y, Terauchi R (2010) Use of TILLING for reverse and forward genetics of rice. In: Meksem K, Kahl G (eds) The handbook of plant mutation screening: mining of natural and induced alleles. Wiley-VCH Verlag GmbH & Co. KGaA, Weinheim. doi:10.1002/9783527629398.ch11

Rostoks N, Ramsay L, MacKenzie K, Cardle L, Bhat PR, Roose ML, Svensson JT, Stein N, Varshney RK, Marshall DF, Graner A, Close TJ, Waugh R (2006) Recent history of artificial outcrossing facilitates whole-genome association mapping in elite inbred crop varieties. Proc Natl Acad Sci USA 103:18656–18661

SAS Institute (1999) SAS/STAT user's guide. Version 8 SAS Institute, Inc, Cary, NC

Searle SR (1987) Linear models for unbalanced data. Wiley, New York

Spielman RS, McGinnes RE, Ewens WJ (1993) Transmission test for linkage disequilibrium: the insulin gene region and insulin-dependent diabetes mellitus (IDDM). Am J Hum Genet 52:506–516

Springer NM, Ying K, Fu Y, Ji T, Yeh CT, Jia Y, Wu W, Richmond T, Kitzman J, Rosenbaum H, Iniguez AL, Barbazuk WB, Jeddeloh JA, Nettleton D, Schnable PS (2009) Maize inbreds exhibit high levels of copy number variation (CNV) and presence/absence variation (PAV) in genome content. PLoS Genet 5(11):e1000734

Stich B, Melchinger AE, Piepho HP, Heckenberger M, Maurer HP, Reif JC (2006) A new test for family-based association mapping with inbred lines from plant breeding programs. Theor Appl Genet 113:1121–1130

Stich B, Piepho HP, Schulz B, Melchinger AE (2008) Multi-trait association mapping in sugar beet (*Beta vulgaris* L.). Theor Appl Genet 117:947–954

Stuber CW, Lincoln SE, Wolff DW, Helentjaris T, Lander ES (1992) Identification of genetic factors contributing to heterosis in a hybrid from elite maize inbred lines using molecular markers. Genetics 132:823–829

Stuber CW, Polacco M, Lynn M (1999) Synergy of empirical breeding, marker assisted selection and genomics to increase crop yield potential. Crop Sci 39:1571–1583

Sun FZ, Flanders WD, Yang QH, Zhao HY (2000) Transmission/disequilibrium tests for quantitative traits. Ann Hum Genet 64:555–565

Tenaillon MI, Sawkins MC, Long AD, Gaut RL, Doebley JF, Gaut BS (2001) Patterns of DNA sequence polymorphism along chromosome 1 of maize (*Zea mays* ssp. mays L.). Proc Natl Acad Sci USA 98:9161–9166

Thornsberry JM, Goodman MM, Doebley J, Kresovich S, Nielsen D, Buckler ES 4th (2001) Dwarf8 polymorphisms associate with variation in flowering time. Nat Genet 28:286–289

Tian F, Bradbury PJ, Brown PJ, Hung H, Sun Q, Flint-Garcia S, Rocheford TR, McMullen MD, Holland JB, Buckler ES (2011) Genome-wide association study of leaf architecture in the maize nested association mapping population. Nat Genet 43:159–162

Varshney RK, Nayak SN, May GD, Jackson SA (2009) Next-generation sequencing technologies and their implications for crop genetics and breeding. Trends Biotechnol 27:522–530

Virk PS, Ford-Lloyd BV, Jackson MT, Pooni HS, Clemeno TP, Newbury HJ (1996) Predicting quantitative variation within rice germplasm using molecular markers. Heredity 76:296–304

Waugh R, Marshall D, Thomas B, Comadran J, Russell J, Close T, Stein N, Hayes P, Muehlbauer G, Cockram J, O'Sullivan D, Mackay I, Flavell A, Agoueb A, Barleycap, Ramsay L (2010) Whole-genome association mapping in elite inbred crop varieties. Genome 53: 967–972

Weigel D, Mott R (2009) The 1001 genome project for *Arabidopsis thaliana*. Genome Biol 10:107

Wen W, Araus JL, Shah T, Cairns J, Mahuku G, Bänziger M, Torres JL, Sánchez C, Yan J (2011) Molecular characterization of a diverse maize inbred line collection and its potential utilization for stress tolerance improvement. Crop Sci. doi:10.2135

Xia X, Melchinger AE, Kuntze L, Lübberstedt T (1999) Quantitative trait loci mapping of resistance to sugarcane mosaic virus in maize. Phytopathology 89:660–667

Xu ML, Melchinger AE, Xia XC, Lübberstedt T (1999) High-resolution mapping of loci conferring resistance to sugarcane mosaic virus in maize using RFLP, SSR, and AFLP markers. Mol Gen Genet 261:574–581

Xue W, Xing Y, Weng X, Zhao Y, Tang W, Wang L, Zhou H, Yu S, Xu C, Li X, Zhang Q (2008) Natural variation in Ghd7 is an important regulator of heading date and yield potential in rice. Nat Genet 40:761–767

Yan J, Bermudez Kandianis C, Harjes CE, Bai L, Kim E, Yang X, Skinner DJ, Fu Z, Mitchell SE, Li Q, Salas Fernandez MG, Zaharieva M, Babu R, Fu Y, Palacios N, Li J, DellaPenna D, Brutnell T, Buckler ES, Warburton ML, Rocheford T (2010) Rare genetic variation at Zea mays crtRB1 increases ß-carotene in maize grain. Nat Genet 42:322–327

Yang X, Gao S, Xu S, Zhang Z, Prasanna BM, Li L, Li J, Yan J (2011) Characterization of a global germplasm collection and its potential utilization for analysis of complex quantitative traits in maize. Mol Breed 28:511–526

Yu J, Buckler ES (2006) Genetic association mapping and genome organization of maize. Curr Opin Biotechnol 17:155–160

Zhang Z, Ersoz E, Lai CQ, Todhunter RJ, Tiwari HK, Gore MA, Bradbury PJ, Yu J, Arnett DK, Ordovas JM, Buckler ES (2010) Mixed linear model approach adapted for genome-wide association studies. Nat Genet 42:355–360

Zhang YX, Wang Q, Jiang L, Liu LL, Wang BX, Shen YY, Cheng XN (2011) Fine mapping of qSTV11(KAS), a major QTL for rice stripe disease resistance. Theor Appl Genet 122: 1591–1604

Zhao K, Tung CW, Eizenga GC, Wright MH, Ali ML, Price AH, Norton GJ, Islam MR, Reynolds A, Mezey J, McClung AM, Bustamante CD, McCouch SR (2011) Genome-wide association mapping reveals a rich genetic architecture of complex traits in *Oryza sativa*. Nat Commun 2:467

Zhu C, Gore M, Buckler ES, Yu J (2008) Status and prospects of association mapping in plants. Plant Genome 1:5–20

Zondervan KT, Cardon LR (2007) Designing candidate gene and genome-wide case-control association studies. Nat Protoc 2:2492–2501

Part III
Validation of QTPs

Chapter 7
TILLING and EcoTILLING

Gunter Backes

Why TILLING?

There are two different ways to link a certain allelic state of an individual's genotype (*e.g.*, a specific mutation) to a trait or phenotype: *forward* and *reverse* genetics. In *forward* genetics, the starting point is a difference in the phenotype and the aim is to find the sequence variation responsible for this difference. *Reverse* genetics starts with an alteration of a DNA sequence and aims to find the change(s) in phenotype, caused by this alteration. Mutants, either naturally occurring or artificially evoked are important in both approaches, as they might represent extreme phenotypes which are easier to differentiate from the wild type than more subtle changes in phenotype, predominant in natural populations of a species. Historically, *forward* genetics has been the predominant approach in genetics for a long time. However, the more recent doubling of sequence information every 9 months (Kahn 2011) favours application of *reverse* genetics to assign functions to genes predicted from sequence annotation. In addition, *forward* genetics is not practical for genome-wide analyses due to the effort and time involved to identify each gene coding for a particular phenotype (Brady and Provart 2007). Further limitations of *forward* genetics in comparison to *reverse* genetics are (a) that many mutations are essentially undetectable in typical phenotypic screens, (b) that rare phenotypes or rare mutations that produce a given phenotype can be missed simply because of the vast number of individuals that need to be screened and (c) that many mutations are silent due to heterozygosity and polyploidy (Stemple 2004).

G. Backes (✉)
Faculty of Organic Agricultural Sciences, Department of Organic Breeding and Agrobiodiversity, University of Kassel, Nordbahnhofstr. 1a, 37213 Witzenhausen, Germany
e-mail: gunter.backes@uni-kassel.de

Table 7.1 Reverse genetic approaches (Henikoff and Comai 2003)

Targeted mutagenesis		
Post-translational gene silencing (iRNA)	Chuang and Meyerowitz (2000)	GMO
Homologous recombination	Struhl et al. (1979)	GMO
Chimeric RNA/DNA oligonucleotides	Beetham et al. (1999)	GMO
Zinc finger nucleases mediated introduction of InDels	Klug (2010)	(GMO)
Genome-wide mutagenesis and specific screening		
Insertional mutagenesis (T-DNA)	Krysan et al. (2002)	GMO
Fast-neutron mutagenesis and size-selection	Li et al. (2002b)	Non-GMO
TILLING	McCallum et al. (2000)	Non-GMO

The reverse-genetics approaches can be divided into two groups (Table 7.1): on the one hand approaches that rely on the mutagenesis of a specific gene and subsequent observation of the resulting phenotype, and on the other hand approaches where undirected mutagenesis across the whole-genome is followed by gene specific screening (Henikoff and Comai 2003). The most frequently applied approach of the first group is Post-Translational Gene Silencing (PTGS), also called RNA interference or RNAi. PTGS exploits defence mechanism against viral double-stranded RNA and is based on the expression of transcripts that results in short two-strand RNA molecules by loop formation (Chuang and Meyerowitz 2000; Waterhouse et al. 1998). Drawbacks are that the results of silencing can vary and that the necessary transformations for each target gene represent both a bottleneck lead to the classification of the product as genetically modified (Henikoff and Comai 2003). Another gene-directed approach is Homologous Recombination, the disruption of specific genes by reciprocal exchange of DNA (Struhl et al. 1979). The technique was mainly applied in microorganisms, fruit flies and mice, but was also used in rice (Iida and Terada 2004). A third technique in this group of gene directed mutations employs chimeric RNA/DNA oligonucleotids (Beetham et al. 1999). The latter two techniques share the need for transformation with PTGS and thereby also the drawbacks mentioned above. A further technique is the use of Zinc-finger nucleases, custom-synthesized enzymes that combine a non-specific DNA nuclease with tandem arrayed Zinc-finger structures (Klug 2010). As each of the Zinc fingers recognizes a certain base-triplet, a specific complex recognition sequence can be constructed. When applied *in vivo*, the respective cleaving site will be repaired and cut again until an imperfect ligation emerges, leaving an InDel at the cleaving site. The mutant plants obtained through this method are also considered to be genetically modified organisms in Europe, as site-specific mutations have been applied.

One approach of the second group, that is based on genome-wide mutation and subsequent specific screening is insertional mutagenesis by transposons. This method has been especially successful in *Arabidopsis thaliana* (L.), where a widely used stock of T-DNA insertion lines has been established (Krysan et al. 2002).

The reason for this success had been the early accessibility of the full *Arabidopsis thaliana* (L.) sequence that enables the identification of the exact position of the insertion point of the transposon based on sequencing of the border regions of the insertion. Further, the availability of an efficient transformation system in this plant (Gelvin 2003) contributed to the fast development of this resource. For this system to be just as effective in other plants, similar conditions have to be met. In rice, the Tos-17 transposon has been used (Hirochika 1997) but problems arose due to background-mutations in tissue culture. Further, the vector construction and plant transformation is time-consuming (Parry et al. 2009). Also for maize, a knockout-population based on the Mu-element has been established and it is offered as a resource to the maize genetic research community (May et al. 2003). A second approach is the introduction of deletions by fast-neutron mutagenesis followed by a size screening of gene-specific PCR-products as described in the "Delete-a-gene" procedure for *Arabidopsis thaliana* (L.) (Li et al. 2002b). With this technique, that also has been used in *Medicago truncatula* (Rogers et al. 2009), deletion mutants can be identified in DNA pools from thousand mutants, but need then to be deconvoluted subsequently using smaller pools. The technique needs access to a fast-neutron source, but on the other hand, it is non-GMO, which is also true for TILLING. In contrast to the fast-neutron technique, TILLING introduces single-base mutations into the sequence of plant lines. Consequently, fragment size screening is no longer feasible, but other techniques have to be used to identify the lines with mutations in the targeted sequences.

What is TILLING?

The development of TILLING began in the late 1990s when Claire McCallum and some of her collaborators worked on characterizing two chromo-methylase genes in *Arabidopsis thaliana* (L.) (Barkley and Wang 2008). They first tried unsuccessfully to apply different approaches such as T-DNA lines and antisense-RNA. At last they came up with TILLING as a new approach. This new technique is defined by combining high density point-mutations provided by chemical mutagenesis with rapid mutational screening in pools of DNA. In their first paper on TILLING (McCallum et al. 2000), they used Ethyl methanesulfonate (EMS) as chemical mutagen and Denaturing High-Performance Liquid Chromatography (DHPLC) as detection method. Variations of TILLING are described below.

Steps in the TILLING Procedure

In the following, the different steps of the TILLING procedure are described with the possible choices that can be applied as summarized in Table 7.2.

Table 7.2 Steps and related choices in the TILLING process

Steps	Choices
Mutagenizing	Choice of wild type
	Choice of mutation target
	Choice of mutagen
	Choice of mutagen dose
Creating population	Partially determined by choices in previous step
	Choice of population structure
DNA isolation	Choice of isolation method
	Choice of concentration equalizing
	Choice of pooling strategy
Target sequence selection	Choice of target gene
	Choice of target region within gene
Mutation detection	Choice of general methods
	Choice of details within general method
Phenotyping	Choice of phenotyping method

Mutagenesis

Before the actual step of mutagenesis, a genotype to be mutagenized, the "wild type", has to be chosen. Two main criteria should be applied for this important choice (Comai and Henikoff 2006): it should be possible to produce several thousand genotypically identical individuals from this 'mother genotype' in one or two steps and whenever possible the individual with the highest amount of sequence information should be chosen. Where it is not possible to start with a homozygous genotype (e.g., due to dioecy), the TILLING procedure has to be adapted (Wienholds et al. 2003).

The next important choice is the plant organ which should be the target of the mutation. In most cases, seeds are chosen as they are easy to handle in the mutation process, as they simply have to be soaked in the mutagenizing chemical. However, out of these M_0 seeds, chimeric M_1 plants grow as single cells of the embryos had been mutagenized (Henikoff and Comai 2003). Thereby, mutations can be lost in the next generation. An alternative, where all mutations are passed to the M_1-generation is the mutagenesis of gametes, such as pollen (Till et al. 2004b). This requires treating of mature pollen with a mutagen solution (Okagaki et al. 1991).

A further decision to take is, which mutagen to use. In the original TILLING paper (McCallum et al. 2000), ethyl methanesulfonate (EMS, $CH_3SO_3C_2H_5$) was used and is still the most often used mutagen for TILLING, as it seems to be fairly constant in the level of mutation rates achieved (Henikoff and Comai 2003). The mutagen works through the reaction of the ethyl group of EMS with guanine in DNA, forming the abnormal base O-6-ethylguanine, which pairs with Thymin and will be repaired to Adenin (Henikoff and Comai 2003). Therefore, around 90% of the mutations after EMS treatment are G/C to A/T transitions (Gilchrist et al. 2006b; Winkler et al. 2005). In Arabidopsis, maize and wheat, up to of 99% of the mutations were G/C to A/T transitions (Greene et al. 2003; Slade et al. 2005). The differences to 100% might be due to 'wild' mutations or contaminations (Greene et al. 2003). In addition, biases in the neighbouring bases have been observed with AGA as the combination with the highest frequency and TGC as the combination with the lowest frequency. Reasons for this imbalance are most likely due to preference of the repair mechanisms. Other possible reasons might be based on preferences of EMS itself or of the detection system (Greene et al. 2003).

N-ethyl-N-nitrosourea (ENU, $C_3H_7N_3O_2$) is less specific in the base transitions it creates. Nevertheless, it shows preference for A>T base transversions and also for AT>GC transitions (Nolan et al. 2002). The mutagen acts by transferring the ethyl group of ENU to nucleobases (usually thymine) in nucleic acids and had been used in mice (Nolan et al. 2002). N-Nitroso-N-methylurea (NMU, $C_2H_5N_3O_2$) has been used in TILLING in soybean (Cooper et al. 2008). NMU is an alkylating agent, and exhibits its toxicity by transferring its methyl group to nucleobases in nucleic acids. Sodium azide (NaN_3) is highly mutagenic in some organisms such as barley and some bacteria, but much less so in other organisms (Al-Qurainy and Khan 2009). The reason for its species-specificity is that the mutagen is metabolized to beta-azidoanaline which then interacts with the DNA (Owais and Kleinhofs 1988). Sodium azide has been applied for TILLING in barley (Lababidi et al. 2009) and a combination of NMU and sodium azide has been used in rice (Till et al. 2007). Further chemical mutagens that have been applied in plants are methyl-methane sulphonate (MMS) and hydrogen fluoride (HF).

The mutagenizing effect of the chosen chemical mutagen is determined by the species that is mutagenized, the plant tissue it is applied to (mostly kernel or pollen, see above), the concentration of the mutagen, and the time of exposure. The optimal effect is the balance between a high density of mutations on the one side, and an acceptable survival and fertility of the M_1-plants on the other side. Therefore, determination of the optimal concentration and time of exposure will typically involve some trial and error. In these trials, the effect of the mutagenesis is measured indirectly by the germination rates of tilled seeds (Hohmann et al. 2005), the fertility rate of M_1-plants (Henikoff and Comai 2003), the appearance of easily detectable mutants such as albinos (Le Signor et al. 2009) or directly by sequencing or TILLING some 'test genes' (Gilchrist et al. 2006b). The optimal frequency of mutations is much higher for polyploid than diploid species, as polyploid species have an up to ten times higher tolerance for mutations (Parry et al. 2009) due to the genetic buffering in the homologous (autoploid) or homoeologous (allopolyploid) chromosomes. Therefore, smaller TILLING populations are sufficient in polyploid species to reach an acceptable probability of having mutations in the respective target genes.

Creating a TILLING Population

The procedure of creating a TILLING population is largely influenced by the choice, whether pollen or seed is used for mutagenesis (see previous section). Its principles

Table 7.3 Steps to create a TILLING population (Comai and Henikoff 2006)

M_0 seeds	Mutagensis of pollen	
M_1 plants	Fertilization and production of seed (Heterozygous) DNA isolation for TILLING	Mutagenesis of seed (Chimeric and heterozygous)
Selfing		
M_2 seeds	Storage	
M_2 plants	(Segregating) Phenotyping	(Segregating) DNA isolation for TILLING
Selfing		
M_3 seeds		Storage
M_3 plants		(Segregating) Phenotyping

are shown in Table 7.3 (see also Comai and Henikoff 2006). The production of M_0 seed involves either the application of the mutagen to seed or to pollen with subsequent pollination of the female organs of another plant of the wild type. The resulting kernels are grown to M_1-plants. These plants are either heterozygous and chimeric for the mutation in case of mutagenesis of seed or heterozygous in the case of pollen mutagenesis. Therefore, in case of pollen mutagenesis, DNA can be directly isolated from M_1-plants for the TILLING procedure, while a further selfing step is required for mutagenesis of seed before DNA can be isolated from M_2-plants. In this context, it is important to state that in case of pollen mutagenesis, all (M_1) plants used for TILLING bear all mutations introduced in pollen, while in case of seed mutagenesis, (M_2) plants have lost part of their mutations either due to the chimeric nature of the mutation or due to segregation which results either in homozygous mutations and wild types (with a frequency of 0.25 each) or plants heterozygous for the mutation (with a frequency of 0.5). Therefore, the M_2 plants coming from one M_1-plant are not identical in terms of mutations, but more similar than plants coming from different M_1-plants. This raises the question of the strategy of composing the M_2-population, where either a broad strategy with many M_1-plants with few derived M_2-plants per M_1-plant, or a deep strategy with few M_1-plants with many derived M_2-plants can be chosen. In *Medicago trunculata* two populations with similar number of M_2-plants and the same wild type, but different structure were analysed. The population with one M_2-plant per M_1-plant had a mutation frequency that was twice as high as a population with nine M_2-plants per M_1 plant (Le Signor et al. 2009). The authors propose two M_2-plants per M_1-plant. In reality the choice of the M_2-strategy will mostly depend on the number of available M_1-plants and the size of the TILLING population targeted. Sreelakshmi et al. (2010) propose that >5,000 individuals in M_1-populations should be chosen for a broad strategy, while in M_1-populations <1,000 individuals, a deep M_2-strategy (8–9 M_2-plants per M_1-plant) should be selected.

DNA Isolation

The most important requirement to DNA isolation for TILLING is a procedure that is cost-effective and high-throughput, as a large number of M_2 (M_1) individuals have to be isolated. Furthermore, the DNA should be of good average size and stable under standard storage conditions. Finally, a low amount of carry-overs hampering the sub-sequent reactions is required (Comai and Henikoff 2006). Sreelakshmi et al. (2010) tested several procedures for DNA isolation for their suitability for TILLING. The CTAB method gave good results but was rather difficult to realize in a 96-well format. Column based methods are quick and clean, but are relatively expensive and generally result in low DNA yield. The optimal method they found was a by Chao and Somers 2012 (Palotta et al. 2003), with minor improvements in the composition of the isolation buffer. Generally, young tissue showed better results than older tissue.

Due to advantages in long term storage, TE buffer has shown to be better than water for dissolving the DNA. The final step of DNA isolation is the normalization to similar concentrations, which is required by most of the detection techniques mentioned below. For this purpose, measuring DNA concentrations, either photometrical or on an Agarose gel is necessary (Comai and Henikoff 2006).

Target Sequence Selection

It should be avoided to choose genes which occur in more than one copy. If this is not possible, PCR primers should be chosen that amplify a specific copy (Barkley and Wang 2008). The most interesting mutations are nonsense mutations, *i.e.*, mutations that cause a premature stop codon, and missense mutations (mutations that produce an altered amino acid in the protein product), which cause a measurable change in the phenotype of the plant. In the Arabidopsis Tilling project, 4.5% of the mutations observed were nonsense mutations, while about 50% were missense mutations, as would be expected from random mutations in the respective genes (Till et al. 2003). As the frequency of nonsense mutations that lead to a clear loss of function is rather low, it is important to choose fragments of the gene for TILLING where missense mutations have a high chance to produce a change in function. Software like CODDLE (http://www.proweb.org/coddle/) uses alignments of related sequences and takes into account both the mutagen used and the coding sequence to identify regions of the target gene most likely to generate deleterious mutations (Henikoff and Comai 2003). For TILLING, a product length of 800–1,200 bp is optimal, dependent on the frequency of mutations. A larger PCR product gives a higher chance of polymorphism, while a single SNP gives a clearer signal in the chosen detection method than several SNPs in one fragment (Barkley and Wang 2008). Further, in 100–200 bp at both ends, the detection of polymorphisms is difficult. This can be circumvented by placing primers in exons or by choosing primers for overlapping PCR products (Barkley and Wang 2008).

Table 7.4 Methods to identify mutations in TILLING (See also Yeung et al. 2005)

On single individuals	Resequencing		
	Single Strand Conformation Polymorphism (SSCP)		Kuhn et al. (2005)
	Mass spectrometry		Stanssens et al. (2004)
On pools of individuals	Melting-temperature based	Denaturing High-Performance Liquid Chromatography (DHPLC)	Hecker et al. (1999)
		3′-Minor Groove Binding (MGB)	Kutyavin (2000)
		Temperature Gradient Capillary Electrophoresis (TGCE)	Li et al. (2002a)
		High-Resolution Melting (HRM)	Gady et al. (2009)
	Secondary structure based	Heteroduplex analysis	Nataraj et al. (1999)
		Conformation Sensitive Capillary Electrophoresis (CSCE)	Gady et al. (2009)
	Chemical cleavage of mismatch	CCM in solution	Cotton et al. (1988)
		CCM with fixed DNA	Bui et al. (2003)
	Enzymatic cleavage of mismatches	DNA N-glycolase based	Lu and Hsu (1992)
		Phage resolvases	Mashal et al. (1995)
		Combination of endonuclease and ligase	Huang et al. (2002)
		Single-strand specific nucleases (S1, P1, Mungbean nuclease, CEL 1, Endo 1)	Desai and Shankar (2003)

Mutation Detection

The aim of the techniques used in the step of mutation detection of TILLING is to identify genotypes with a mutation in a certain gene or gene region. This technique needs to be high-throughput and reliable. In addition, it is desirable that the technique can distinguish genotypes homozygous and heterozygous for the trait and that it delivers information on the position of the mutation in the sequence. The different approaches are outlined in Table 7.4 (see also Yeung et al. 2005).

These approaches can be divided into techniques carried out on single individuals or on pools of individuals. Of the techniques established for single individuals, *simple re-sequencing* is the most straightforward approach. Even though it might be not sufficiently cost-efficient and high-throughput at the moment, the rapidly increasing progress in sequencing techniques can change this condition soon (Shendure and Ji 2008). Another approach is *Single Strand Conformation Polymorphism*

(SSCP, Kuhn et al. 2005). It relies on differences in the secondary structure of DNA that results in different runs on an polyacrylamide gel or capillaries. As these differences occur under very narrowly defined running conditions, the technique needs extensive optimization for every PCR product. Furthermore, it does not give the position of the mutation. *Mass spectrometry for mutation detection* (Stanssens et al. 2004) is a rather complicated technique with many steps, such as transcription of DNA into RNA, the cleavage with four different base-specific RNases and subsequent MALDI-TOF mass spectrometry.

The methods carried out on pooled genotypes are all based on the mixing of DNA of the different mutant lines into pools, followed by a PCR of the pooled DNA and the creation of heteroduplices of PCR products of those different genotypes. In principle, mixing of fragments after PCR is possible, but this would increase sample numbers and work load substantially. The different techniques are based on differing melting temperatures, secondary structure, chemical cleavage of mismatches, and enzymatic approaches (Table 7.1).

The mutation detection approach of the first TILLING paper (McCallum et al. 2000) was *Denaturing High-Performance Liquid Chromatography* (DHPLC, Hecker et al. 1999), a melting-temperature based approach. DHPLC for TILLING is based on the fact that at a certain temperature range, heteroduplices with mismatches show more single-strands than base-pairings without mismatches. Due to their lower hydrophobicity, melted strands are retained longer in a column of a reverse-phase liquid chromatography. Drawbacks of this method are the relatively high costs and the limited pooling potential (up to fivefold). Other melting temperature-based approaches are: (a) *3'-Minor Groove Binding* (MGB, Kutyavin 2000), where short DNA (12–20mer) probes are conjugated with minor groove structures and show thus higher sequence specificity than unbound DNA probes. Therefore, they can detect single base mismatches that occur within the probe sequence. The relative short probe length restricts the potential MGB in TILLING. (b) *Temperature Gradient Capillary Electrophoresis* (TGCE, Li et al. 2002a) that applies a temperature gradient that covers all possible temperatures of 50% melting equilibrium for the samples during capillary electrophoresis. As heteroduplices have different melting temperatures, they change their running characteristics in different positions of the gradient as homoduplices. (c) *High-Resolution Melting* (Gady et al. 2009) is a recent high throughput technique. In the PCR reaction, an intercalating dye attaches to the PCR product, which only emits fluorescence when captured in the molecule. Afterwards, the PCR product is slowly heated and fluorescence observed. Heteroduplices with their lower melting temperature loose fluorescence earlier than homoduplices. All four melting-temperature based approaches mentioned above only show the presence of a mutation in DNA sequence, but not the position.

This is also true for the next group of methods that are based on the fact that the non-denatured heteroduplex DNA molecules form a different secondary structure than homoduplices. *Heteroduplex analysis* (HA, Nataraj et al. 1999) uses non-degenerating polyacrylamide-gels to show differences in running speed of different secondary structures. *Conformation Sensitive Capillary Electrophoresis* (CSCE, Gady et al. 2009) uses differences in running speed in a semi-denaturing polymer. Peaks with heteroduplices show a different shape than homoduplex peaks.

Unpaired bases show a different reactivity to certain chemicals than paired bases. The method of *chemical cleavage of mismatch DNA* (CCM) uses this difference to treat the DNA with Osmium tetroxide or $KMNO_4$ (both oxidizing) or hydroxylamine to cleave the DNA at the mismatch position. This approach was first applied with DNA in solution (Cotton et al. 1988), but later also with DNA fixed to a solid support (Bui et al. 2003).

The third and last group of methods to detect mutations in DNA heteroduplices are methods based on the *enzymatic cleavage of mismatches*. Criteria for the best suited approach are that the enzyme cleaves all mismatches present in the heteroduplex, that there are no or only few non-specific cleavages and that the reaction conditions are neither damaging the DNA nor disturbing the subsequent steps of fragment detection. *DNA N-glycolases* are mismatch repair enzymes that recognize specific mismatches and create apurinic or apyrimidinic sites which can be cleaved (Lu and Hsu 1992). Their use for mutation detection in TILLING is restricted by their specificity and the fact that, currently, only N-glycolases that cleave TG and AG mismatches are known. *Phage resolvases* are enzymes in phages that cut at holliday junctions (cross-shaped structures of single strand DNA that forms during the process of genetic recombination). They can also be used in mismatch cleaving of heteroduplices, but tend to show a high percentage of non-specific cleavages (Mashal et al. 1995). Also *combinations of endonucleases and ligases* have been used for the cleavage of mismatches (Huang et al. 2002). In this approach nicks are introduced by the endonuclease Endo V that cleaves preferentially at mismatches, followed by repairing non-specific cleavage points by the action of a DNA-ligase.

Single strand (SS) specific nucleases are the group of enzymes that is most frequently used in TILLING. Historically, S1 from *Aspergillus oryzae*, P1 from *Penicillium citirinum* and Mungbean nuclease from *Vigna radiate* had been the ones first applied to cleave single strands selectively, e.g., overhanging single strand ends of DNA double strands. These SS-specific endonuclease have also been used for mutation detection (Howard et al. 1999), but show the disadvantage that not all mismatches are recognized and cleaved and that they are working in a relatively acidic pH of around 5.0 (Yeung et al. 2005).

The function of CEL I, the single-strand specific endonuclease that is most frequently used in TILLING, was published by Oleykowski et al. (1998) and the enzyme was cloned and characterized by Yang et al. (2000). Advantages compared to other endocnucleases are the high sensitivity, a neutral pH range of optimal conditions and the fact that CEL I is a very stable enzyme during purification, storage and assay (Oleykowski et al. 1998). The enzyme needs a divalent metal ion (Mg^{2+}, Ca^{2+} or Zn^{2+}) for its function. Normally, only one strand at the 3'-end of the mismatch is cleaved, but double strand cuts can occur, which is likely explained by two single strand cuts on both strands and is favored by higher enzyme concentration and longer incubation time (Yeung et al. 2005). A cross-reaction of CEL I with polymerases has been observed, leading to an up to 30-fold increase in activity by the presence of ligases and polymerases (Yeung et al. 2005), but the stimulating effect of the polymerase is not effective, when the PCR-product is dye-labelled. Problems of the application of the enzyme are a base preference in the cleavage sites (C/C > C/A = C/T > G/G > A/C = A/A = T/C > T/G = G/T = G/A = A/G > T/T) that might reflect incomplete cleavage of mismatches (Oleykowski et al. 1998) and a $5'-3'$ exonuclease activity of the enzyme (Brady and Provart 2007; Yeung et al. 2005). Digested fragments are better protected than the full-length product of the sequence, but generally over-digestion with CEL I leads to loss of product. Therefore, an optimisation of CEL I concentration and incubation time is necessary. An optimal band to background relationship is reached when a larger part of the full-length product is digested (Till et al. 2004a). The incubation time varies normally from 15–60 min at 45°C and is stopped by adding EDTA (Till et al. 2006). Further background noise is observed by unspecific restriction, which is likely caused by mismatches introduced by PCR (Yeung et al. 2005), but DNA polymerases with $3'-5'$ exonuclease proofreading activity cannot be used for mutation detection as they are degrading the cleavage product of CEL I (Oleykowski et al. 1998). Originally the enzyme was isolated from celery juice through a laborious procedure, but it was shown that a crude juice extracts works as well as the pure form of the enzyme (Till et al. 2004a).

Looking for alternatives to CEL I, ENDO1 was found in *Arabidopsis thaliana* (L.) using common sequence features of S1 and P1 nucleases (Triques et al. 2007). ENDO1 was one of the two detected enzymes that showed mismatch cleaving, and it had the highest sequence similarity to CEL I. ENDO1 shows no base preferences at the mismatch cleaving point and produces more consistent results than CEL I (Parry et al. 2009). However, the crude extract loses quickly its activity and must, therefore, be purified (Triques et al. 2008). In order to achieve a higher yield of the enzyme, the respective gene was transiently transferred to *Nicotiana benthamiana* (L.) and overexpressed (Triques et al. 2008).

The cleavage of the mismatch in the heteroduplex is followed by fragment analysis. In the most common TILLING protocol this is done by polyacrylamide gel electrophoresis (PAGE) with automatic detection of dye-labeled fragments (Till et al. 2006). Alternatives are capillary based systems that provide better automation and a more sensitive detection (Igarashi et al. 2000; Suzuki et al. 2008). For both systems, dye-labelling is important. Normally, the forward and reverse primers are differentially $5'$-labelled. This avoids false positives, which are either caused by the fact that certain homoduplex sites are especially sensitive to variability in CEL I digestion, causing bands to appear in multiple lanes above the background pattern or caused by mis-priming that leads to a large amount of double-end-labelled products of a single size, causing the appearance of the same band for both dyes (Igarashi et al. 2000). Unfortunately, the $5'$-exonuclease activity of CEL I cleaves the $5'$-label from the full-size product. The labelling itself seems to accelerate this process by destabilizing the double strand (Cross et al. 2008). One way to circumvent this problem is the internally labelling of the PCR product, a derivate of TILLING, described as EMAIL (Endonuclolytic Mutation Analysis by Internal Labeling,

Cross et al. 2008). The authors propose an internal labelling with dUTP[R110]. The advantages of this method are the improvement of the signal-to-noise ratio and the cost reduction for PCR primers. A disadvantage is the reduced potential of EMAIL to filter for false positives by the application of two different dyes.

An alternative to the PAGE and capillary systems using dye-labels is a more simplified (and less expensive) fragment detection by agarose gel electrophoresis (Raghavan et al. 2006). The drawback of this method is a lower sensitivity and, as only about 20% of the full-length product is digested, the initial amount of full length products need to be high. Furthermore, it requires an effective reduction of unspecific PCR products to reach a high signal-to-noise ratio. A relatively new development offering a further alternative for detection of TILLING fragments is a capillary electrophoresis with detection based on intercalating dyes (Dibya 2010).

Pooling has two functions: it enables the formation of heteroduplices and it increases throughput. Thus, a high depth of pooling is increasing the throughput of the approach and is, therefore, desirable. The possible depth of pooling depends on the sensitivity of the mutation detection method applied and the expected number of mutations. Using DHPLC, a 1:9 ratio of heteroduplex to homoduplex was detectable. Since many mutations are in heterozygous state, fivefold pooling is recommended (Oleykowski et al. 1998). With CSCE, the optimal pooling-depth is fivefold (Gady et al. 2009). Using CEL I, eightfold pooling is optimal when using PAGE or capillary systems for fragment detection (Colbert et al. 2001), while only fourfold pooling can be achieved when agarose gels are applied for fragment detection (Raghavan et al. 2006). High Resolution Melting (HRM) can also detect a heterozygous mutant in an eightfold pool (Gady et al. 2009). The expected number of mutations depends on the size of the full-length product and frequency of mutations. As the mutation frequency can be much higher in polyploid than in diploid species, in hexaploid wheat only twofold and fourfold pools are applied (Slade et al. 2005). Especially, when eightfold pooling is used, 2-D pooling is advantageous, both in terms of speed of detection and confidence in detected bands (Gilchrist et al. 2006b). When only one-dimensional pooling is applied, a further mutation-detection step in a mixture of the DNA of the wild type and each of the mutant lines of a pool with a detected mutation has to be carried out. A direct 2-D pooling of tissue before DNA isolation was proposed by Sreelakshmi et al. (2010) to save time in DNA isolation and to prevent contamination.

Phenotyping the Mutant(s)

When a mutant is detected, the next step would be to examine the phenotype of this genotype. However, as most mutations are recessive, a genotype homozygous for the mutation will normally be needed to see the effect. In M_1-plants after pollen mutagenesis, each genotype will be heterozygous (Table 7.3). In the case of mutant detection in M_2-plants after kernel mutagenesis, homozygosity or heterozygosity can be distinguished by running the mutation-detection step on the unpooled mutant-DNA. If the TILLING-band also shows up in this case, a heterozygous genotype was detected (Colbert et al. 2001). In this case, a homozygous genotype can most likely be identified in the next generation of plants, which is available as seed (see Table 7.3 and section "Creating a TILLING Population"). For polyploid individuals, the situation is more complicated. In case of autopolyploidy, complete homozygosity must be reached. In case of allopolyploidy, it might be necessary to go through several crosses and selfing steps to prevent that gene-copies from homoeologous chromosomes cover the effect of the mutation (Uauy et al. 2009). When a change in phenotype has been identified, it is important to exclude that other background mutations caused the observed phenotype. One possibility to clarify this question is to cross two mutants with changes in the same gene and observe the segregation ratios in the resulting population (Slade and Knauf 2005). Another way is to observe the segregation of the respective trait in relation to the mutation in the M_2-population in the case of pollen mutagenesis or the M_3-population in the case seed mutagenesis (Henikoff and Comai 2003).

Application of TILLING

TILLING can either be used as a reverse-genetics tool to connect a known DNA sequence with a phenotype, or as a molecular breeding tool (Slade et al. 2005). The broadest application of TILLING as a reverse-genetics tools can be found in *Arabidopsis thaliana* (L.) were the Arabidopsis TILLING project (ATP) offers TILLING services to the research community (Greene et al. 2003). In cabbage (*Brassica oleracera* L.), the function of genes related to abiotic stress were studied with TILLING (Himelblau et al. 2009) In a model species for legumes, *Lotus japonicus*, published TILLING results based on reverse genetics, deal mostly with genes involved in the nodulation process (Heckmann et al. 2006; Horst et al. 2007; Perry et al. 2003). Also in *Medicago trunculata* (Gaertn.), another model species for legumes, the functions of several genes were analysed with TILLING (Lefebvre et al. 2001). As a proof-of-concept, the functions of four candidate genes for internode length were investigated by TILLING in pea (*Pisum sativum*). In grasses, TILLING projects were carried out for reverse-genetics purposes mostly in the already sequenced species: In rice two different TILLING populations have been established and are in use (Suzuki et al. 2008; Till et al. 2007). For maize a Maize TILLING project similar to the Arabidopsis TILLING projects offers its service to the community (Till et al. 2004b; Weil 2009). For sorghum TILLING results of candidate genes for lignin-synthesis have been published (Xin et al. 2008). Besides for these species with full sequences available, TILLING has been applied in two further economically important species: for barley, proof-of-concept papers have been published for TILLING as reverse-genetics tool (Caldwell et al. 2004;

Lababidi et al. 2009; Talamè et al. 2008; Gottwald et al. 2009); and in wheat, genes for different enzymes related to starch synthesis were associated with mutant genotypes as a proof-of-concept (Uauy et al. 2009).

When TILLING is used as a molecular breeding tool, it is mostly applied when the genetic variation needed is not accessible in the gene pool of the respective species or as an alternative to use related wild species for this purpose, which is often difficult to handle in breeding. Mostly, this method will be applied to knock out genes, but also the creation of new allelic variation for certain genes might be a possible aim. As no transformation of DNA is involved, TILLING is a non-genetic modified organism (GMO) method and the restrictions to GMO crops do not apply to genotypes obtained by TILLING (Slade and Knauf 2005). In wheat, starch composition could be changed by knocking out genes for granule-bound starch synthase (Slade et al. 2005). Sestili et al. (2010) knocked out the gene *Sgp-1* responsible for Starch synthase II in wheat for all three homoeologous chromosomes. In rapeseed (*Brassica napus*) the fatty acid elongase 1 (FAE1), a key gene in erucid acid biosynthesis was the target for a knock-out by TILLING. When a mutant with the expected phenotype is obtained through TILLING, the next steps would be backcrossing this genotype to the wild type or diverse high-yielding, high-quality varieties in order to eliminate background mutations (Slade and Knauf 2005).

EcoTILLING

EcoTILLING is a technique derived from TILLING to discover variation in natural populations, in contrast to mutant population used for TILLING. Just as for TILLING, it was originally developed in Arabidopsis to detect variation in Arabidopsis eco-types (Comai et al. 2004). Thus, no mutant population is developed in EcoTILLING. Instead, different existing genotypes representing the genetic variation of a species or a certain group within the species are collected. In the mutation detection approach of EcoTILLING, DNA from only two individuals is pooled: DNA from a reference-genotype and from the queried genotype. A higher pooling depth than twofold pooling could be applied, but the expected high diversity in natural populations used for EcoTILLING prevents the usefulness of this course of action in most cases. For EcoTILLING, most modifications of the procedure described above for TILLING apply too. The lower pooling depth (only one individual is characterized per reaction) is reducing the advantage of the method compared to full re-sequencing of all genotypes. This is especially true as it is necessary to re-sequence anyway the allelic variation detected by EcoTILLING in order to get the precise location of the differing bases. But as it can be expected that the same changes in coding sequence can be found in many genotypes of a species, EcoTILLING has still a relative advantage compared to full re-sequencing (Garvin and Gharett 2007). The comparative advantage is more obvious in cases were the genetic variation of a certain gene in a certain population is low.

Besides in Arabidopsis, EcoTILLING has been applied in common bean (*Phaseolus vulgaris* (L.), Galeano et al. 2009), melon (*Cucumis melo* (L.), Nieto et al. 2007), poplar (*Populus trichocarpa* (Torr. & A.Gray), Gilchrist et al. 2006a), pickerelweed (*Monochoria vaginalis* (C. Presl ex Kunth), Wang et al. 2007, 2008b), rice (*Oryza sativa* (L.), Kadaru et al. 2006; Raghavan et al. 2006), wheat (*Triticum aestivum* (L.), Wang et al. 2008b), barley (*Hordem vulgare* (L.), Mejlhede et al. 2006) and sugarcane (*Saccharum spec.*, Hermann et al. 2006). In contrast to TILLING, a broader choice of applications can be adopted. The most straight-forward application is the survey of the genetic variation in a plant population by examining the allelic variation at several genes. Gilchrist et al. (2006a) used EcoTILLING to catalogue the level of diversity in natural populations of poplar collect in the western part of Canada and the USA. EcoTILLING can also be used to find Single Nucleotide Polymorphisms (SNPs) in certain genes, e.g., in resistance gene analogs in sugarcane (Hermann et al. 2006). By scanning many individuals, the SNPs with the highest diversity or with diversity between certain groups of genotypes can be selected. The SNPs detected this way can either be transformed into markers with easier detection methods or EcoTILLING can also be applied to call the alleles of the SNPs. This is especially useful when dealing with highly polymorphic sequences with several polymorphic sites, as multiple sequence polymorphisms can be detected in one EcoTILLING reaction (Mejlhede et al. 2006). When used as a marker, often simplified methods are applied, such as the replacement of polyacrylamide gel or capillary electrophoresis by agarose gel electrophoresis (Galeano et al. 2009; Garvin and Gharett 2007). In case of polyploid and/or outbreeding species, no pooling is necessary, as different alleles can be found within the same genotype (Hermann et al. 2006). This is sometimes referred to as Self-ecoTILLING (Wang et al. 2008a). Also the presence of several copies of the same gene can lead to bands in the EcoTILLING procedure without the need of pooling. Thus, in barley 13 different allelic groups of the highly diverse *Mla*-gene conferring gene-for-gene resistance to powdery mildew were identified by EcoTILLING (Mejlhede et al. 2006). When allelic differences for a gene are identified via EcoTILLING between two parental lines of a segregating population, the gene can also be localised in this population by pooling the DNA of each descendant line with one of the parents or in two pools with each of the parents (Raghavan et al. 2006). Furthermore, the results of an EcoTILLING experiment can be used to associate the allelic variation at specific sequence sites with phenotypic variation of the respective genotypes. This has been applied to identify causal base changes for virus resistance in common bean (Galeano et al. 2009) and melon (Nieto et al. 2007), for herbicide resistance in pickerelweed (Wang et al. 2007, 2008a, b) and for kernel hardness in wheat (Wang et al. 2008b). Finally, EcoTILLING can be used to search new naturally occurring variation in certain genes to be used in a plant breeding context. This has been the case in the search of new alleles for puroindoline genes in wheat (Wang et al. 2008b), for new resistance genes in sugarcane (Hermann et al. 2006) and new virus resistance alleles in melon based on allelic variation of the eIF4E transcription factor (Nieto et al. 2007).

Further Developments

TILLING has proven to be a relevant reverse-genetic technique, complementing the spectrum of available methods and a useful non-GMO method for molecular breeding, especially when the respective aim can be reached with gene knockouts. EcoTILLING has many different applications that are interesting for a wide range of scientific questions and practical breeding challenges. Nevertheless, with the increasing number of bases per read and decreasing costs (Shendure and Ji 2008), second generation sequencing could get sufficiently cost-effective and quick to replace TILLING (and EcoTILLING) as described above. SequeTILLING has been proposed as an extension of TILLING with the help of second generation sequencing techniques (Weil 2009). The proposed technique includes the two-dimensional 48-fold pooling of the DNA of the mutant lines, the amplification of the target genes for each of the pools, shearing the pooled amplicons into random sets of fragments of about 100 bp, the barcoding of these fragments by ligating row and column barcodes to the different ends of the DNA and finally the sequencing and reassembling of DNA. Thereby, advantages of TILLING and second generation sequencing can be combined and TILLING can be promoted to the next level.

References

Al-Qurainy F, Khan S (2009) Mutagenic effects of sodium azid and its application in crop improvement. World Appl Sci J 6:1589–1601

Barkley NA, Wang ML (2008) Application of TILLING and EcoTILLING as reverse genetic approaches to elucidate the function of genes in plants and animals. Curr Genomics 9: 212–226. doi:10.2174/138920208784533656

Beetham PR, Kipp PB, Sawycky XL, Arntzen CJ, May GD (1999) A tool for functional plant genomics: chimeric RNA/DNA oligonucleotides cause in vivo gene-specific mutations. Proc Natl Acad Sci USA 96:8774–8778. doi:10.1073/pnas.96.15.8774

Brady SM, Provart NJ (2007) Extreme breeding: leveraging genomics for crop improvement. J Sci Food Agric 87:925–929. doi:10.1002/jsfa.2763

Bui CT, Lambrinakos A, Babon JJ, Cotton RG (2003) Chemical cleavage reactions of DNA on solid support: application in mutation detection. BMC Chem Biol 3:1. doi:10.1186/1472-6769-3-1

Caldwell DG, McCallum N, Shaw P, Muehlbauer GJ, Marshall DF, Waugh R (2004) A structured mutant population for forward and reverse genetics in Barley (*Hordeum vulgare* L.). Plant J 40:143–150. doi:10.1111/j.1365-313X.2004.02190.x

Chao SM, Somers D (2012) Wheat and barley DNA extraction in 96-well plates. In: MAS wheat. http://maswheat.ucdavis.edu/protocols/general_protocols/DNA_extraction_003.htm. Accessed 3 Jan 2012

Chuang CF, Meyerowitz EM (2000) Specific and heritable genetic interference by double-stranded RNA in *Arabidopsis thaliana*. Proc Natl Acad Sci USA 97:4985–4990. doi:10.1073/pnas.060034297

Colbert T, Till BJ, Tompa R, Reynolds S, Steine MN, Yeung AT, McCallum CM, Comai L, Henikoff S (2001) High-throughput screening for induced point mutations. Plant Physiol 126:480–484. doi:10.1104/pp. 126.2.480

Comai L, Henikoff S (2006) TILLING: practical single-nucleotide mutation discovery. Plant J 45:684–694. doi:10.1111/j.1365-313X.2006.02670.x

Comai L, Young KJ, Codomo CA, Enns LC, Johnson JE, Burtner C, Odden AR, Henikoff S, Reynolds SH, Greene EA (2004) Efficient discovery of DNA polymorphisms in natural populations by Ecotilling. Plant J 37:778–786. doi:10.1111/j.1365-313X.2003.01999.x

Cooper J, Till B, Laport R, Darlow M, Kleffner J, Jamai A, El-Mellouki T, Liu S, Ritchie R, Nielsen N, Bilyeu K, Meksem K, Comai L, Henikoff S (2008) TILLING to detect induced mutations in soybean. BMC Plant Biol 8:9. doi:10.1186/1471-2229-8-9

Cotton RG, Rodrigues NR, Campbell RD (1988) Reactivity of cytosine and thymine in single-base-pair mismatches with hydroxylamine and osmium tetroxide and its application to the study of mutations. Proc Natl Acad Sci USA 85:4397–4401. doi:10.1073/pnas.85.12.4397

Cross MJ, Waters DLE, Lee LS, Henry RJ (2008) Endonucleolytic mutation analysis by internal labeling (EMAIL). Electrophoresis 29:1291–1301. doi:10.1002/elps.200700452

Desai NA, Shankar V (2003) Single-strand-specific nucleases. FEMS Microbiol Rev 26:457–491. doi:10.1111/j.1574-6976.2003.tb00626.x

Dibya D (2010) High-throughput mutation screening on a TILLING platform using the AdvanCE™ FS96 system. Advanced Analytical, Ames

Gady AL, Hermans FW, Van de Wal MH, van Loo EN, Visser RG, Bachem CW (2009) Implementation of two high through-put techniques in a novel application: detecting point mutations in large EMS mutated plant populations. Plant Methods 5:13. doi:10.1186/1746-4811-5-13

Galeano CH, Gomez M, Rodriguez LM, Blair MW (2009) CEL I nuclease digestion for SNP discovery and marker development in common bean (*Phaseolus vulgaris* L.). Crop Sci 49: 381–394. doi:10.2135/cropsci2008.07.0413

Garvin MR, Gharett AJ (2007) DEco-TILLING: an inexpensive method for single nucleotide polymorphism discovery that reduces ascertainment bias. Mol Ecol Notes 7:735–746. doi:10.1111/j.1471-8286.2007.01767.x

Gelvin SB (2003) Agrobacterium-mediated plant transformation: the biology behind the "gene-jockeying" tool. Microbiol Mol Biol Rev 67:16–37. doi:10.1128/MMBR.67.1.16-37.2003

Gilchrist EJ, Haughn GW, Ying CC, Otto SP, Zhuang J, Cheung D, Hamberger B, Aboutorabi F, Kalynyak T, Johnson L, Bohlmann J, Ellis BE, Douglas CJ, Cronk QCB (2006a) Use of Ecotilling as an efficient SNP discovery tool to survey genetic variation in wild populations of *Populus trichocarpa*. Mol Ecol 15:1367–1378. doi:10.1111/j.1365-294X.2006.02885.x

Gilchrist EJ, O'Neil NJ, Rose AM, Zetka MC, Haughn GW (2006b) TILLING is an effective reverse genetics technique for *Caenorhabditis elegans*. BMC Genomics 7:262. doi:10.1186/1471-2164-7-262

Gottwald S, Bauer P, Komatsuda T, Lundqvist U, Stein N (2009) TILLING in the two-rowed barley cultivar "Barke" reveals preferred sites of functional diversity in the gene HvHox1. BMC Res Notes 2:258. doi:10.1186/1756-0500-2-258

Greene EA, Codomo CA, Taylor NE, Henikoff JG, Till BJ, Reynolds SH, Enns LC, Burtner C, Johnson JE, Odden AR, Comai L, Henikoff S (2003) Spectrum of chemically induced mutations from a large-scale reverse-genetic screen in Arabidopsis. Theor Appl Genet 164:731–740

Hecker KH, Taylor PD, Gjerde DT (1999) Mutation detection by denaturing DNA chromatography using fluorescently labeled polymerase chain reaction products. Anal Biochem 272:156–164. doi:10.1006/abio.1999.4171

Heckmann AB, Lombardo F, Miwa H, Perry JA, Bunnewell S, Parniske M, Wang TL, Downie JA (2006) Lotus japonicus nodulation requires two GRAS domain regulators, one of which is functionally conserved in a non-legume. Plant Physiol 142:1739–1750. doi:10.1104/pp.106.089508

Henikoff S, Comai L (2003) Single-nucleotide mutations for plant functional genomics. Annu Rev Plant Biol 54:375–401. doi:10.1143/annurev.arplant.54.031902.135009

Hermann S, Brumbley S, McIntyre CL (2006) Analysing diversity in sugarcane resistance gene analogues. Australas Plant Pathol 35:631. doi:10.1071/AP06066

Himelblau E, Gilchrist EJ, Buono K, Bizzell C, Mentzer L, Vogelzang R, Osborn T, Amasino RM, Parkin IAP, Haughn GW (2009) Forward and reverse genetics of rapid-cycling *Brassica oleracea*. Theor Appl Genet 118:953–961. doi:10.1007/s00122-008-0952-7

Hirochika H (1997) Retrotransposons of rice: their regulation and use for genome analysis. Plant Mol Biol 35:231–240. doi:10.1023/A:1005774705893

Hohmann U, Jacobs G, Jung C (2005) An EMS mutagenesis protocol for sugar beet and isolation of non-bolting mutants. Plant Breed 124:317–321. doi:10.1111/j.1439-0523.2005.01126.x

Horst I, Welham T, Kelly S, Kaneko T, Sato S, Tabata S, Parniske M, Wang TL (2007) TILLING mutants of Lotus japonicus reveal that nitrogen assimilation and fixation can occur in the absence of nodule-enhanced sucrose synthase. Plant Physiol 144:806–820. doi:10.1104/pp.107.097063

Howard JT, Ward J, Watson JN, Roux KH (1999) Heteroduplex cleavage analysis using S1 nuclease. Biotechniques 27:18–19

Huang J, Kirk B, Favis R, Soussi T, Paty P, Cao W, Barany F (2002) An endonuclease/ligase based mutation scanning method especially suited for analysis of neoplastic tissue. Oncogene 21:1909–1921. doi:10.1038/sj.onc.1205109

Igarashi H, Nagura K, Sugimura H, Kulinski J, Besack D, Oleykowski CA, Godwin AK, Yeung AT (2000) CEL I enzymatic mutation detection. Biotechniques 29:44–48

Iida S, Terada R (2004) A tale of two integrations, transgene and T-DNA: gene targeting by homologous recombination in rice. Curr Opin Biotechnol 15:132–138. doi:10.1016/j.copbio.2004.02.005

Kadaru SB, Yadav AS, Fjellstrom RG, Oard JH (2006) Alternative ecotilling protocol for rapid, cost-effective single-nucleotide polymorphism discovery and genotyping in rice (*Oryza sativa* L.). Plant Mol Biol Rep 24:3–22. doi:10.1007/BF02914042

Kahn SD (2011) On the future of genomic data. Science 331:728–729. doi:10.1126/science.1197891

Klug A (2010) The discovery of Zinc figures and their applications in gene regulation and genome manipulation. Annu Rev Biochem 79:213–231. doi: 10.1146/annurev-biochem-010909-095056

Krysan PJ, Young JC, Jester PJ, Monson S, Copenhaver G, Preuss D, Sussman MR (2002) Characterization of T-DNA insertion sites in *Arabidopsis thaliana* and the implications for saturation mutagenesis. OMICS 6:163–174. doi:10.1089/153623102760092760

Kuhn DN, Borrone J, Meerow AW, Motamayor JC, Brown JS, Schnell RJ (2005) Single-strand conformation polymorphism analysis of candidate genes for reliable identification of alleles by capillary array electrophoresis. Electrophoresis 26:112–125. doi:10.1002/elps.200406106

Kutyavin IV (2000) 3′-Minor groove binder-DNA probes increase sequence specificity at PCR extension temperatures. Nucleic Acids Res 28:655–661. doi:10.1093/nar/28.2.655

Lababidi S, Mejlhede N, Rasmussen SK, Backes G, Al-Said W, Baum M, Jahoor A (2009) Identification of barley mutants in the cultivar "Lux" at the Dhn loci through TILLING. Plant Breed 128:332–336. doi:10.1111/j.1439-0523.2009.01640.x

Le Signor C, Savois V, Aubert G, Verdier J, Nicolas M, Pagny G, Moussy F, Sanchez M, Baker D, Clarke J, Thompson R (2009) Optimizing TILLING populations for reverse genetics in *Medicago truncatula*. Plant Biotechnol J 7:430–441. doi:10.1111/j.1467-7652.2009.00410.x

Lefebvre V, Goffinet B, Chauvet JC, Caromel B, Signoret P, Brand R, Palloix A (2001) Evaluation of genetic distances between pepper inbred lines for cultivar protection purposes: comparison of AFLP, RAPD and phenotypic data. Theor Appl Genet 102:741–750. doi:10.1007/s001220051705

Li Q, Liu Z, Monroe H, Culiat CT (2002a) Integrated platform for detection of DNA sequence variants using capillary array electrophoresis. Electrophoresis 23:1499–1511. doi:10.1002/1522-2683(200205)23:10<1499::AID-ELPS1499>3.0.CO;2-X

Li X, Lassner M, Zhang Y (2002b) Deleteagene: a fast neutron deletion mutagenesis-based gene knockout system for plants. Comp Funct Genomics 3:158–160. doi:10.1002/cfg.148

Lu A-L, Hsu I-C (1992) Detection of single DNA base mutations with mismatch repair enzymes. Genomics 14:249–255. doi:10.1016/S0888-7543(05)80213-7

Mashal RD, Koontz J, Sklar J (1995) Detection of mutations by cleavage of DNA heteroduplexes with bacteriophage resolvases. Nat Genet 9:177–183. doi:10.1038/ng0295-177

May BP, Liu H, Vollbrecht E, Senior L, Rabinowicz PD, Roh D, Pan X, Stein L, Freeling M, Alexander D, Martienssen R (2003) Maize-targeted mutagenesis: a knockout resource for maize. Proc Natl Acad Sci USA 100:11541–11546. doi:10.1073/pnas.1831119100

McCallum CM, Comai L, Greene EA, Henikoff S (2000) Targeted screening for induced mutations. Nat Biotechnol 18:455–457

Mejlhede N, Kyjovska Z, Backes G, Burhenne K, Rasmussen SK, Jahoor A (2006) EcoTILLING for the identification of allelic variation in the powdery mildew resistance genes mlo and Mla of barley. Plant Breed 125:461–467. doi:10.1111/j.1439-0523.2006.01226.x

Nataraj AJ, Olivos-Glander I, Kusukawa N, Highsmith WE (1999) Single-strand conformation polymorphism and heteroduplex analysis for gel-based mutation detection. Electrophoresis 20:1177–1185. doi:10.1002/(SICI)1522-2683(19990101)20:6<1177::AID-ELPS1177>3.0.CO;2-2

Nieto C, Piron F, Dalmais M, Marco CF, Moriones E, Gómez-Guillamon ML, Truniger V, Gómez P, Garcia-Mas J, Aranda MA, Bendahmane A (2007) EcoTILLING for the identification of allelic variants of melon eIF4E, a factor that controls virus susceptibility. BMC Plant Biol 7:34. doi:10.1186/1471-2229-7-34

Nolan PM, Hugill A, Cox RD (2002) ENU mutagenesis in the mouse: application to human genetic disease. Brief Funct Genomic Proteomic 1:278–289. doi:10.1093/bfgp/1.3.278

Okagaki RJ, Neuffer MG, Wessler SR (1991) A deletion common to two independently derived waxy mutations of maize. Genetics 128:425–431

Oleykowski CA, Mullins CRB, Godwin AK, Yeung AT (1998) Mutation detection using a novel plant endonuclease. Nucleic Acids Res 26:4597–4602. doi:10.1093/nar/26.20.4597

Owais WM, Kleinhofs A (1988) Metabolic activation of the mutagen azide in biological systems. Mutat Res 197:313–323. doi:10.1016/0027-5107(88)90101-7

Palotta MA, Warner P, Fox RL, Kuchel H, Jefferies SP, Langridge P (2003) Marker assisted wheat breeding in the southern region of Australia. In: Proceedings of the tenth international wheat genetics symposium, Paestum, 1–6 September 2003, pp 789–791

Parry MAJ, Madgwick PJ, Bayon C, Tearall K, Hernandez-Lopez A, Baudo M, Rakszegi M, Hamada W, Al-Yassin A, Ouabbou H, Labhilili M, Phillips AL (2009) Mutation discovery for crop improvement. J Exp Bot 60:2817–2825. doi:10.1093/jxb/erp189

Perry JA, Wang TL, Welham TJ, Gardner S, Pike JM, Yoshida S, Parniske M (2003) A TILLING reverse genetics tool and a web-accessible collection of mutants of the legume *Lotus japonicus*. Plant Physiol 131:866–871. doi:10.1104/pp. 102.017384

Raghavan C, Naredo MEB, Wang H, Atienza G, Liu B, Qiu F, McNally KL, Leung H (2006) Rapid method for detecting SNPs on agarose gels and its application in candidate gene mapping. Mol Breed 19:87–101. doi:10.1007/s11032-006-9046-x

Rogers C, Wen JQ, Chen RJ, Oldroyd G (2009) Deletion-based reverse genetics in *Medicago truncatula*. Plant Physiol 151:1077–1086. doi:10.1104/pp. 109.142919

Sestili F, Botticella E, Bedo Z, Phillips A (2010) Production of novel allelic variation for genes involved in starch biosynthesis through mutagenesis. Mol Breed 25:145–154. doi:10.1007/s11032-009-9314-7

Shendure J, Ji HL (2008) Next-generation DNA sequencing. Nat Biotechnol 26:1135–1145. doi:10.1038/nbt1486

Slade AJ, Knauf VC (2005) TILLING moves beyond functional genomics into crop improvement. Transgenic Res 14:109–115. doi:10.1007/s11248-005-2770-x

Slade AJ, Fuerstenberg S, Loeffler D, Steine MN, Facciotti D (2005) A reverse genetic, nontransgenic approach to wheat crop improvement by TILLING. Nat Biotechnol 23:75–81. doi:10.1038/nbt1043

Sreelakshmi Y, Gupta S, Bodanapu R, Chauhan VS, Hanjabam M, Thomas S, Mohan V, Sharma S, Srinivasan R, Sharma R (2010) NEATTILL: a simplified procedure for nucleic acid extraction from arrayed tissue for TILLING and other high-throughput reverse genetic applications. Plant Methods 6:3. doi:10.1186/1746-4811-6-3

Stanssens P, Zabeau M, Meersseman G, Remes G, Gansemans Y, Storm N, Hartmer R, Honisch C, Rodi CP, Böcker S, van den Boom D (2004) High-throughput MALDI-TOF discovery of genomic sequence polymorphisms. Genome Res 14:126–133. doi:10.1101/gr.1692304

Stemple DL (2004) TILLING – a high-throughput harvest for functional genomics. Nat Rev Genet 5:145–150. doi:10.1038/nrg1273

Struhl K, Stinchcomb DT, Scherer S, Davis RW (1979) High-frequency transformation of yeast: autonomous replication of hybrid DNA molecules. Proc Natl Acad Sci USA 76:1035–1039. doi:10.1073/pnas.76.3.1035

Suzuki T, Eiguchi M, Kumamaru T, Satoh H, Matsusaka H, Moriguchi K, Nagato Y, Kurata N (2008) MNU-induced mutant pools and high performance TILLING enable finding of any gene mutation in rice. Mol Genet Genomics 279:213–223. doi:10.1007/s00438-007-0293-2

Talamè V, Bovina R, Sanguineti MC, Tuberosa R, Lundqvist U, Salvi S (2008) TILLMore, a resource for the discovery of chemically induced mutants in barley. Plant Biotechnol J 6: 477–485. doi:10.1111/j.1467-7652.2008.00341.x

Till BJ, Reynolds SH, Greene EA, Codomo CA, Enns LC, Johnson JE, Burtner C, Odden AR, Young K, Taylor NE, Henikoff JG, Comai L, Henikoff S (2003) Large-scale discovery of induced point mutations with high-throughput TILLING. Genome Res 13:524–530. doi:10.1101/gr.977903

Till BJ, Burtner C, Comai L, Henikoff JG (2004a) Mismatch cleavage by single-strand specific nucleases. Nucleic Acids Res 32:2632–2641. doi:10.1093/nar/gkh599

Till BJ, Reynolds SH, Weil C, Springer N, Burtner C, Young K, Bowers E, Codomo CA, Enns LC, Odden AR, Greene EA, Comai L, Henikoff S (2004b) Discovery of induced point mutations in maize genes by TILLING. BMC Plant Biol 4:12. doi:10.1186/1471-2229-4-12

Till BJ, Zerr T, Comai L, Henikoff S (2006) A protocol for TILLING and Ecotilling in plants and animals. Nat Protoc 1:2465–2477. doi:10.1038/nprot.2006.329

Till B, Cooper J, Tai T, Colowit P, Greene E, Henikoff S, Comai L (2007) Discovery of chemically induced mutations in rice by TILLING. BMC Plant Biol 7:19. doi:10.1186/1471-2229-7-19

Triques K, Sturbois B, Gallais S, Dalmais M, Chauvin S, Clepet C, Aubourg S, Rameau C, Caboche M, Bendahmane A (2007) Characterization of *Arabidopsis thaliana* mismatch specific endonucleases: application to mutation discovery by TILLING in pea. Plant J 51:1116–1125. doi:10.1111/j.1365-313X.2007.03201.x

Triques K, Piednoir E, Dalmais M, Schmidt J, Le Signor C, Sharkey M, Caboche M, Sturbois B, Bendahmane A (2008) Mutation detection using ENDO1: application to disease diagnostics in humans and TILLING and Eco-TILLING in plants. BMC Mol Biol 9:42. doi:10.1186/1471-2199-9-42

Uauy C, Paraiso F, Colasuonno P, Tran R, Tsai H, Berardi S, Comai L, Dubcovsky J (2009) A modified TILLING approach to detect induced mutations in tetraploid and hexaploid wheat. BMC Plant Biol 9:115. doi:10.1186/1471-2229-9-115

Wang G-X, Tan M-K, Rakshit S, Saitoh H, Terauchi R, Imaizumi T, Ohsako T, Tominaga T (2007) Discovery of single-nucleotide mutations in acetolactate synthase genes by Ecotilling. Pestic Biochem Physiol 88:143–148. doi:16/j.pestbp.2006.10.006

Wang G-X, Imaizumi T, Li W, Saitoh H, Terauchi R, Ohsako T, Tominaga T (2008a) Self-EcoTILLING to identify single-nucleotide mutations in multigene family. Pestic Biochem Physiol 92:24–29. doi:10.1016/j.pestbp. 2008.05.001

Wang J, Sun JZ, Liu DC, Yang WL, Wang DW, Tong YP, Zhang AM (2008b) Analysis of Pina and Pinb alleles in the micro-core collections of Chinese wheat germplasm by Ecotilling and identification of a novel Pinb allele. J Cereal Sci 48:836–842. doi:10.1016/j.jcs.2008.06.005

Waterhouse PM, Graham MW, Wang M-B (1998) Virus resistance and gene silencing in plants can be induced by simultaneous expression of sense and antisense RNA. Proc Natl Acad Sci USA 95:13959–13964. doi:10.1073/pnas.95.23.13959

Weil CF (2009) TILLING in grass species. Plant Physiol 149:158–164. doi:10.1104/pp. 108.128785

Wienholds E, van Eeden F, Kosters M, Mudde J, Plasterk RHA, Cuppen E (2003) Efficient target-selected mutagenesis in zebrafish. Genome Res 13:2700–2707. doi:10.1101/gr.1725103

Winkler S, Schwabedissen A, Backasch D, Bökel C, Seidel C, Bönisch S, Fürthauer M, Kuhrs A, Cobreros L, Brand M, González-Gaitán M (2005) Target-selected mutant screen by TILLING in Drosophila. Genome Res 15:718–723. doi:10.1101/gr.3721805

Xin Z, Li Wang M, Barkley N, Burow G, Franks C, Pederson G, Burke J (2008) Applying genotyping (TILLING) and phenotyping analyses to elucidate gene function in a chemically induced sorghum mutant population. BMC Plant Biol 8:103. doi:10.1186/1471-2229-8-103

Yang B, Wen X, Kodali NS, Oleykowski CA, Miller CG, Kulinski J, Besack D, Yeung JA, Kowalski D, Yeung AT (2000) Purification, cloning, and characterization of the CEL I nuclease. Biochemistry 39:3533–3541. doi:10.1021/bi992376z

Yeung AT, Hattangadi D, Blakesly L, Nicolas E (2005) Enzymatic mutation detection technologies. Biotechniques 38:749–758. doi:10.2144/05385RV01

Chapter 8
Gene Replacement

Sylvia de Pater and Paul J.J. Hooykaas

Introduction

Genetic modification of plants is now routinely performed. Transformation can be done by various methods and vectors including *Agrobacterium tumefaciens*. It has been observed that transgenes integrate at fairly random positions via non-homologous end-joining (NHEJ) in variable copy numbers in the plant genome. This may cause position effects (such as silencing of transgenes) and mutation of genes at the integration site. Therefore, it would be an advantage, if integration could be targeted to a specific locus. Targeted DNA integration via gene-targeting (GT) by homologous recombination (HR) is efficient in yeast but a very rare event in somatic cells of higher eukaryotes, like animals and plants. The main steps of the NHEJ and HR DNA repair pathways are shown in Fig. 8.1. In animal systems, the use of specific cells has led to significant progress. Using mouse embryonic stem cells, efficient GT methods have been developed and large-scale knockout programmes have already resulted in 9,000 conditional targeted alleles (Skarnes et al. 2011). In plants, the GT frequency varies considerably depending on the plant species. In lower plants such as the moss *Physcomitrella patens*, integration of foreign DNA predominantly occurs via homologous recombination. Stretches of 50–200 bp homology resulted in high GT frequencies (Schaefer 2002). In higher plants, however, integration occurs by non-homologous recombination, even when much larger homologous sequences are used from one to several kb. Estimates of GT frequencies in several plant species vary from 10^{-4} to 10^{-6} (Halfter et al. 1992; Hanin et al. 2001; Hrouda and Paszkowski 1994; Lee et al. 1990; Miao and Lam 1995; Offringa et al. 1990; Paszkowski et al. 1988; Risseeuw et al. 1995). Such low

S. de Pater (✉) • P.J.J. Hooykaas
Department of Molecular and Developmental Genetics, Institute of Biology, Sylvius Laboratory, Leiden University, Sylviusweg 72, 2333 BE Leiden, The Netherlands
e-mail: b.s.de.pater@biology.leidenuniv.nl

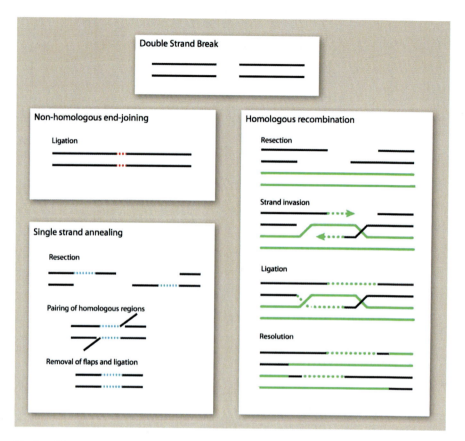

Fig. 8.1 DSB repair pathways. DSBs can be repaired via NHEJ, HR or single strand annealing (SSA). NHEJ does not require any homologous sequence. The ends are directly ligated, but this is often accompanied by small deletions and insertions. HR uses homologous sequences to repair the break. First the ends are resectioned to create 3′ overhangs, followed by strand invasion of the homologous sequence, DNA replication and finally second-end capture and ligation. In the last step the Holliday junctions are dissolved (*non-crossover*; not shown) or resolved (*crossover*). After resection, the single strand ends may contain homologous regions, which may pair, followed by removal of the flaps, gap filling and ligation. This single strand annealing (SSA) pathway will result in deletion of one of the regions of homology and the region in between the homologous sequences

frequencies make gene modification in model plants for answers on fundamental questions and in crop plants for agricultural applications very impractical and not cost effective. For more than two decades, researchers have looked for ways to improve GT. Using strong positive-negative selection in rice, GT events could be enriched (Endo et al. 2007; Johzuka-Hisatomi et al. 2008; Saika et al. 2011; Yamauchi et al. 2009) to an estimated frequency of up to 2% in the surviving calli (Terada et al. 2007). In other plant species such positive-negative selection schemes have not been that successful (Iida and Terada 2005). Two other approaches for developing efficient methods for targeted mutagenesis and gene modification have, therefore, been tested, which will be discussed below. In this chapter, we focus mainly on one of these, for which very promising results have been obtained, i.e., introduction of double strand breaks (DSBs) at the site of the desired recombination event using artificial zinc finger nucleases (ZFNs).

Effects of Mutations in Non-homologous End-Joining DNA Repair Pathways on Gene-Targeting

Agrobacterium tumefaciens is not only able to transform plant cells, but also non-plant hosts like yeast and fungi (Bundock et al. 1995; de Groot et al. 1998; Michielse et al. 2005). DNA repair mutants of the model yeast *Saccharomyces cerevisiae* were used to identify the genes involved in integration of T-DNA. Genes like Ku70, Ku80, and Lig4, which are involved in NHEJ (Van Attikum et al. 2001) and conserved in all eukaryots, are required for integration of T-DNA by non-homologous recombination. Genes involved in homologous recombination, like Rad51 and Rad52, are required for targeted integration of T-DNA in yeast (Van Attikum and Hooykaas 2003). The same genes are responsible for chromosomal integration of genes that are introduced by other methods of transformation. Inactivation of NHEJ prevents stable transformation, unless T-DNA shows homology to the yeast genome. In this case, integration occurs exclusively by HR. A generic method for GT in yeast and fungi was developed by transient or stable inactivation of the NHEJ pathway (Hooykaas et al. 2001). Proof of principle was obtained in *Kluyveromyces lactis* (Kooistra et al. 2004) and *Neurospora crassa* (Ninomiya et al. 2004), where extremely high (80–100%) GT frequencies were obtained in NHEJ mutants. Since this approach has been very successful, it has been applied to other fungal species (Bhadauria et al. 2009; de Boer et al. 2010; Meyer et al. 2007).

Since GT was significantly increased in NHEJ mutants in yeast and fungi, we tested several NHEJ mutants of *Arabidopsis thaliana* to study their GT frequency. The transformation frequency using the floral dip method dropped significantly (5–10 fold) in such Arabidopsis mutants (Jia et al. 2012). However, in these NHEJ mutants, integration of T-DNA still occurred at mostly random positions. Thus, alternative (back-up) end-joining pathways must be functional in NHEJ mutants of plants. In mammals, back-up (B)-NHEJ pathways have been identified, which become active in the absence of the core NHEJ factors, such as Ku and Lig4 (Nussenzweig and Nussenzweig 2007). Factors required for B-NHEJ are among others PARP1, Lig3, and XRCC1 (Audebert et al. 2004). Orthologs for Parp1 and Xrcc1 have been identified in Arabidopsis and mutants have been tested in floral dip transformation. No significant change in transformation frequency was observed in the *parp1parp2* double mutant (Jia 2011) or the *xrcc1* mutant (de Pater, unpublished results). When both DNA repair pathways were inactivated in the *parp1parp2ku80* triple mutant, the transformation frequency was further decreased, but the integration still occurred at random positions, suggesting that a third end-joining pathway is active in this triple mutant. Recently, similar results were reported with the *ku80xrcc1* mutant, which still showed DNA repair, although at a slow speed (Charbonnel et al. 2010). Inactivation of lig4 and lig3 in mammals also did not completely abolish end-joining, indicating that a third ligase may be employed for ligation in an alternative end-joining pathway (Simsek et al. 2011). In conclusion, GT frequency was not significantly improved in these different Arabidopsis end-joining mutants.

An Arabidopsis T-DNA insertion mutant of Mre11, a protein involved in several DNA repair pathways (Stracker and Petrini 2011), has also been tested for transformation and GT frequencies. This resulted in 5-fold lower transformation frequency and three GT events in about 3,600 transformants, whereas no GT events were found in about 11,000 wild-type transformants (Jia et al. 2012). Thus, mutation of Mre11 seems to be a way to improve GT. The *mre11* mutant that was used for these GT experiments had a mild phenotype. However, other *mre11* mutants often have severe phenotypes and are often sterile (Bundock and Hooykaas 2002; de Pater, unpublished results), making them unsuitable for GT experiments via floral dip transformation.

Besides the inactivation of the end-joining pathways, stimulation of the HR pathway may decrease the relative contribution of end-joining to DNA repair and integration. Overexpression of yRAD54, involved in chromatin remodelling during D-loop formation indeed stimulated GT in Arabidopsis (Shaked et al. 2005).

Double Strand Break-Mediated Mutagenesis and Gene-Targeting

Using artificial constructs, introduction of a targeted DNA double strand break (DSB) near the site of the desired recombination event dramatically improved the frequency of GT in animals (Choulika et al. 1995; Jasin 1996; Rouet et al. 1994) and in plants (Puchta et al. 1996). Methods to introduce DSBs at the sites of interest in the genome were lacking until more recently, when artificial nucleases were developed. Very promising results have been obtained with such artificial nucleases that can induce DSBs at any selected site, as will be discussed in the next paragraphs.

Eukaryotic cells repair DSBs primarily via two mechanisms: NHEJ and HR (Fig. 8.1). In animals and plants, DSBs are mainly repaired via NHEJ, for which no homologous sequences are required. The balance between NHEJ and HR shifts during the cell cycle (Hiom 2010). In the G1 phase and the beginning of the S phase DSBs will mainly be repaired via NHEJ, whereas during S and G2 phases of the cell cycle HR is upregulated, when sister chromatids are available. The choice of DNA repair pathway may be regulated by one or more proteins that act in both pathways, including the damage response factor ATM (Shrivastav et al. 2008). This choice will determine the outcome of the repair. During NHEJ the majority of compatible ends are precisely ligated but the presence of incompatible ends leads to small deletions or insertions, so-called indels. Even with compatible ends this will be the case when the nuclease remains present and precise repair leads to restoration of the recognition site, which will be cut again until imprecise repair results in the mutation of the recognition site. There may also be species differences in the NHEJ pathway: in tobacco, protoplasts joining without sequence alterations were shown to be rare (Gorbunova and Levy 1997). HR will lead to precise restoration by using the sister chromatid as a template for repair. By using an introduced transgene as a template for repair, targeted integration can be accomplished.

Required steps for DSB-induced mutagenesis and GT are shown in Fig. 8.2 and will be discussed below.

Site-Specific Double Strand Breaks

Site-specific DSBs can be introduced by endonucleases, which have long and, therefore, rare target recognition sites in eukaryotic genomes. These include the naturally occurring homing endonucleases or meganucleases, such as I-*Sce*I (Stoddard 2011). I-*Sce*I, encoded by a mitochondrial intron of *Saccharomyces cerevisiae*, helps to convert intronless alleles into alleles with an intron. Meganucleases (MNs) recognize long DNA stretches (12–40 bp) and cleave double-stranded DNA with high specificity in the presence of divalent metal ions. In plants, this was demonstrated by using artificially introduced gene constructs, for the first time through the use of the rare cutting meganuclease I-*Sce*I, resulting in an increase in GT frequency by two orders of magnitude (Puchta et al. 1996). Though GT frequencies can be increased considerably by the use of the natural homing endonucleases, its application was so far limited to artificially introduced target sites. Recently, however, a combinatorial approach was reported to redesign homing endonucleases to match with target sites that are naturally present in the genome (Arnould et al. 2011; Grizot et al. 2010). It is to be expected that such collections of meganucleases will expand their use in the near future.

Zinc finger nucleases (ZFNs) are artificial nucleases, which emerged as the tools of choice to create DSBs at any desired site in the genome (Klug 2010). The current generation of ZFNs combines the nonspecific cleavage nuclease domain of the *Fok*I restriction enzyme with a specific DNA binding domain with several C_2H_2 zinc fingers (ZFs) to provide cleavage specificity. Each individual ZF present within a polydactyl zinc finger domain consists of a stretch of $\sim$30 amino acids, stabilized by a zinc ion, which binds a particular three-base DNA sequence (triplet). Efficient cleavage of the target site requires dimerization of the *Fok*I cleavage domain (Bitinaite et al. 1998; Mani et al. 2005; Smith et al. 2000). Therefore, two ZFN subunits are typically designed to recognize the target sequence in a tail-to-tail conformation and the DSB is then introduced within a 4–7 base pair DNA sequence, which is located in between the binding sites of the ZFN subunits. Series of ZF modules have been created for recognition of most of the 64 possible triplets (Blancafort et al. 2003; Dreier et al. 2000, 2001, 2005; Liu et al. 2002; Segal et al. 1999). In general, 3–6 ZFs are linked together, thereby obtaining a recognition

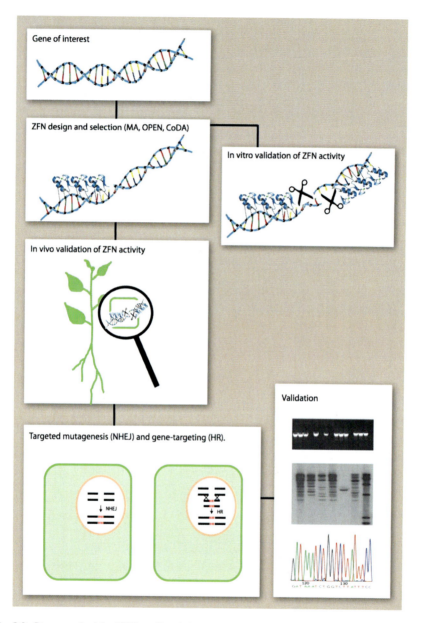

Fig. 8.2 Steps required for ZFN-mediated site-specific mutagenesis and GT. After identification of a suitable target site in the gene of interest, ZF domains can be selected via one of the described methods (modular assembly, OPEN selection or CoDA). Subsequently, the ZF domains are linked to (obligate heterodimer) *Fok*I nuclease domains. The resulting ZFNs can be tested *in vitro* for nuclease activity on their target sites present on plasmids or DNA fragments. *In vivo* activity on chromosomal target sites may result in small deletions or insertions after inaccurate repair via NHEJ (targeted mutagenesis). When a repair template is provided together with the ZFNs (or in two consecutive steps), GT may occur resulting in replacement of endogenous sequences. The validation can be done by PCR, Southern blotting and sequencing

sequence for the ZFN pair of 18–36 bps. ZFNs have been used for targeted mutagenesis, targeted integration and GT for a wide variety of animal species, like *Drosophila* (Beumer et al. 2006; Bibikova et al. 2002), *Xenopus* (Bibikova et al. 2001), *Caenorhabditis elegans* (Morton et al. 2006), zebrafish (Doyon et al. 2008; Meng et al. 2008), rat (Jacob et al. 2010; Rémy et al. 2010), hamster (Santiago et al. 2008) and human cells (Herrmann et al. 2011; Lombardo et al. 2007; Moehle et al. 2007; Urnov et al. 2005) and plants (see below), showing that it is possible to obtain active ZFNs for many target sequences and genes. However, not all interactions are robust enough for efficient binding (Ramirez et al. 2008). Especially for T-rich sequences, it is hard to obtain active ZFNs (Sander et al. 2010a).

Design and Structure of ZFNs

The efficiency of ZFNs for DSB formation depends on the three domains present in the ZFN: the ZF DNA binding domain, the nuclease domain and the linker that connects the two. Important determinants of these domains are the binding affinity of the ZFs to the DNA, dimerization and nuclease activity of the nuclease domain and the length and sequence of the linker. Each of these three domains needs to be optimized for high activity of the ZFNs.

The C_2H_2 zinc finger domain consists of an antiparallel β-sheet, which contains the two cysteine residues and an α-helix containing the two histidine residues. These two structural units form a finger-like structure in the presence of a Zn ion. The α-helix of the ZF recognizes a DNA triplet and the specificity lies in the amino acids at positions $-1, 3,$ and 6 (Klug 2010). In polydactyl ZFNs, the situation is, however, more complex as the amino acid at position 2 interacts with the neighbouring base pair of the adjacent triplet (Isalan et al. 1997). Therefore, the construction of polydactyl ZF DNA binding domains via modular assembly (Beerli and Barbas 2002; Wright et al. 2006; Bhakta and Segal 2010), is not always successful. In order to increase the success rate for obtaining more optimal polydactyl ZFs, these may be selected from combinatorial libraries on the basis of their DNA binding in vivo (Maeder et al. 2008). In the Oligomerized Pool ENgineering (OPEN) method, libraries of 3 ZF modules each binding a triplet are used for selection of binding to the target half sites. Recently, a platform for Context Dependent Assembly (CoDA) was developed for assembling three-finger arrays using N- and C-terminal fingers that were previously identified in other arrays containing a common middle finger (Sander et al. 2010b).

A target site thus consists of 2×9 bp, which may not be completely unique especially in organisms with large genomes. Longer DNA binding domains with 4–6 ZF modules have, therefore, also been employed (Cai et al. 2009; de Pater et al. 2009; Shukla et al. 2009) . Neuteboom et al. (2006) reported high binding activity only with 6 ZF modules, whereas Shimizu et al. (2011) found that increasing the number of ZF modules may reduce the activity of the ZFNs. These contradictory observations probably resulted from the fact that Shimizu et al. (2011)

performed *in vitro* experiments on plasmid DNA and Neuteboom et al. (2006) used chromosomal chromatin-embedded DNA sequences in the yeast *Saccharomyces cerevisiae*. An important issue is the specificity and the toxicity of ZFNs. If ZFNs are not extremely specific, and can cut the genome at off-target sites, they will cause instability of the genome and be toxic (Cornu et al. 2008). This can happen more easily when the two ZFNs not only form heterodimers, but also form homodimers. To reduce toxicity, the specificity of ZFNs needs to be improved by optimizing the design of the ZFNs' structure or regulating the protein level and duration of the presence of ZFNs in the target cells (Pruett-Miller et al. 2009). Information on ZFNs, including programs for target site selection and ZFN design, can be found on the websites of the Zinc Finger Consortium, a consortium of several researchers in the ZFN field (www.zincfingers.org) and the Barbas lab (www.zincfingertools.org).

As mentioned above, activity of the nuclease domain of the *Fok*I enzyme requires dimerization. For cleavage of non-palindromic target sites two different ZFNs are necessary to heterodimerize to form a catalytically active nuclease complex. However, two different ZFNs with the wild-type *Fok*I nuclease domains can also form homodimers, which may create DSBs at off target sites, which may be toxic for the cell. In order to prevent undesired homodimerization, the ZFN dimer interface has been engineered to produce *Fok*I nuclease domains that function as obligate heterodimers and thus increase specificity (Miller et al. 2007; Szczepek et al. 2007). The first generation of these obligate heterodimer *Fok*I domains exhibited a reduced cutting rate. Via an *in vivo* evolution-based approach (Guo et al. 2010) or a combination of cold-sensitive mutations and rational design (Doyon et al. 2010), improved obligate heterodimeric *Fok*I domains were obtained, that have high activity. Further engineering resulted in autonomous ZFN pairs that can be directed simultaneously to two different sites to induce targeted chromosomal deletions (Şöllü et al. 2010).

The ZF DNA binding domain and the *Fok*I nuclease domain are linked by a short stretch of 4–20 amino acids. The length and amino acid composition of this linker is also an important determinant of the activity of the ZFNs. Longer linker sequences allowed the ZFNs to induce DSBs on target sites with longer spacers between the ZF recognition half sites. With linkers of 4–8 amino acids efficient cleavage was seen at target sites with 6 bp spacers, whereas linkers of 9–16 amino acids enabled ZFN activity on target sites with 7 or 8 bp spacers (Händel et al. 2009).

Upon transformation, equimolar amounts of both ZFN partners of the heterodimer need to be produced. For coordinated expression of both ZFNs the open reading frames can be cloned under control of one promoter, whereby the open reading frames are linked by the 2A ribosomal stuttering signal (de Felipe et al. 2006). Another possibility is the construction of single-chain ZFNs. Such ZFNs have been constructed with two *Fok*I nuclease domains separated by a flexible protein linker (Mino et al. 2009; Minczuk et al. 2008). This approach extended the number of target sites in the genome and was successful for mitochondrial sequences for which no conventional ZFN pair could be found (Minczuk et al. 2008).

In Vitro and In Vivo Activity of ZFNs

Whether constructed by modular assembly or selected by the OPEN method, novel ZFNs have to be validated before they are used for mutagenesis or GT experiments. Several *in vitro* and *in vivo* assays have been developed to assess the binding and nuclease activity. As part as the OPEN procedure, binding of ZF arrays to target half sites is being tested in bacterial two hybrid assays (Maeder et al. 2009). Alternatively, purified ZF proteins have been used for *in vitro* binding assays (Mino et al. 2009) and *in vitro* cleavage assays (Guo et al. 2010; Szczepek et al. 2007). Assays for assessing nuclease activity are often performed with proteins synthesized *in vitro* (Cathomen and Şöllü 2010). *In vivo* cleavage of DNA by ZFN pairs can be indirectly assayed by analysis of footprints (indels), resulting from imperfect repair of DSBs via NHEJ. Footprints can be detected via the surveyor nuclease mismatch assay (Guschin et al. 2010) or via loss of a restriction enzyme recognition sequence (Hoshaw et al. 2010). For easy detection of ZFN activity, reporter constructs to detect cleavage of plasmid-derived or chromosomal target sites *in vivo* have been developed (Tovkach et al. 2009; Zhang et al. 2010). These reporter constructs are based on repair via NHEJ, resulting in reconstitution of a mutated reporter gene or repair via single strand annealing (SSA) (Fig. 8.1) of two repeats flanking the target site.

ZFN-Mediated Gene Modification in Plants

For site-directed mutagenesis (SDM) in animals and plants, one can exploit the fact that DSBs are mainly repaired via NHEJ, which is intrinsically error prone. When active ZFNs are present, the cycle of cutting and repairing the ZFN target site continues until an imperfect NHEJ-mediated repair event results in a mutation or footprint within the target site, which prevents recognition and subsequent cleavage by the ZFNs. The mutations that are mainly obtained in this way are small deletions, sometimes accompanied by insertions. Such a NHEJ-based mutagenesis strategy was developed in Drosophila (Bibikova et al. 2002) and was also shown to be an efficient mutagenesis method not only in the model plants Arabidopsis (Lloyd et al. 2005; Osakabe et al. 2010; Zhang et al. 2010; de Pater et al. 2009) and tobacco (Cai et al. 2009; Maeder et al. 2008; Tovkach et al. 2009; Townsend et al. 2009) but also in crop plants, like maize (Shukla et al. 2009) and soybean (Curtin et al. 2011).

ZFN-mediated GT was demonstrated in plants initially by precise repair of defective reporter genes (Cai et al. 2009; Wright et al. 2005) or by exchange of two different reporter genes (de Pater et al. 2009). Later, it was demonstrated that ZFN technology can be instrumental for HR-mediated modification of endogenous genes in maize (Shukla et al. 2009) and tobacco (Townsend et al. 2009).

In most studies, the GT repair constructs and ZFNs expression constructs were co-delivered to cell suspensions (Cai et al. 2009; Shukla et al. 2009) or protoplasts (Townsend et al. 2009; Wright et al. 2006) via a variety of direct DNA transformation procedures. DNA transfer via the widely used bacterial vector *Agrobacterium tumefaciens*, was also successful for tobacco cell cultures (Cai et al. 2009) and Arabidopsis plants (de Pater et al. 2009; Zhang et al. 2010). ZFN genes that are stably integrated in plant genomes can be removed via segregation after the required modifications have been introduced.

Unfortunately, cells with site-specific mutations or derived plants are considered to be genetically modified organisms (GMOs) in Europe even if introduced transgenes have been lost from the plant genome. Such a "GMO label" increases the administrative burden in the regulation processes needed for marketing of the products, and eventual acceptance is far from certain. Therefore, it would be a great advantage if the required MNs or ZFNs, could be introduced directly as proteins, so that the plant is not classified as GMO. This can be achieved by protein transfer via the *Agrobacterium* type four secretion system (T4SS), which transfers not only DNA, but also a range of specific effector proteins into recipient cells. Transfer of these effector proteins occurs independently of transfer of the T-DNA (Vergunst et al. 2000). Heterologous proteins can be translocated into plants cells when they are N-terminally fused to one of the Vir proteins or the C-terminal part thereof, containing the T4SS transfer signal (Vergunst et al. 2000, 2005). Recently, our laboratory showed that homing endonucleases and ZFNs can be introduced into recipient plant cells via the T4SS (van Kregten 2011). This opens possibilities for non-transgenic methods of genome modification. The use of viral RNA vectors for transient expression of ZFNs was also successful (Marton et al. 2010). This method was presented as a non-transgenic method for genome modification of plant cells. However, to date is not clear whether this latter method will indeed be classified as non-GMO, as it is based on the introduction of a nucleic acid.

Validation of Introduced Modifications

After introduction and expression of ZFNs, the presence of the desired gene modifications at the target site in the genome needs to be verified. The surveyor nuclease mismatch assay or the analysis of the loss of a restriction enzyme recognition sequence, are often used for the initial analysis of the modified DNA locus. However, when less than 2% of the sequences contain footprints or no convenient restriction site is present in the target locus, other methods are needed. One of the strategies that will give the best results is deep sequencing of PCR products of the region surrounding the target site by next generation sequencing (NGS). This method has been used by Shukla et al. (2009) and is expected to be the method of choice due to further decreasing costs for NGS in the future.

Evaluation of GT events mostly has three stages: selection via repair of reporter constructs, PCR analysis, and finally Southern blotting. This last method is necessary in order to determine whether HR has occurred at the endogenous locus and to find out whether extra integrations have occurred. For instance, Wright et al. (2005) showed that of the 12 plants analysed, one represented a clean GT event without any additional integration of T-DNA. Two other GT events had additional random T-DNA integrations and in the rest of the plants integration probably had occurred via NHEJ. Shukla et al. (2009) obtained a high frequency of targeted integration without random integration after introduction of ZFN-mediated DSBs. De Pater et al. (2009) obtained two so-called true GT events, but these plants contained extra randomly integrated T-DNAs. However, in the next generation homozygous plants were obtained of which one had lost the extra copies of the repair construct and the ZFN-containing T-DNA.

Conclusions and Future Prospects

ZFN technology has been successfully applied for precise engineering of plant genomes. GT frequencies have increased considerably through induction of DSBs. However, knowledge on ZFN specificity and toxicity, methods for delivery of ZFNs and repair constructs, and ZFN expression, need to be further extended and optimized in order to obtain reliable and highly efficient methods for targeting native sequences in model and crop species. Since the majority of the desirable mutations or GT events will lead to non-selectable genome changes, high throughput DNA isolation combined with NGS will be the methods of choice to identify these modifications.

Recently, a novel DNA binding domain, derived from transcription-activator like effectors (TALEs), was used in combination with the *Fok*I nuclease domain as artificial rare cutting enzymes for gene modification. TALE proteins are naturally produced by plant pathogens and transported to the plant via the type three secretion system (T3SS), where they contribute to disease or trigger resistance by binding to DNA and turning on TALE-specific host genes (Bogdanove et al. 2010). The TALE DNA binding domain consists of up to 30 nearly identical repeats of ~33–35 amino acids, each binding to a single nucleotide. How these repeats determine binding specificity was recently elucidated (Boch et al. 2009). TALE nucleases (TALENs) have been successfully used for gene modification in human cells (Cermak et al. 2011; Zhang et al. 2011), yeast (Christian et al. 2010; Li et al. 2011) and tobacco leaves (Mahfouz et al. 2011). The use of these TALEN proteins opens up possibilities for designing and selecting proteins that recognize DNA targets for which so far no active ZFNs could be obtained, since TALENs seem to have fewer limitations for target site sequences compared to ZFNs.

References

Arnould S, Delenda C, Grizot S, Desseaux C, Pâques F, Silva GH, Smith J (2011) The I-CreI meganuclease and its engineered derivatives: applications from cell modification to gene therapy. Protein Eng Des Sel 24:27–31

Audebert M, Salles B, Calsou P (2004) Involvement of poly(ADP-ribose) polymerase-1 and XRCC1/DNA ligase III in an alternative route for DNA double-strand breaks rejoining. J Biol Chem 279:55117–55126

Beerli RR, Barbas CF (2002) Engineering polydactyl zinc-finger transcription factors. Nat Biotechnol 20:135–141

Beumer K, Bhattacharyya G, Bibikova M, Trautman JK, Carroll D (2006) Efficient gene targeting in Drosophila with zinc-finger nucleases. Genetics 172:2391–2403

Bhadauria V, Banniza S, Wei Y, Peng YL (2009) Reverse genetics for functional genomics of phytopathogenic fungi and oomycetes. Comp Funct Genom. doi:10.1155/2009/380719

Bhakta MS, Segal DJ (2010) The generation of zinc finger proteins by modular assembly. In: Mackay JP, Segal DJ (eds) Engineered zinc finger proteins, Methods of molecular biology 649. Springer, New York, pp 3–30

Bibikova M, Carroll D, Segal DJ, Trautman JK, Smith J, Kim YG, Chandrasegaran S (2001) Stimulation of homologous recombination through targeted cleavage by chimeric nucleases. Mol Cell Biol 21:289–297

Bibikova M, Golic M, Golic KG, Carroll D (2002) Targeted chromosomal cleavage and mutagenesis in Drosophila using zinc-finger nucleases. Genetics 161:1169–1175

Bitinaite J, Wah DA, Aggarwal AK, Schildkraut I (1998) *FokI* dimerization is required for DNA cleavage. Proc Natl Acad Sci USA 95:10570–10575

Blancafort P, Magnenat L, Barbas CF (2003) Scanning the human genome with combinatorial transcription factor libraries. Nat Biotechnol 21:269–274

Boch J, Scholze H, Schornack S, Landgraf A, Hahn S, Kay S, Lahaye T, Nickstadt A, Bonas U (2009) Breaking the code of DNA binding specificity of TAL-type III effectors. Science 326:1509–1512

Bogdanove AJ, Schornack S, Lahaye T (2010) TAL effectors: finding plant genes for disease and defense. Curr Opin Plant Biol 13:394–401

Bundock P, Hooykaas P (2002) Severe developmental defects, hypersensitivity to DNA-damaging agents, and lengthened telomeres in Arabidopsis MRE11 mutants. Plant Cell 14:2451–2462

Bundock P, den Dulk-Ras A, Beijersbergen A, Hooykaas PJJ (1995) Trans-kingdom T-DNA transfer from *Agrobacterium tumefaciens* to *Saccharomyces cerevisiae*. EMBO J 14:3206–3214

Cai CQ, Doyon Y, Ainley WM, Miller JC, DeKelver RC, Moehle EA, Rock JM, Lee Y-L, Garrison R, Schulenberg L, Blue R, Worden A, Baker L, Faraji F, Zhang L, Holmes MC, Rebar EJ, Collingwood TN, Rubin-Wilson B, Gregory PD, Urnov FD, Petolino JF (2009) Targeted transgene integration in plant cells using designed zinc finger nucleases. Plant Mol Biol 69:699–709

Cathomen T, Şöllü C (2010) In vitro assessment of zinc finger nuclease activity. In: Mackay JP, Segal DJ (eds) Engineered zinc finger proteins, Methods in molecular biology 649. Springer, New York, pp 227–235

Cermak T, Doyle EL, Christian M, Wang L, Zhang Y, Schmidt C, Baller JA, Somia NV, Bogdanove AJ, Voytas DF (2011) Efficient design and assembly of custom TALEN and other TAL effector-based constructs for DNA targeting. Nucl Acid Res. doi:10.1093/nar/gkr218

Charbonnel C, Gallego ME, White CI (2010) Xrcc1-dependent and Ku-dependent DNA double-strand break repair kinetics in Arabidopsis plants. Plant J 64:280–290

Choulika A, Perrin A, Dujon B, Nicolas J-F (1995) Induction of Homologous recombination in mammalian chromosomes by using the I-SceI system of *Saccharomyces cerevisiae*. Mol Cell Biol 15:1968–1973

Christian M, Cermak T, Doyle EL, Schmidt C, Zhang F, Hummel A, Bogdanove AJ, Voytas DF (2010) Targeted DNA double-strand breaks with TAL effector nucleases. Genetics 186:757–761

Cornu TI, Thibodeau-Begganny S, Guhl E, Alwin S, Eichtinger M, Joung JK, Cathomen T (2008) DNA-binding specificity is a major determinant of the activity and toxicity of zinc-finger nucleases. Mol Ther 16:352–358

Curtin SJ, Zhang F, Sander JD, Haun WJ, Starker C, Baltes NJ, Reyon D, Dahlborg EJ, Goodwin MJ, Coffman AP, Dobbs D, Joung JK, Voytas DF, Stupar RM (2011) Targeted mutagenesis of duplicated genes in soybean with zinc-finger nucleases. Plant Phys 156:466–473

De Boer P, Bastiaans J, Touw H, Kerkman R, Bronkhof J, van den Berg M, Offringa R (2010) Highly efficient gene targeting in *Penicillium chrysogenum* using the bi-partite approach in Δ*lig4* or Δ*ku70* mutants. Fung Genet Biol 47:839–846

De Felipe P, Luke GA, Hughes LE, Gani D, Halpin C, Ryan MD (2006) E unum pluribus: multiple proteins from a self-processing polyprotein. Trends Biotechnol 24:68–75

De Groot MJA, Bundock P, Beijersbergen AGM, Hooykaas PJJ (1998) *Agrobacterium tumefaciens*-mediated transformation of filamentous fungi. Nat Biotechnol 16:839–842

De Pater S, Neuteboom LW, Pinas JE, Hooykaas PJJ, van der Zaal BJ (2009) ZFN-induced mutagenesis and gene-targeting in Arabidopsis through *Agrobacterium*-mediated floral dip transformation. Plant Biotechnol J 7:821–835

Doyon Y, McCammon JM, Miller JC, Faraji F, Ngo C, Katibah GE, Amora R, Hocking TD, Zhang L, Rebar EJ, Gregory PD, Urnov FD, Amacher SL (2008) Heritable targeted gene disruption in zebrafish using designed zinc-finger nucleases. Nat Biotechnol 26:702–708

Doyon Y, Vo TD, Mendel MC, Greenberg SG, Wang J, Xia DF, Miller JC, Urnov FD, Gregory PD, Holmes MC (2010) Enhancing zinc-finger-nuclease activity with improved obligate heterodimeric architectures. Nat Method 8:74–79

Dreier B, Segal DJ, Barbas CF (2000) Insights into the molecular recognition of the 5′-GNN-3′ family of DNA sequences by zinc finger domains. J Mol Biol 303:489–502

Dreier B, Beerli RR, Segal DJ, Flippin JD, Barbas CF (2001) Development of zinc finger domains for recognition of the 5′-ANN-3′ family of DNA sequences and their use in the construction of artificial transcription factors. J Biol Chem 276:29466–29478

Dreier B, Fuller RP, Segal DJ, Lund CV, Blancafort P, Huber A, Koksch B, Barbas CF (2005) Development of zinc finger domains for recognition of the 5′-CNN-3′ family DNA sequences and their use in the construction of artificial transcription factors. J Biol Chem 280: 35588–35597

Endo M, Osakabe K, Ono K, Handa H, Shimizu T, Toki S (2007) Molecular breeding of a novel herbicide-tolerant rice by gene targeting. Plant J 52:157–166

Gorbunova V, Levy AA (1997) Non-homologous DNA end joining in plant cells is associated with deletions and filler DNA insertions. Nucl Acids Res 25:4650–4657

Grizot S, Epinat J-C, Thomas S, Duclert A, Rolland S, Pâques F, Duchateau P (2010) Generation of redesigned homing endonucleases comprising DNA-binding domains derived from two different scaffolds. Nucl Acids Res 38:2006–2018

Guo J, Gaj T, Barbas CF (2010) Directed evolution of an enhanced and highly efficient FokI cleavage domain for zinc finger nucleases. J Mol Biol 400:96–107

Guschin DY, Waite AJ, Katibah GE, Miller JC, Holmes MC, Rebar EJ (2010) A rapid and general assay for monitoring endogenous gene modification. In: Mackay JP, Segal DJ (eds) Engineered zinc finger proteins, Methods in molecular biology 649. Springer, New York, pp 247–256

Halfter U, Morris PC, Willmitzer L (1992) Gene targeting in *Arabidopsis thaliana*. Mol Gen Genet 231:184–193

Händel E-M, Alwin S, Cathomen T (2009) Expanding or restricting the target site repertoire of zinc-finger nucleases: the inter-domain linker as a major determinant of target site selectivity. Mol Ther 17:104–111

Hanin M, Volrath S, Bogucki A, Briker M, Ward E, Paszkowski J (2001) Gene targeting in *Arabidopsis*. Plant J 28:671–677

Herrmann F, Garriga-Canut M, Baumstark R, Fajardo-Sanchez E, Cotterell J, Minoche A, Himmelbauer H, Isalan M (2011) p53 gene repair with zinc finger nucleases optimised by yeast 1-hybrid and validated by Solexa sequencing. PLoS One 6:e20913

Hiom K (2010) Coping with DNA double strand breaks. DNA Repair 9:1256–1263

Hooykaas PJJ, van Attikum H, Bundock P (2001) Nucleic acid integration in eukaryotes. Patent EP 00204693.6

Hoshaw JP, Unger-Wallace E, Zhang F, Voytas DF (2010) A transient assay for monitoring zinc finger nuclease activity at endogenous plant gene targets. In: Mackay JP, Segal DJ (eds) Engineered zinc finger proteins, Methods in molecular biology 649. Springer, New York, pp 299–314

Hrouda M, Paszkowski J (1994) High fidelity extrachromosomal recombination and gene targeting in plants. Mol Gen Genet 243:106–111

Iida S, Terada R (2005) Modification of endogenous natural genes by gene targeting in rice and other higher plants. Plant Mol Biol 59:205–219

Isalan M, Choo Y, Klug A (1997) Synergy between adjacent zinc fingers in sequence-specific DNA recognition. Proc Natl Acad Sci USA 94:5617–5621

Jacob HJ, Lazar J, Dwinell MR, Moreno C, Geurts AM (2010) Gene targeting in the rat: advances and opportunities. Trends Genet 26:510–518

Jasin M (1996) Genetic manipulation of genomes with rare-cutting endonucleases. Trends Genet 12:224–228

Jia Q (2011) DNA repair and gene-targeting in plant end-joining mutants. Dissertation, Leiden University

Jia Q, Bundock P, Hooykaas PJJ, de Pater S (2012) *Agrobacterium tumefaciens* T-DNA integration and gene targeting in *Arabidopsis thaliana* non-homologous end-joining mutants. J Bot. doi:10.1155/2012/989272

Johzuka-Hisatomi Y, Terada R, Iida S (2008) Efficient transfer of base changes from a vector to the rice genome by homologous recombination: involvement of heteroduplex formation and mismatch correction. Nucl Acids Res 36:4727–4735

Klug A (2010) The discovery of zinc fingers and their applications in gene regulation and genome manipulation. Ann Rev Biochem 79:213–231

Kooistra R, Hooykaas PJJ, Steensma HY (2004) Efficient gene targeting in *Kluyveromyces lactis*. Yeast 21:781–792

Lee KY, Lund P, Lowe K, Dunsmuir P (1990) Homologous recombination in plant cells after *Agrobacterium*-mediated transformation. Plant Cell 2:415–425

Li T, Huang S, Zhao X, Wright DA, Carpenter S, Spalding MH, Weeks DP, Yang B (2011) Modularly assembled designer TAL effector nucleases for targeted gene knockout and gene replacement in eukaryotes. Nucl Acids Res. doi:10.1093/nar/gkr188

Liu Q, Xia Z, Case CC (2002) Validated zinc finger protein designs for all 16 GNN DNA triplet targets. J Biol Chem 277:3850–3856

Lloyd A, Plaisier CL, Carroll D, Drews GN (2005) Targeted mutagenesis using zinc-finger nucleases in Arabidopsis. Proc Natl Acad Sci USA 102:2232–2237

Lombardo A, Genovese P, Beausejour CM, Colleoni S, Lee Y-L, Kim KA, Ando D, Urnov FD, Galli C, Gregory PD, Holmes MC, Naldini L (2007) Gene editing in human stem cells using zinc finger nucleases and integrase-defective lentiviral vector delivery. Nat Biotechnol 25:1298–1306

Maeder ML, Thibodeau-Beganny S, Osiak A, Wright DA, Reshma MA, Eichtinger M, Jiang T, Foley JE, Winfrey RJ, Townsend JA, Unger-Wallace E, Sander JD, Müller-Lerch F, Fu F, Pearlberg J, Göbel C, Dassie JP, Pruett-Miller SM, Porteus MH, Sgroi DC, Iafrate AJ, Dobbs D, McCray PB Jr, Cathomen T, Voytas DF, Joung JK (2008) Rapid "open-source" engineering of customized zinc-finger nucleases for highly efficient gene modification. Mol Cell 31:294–301

Maeder ML, Thibodeau-Beganny S, Sander JD, Voytas DF, Joung JK (2009) Oligomerized pool engineering (OPEN): an 'open-source' protocol for making customized zinc-finger arrays. Nat Prot 4:1471–1501

Mahfouz MM, Li L, Shamimuzzaman M, Wibowo A, Fang X, Zhu J-K (2011) De novo-engineered transcription activator-like effector (TALE) hybrid nuclease with novel DNA binding specificity creates double-strand breaks. Proc Natl Acad Sci USA 108:2623–2628

Mani M, Smith J, Kandavelou K, Berg JM, Chandrasegaran S (2005) Binding of two zinc finger nuclease monomers to two specific sites is required for effective double-strand DNA cleavage. Biochem Biophys Res Commun 334:1191–1197

Marton I, Zuker A, Shklarman E, Zeevi V, Tovkach A, Roffe S, Ovadis M, Tzfira T, Vainstein A (2010) Nontransgenic genome modification in plant cells. Plant Phys 154:1079–1087

Meng X, Noyes MB, Zhu LJ, Lawson ND, Wolfe SA (2008) Targeted gene inactivation in zebrafish using engineered zinc-finger nucleases. Nat Biotechnol 26:695–701

Meyer V, Arentshorst M, El-Ghezal A, Drews A-C, Kooistra R, van den Hondel CAMJJ, Ram AFJ (2007) Highly efficient gene targeting in the *Aspergillus niger kusA* mutant. J Biotechnol 128:770–775

Miao ZH, Lam E (1995) Targeted disruption of the TGA3 locus in *Arabidopsis thaliana*. Plant J 7:359–365

Michielse CB, Hooykaas PJJ, van den Hondel CAMJJ, Ram AFJ (2005) *Agrobacterium*-mediated transformation as a tool for functional genomics in fungi. Curr Genet 48:1–17

Miller JC, Holmes MC, Wang J, Guschin DY, Lee Y-L, Rupniewski I, Beausejour CM, Waite AJ, Wang NS, Kim KA, Gregory PD, Pabo CO, Rebar EJ (2007) An improved zinc-finger nuclease architecture for highly specific genome editing. Nat Biotechnol 25:778–785

Minczuk M, Papworth MA, Miller JC, Murphy MP, Klug A (2008) Development of a single-chain, quasi-dimeric zinc-finger nuclease for the selective degradation of mutated human mitochondrial DNA. Nucl Acids Res 36:3926–3938

Mino T, Aoyama Y, Sera T (2009) Efficient double-stranded DNA cleavage by artificial zinc-finger nucleases composed of one zinc-finger protein and a single-chain FokI dimer. J Biotechnol 140:156–161

Moehle E, Rock JM, Lee YL, Jouvenot Y, DeKelver RC, Gregory PD, Urnov FD, Holmes MC (2007) Targeted gene addition into a specified location in the human genome using designed zinc finger nucleases. Proc Natl Acad Sci USA 104:3055–3060

Morton J, Davis MW, Jorgensen EM, Carroll D (2006) Induction and repair of zinc-finger nuclease-targeted double-strand breaks in *Caenorhabditis elegans* somatic cells. Proc Natl Acad Sci USA 103:16370–16375

Neuteboom LW, Lindhout BI, Saman IL, Hooykaas PJJ, van der Zaal BJ (2006) Effects of different zinc finger transcription factors on genomic targets. Biochem Biophys Res Commun 339:263–270

Ninomiya Y, Suzuki K, Ishii C, Inoue H (2004) Highly efficient gene replacements in *Neurospora* strains deficient for nonhomologous end-joining. Proc Natl Acad Sci USA 101:12248–12253

Nussenzweig A, Nussenzweig MC (2007) A backup DNA repair pathway moves to the forefront. Cell 131:223–225

Offringa R, de Groot MJA, Haagsman HJ, Does MP, van den Elzen PJM, Hooykaas PJJ (1990) Extrachromosomal homologous recombination and gene targeting in plant cells after *Agrobacterium* mediated transformation. EMBO J 9:3077–3084

Osakabe K, Osakabe Y, Toki S (2010) Site-directed mutagenesis in *Arabidopsis* using custom-designed zinc finger nucleases. Proc Natl Acad Sci USA 107:12034–12039

Paszkowski J, Baur M, Bogucki A, Potrykus I (1988) Gene targeting in plants. EMBO J 7:4021–4026

Pruett-Miller SM, Reading DW, Porter SN, Porteus MH (2009) Attenuation of zinc finger nuclease toxicity by small-molecule regulation of protein levels. PLoS Genet 5:e1000376

Puchta H, Dujon B, Hohn B (1996) Two different but related mechanisms are used in plants for the repair of genomic double-strand breaks by homologous recombination. Proc Natl Acad Sci USA 93:5055–5060

Ramirez CL, Foley JE, Wright DA, Müller-Lerch F, Rahman SH, Cornu TI, Winfrey RJ, Sander JD, Fu F, Townsend JA, Cathomen T, Voytas DF, Joung JK (2008) Unexpected failure rates for modular assembly of engineered zinc fingers. Nat Method 5:374–375

Rémy S, Tesson L, Ménoret S, Usal C, Scharenberg AM, Anegon I (2010) Zinc-finger nucleases: a powerful tool for genetic engineering of animals. Transgenic Res 19:363–371

Risseeuw E, Offringa R, Franke-van Dijk MEI, Hooykaas PJJ (1995) Targeted recombination in plants using *Agobacterium* coincides with additional rearrangements at the target locus. Plant J 7:109–119

Rouet P, Smih F, Jasin M (1994) Introduction of double-strand breaks into the genome of mouse cells by expression of a rare-cutting endonuclease. Mol Cell Biol 14:8096–8106

Saika H, Oikawa A, Matsuda F, Onodera H, Saito K, Toki S (2011) Application of gene targeting to designed mutation breeding of high-tryptophan rice. Plant Phys 156:1269–1277

Sander JD, Dahlborg EJ, Goodwin MJ, Cade L, Zhang F, Cifuentes D, Curtin SJ, Blackburn JS, Thibodeau-Beganny S, Qi Y, Pierick CJ, Hoffman E, Maeder ML, Khayter C, Reyon D, Dobbs D, Langenau DM, Stupar RM, Giraldez AJ, Voytas DF, Peterson RT, Yeh J-RJ, Joung JK (2010a) Selection-free zinc-finger-nuclease engineering by context-dependent assembly (CoDA). Nat Method 8:67–69

Sander JD, Reyon D, Maeder ML, Foley JE, Thibodeau-Beganny S, Li X, Regan MR, Dahlborg EJ, Goodwin MJ, Fu F, Voytas DF, Joung JK, Dobbs D (2010b) Predicting success of oligomerized pool engineering (OPEN) for zinc finger target site sequences. BMC Bioinforma 11:543

Santiago Y, Chan E, Liu P-Q, Orlando S, Zhang L, Urnov FD, Holmes MC, Guschin D, Waite A, Miller JC, Rebar EJ, Gregory PD, Klug A, Collingwood TN (2008) Targeted gene knockout in mammalian cells by using engineered zinc-finger nucleases. Proc Natl Acad Sci USA 105:5809–5814

Schaefer DG (2002) A new moss genetics: targeted mutagenesis in *Physcomitrella patens*. Ann Rev Plant Biol 53:477–501

Segal DJ, Dreier B, Beerli RR, Barbas CF (1999) Toward controlling gene expression at will: selection and design of zinc finger domains recognizing each of the 5′-GNN-3' DNA target sequences. Proc Natl Acad Sci USA 96:2758–2763

Shaked H, Melamed-Bessudo C, Levy AA (2005) High-frequency gene targeting in *Arabidopsis* plants expressing the yeast RAD54 gene. Proc Natl Acad Sci USA 102:12265–12269

Shimizu Y, Şöllü C, Meckler JF, Adriaenssens A, Zykovich A, Cathomen T, Segal DJ (2011) Adding fingers to an engineered zinc finger nuclease can reduce activity. Biochemistry 50:5033–5041

Shrivastav M, De Haro LP, Nickoloff JA (2008) Regulation of DNA double-strand break repair pathway choice. Cell Res 18:134–147

Shukla VK, Doyon Y, Miller JC, DeKelver RC, Moehle EA, Worden SE, Mitchell JC, Arnold NL, Gopalan S, Meng X, Choi VM, Rock JM, Wu Y-Y, Katibah GE, Zhifang G, McCaskill D, Simpson MA, Blakeslee B, Greenwalt SA, Butler HJ, Hinkley SJ, Zhang L, Rebar EJ, Gregory PD, Urnov FD (2009) Precise genome modification in the crop species *Zea mays* using zinc-finger nucleases. Nature 459:437–441

Simsek D, Brunet E, Wong SY-W, Katyal S, Gao Y, McKinnon PJ, Lou J, Zhang L, Li J, Rebar EJ, Gregory PD, Holmes MC, Jasin M (2011) DNA ligase III promotes alternative nonhomologous end-joining during chromosomal translocation formation. PLoS Genet 7:e1002080

Skarnes WC, Rosen B, West AP, Koutsourakis M, Bushell W, Iyer V, Mujica AO, Thomas M, Harrow J, Cox T, Jackson D, Severin J, Biggs P, Fu J, Nefedov M, de Jong PJ, Stewart AF, Bradley A (2011) A conditional knockout resource for the genome-wide study of mouse gene function. Nature 474:337–342

Smith J, Bibikova M, Whitby FG, Reddy AR, Chandrasegaran S, Carroll D (2000) Requirements for double-strand cleavage by chimeric restriction enzymes with zinc finger DNA-recognition domains. Nucl Acids Res 28:3361–3369

Şöllü C, Pars K, Cornu TI, Thibodeau-Beganny S, Maeder ML, Joung JK, Heilbronn R, Cathomen T (2010) Autonomous zinc-finger nuclease pairs for targeted chromosomal deletion. Nucl Acids Res 38:8269–8276

Stoddard BL (2011) Homing endonucleases: from microbial genetic invaders to reagents for targeted DNA modification. Structure 19:7–15

Stracker TH, Petrini JHJ (2011) The MRE11 complex: starting from the ends. Nat Rev Mol Cell Biol 12:90–103

Szczepek M, Brondani V, Büchel J, Serrano L, Segal DJ, Cathomen T (2007) Structure-based redesign of the dimerization interface reduces the toxicity of zinc-finger nucleases. Nat Biotechnol 25:786–793

Terada R, Johzuka-Hisatomi Y, Saitoh M, Asao H, Iida S (2007) Gene targeting by homologous recombination as a biotechnological tool for rice functional genomics. Plant Phys 144:846–856

Tovkach A, Zeevi V, Tzfira T (2009) A toolbox and procedural notes for characterizing novel zinc finger nucleases for genome editing in plant cells. Plant J 57:747–757

Townsend JA, Wright DA, Winfrey RJ, Fu F, Maeder ML, Joung JK, Voytas DF (2009) High-frequency modification of plant genes using engineered zinc-finger nucleases. Nature 459: 442–445

Urnov FD, Miller JC, Lee Y-L, Beausejour CM, Rock JM, Augustus S, Jamieson AC, Porteus MH, Gregory PD, Holmes MC (2005) Highly efficient endogenous human gene correction using designed zinc-finger nucleases. Nature 435:646–651

Van Attikum H, Hooykaas PJJ (2003) Genetic requirements for the targeted integration of *Agrobacterium* T-DNA in *Saccharomyces cerevisiae*. Nucl Acids Res 31:826–832

Van Attikum H, Bundock P, Hooykaas PJJ (2001) Non-homologous end-joining proteins are required for *Agrobacterium* T-DNA integration. EMBO J 20:6550–6558

Van Kregten M (2011) VirD2 of *Agrobacterium tumefaciens*: functional domains and biotechnological applications. Dissertation, Leiden University

Vergunst AC, Schrammeijer B, den Dulk-Ras A, de Vlaam CMT, Regensburg-Tuink TJG, Hooykaas PJJ (2000) VirB/D4-dependent protein translocation from *Agrobacterium* into plant cells. Science 290:979–982

Vergunst AC, van Lier MCM, den Dulk-Ras A, Grosse Stüve TA, Ouwehand A, Hooykaas PJJ (2005) Positive charge is an important feature of the C-terminal transport signal of the VirB/D4-translocated proteins of *Agrobacterium*. Proc Natl Acad Sci USA 102:832–837

Wright DA, Townsend JA, Winfrey RJ, Irwin PA, Rajagopal J, Lonosky PM, Hall BD, Jondle MD, Voytas DF (2005) High-frequency homologous recombination in plants mediated by zinc-finger nucleases. Plant J 44:693–705

Wright DA, Thibodeau-Beganny S, Sander JD, Winfrey RJ, Hirsh AS, Eichtinger M, Fu F, Porteus MH, Dobbs D, Voytas DF, Joung JK (2006) Standardized reagents and protocols for engineering zinc finger nucleases by modular assembly. Nat Prot 1:1637–1652

Yamauchi T, Johzuka-Hisatomi Y, Fukada-Tanaka S, Terada R, Nakamura I, Iida S (2009) Homologous recombination-mediated knock-in targeting of the *MET1a* gene for a maintenance DNA methyltransferase reproducibly reveals dosage-dependent spatiotemporal gene expression in rice. Plant J 60:386–396

Zhang F, Maeder ML, Unger-Wallace E, Hoshaw JP, Reyon D, Christian M, Li X, Pierick CJ, Dobbs D, Peterson T, Joung JK, Voytas DF (2010) High frequency targeted mutagenesis in *Arabidopsis thaliana* using zinc finger nucleases. Proc Natl Acad Sci USA 107:12028–12033

Zhang F, Cong L, Lodato S, Kosuri S, Church GM, Arlotta P (2011) Efficient construction of sequence-specific TAL effectors for modulating mammalian transcription. Nat Biotechnol 29:149–154

Part IV
Conversion of QTPs into Functional Markers

Chapter 9
SNP Genotyping Technologies

Bruno Studer and Roland Kölliker

Preface

In the recent years, single nucleotide polymorphism (SNP) markers have emerged as the marker technology of choice for plant genetics and breeding applications. Besides the efficient technologies available for SNP discovery even in complex genomes, one of the main reasons for this is the availability of high-throughput platforms for multiplexed SNP genotyping. Advancements in these technologies have enabled increased flexibility and throughput, allowing for the generation of adequate SNP marker data at very competitive cost per data point.

Starting with a technical description of the most widely used SNP genotyping platforms, this chapter aims at discussing potentials and limitations for each technology, thereby providing a basis for the selection of the platform of choice for a specific biological application under technical and economical considerations.

Introduction

Precise molecular marker data with high density at the genomic location under investigation is a basic prerequisite for the molecular dissection of complex traits and the diagnostic application of molecular tools in plant breeding and research. Technological advances in methods for high-throughput genotyping of SNP markers

B. Studer (✉)
Institute of Agricultural Sciences, Forage Crop Genetics, ETH Zurich, Universitätsstrasse 2, 8092 Zurich, Switzerland
e-mail: bruno.studer@usys.ethz.ch

R. Kölliker
Molecular Ecology, Agroscope Reckenholz-Tänikon Research Station ART, Reckenholzstrasse 191, 8046 Zurich, Switzerland

have initiated a novel era of using molecular markers in numerous fields such as genetic linkage analysis and trait mapping, diversity analysis, association studies and single marker or genome-wide marker assisted selection (MAS) (Varshney et al. 2009). The potential of SNPs has been impressively demonstrated in human and animal genetics, as well as in model plant species such as *Arabidopsis thaliana*, rice (*Oryza sativa* L.) and maize (*Zea mays* L.), where fully sequenced genomes resulted in the identification of millions of SNPs suitable for genome-wide association studies (GWAS) and molecular breeding concepts such as genomic selection (GS) (Morrell et al. 2012).

Definition and Characteristics of Single Nucleotide Polymorphism (SNP) in Plants

A SNP represents a nucleotide difference between alleles at a specific locus in the genome. As single nucleotides are the smallest unit of inheritance, SNPs represent the most basic and abundant form of genetic sequence variation in genomes occurring at frequencies of up to one SNP per 21 bp in plant genomes (Edward et al. 2008). SNPs can be divided into three different forms; nucleotide transitions (a point mutation that changes a purine nucleotide to another purine, or a pyrimidine nucleotide into another pyrimidine i.e. $G \Longleftrightarrow A$ or $C \Longleftrightarrow T$), nucleotide transversions (a point mutation that changes a purine nucleotide into a pyrimidine nucleotide or vice versa) or single nucleotide insertions/deletions (indels) (Edwards et al. 2007). Approximately two out of three SNPs are transitions (Collins and Jukes 1994). This chapter will mostly focus on single nucleotide exchanges (transitions and transversions), as indels are discussed in Chap. 10.

Theoretically, a SNP polymorphism can involve the four different nucleotide variants A, T, C and G. In practice, SNPs are generally biallelic and the different variants occur at different frequencies (Schmid et al. 2003). The limited polymorphic information content of SNP markers is compensated by their frequent occurrence in the plant genome, making them valuable for targeting any gene of interest, for high density genotyping as well as for haplotyping (International HapMap Consortium 2007). SNPs, however, are not evenly distributed across the genome and occur at lower frequency in coding when compared to non-coding or intergenic regions (Choi et al. 2007). In coding sequences, SNPs do not necessarily change the amino acid sequence of the produced protein, they are often synonymous and, thus, non-functional. If a nucleotide change is non-synonymous and affects the amino acid sequence, missense polymorphisms resulting in different amino acid compositions and nonsense polymorphisms resulting in a premature stop codon can be distinguished. From an evolutionary point of view, SNPs are stable, not changing significantly from one generation to another and, due to the low mutation rate, excellent markers for studying genome evolution (Syvanen 2001).

SNP genotyping refers to the process of assigning the SNP variant to one of the four nucleotides, thereby discriminating alleles at a particular locus. Recently, a large number of different techniques, chemistries and allele discrimination methods from low- to ultra-high-throughput have been developed and discussed in numerous review articles and books (Syvanen 2001, 2005; Perkel 2008; Ragoussis 2009; Gupta et al. 2008; Bayes and Gut 2011; Chagné et al. 2007).

SNP Genotyping Technologies

New technologies for SNP genotyping are under continuous development and dozens of different platforms are available to date. Here, we focus on the most widely used platforms for plant breeding applications, which we suggest to divide into hybridization-based technologies, enzyme-based technologies and technologies based on physical properties of DNA. The following section will not attempt to provide a fully comprehensive overview of all SNP genotyping technologies available, but will aim to describe the key features of technologies that appear promising for plant breeding applications.

Hybridization-Based Technologies

SNP genotyping technologies based on hybridization of DNA probes complementary to the SNP sites include dynamic allele-specific hybridization (DASH) (Podder et al. 2008), molecular beacons (Mhlanga and Malmberg 2001; Täpp et al. 2000) and microarrays (Nazar and Robb 2011). The challenge with these approaches is to minimize cross-hybridization between allele-specific probes, which can be overcome by optimizing hybridization stringency conditions (Nazar and Robb 2011).

SNP Microarrays

The basic principle of a SNP microarray is the convergence of solid surface DNA capture, DNA hybridization, and fluorescence microscopy (Fig. 9.1a). On high density oligonucleotide SNP arrays, allele-specific oligonucleotide probes (AOP) are immobilized at high density on a small chip, allowing for hundreds of thousands of SNPs to be interrogated simultaneously. Each SNP interacts with different oligonucleotide probes. These probes contain the SNP site at several positions, some of them with mismatches to the SNP variant. For efficient hybridization to immobilized probes, the complexity of the genomic DNA must be reduced through digestion with restriction endonucleases (Kennedy et al. 2003). The comparison of hybridization efficiencies of the SNP to each of these redundant oligonucleotide

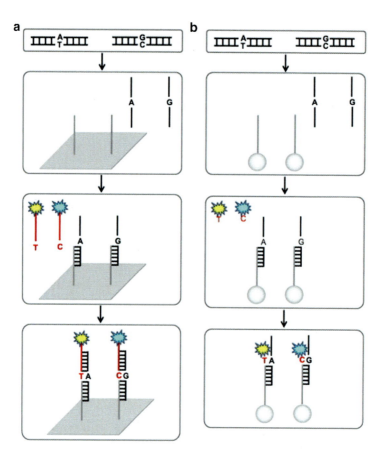

Fig. 9.1 **Array-based SNP genotyping platforms (Paux et al.** 2012). (**a**) Affymetrix Axiom. On high density oligonucleotide arrays containing 50 bp nucleotide sequence upstream the SNP, fragmented genomic DNA is hybridized. Differentially labeled probes corresponding to the four different nucleotides with downstream sequence of the SNP are added and ligated. Arrays are then scanned to interrogate each individual SNP. (**b**) Illumina Infinium. A whole-genome amplification step is followed by hybridization of the amplified DNA to array-bound target sequences that correspond to the 50 bases directly upstream of the SNP. Following hybridization, a single nucleotide extension reaction with hapten-labeled ddNTPs is performed in parallel for all SNPs. Genotype calls are derived from the relative intensity of the fluorescent signals

probes makes it possible to identify specific homozygous and heterozygous alleles (Nazar and Robb 2011). Because AOPs differ only in one nucleotide, the target DNA may hybridize to mismatched probes. The comparatively limited specificity and sensitivity of a microarray approach is compensated by the number of SNPs that can be interrogated simultaneously. High density SNP microarrays have mainly been applied in humans, or in major crop plant species such as rice (www.ricearray.org), where SNP arrays are commercially available (www.affymetrix.com). However, the Affymetrix Axiom™ myDesign™ genotyping arrays allow the development of fully customized SNP genotyping arrays containing from 1,500 up to 2.6 million SNPs. Microarray applications are limited to whole genome SNP interrogation at high density and offer only limited flexibility for targeting only selected genes or genomic regions. Although microarray SNP genotyping data achieve pass rates of >95% and accuracy of >99% (Matsuzaki et al. 2004), the total equipment required is rather expensive (Ragoussis and Elvidge 2006).

Enzyme-Based Technologies

The most widely used SNP genotyping technologies are based on a broad range of enzymes including DNA polymerases, ligases, and nucleases. The majority of the technologies presented follow a polymerase or ligase-based primer extension approach. Allele-specific PCR to selectively amplify one of the SNP variant is also known as amplification refractory mutation system (ARMS) (Newton et al. 1989). Two main approaches can be distinguished: Firstly, a single nucleotide primer extension, where a primer hybridizes immediately upstream of the SNP and a DNA polymerase incorporates a fluorescently labeled dideoxynucleotide (ddNTP) that is complementary to the SNP variant, and, secondly, allele-specific primer extension, where a primer perfectly matching the SNP variant with its 3′ end is hybridized and extended by PCR. Generally, DNA polymerase-based extension is very reliable due to the high sequence specificity and fidelity of DNA polymerases. Moreover, it allows a high degree of flexibility and multiplexing as PCR for primer extension can be performed under very similar reaction conditions, making such methods amenable to high-throughput (Syvanen 2001, 2005). SNP detection following primer extension involves a wide range of methods such as DNA sequencing (Ekstrom et al. 2000), MALDI-TOF mass spectrometry and fluorescent analysis (Tang et al. 2004).

SNuPE™ and SNaPshot™ Single Nucleotide Primer Extension

Low-scale but highly flexible systems based on single nucleotide primer extension (also called mini-sequencing) are the MegaBACE™ SNuPe™ system (GE Healthcare, formerly Amersham Biosciences AB) and the ABI PRISM® SNaPshot™ [Applied Biosystems (Life Technologies)], combining DNA polymerase-based primer extension using fluorescently labeled ddNTPs (terminators) with capillary electrophoresis in a single-tube reaction premix. The product size is determined by the length of the initial primer plus one labeled nucleotide. Multiplexing up to ten primers is possible by using primers of different length (Suharyanto and Shiraishi 2011).

Besides the reaction kits supplied by commercial manufacturers, a capillary-based electrophoresis instrument is required. Sequence data need to be analyzed using softwares for allele calling, editing, and verification (Torjek et al. 2003).

Assay workflow time is around 8 h, with a call- and accuracy rate of >95% and >99.9%, respectively. Genomic DNA quantity and quality requirements are low (3 ng per PCR reaction).

APEX (Arrayed Primer Extension) Technology

APEX is a mini-sequencing methodology based on oligonucleotide probes arrayed on slides that are used for primer extension (Shumaker et al. 1996; Pastinen et al. 1997; Kurg et al. 2000). The locus-specific PCR amplification is followed by a fragmentation of PCR products based on uracil N-glycosylase. Fragmented PCR products are then denatured and hybridized to complementary oligonucleotides that have been immobilized on a glass array in a reaction mixture. The buffered reaction mixture contains DNA polymerase and four different terminators (ddNTP), each labeled with an individual fluorescent dye. PCR primers are extended under elevated temperature in order to avoid secondary structures of the oligos. After stringent washing, detection is based on imaging using a microarray reader. Imaging is followed by data analysis to convert the fluorescence information into sequence data (Syvanen 2001). APEX can interrogate hundreds to thousands of SNPs in a single multiplexed reaction simultaneously. As the genotype information is obtained by single base extension performed by a specific DNA polymerase, this approach has a higher discrimination power but a lower throughput per run when compared to methods based on allele-specific oligonucleotide hybridization (microarrays) (Pastinen et al. 1997).

iPLEX® Gold MassARRAY SNP Genotyping

MassARRAY SNP genotyping (Sequenom) combines highly specific single nucleotide primer extension using ddNTPs with MALDI-TOF mass spectrometry. First, a locus specific PCR is used to amplify the SNP target region (Fig. 9.2, blue). Following amplification, a third primer (Fig. 9.2, red) anneals upstream with the 3' end directly flanking the SNP. This primer is then extended by PCR according to the template sequence, resulting in an allele-specific difference in mass between extension products. SNP variants are detected on the actual mass of the extension product determined by MALDI-TOF MS (matrix-assisted laser desorption/ionization time-of-flight mass spectrometry) (Sauer et al. 2000). Up to 40plex reactions in 384 well formats allow a single person to generate 100,000 data points per day. MassARRAY is flexible and suitable to generate both small and large marker numbers per sample (Jones et al. 2007). This method is for medium to high-throughput, and is not intended for whole genome scanning. The main advantage of the MassARRAY system is the low cost per data point at its given flexibility (Bagge and Lübberstedt 2008). However, it requires an expensive and rather complex instrumentation.

Fig. 9.2 MassARRAY SNP genotyping (Paux et al. 2012). The iPLEX platform combines highly specific single nucleotide primer extension using ddNTPs with MALDI-TOF mass spectrometry

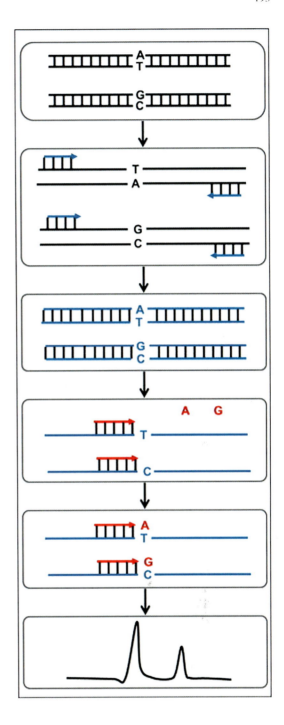

SNPlex™ Genotyping System

Based on oligonucleotide ligation, PCR and capillary electrophoresis, the SNPlex genotyping system [Applied Biosystems (Life Technologies)] allows to detect up to 48 bi-allelic SNP genotypes in a single reaction (Tobler et al. 2005). First, allele- and locus-specific oligonucleotide probes are hybridized to the target sequence. Successful hybridization leads then to ligation of the two probes and simultaneous ligation of two universal linkers. In order to encode the genotype information of each SNP, a unique ZipCode sequence is added to each allele-specific probe. Genomic DNA, unligated probes and linkers are removed using exonuclease digestion, and all 96 ligation products are amplified simultaneously using a single pair of PCR primers. Single-stranded amplicons are produced by binding biotinylated amplicons to the well of a streptavidin-coated microtiter plate and by removing non-biotinylated strands. Universal ZipChute probes containing a sequence complementary to the unique ZipCode sequence in each allele-specific probe, a fluorescent label and a mobility modifier are hybridized to the bound single-stranded amplicons. Finally, amplicons are separated by capillary electrophoresis where SNP genotypes are assigned based on the rate of mobility determined by the mobility modifier of the allele-specific ZipChute probe. SNP detection using the SNPlex system can be completed in 2 days and is amenable to automation, making it a medium to high-throughput system with more than one million data points in a week (Tobler et al. 2005) The protocol is based on standardized hybridization and amplification and does not require optimization for individual SNPs to be genotyped. For designing SNPlex assays, an automated high-throughput pipeline is available which consists of screening the SNP sequence against the target genome, selecting and designing SNP specific ligation probes, assignment of ZipCode sequences and separating assays into compatible multiplex pools (De la Vega et al. 2005). SNPlex has proven to be particularly powerful in investigations, where several hundred of SNPs are genotyped in a hundred or more samples, a situation often encountered in projects aiming at the development of tools for MAS in plants.

Molecular Inversion Probes (MIP) Assay

The MIP assay is a large-scale technology from Affymetrix and uses inverted oligonucleotide probes that contain the sequence information of the SNP and its surrounding sequence, and transfer this information into tags analyzed on DNA microarrays (Fig. 9.3a). MIPs originated from padlock probes (Nilsson et al. 1994) which were modified for SNP genotyping in a way to form gaps at the SNP position when the probe is hybridized to the target region. MIPs are circularizable, single-stranded DNA molecules containing two regions complementary to the DNA sequence flanking the target SNP (Fig. 9.3a, red), universal primer sequences (Fig. 9.3a, blue) that are separated by ribonuclease recognition site (Fig. 9.3a, orange), and a 20 bp sequence tag (Fig. 9.3a, green) (Hardenbol et al. 2003). During the assay, the probes are circularized around the target SNP, complemented

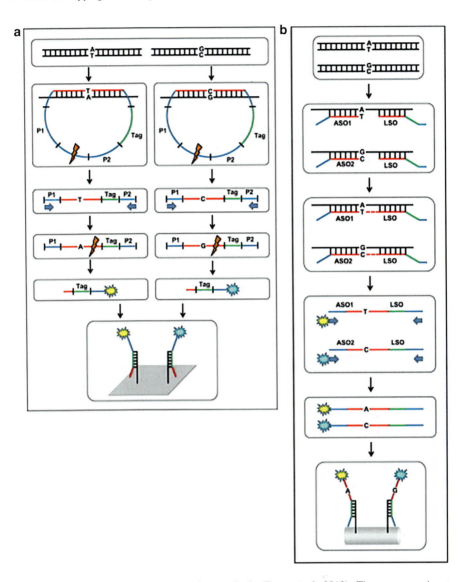

Fig. 9.3 Ligation-based SNP genotyping methods (Paux et al. 2012). The most prominent examples are (**a**) Affymetrix' molecular inversion probe (MIP) and (**b**) Illumina's GoldenGate assay

with the nucleotides corresponding to the SNPs in four separate allele-specific polymerizations (A, C, G, and T) and ligation reactions. The resulting circular molecule is cleaved between the PCR primers before and after PCR, followed by fluorescent labeling. The labeled molecules are captured on glass microarrays carrying complementary tag sequences for fluorescence detection (Paux et al. 2012).

With this technology, multiplex analysis of 3,000 up to 50,000 SNPs can be achieved in a single tube in parallel. Conversion rates are reported to be >80%, with pass rates of >98% and accuracy >99% (Hardenbol et al. 2005). The hardware required for MIP assays is similar to that used for Affymetrix gene chips, apart from the requirement for a four-color scanner.

Illumina GoldenGate Assays

Illumina's GoldenGate (GG) genotyping assay is an example of a ligation based primer extension using the BeadArray (Fan et al. 2006), or VeraCode technology. Technically, GG combines oligonucleotide ligation and allele-specific extension PCR. The assay is based on three primers per SNP, one locus-specific (LSO, Fig. 9.3b) and two allele-specific primers (ASO1, ASO2, Fig. 9.3b), directly annealing to genomic DNA. This is followed by an extension reaction at the ASO towards the LSO situated a few nucleotides farther from its 3′end. A ligation reaction links the successfully extended allele-specific product to the LSO, a reaction that gives very high specificity to the assay. As both, ASOs and LSO contain universal primer tails (Fig. 9.3b, blue), the successfully extended and ligated products are amplified by PCR with fluorescently labeled primers. Denatured PCR products are then hybridized to an array of beads (Sentrix Array) carrying sequences complementary to locus-specific tags located in the LSO sequence (Fig. 9.3b, green). For imaging, Illumina's iScan array scanner is being used for the BeadArray technology. GG genotyping using the BeadArray technology allows levels of multiplexing of 96-, 192-, 384-, 768-, 1,536- and 3,072-SNPs that can be assayed on 32 samples in parallel (Fan et al. 2003). Genotyping with the VeraCode technology with its increased flexibility is available with the GG chemistry at 48-, 96-, 144-, 192-, and 384-plex format. The VeraCode technology can be detected by Illumina's BeadXpress reader system, thereby increasing sample throughput.

Generally, the GG platform is very flexible, protocols can be performed manually or can be easily automated and throughput is high (up to 300,000 genotypes per six hands-on hours). Customized oligo pool assays can be designed for many species, but this is generally laborious. GG genotyping has demonstrated to produce highly reproducible results with a high call rate and accuracy (>99%) (Shen et al. 2005). Despite of the high initial cost, the cost per data point is competitive, especially for highly multiplexed chips.

Illumina's Infinium iSelect HD Custom Genotyping Beadchips

For genome-wide marker profiling, Illumina's Infinium assay allows to simultaneously genotype 3,072–1,000,000 SNPs in customized panels. The assay includes first a whole-genome amplification step, followed by enzymatic fragmentation and hybridization to bead arrays of 50-bp-long capture probes (Fig. 9.1b). The assay uses a single bead type and dual color channel approach, i.e. one color for A and T, another for G and C. After hybridization, allelic specificity is conferred by enzymatic base extension. The hapten-labelled nucleotides are recognized by anti-bodies, that are coupled to a detectable signal (Gunderson et al. 2006). The BeadChips can be deployed on the 24-sample format (3,072–90,000 attempted bead types), the 12-sample format (90,001–250,000 attempted bead types), or the 4-sample format (250,001–1,000,000 attempted bead types). The Infinium HD BeadChips offer the ability to interrogate virtually any SNP for any species, however, has only been used so far in sequenced model crop species such as soybean (*Glycine max* L.) (Haun et al. 2011), maize and loblolly pine (*Pinus taeda* L.). An Infinium assay for tetraploid and hexaploid wheat is on its way (Paux et al. 2012). The two-color system of the Infinium assay restricts somewhat the classes of SNPs that can be genotyped, but high pass rates and accuracy (>99.9) are performance characteristic (Steemers et al. 2006). One of the advantages of this system is that it allows simultaneous measurement of both signal intensity variations as well as changes in allelic composition (Gupta et al. 2008). DNA requirements are low, ranging from 200 ng for 3,072–250,000 SNPs and 400 ng for 250,001–1,000,000 SNPs.

Invader® Assay

The basic Invader® assay is based on the hybridization of two oligonucleotide probes and subsequent cleavage using thermostable flap endonucleases (FEN) (Olivier 2005). An allele-specific probe together with an Invader oligonucleotide which overlaps the SNP site with a non matching probe form a three-dimensional invader structure that can be recognized by an FEN cleavase. The fluorophore attached to the allele-specific probe is separated from its quencher resulting in a measurable fluorescent signal. This initial assay requires a substantial amount of target DNA and only allows for a single allele to be detected in one assay. Consequently, a biallelic assay was developed based on two subsequent invasive amplification reactions (Olivier 2005). Although the technique is highly reliable and multiplex systems have been developed which allow for more than 20 SNPs to be genotyped simultaneously (Nakahara et al. 2010), possibilities for high-throughput platforms are limited and the assay is rather suitable for specific applications than for general large-scale SNP genotyping.

TaqMan™ Assays

The TaqMan™ assays (Holland et al. 1991) is based on the 5′-nuclease activity of the Taq DNA polymerase and can be analysed using real-time PCR (McGuigan and Ralston 2002). The assay requires forward (FP) and reverse (RP) PCR primers that are used to amplify the region including the SNP (Fig. 9.4a). SNP variants are determined with two fluorescence resonance energy transfer (FRET) oligonucleotides (also called TaqMan probes) that hybridize to the SNP. The probes

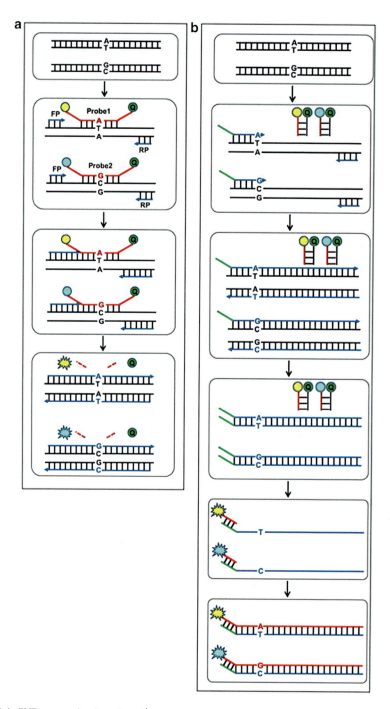

Fig. 9.4 SNP genotyping based on 5′-nuclease activity of Taq DNA polymerase. This illustration from Paux et al. (2012) compares (**a**) the TaqMan assay supplied by Applied Biosystems (Life Technologies) with the slightly modified method of KBiosciences KASPar (**b**)

are fluorescently labeled at their 5' end and contain a quencher molecule (Q) linked to their 3' end. During PCR amplification, the allele-specific probe complementary to the SNP allele binds to the target DNA strand and gets degraded by 5'-nuclease activity of the Taq DNA polymerase. Upon cleavage of the probe, the separation of the fluorescent dye from the quencher molecule is generating a detectable signal with measurable intensity (Fig. 9.4a). If the allele-specific probe is not complementary to the target SNP, it will have lower melting temperature and will not perfectly match the SNP site, preventing the nuclease to act on the probe. The use of sterically modified locked nucleic acids for probe design allowed to use shorter probes, thereby optimizing hybridization kinetics and improving detection sensitivity (Kennedy et al. 2006).

Since the TaqMan assay is based on a single tube PCR, it is relatively simple to implement, fast and has a high sample throughput (Holloway et al. 1999). The TaqMan assay using well performing probes under optimized reaction conditions can also be multiplexed by combining the detection of up to seven SNPs in one reaction. Thus, TaqMan is an ideal method for genotyping a low to medium number of SNP markers on a high number of samples (Syvanen 2001).

Recently, Applied Biosystems (Life Technologies) combined its TaqMan genotyping assays with the OpenArray® technology which uses nanofilter fluidics for massively parallel analysis of large samples at lower costs per data point. Different formats are available ranging from 16 SNPs on 144 samples to 256 SNPs on 12 samples. One person can run up to 24 plates per day without need of automation and generate more than 70,000 genotyping points (256 SNPs × 12 samples × 24 plates). The use of this technology has been recently applied in maize where a set of 162 gene-based SNPs were converted successfully from Illumina GoldenGate to TaqMan assays (Mammadov et al. 2012).

A similar invention is the recently launched Fluidigm Dynamic Array Integrated Fluidic Circuit (IFC), providing an interesting solution to run the Taqman assay on a high sample throughput (http://www.fluidigm.com/snp-genotyping.html). This micro-fluidic PCR system allows for parallel amplifications in separated nano-volumes and thus the interrogation of multiple SNP on up to 192 samples. Main advantages are the comparably easy workflow (reduced number of pipetting steps) and significantly reduced reagent usage at high data quality, resulting in lower costs and high sample throughput.

KASPar (KBioScience Allele-Specific Polymorphism) Assays

KASPar provided by K-Biosciences (http://www.kbioscience.co.uk/index.html) is a slightly modified method to TaqMan™, still using FRET quencher cassette oligos but a unique form of allele-specific PCR; allele-specific primers complementary to the region upstream of the SNP hybridize in a way that the 3' end perfectly matches the SNP variant. Each allele-specific primer contains a unique tail sequence at the 5' end (Fig. 9.4b, green). A common reverse primer complementary of the downstream sequence of the SNP (Fig. 9.4b, blue) and two additional 5' labeled primers complementary to the allele-specific tail referred to as reporters (Fig. 9.4b, red) that are matching with oligos containing the quenchers, are added to the PCR reaction. In a first step, allele-specific and common reverse primers bind to genomic DNA and generate a product without separating reporter and quencher. Subsequent PCR steps incorporate the labeled reporter complementary to the tail of the allele-specific primer and separate the quencher, thereby generating a fluorescence signal.

As a monoplex, single-step and closed tube system, KASPar is a simple and highly flexible platform on 96, 384 or even 1,536-well plate format. Assay design is easy, but time consuming with increasing number of SNPs. It has a higher SNP conversion rate when compared to TaqMan.

SNP Interrogation Based on Physical Properties of PCR Amplified DNA

The physical characteristics of DNA, i.e. the melting temperature and single strand conformation, are the basis for SNP allele discrimination in this group of SNP genotyping technologies. Methods such as single strand conformation polymorphism (SSCP) (Orita et al. 1989) or temperature gradient gel electrophoresis (TGGE) often require optimized conditions to achieve a high reaction specificity.

High Resolution Melting Curve Analysis (HRM)

High resolution melting curve analysis (HRM) recently emerged as a simple, high-throughput single marker system in plants. HRM measures dissociation of double stranded (ds) DNA of a PCR product amplified in the presence of a saturating fluorescence dye such as LCgreen (Idaho Technology, Salt Lake City, UH) or EvaGreen® (Biotium, Inc. Hayward, CA). The dye integrates itself into the dsDNA of the PCR product. Following PCR, the dissociation of the amplified dsDNA can be monitored with a CCD camera. The shape of the resulting melting curve is used to differentiate the SNP variants.

Two different HRM approaches can be applied for SNP genotyping: the first is a probe-based approach where an unlabeled oligonucleotide probe (also called luna probe) is included in the PCR reaction to interrogate the SNP by post-amplification melting. The probe fully hybridizes with one form of the allele and thus will have a 1 bp mismatch to the alternative SNP variant. During melting, the probe queries the SNP and the genotype is determined by the melting curve shape. The high Tm signal is obtained from the fully hybridized probe, while a lower Tm signal is obtained from the mismatched probe.

The second approach refers to short amplicon genotyping and queries the SNP directly in the PCR product without the need of a probe (Liew et al. 2004). Here, primers are designed to directly flank the SNP in order to minimize the chances of amplifying additional polymorphisms. Therefore the entire amplicon may be as short as 37–44 bp. The melting curves of most homozygotes are sufficiently different, the heterozygotes are even easier to differentiate because they form heteroduplexes, broadening the melt transition and usually give two discernible peaks. HRM is simple, highly effective, and cheap. Multiplexing up to 3 SNPs is possible. Unique to this technology is, that it can also be used for genotyping any other type of DNA sequence polymorphism present in the amplified fragment (Studer et al. 2009).

Considerations for Selecting SNP Technologies for Breeding Applications

The selection of a suitable technology for SNP genotyping mainly depends on the specific biological question, and, consequently, a bottom-up approach in the decision process is suggested. For example, gene-targeting and marker assisted backcrossing may call for a low to moderate number of markers to be screened in hundreds or thousands of samples. Thus, technologies such as HRM, SNuPE/SNaPshot, TaqMan, KASPar, or Invader might be interesting. For genetic diversity studies, the selection of parental lines or trait mapping, a moderate to high number of markers are genotyped on a moderate number of samples, which favors platforms such as iPLEX, SNPlex, APEX or GoldenGate. For GWAS and GS, ultra-high density platforms such as MIP, Infinium, Axiom, or genotyping by sequencing (GBS) might be necessary (Table 9.1, Fig. 9.5). Within these three groups, the technology of choice mainly depends on technical and economical considerations.

Technical Considerations

The relationship between the number of SNPs and the number of samples under investigation constitutes an important technical issue and has been described earlier (Bagge and Lübberstedt 2008). As these two factors may vary substantially from one SNP genotyping project to another, the flexibility of an assay, i.e. the ability to adapt a technology for a specific number of SNPs and samples, determine a technology's applicability (Fig. 9.5).

The challenge of successful assay design adds another technical dimension. It has been shown that the genomic DNA sequence flanking the target SNP has a high impact on genotyping performance (Grattapaglia et al. 2011). Some alleles may have mutations or additional polymorphisms in primer binding sites leading to null alleles. This can be problematic in allogamous plant species where SNPs generally occur at higher frequencies. Further characterization of the target SNP and the flanking region can be achieved by allele sequencing prior to SNP genotyping assay design. However, allele sequencing might be limited to low and medium SNP density assays.

Table 9.1 Genotyping scales and potential breeding applications of selected SNP genotyping platforms. Main characteristics of each platform in terms of the providing company, the assay type and the detection technology are summarized

Genotyping scale	Single marker assays (low scale)				Medium density assay (medium scale)			(Ultra-) high density (genome wide scale)		
	Gene tagging/marker assisted backcrossing									
	Genetic diversity studies/selection of parental lines									
Breeding application					Linkage mapping			Genome-wide association studies		
								Genomic selection		
Platform	HRM	SNuPE/SNaPshot	TaqMan	KASPar	iPlex	SNPlex	GoldenGate	Infinium	Axiom	GBS
Provider	Several[a]	GE Healthcare/Applied Biosystems (Life Technologies)	Applied Biosystems (Life Technologies)	KBioscience	Sequenom	Applied Biosystems (Life Technologies)	Illumina	Illumina	Affymetrix	Illumina/Applied Biosystems (Life Technologies)
Assay type	PCR/melting	Single nucleotide extension/PCR	5′exonuclease/PCR	5′exonuclease/PCR	Single nucleotide extension/PCR	Ligation/PCR	Ligation/PCR	Hybridization	Hybridization	Sequencing by synthesis
Detection technology	LightScanner/RT-PCR instrument	Capillary electrophoresis	RT-PCR instrument	RT-PCR instrument	MALDI-TOF Mass spectrometry	Capillary electrophoresis	VeraCode BeadArray	BeadArray	Oligonucleotide array	Flow cell imaging

[a] Several companies such as Idaho Technology, Bio-Rad and Applied Biosystems (Life Technologies) are supplying HRM solutions

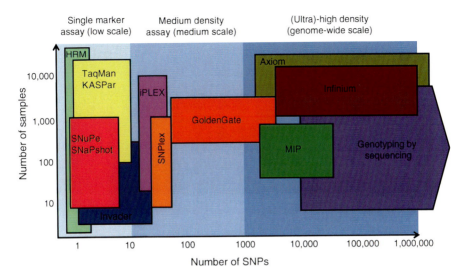

Fig. 9.5 Applicability of SNP genotyping platforms as a function of SNP number (x-axis, log scale) and sample number (y-axis, log scale). SNP genotyping platforms were divided into three main groups (*white-blue*, *light-blue* and *blue*) representing single marker, medium density and (ultra)-high density SNP assays, respectively. In order to increase readability, SNPlex and iPLEX were assigned to medium density assays, but can technically also handle single SNPs

DNA pooling has been suggested as a practical way to reduce SNP genotyping costs but requires the ability to measure allele frequencies in pools of individuals (quantitative genotyping). Additional technical factors are, among others, (1) achieved call rates and accuracy, (2) the DNA quantity and quality required, and (3) the degree of automation. Main technical characteristics of selected SNP genotyping platforms are summarized in Table 9.2.

Economic Considerations

Cost effectiveness, both in terms of initial investment and cost per data point is a major factor for the deployment of a SNP genotyping technology. As shown earlier, it includes salary, fixed costs per SNP as well as consumables (Bagge and Lübberstedt 2008). However, for several reasons, it might be rather difficult – if not impossible – to capture all the different economical levels and combine them in a global concept. Firstly, instrument, consumables and labor cost are subjected to major fluctuations in an economically changing environment, they vary between companies, countries, and currencies, as well as over time. Secondly, with increasing flexibility of SNP genotyping technology in terms of SNP and sample numbers, the cost ranges of a particular technology vary extensively for the different multiplex-levels and increasingly overlap with the cost ranges of other platforms, making it meaningless

Table 9.2 Main technical characteristics of selected SNP genotyping platforms applied in plant breeding

Platform	HRM	SNuPE SNaPshot	TaqMan	KASPar	iPlex	SNPlex	GoldenGate	Infinium	Axiom	GBS
Multiplexing	1–3	1–10	1–10	1–10	1–40	1–48	96–3,072	3,072–1 m	1,500–2.6 m	[a]
Sample sizes	Low-high	Low-high	Low-high	Low-high	Medium	Medium	Medium-high	Medium-high	Medium-high	Low-medium
Throughput[b]	2	1	2	2	6	5	7	9	9	10
Flexibility[c]	****	***	***	***	**	**	**	*	*	***
Allele frequencies	Yes						Maybe	Yes		Yes
Call rates and accuracy	Medium	Medium	High	High	High	Medium	High	High	Medium	Very high
DNA quantity and quality required	Low	Low	Medium	Medium	Low	Low	Low	Low	High	High
Degree of automation	High	Low	High	High	High	Medium	High	High	High	Medium

[a] Marker scale depends on the genome fraction sequenced
[b] Throughput refers to data points per person per day and is estimated on a scale from 1 (limited throughput) to 10 (highest possible throughput)
[c] Flexibility in terms of the possibility to change the SNP setup after assay design is rated on a scale from *(limited flexibility) to *****(highest possible flexibility)

to compare across different genotyping technologies. Thirdly, it makes a difference whether SNP genotyping is outsourced or (partially) conducted in-house: This again depends on the lab equipment and technical assistance available. All this makes it difficult comparing costs per data point. Therefore, a case-to-case decision rather than a global conclusion will be necessary. In this sense, the following points will rather aim to describe major economical tendencies than providing a global guide to select the most economical platform.

For single marker assays, HRM outperforms all other technologies both in terms of cost for investments as well as cost per data point. Main reasons are (1) low variable cost (PCR chemistry including a saturating fluorescent dye), (2) low initial investments (a PCR thermocycler including an optical unit), (3) the ease of assay-design (primer design flanking a SNP, a single PCR reaction followed by melting of the PCR product), and (4) the speed of the assay (2 h including data analysis) leading to cost savings in terms of labor.

Comparing different low to medium multiplex platforms, Sequenom's MassARRAY platform seems most effective in terms of cost per marker data point (at least for sample numbers close to 384) up to a high SNP number. The SNP number at which a platform with a higher multiplex level such as Illumina GoldenGate can not be generally determined as this depends on the sample number. For other low to medium multiplex platforms (e.g. KASPar, TaqMan, Invader, SNPlex) cost effectiveness depend on the number of markers and the number samples that can be analyzed in parallel.

On the genome-wide scale, GBS will emerge as the technology of choice and further reduction of reagent-consumable and sequencing costs can be expected. In combination with increased throughput of next generation sequencing (NGS) such as Illumina's HiSeq™ 2000 and Applied Bio-systems (Life Technologies)'s SOLiD™ sequencing system, or even third generation single molecule sequencing (Thompson and Milos 2011), a tremendous potential of cost reduction for GBS can be expected (Davey et al. 2011).

Of major importance in economic considerations is the sample number, particularly for technologies with high fixed costs per SNP such as TaqMan or KASPar. Cost per data point will only become competitive when screening a high sample number. Of further consideration are the investments necessary for in-house SNP genotyping. Initial investment cost vary significantly for platforms requiring an RT-PCR instrument, a capillary electrophoresis instrument, a mass spectrometer or a holistic solution to run a NGS instrument. The lab equipment already available will finally drive the decision.

SNP Genotyping for Plant Breeding

For plant breeders, two main scenarios emerged: On the one hand, large breeding companies working with species with an established (or nearly completed) reference genome make use of a very high number of available SNPs for GWAS and

GS (Hamblin et al. 2011). For these purposes, customized Affymetrix Axiom™, Illumina Infinium and Invader arrays, or GBS strategies might be chosen.

On the other hand, small to medium scale breeding companies with a lower budged allocated to molecular breeding approaches might focus on functional markers (Andersen and Lübberstedt 2003) for a few traits only but on a very high scale. In this case, single SNP technologies with an enormous sample throughput are interesting.

Future Prospective of SNP Genotyping in Plants

For single marker approaches, higher sample throughput and further cost reductions can be achieved by reducing the reaction volumes and using higher density plate, chip, or array formats. Nano fluidics such as OpenArray or the Fluidigm Dynamic Array technologies provide interesting solutions for nanoliter scale-PCR on multiple samples. For more complex traits addressed by GWA and improved by GS in the breeding programmes, the detection of haplotypes will be important to understand functional effects of SNPs in cis and meiotic recombination. Thus, the utility of SNP arrays for long range haplotyping in plants will become an important issue in the future.

Looking beyond plant genomics, structural or copy number variation (CNV), defined as genome fragments larger than 1 kb varying in copy number between individuals, has emerged as a significant contributor to human genetic variation in addition to sequence variants (Redon et al. 2006). CNVs are increasingly evaluated for their contribution to phenotypes, prompting a new race to incorporate assays for CNVs within SNP genotyping chips and new analysis algorithms to infer CNVs from SNP genotyping data (Ragoussis 2009). Resequencing strategies will be very promising to assess CNV and genomic regions difficult to address by SNPs (International HapMap Consortium 2007).

Advancements in throughput and multiplexing capacities of NGS technologies offer the opportunity of by-passing the necessity for array-based genotyping by means of sequencing (see Chap. 11). GBS will prove extremely powerful for ultra-high density SNP genotyping applied in GWAS and GS (Hamblin et al. 2011). A major advantage of GBS is the reduced ascertainment bias, i.e. the bias attributable to the fact that different plant material was used for SNP discovery (or SNP ascertainment) and SNP genotyping. Moreover, GBS allows to characterize allele frequencies in genetically heterogeneous plant populations, especially useful for genotyping in crop species where cultivars consist of open-pollinated populations. For small plant genomes such as those of Arabidopsis or rice, for which high-quality reference genome sequences are established, whole-genome resequencing might be the most powerful and straightforward genotyping approach (Huang 2009). For larger and more complex genomes, target enrichment (see Chap. 5) or complexity reduction strategies (see Chap. 11) will allow sequencing only a well distributed portion of the genome. A cost-effective approach of GBS on a small portion of the genome has recently been described and demonstrated in both maize and barley mapping populations (Elshire 2011). However, GBS implies a careful consideration of the experimental setup in order to sequence the right proportion of the genome at sufficient depth to meet SNP number and accuracy for the application in question. The potential of GBS to replace dedicated marker technologies remains to be demonstrated.

Conclusions

Current SNP genotyping technologies offer a wide range of opportunities for using SNP markers as a diagnostic tool in plant breeding. These technologies involve single SNP assays genotyped on thousands of samples, a wide spectrum of multiplexed assays run on several samples in parallel, as well as high-throughput genotyping platforms interrogating millions of markers simultaneously at genome-wide coverage.

A single technology is not applicable to answer all plant breeding related questions and each technology has its own advantages and disadvantages. The selection of a suitable technology for SNP genotyping mainly depends on the biological question that determines SNP and sample number under investigation. Within three main groups (single marker, medium density and high density assays), selection criteria involve technical (flexibility, throughput, success rate, degree of automation) and economic (cost per data point, time) considerations.

SNP genotyping technologies are continuously evolving and the best technology today is likely to become outdated in the near future. For single marker approaches, further sample throughput and cost reductions by minimizing reaction volumes and using nano fluidics can be achieved. For genome-wide SNP genotyping, the ability to detect haplotypes and structural variations is likely to become an important issue. For both, GBS seems promising.

References

Andersen JR, Lübberstedt T (2003) Functional markers in plants. Trends Plant Sci 8(11):554–560

Bagge M, Lübberstedt T (2008) Functional markers in wheat: technical and economic aspects. Mol Breed 22(3):319–328

Bayes M, Gut IG (2011) Overview of genotyping. In: Rapley R, Harbron S (eds) Molecular analysis and genome discovery, 2nd edn. Wiley, Chichester, pp 1–19

Chagné D, Batley J, Edwards D, Forster JW (2007) Single nucleotide polymorphism genotyping in plants. In: Oraguzie NC, Rikkerink EHA, Gardiner SE, Silva HN (eds) Association mapping in plants. Springer, New York, pp 77–94

Choi I-Y, Hyten DL, Matukumalli LK, Song Q, Chaky JM, Quigley CV, Chase K, Lark KG, Reiter RS, Yoon M-S, Hwang E-Y, Yi S-I, Young ND, Shoemaker RC, van Tassell CP, Specht JE, Cregan PB (2007) A soybean transcript map: gene distribution, haplotype and single-nucleotide polymorphism analysis. Genetics 176(1):685–696

Collins DW, Jukes TH (1994) Rates of transition and transversion in coding sequences since the human-rodent divergence. Genomics 20(3):386–396

Davey JW, Hohenlohe PA, Etter PD, Boone JQ, Catchen JM, Blaxter ML (2011) Genome-wide genetic marker discovery and genotyping using next-generation sequencing. Nat Rev Genet 12(7):499–510

De la Vega FM, Lazaruk KD, Rhodes MD, Wenz MH (2005) Assessment of two flexible and compatible SNP genotyping platforms: TaqMan SNP genotyping assays and the SNPlex genotyping system. Mutat Res 573(1–2):111–135

Edward K, Poole R, Barker G (2008) SNP discovery in plants. In: Henry R (ed) Plant genotyping II SNP technology. CABI, Oxfordshire, pp 1–29

Edwards D, Forster JW, Chagné D, Batley J (2007) What are SNPs? In: Oraguzie NC, Rikkerink EHA, Gardiner SE, De Silva HN (eds) Association mapping in plants. Springer, New York, pp 41–52

Ekstrom B, Alderborn A, Hammerling U (2000) Pyrosequencing for SNPs. Prog Biomed Opt Imaging 1:134–139

Elshire RJ (2011) A robust, simple genotyping-by-sequencing (GBS) approach for high diversity species. PLoS One 6:e19379

Fan JB, Oliphant A, Shen R, Kermani BG, Garcia F, Gunderson KL, Hansen M, Steemers F, Butler SL, Deloukas P, Galver L, Hunt S, McBride C, Bibikova M, Rubano T, Chen J, Wickham E, Doucet D, Chang W, Campbell D, Zhang B, Kruglyak S, Bentley D, Haas J, Rigault P, Zhou L, Stuelpnagel J, Chee MS (2003) Highly parallel SNP genotyping. Cold Spring Harb Symp Quant Biol 68:69–78

Fan JB, Gunderson KL, Bibikova M, Yeakley JM, Chen J, Wickham Garcia E, Lebruska LL, Laurent M, Shen R, Barker D, Kimmel A, Oliver B (2006) Illumina universal bead arrays. In: Methods in enzymology, vol 410. Academic, San Diego, pp 57–73

Grattapaglia D, Silva OB, Kirst M, de Lima BM, Faria DA, Pappas GJ (2011) High-throughput SNP genotyping in the highly heterozygous genome of Eucalyptus: assay success, polymorphism and transferability across species. BMC Plant Biol 11:65

Gunderson KL, Steemers FJ, Ren H, Ng P, Zhou L, Tsan C, Chang W, Bullis D, Musmacker J, King C, Lebruska LL, Barker D, Oliphant A, Kuhn KM, Shen R (2006) Whole-genome genotyping. In: Alan K, Brian O (eds) Methods in enzymology, vol 410. Academic, San Diego, pp 359–376

Gupta PK, Rustgi S, Mir RR (2008) Array-based high-throughput DNA markers for crop improvement. Heredity 101(1):5–18

Hamblin MT, Buckler ES, Jannink J-L (2011) Population genetics of genomics-based crop improvement methods. Trends Genet 27(3):98–106

Hardenbol P, Baner J, Jain M, Nilsson M, Namsaraev EA, Karlin-Neumann GA, Fakhrai-Rad H, Ronaghi M, Willis TD, Landegren U, Davis RW (2003) Multiplexed genotyping with sequence-tagged molecular inversion probes. Nat Biotechnol 21(6):673–678

Hardenbol P, Yu FL, Belmont J, MacKenzie J, Bruckner C, Brundage T, Boudreau A, Chow S, Eberle J, Erbilgin A, Falkowski M, Fitzgerald R, Ghose S, Iartchouk O, Jain M, Karlin-Neumann G, Lu XH, Miao X, Moore B, Moorhead M, Namsaraev E, Pasternak S, Prakash E, Tran K, Wang ZY, Jones HB, Davis RW, Willis TD, Gibbs RA (2005) Highly multiplexed molecular inversion probe genotyping: over 10,000 targeted SNPs genotyped in a single tube assay. Genome Res 15(2):269–275

Haun WJ, Hyten DL, Xu WW, Gerhardt DJ, Albert TJ, Richmond T, Jeddeloh JA, Jia G, Springer NM, Vance CP, Stupar RM (2011) The composition and origins of genomic variation among individuals of the soybean reference cultivar Williams 82. Plant Physiol 155(2):645–655

Holland PM, Abramson RD, Watson R, Gelfand DH (1991) Detection of specific polymerase chain reaction product by utilizing the 5′-3′ exonuclease activity of *Thermus aquaticus* DNA polymerase. Proc Natl Acad Sci USA 88(16):7276–7280

Holloway JW, Beghé B, Turner S, Hinks LJ, Day INM, Howell WM (1999) Comparison of three methods for single nucleotide polymorphism typing for DNA bank studies: sequence-specific oligonucleotide probe hybridisation, TaqMan liquid phase hybridisation, and microplate array diagonal gel electrophoresis (MADGE). Hum Mutat 14(4):340–347

Huang X (2009) High-throughput genotyping by whole-genome resequencing. Genome Res 19:1068–1076

International HapMap Consortium (2007) A second generation human haplotype map of over 3.1 million SNPs. Nature 449(7164):851–861

Jones ES, Sullivan D, Bhattramakki D, Smith JSC (2007) A comparison of simple sequence repeat and single nucleotide polymorphism marker technologies for the genotypic analysis of maize (*Zea mays* L.). Theor Appl Genet 115:361–371

Kennedy GC, Matsuzaki H, Dong S, W-m L, Huang J, Liu G, Su X, Cao M, Chen W, Zhang J, Liu W, Yang G, Di X, Ryder T, He Z, Surti U, Phillips MS, Boyce-Jacino MT, Fodor SPA, Jones KW (2003) Large-scale genotyping of complex DNA. Nat Biotechnol 21(10):1233–1237

Kennedy B, Arar K, Reja V, Henry RJ (2006) Locked nucleic acids for optimizing displacement probes for quantitative real-time PCR. Anal Biochem 348(2):294–299

Kurg A, Tõnisson N, Georgiou I, Shumaker J, Tollett J, Metspalu A (2000) Arrayed primer extension: solid-phase four-color DNA resequencing and mutation detection technology. Genet Test 4(1):1–7

Liew M, Pryor R, Palais R, Meadows C, Erali M, Lyon E, Wittwer C (2004) Genotyping of single-nucleotide polymorphisms by high-resolution melting of small amplicons. Clin Chem 50(7):1156–1164

Mammadov J, Chen W, Mingus J, Thompson S, Kumpatla S (2012) Development of versatile gene-based SNP assays in maize (*Zea mays* L.). Mol Breed 29(3):779–790

Matsuzaki H, Dong S, Loi H, Di X, Liu G, Hubbell E, Law J, Berntsen T, Chadha M, Hui H, Yang G, Kennedy GC, Webster TA, Cawley S, Walsh PS, Jones KW, Fodor SPA, Mei R (2004) Genotyping over 100,000 SNPs on a pair of oligonucleotide arrays. Nat Methods 1(2):109–111

McGuigan FEA, Ralston SH (2002) Single nucleotide polymorphism detection: allelic discrimination using TaqMan. Psychiatr Genet 12(3):133–136

Mhlanga MM, Malmberg L (2001) Using molecular beacons to detect single-nucleotide polymorphisms with real-time PCR. Methods 25(4):463–471

Morrell LP, Buckler ES, Ross-Ibarra J (2012) Crop genomics: advances and applications. Nat Rev Genet 13:85–96

Nakahara H, Sekiguchi K, Hosono N, Kubo M, Takahashi A, Nakamura Y, Kasai K (2010) Criterion values for multiplex SNP genotyping by the invader assay. Forensic Sci Int Genet 4:130–136

Nazar RN, Robb J (2011) DNA chip analysis in genome discovery. In: Rapley R, Harbron S (eds) Molecular analysis and genome discovery, 2nd edn. Wiley, Chichester, pp 24–37

Newton CR, Graham A, Heptinstall LE, Powell SJ, Summers C, Kalsheker N, Smith JC, Markham AF (1989) Analysis of any point mutation in DNA. The amplification refractory mutation system (ARMS). Nucleic Acids Res 17(7):2503–2516

Nilsson M, Malmgren H, Samiotaki M, Kwiatkowski M, Chowdhary B, Landegren U (1994) Padlock probes: circularizing oligonucleotides for localized DNA detection. Science 265(5181):2085–2088

Olivier M (2005) The Invader® assay for SNP genotyping. Mutat Res 573:103–110

Orita M, Iwahana H, Kanazawa H, Hayashi K, Sekiya T (1989) Detection of polymorphisms of human DNA by gel electrophoresis as single-strand conformation polymorphisms. Proc Natl Acad Sci USA 86:2766–2770

Pastinen T, Kurg A, Metspalu A, Peltonen L, Syvänen A-C (1997) Minisequencing: a specific tool for DNA analysis and diagnostics on oligonucleotide arrays. Genome Res 7(6):606–614

Paux E, Sourdille P, Mackay I, Feuillet C (2012) Sequence-based marker development in wheat: advances and applications to breeding. Biotechnol Adv 30(5):1071–1088

Perkel J (2008) SNP genotyping: six technologies that keyed a revolution. Nat Methods 5(5):447–453

Podder M, Ruan J, Tripp B, Chu Z, Tebbutt S (2008) Robust SNP genotyping by multiplex PCR and arrayed primer extension. BMC Med Genomics 1(1):5

Ragoussis J (2009) Genotyping technologies for genetic research. Annu Rev Genomics Hum Genet 10(1):117–133

Ragoussis J, Elvidge G (2006) Affymetrix GeneChip® system: moving from research to the clinic. Expert Rev Mol Diagn 6(2):145–152

Redon R, Ishikawa S, Fitch KR, Feuk L, Perry GH, Andrews TD, Fiegler H, Shapero MH, Carson AR, Chen W, Cho EK, Dallaire S, Freeman JL, Gonzalez JR, Gratacos M, Huang J, Kalaitzopoulos D, Komura D, MacDonald JR, Marshall CR, Mei R, Montgomery L, Nishimura K, Okamura K, Shen F, Somerville MJ, Tchinda J, Valsesia A, Woodwark C, Yang F, Zhang J, Zerjal T, Zhang J, Armengol L, Conrad DF, Estivill X, Tyler-Smith C, Carter NP, Aburatani H, Lee C, Jones KW, Scherer SW, Hurles ME (2006) Global variation in copy number in the human genome. Nature 444(7118):444–454

Sauer S, Lechner D, Berlin K, Plançon C, Heuermann A, Lehrach H, Gut IG (2000) Full flexibility genotyping of single nucleotide polymorphisms by the GOOD assay. Nucleic Acids Res 28(23):e100

Schmid KJ, Sörensen TR, Stracke R, Törjék O, Altmann T, Mitchell-Olds T, Weisshaar B (2003) Large-scale identification and analysis of genome-wide single-nucleotide polymorphisms for mapping in *Arabidopsis thaliana*. Genome Res 13(6a):1250–1257

Shen R, Fan J-B, Campbell D, Chang W, Chen J, Doucet D, Yeakley J, Bibikova M, Wickham Garcia E, McBride C, Steemers F, Garcia F, Kermani BG, Gunderson K, Oliphant A (2005) High-throughput SNP genotyping on universal bead arrays. Mutat Res 573(1–2):70–82

Shumaker JM, Metspalu A, Caskey CT (1996) Mutation detection by solid phase primer extension. Hum Mutat 7(4):346–354

Steemers FJ, Chang W, Lee G, Barker DL, Shen R, Gunderson KL (2006) Whole-genome genotyping with the single-base extension assay. Nat Methods 3(1):31–33

Studer B, Jensen LB, Fiil A, Asp T (2009) "Blind" mapping of genic DNA sequence polymorphisms in *Lolium perenne* L. by high resolution melting curve analysis. Mol Breed 24(2):191–199

Suharyanto S, Shiraishi S (2011) An identification system using multiplex allele-specific PCR assay in the Japanese black pine (*Pinus thunbergii* Parl.). Breed Sci 61(3):301–306

Syvanen A-C (2001) Accessing genetic variation: genotyping single nucleotide polymorphisms. Nat Rev Genet 2(12):930–942

Syvanen A-C (2005) Toward genome-wide SNP genotyping. Nat Genet 37:5–10

Tang K, Oeth P, Kammerer S, Denissenko MF, Ekblom J, Jurinke C, van den Boom D, Braun A, Cantor CR (2004) Mining disease susceptibility genes through SNP analyses and expression profiling using MALDI-TOF mass spectrometry. J Proteome Res 3(2):218–227

Täpp I, Malmberg L, Rennel E, Wik M, Syvänen AC (2000) Homogeneous scoring of single-nucleotide polymorphisms: comparison of the 5′-nuclease TaqMan assay and Molecular Beacon probes. Biotechniques 28(4):732–738

Thompson JF, Milos PM (2011) The properties and applications of single-molecule DNA sequencing. Genome Biol 12:217

Tobler AR, Short S, Andersen MR, Paner TM, Briggs JC, Lambert SM, Wu PP, Wang Y, Spoonde AY, Koehler RT, Peyret N, Chen C, Broomer AJ, Ridzon DA, Zhou H, Hoo BS, Hayashibara KC, Leong LN, Ma CN, Rosenblum BB, Day JP, Ziegle JS, De La Vega FM, Rhodes MD, Hennessy KM, Wenz HM (2005) The SNPlex genotyping system: a flexible and scalable platform for SNP genotyping. J Biomol Tech 16(4):398–406

Torjek O, Berger D, Meyer RC, Mussig C, Schmid KJ, Sorensen TR, Weisshaar B, Mitchell-Olds T, Altmann T (2003) Establishment of a high-efficiency SNP-based framework marker set for Arabidopsis. Plant J 36(1):122–140

Varshney RK, Nayak SN, May GD, Jackson SA (2009) Next-generation sequencing technologies and their implications for crop genetics and breeding. Trends Biotechnol 27(9):522–530

Chapter 10
Insertion-Deletion Marker Targeting for Intron Polymorphisms

Ken-ichi Tamura, Jun-ichi Yonemaru, and Toshihiko Yamada

Introduction

Insertion-deletion (indel) polymorphisms are the second most frequent type of polymorphisms after nucleotide substitutions such as single nucleotide polymorphisms (SNPs). Indel polymorphisms are more easily detectable as the difference of fragment length and at lower costs than SNPs, by using the combination of PCR and electrophoresis. Furthermore, wide length variations of indels are useful for studies requiring detection of multi-allelic loci, such as in population genetics, association studies, and for genetic mapping of polyploid species. In this decade, whole-genome sequencing has accelerated the map-based cloning of genes for qualitative and quantitative trait loci (QTL) for agronomic traits in several crops such as rice (*Oryza sativa* L., reviewed by Yamamoto et al. 2009; Miura et al. 2011). In addition to genetic analyses such as QTL analysis, candidate gene approaches based on allele mining in diverse germplasm identified several functional nucleotide polymorphisms (FNPs) by association analysis or comparative genomic approaches. FNPs are directly linked to phenotypes, and can thus be used as diagnostic markers in plant breeding. FNPs are usually located in genic regions such as coding sequences, promoter and intron regions, generating gain or loss of function or

K. Tamura
NARO Hokkaido Agricultural Research Center, National Agriculture and Food Research Organization, Hitsujigaoka 1, Toyohira,
Sapporo 062-8555, Japan

J. Yonemaru
National Institute of Agrobiological Sciences, 2-1-2 Kannondai, Tsukuba,
Ibaraki 305-8602, Japan

T. Yamada (✉)
Field Science Center for Northern Biosphere, Hokkaido University, Kita 11, Nishi 10,
Kita, Sapporo 060-0811, Japan
e-mail: yamada@fsc.hokudai.ac.jp

up- or down regulation of genes involved in agronomic specific traits. Previous reports identified several indels as FNPs for qualitative or quantitative trait in crops (Selinger and Chandler 1999; Ashikari et al. 2002; Palaisa et al. 2003; Guillet-Claude et al. 2004; Kobayashi et al. 2004; Ashikari et al. 2005; Shomura et al. 2008; Zhou et al. 2009; Kawahigashi et al. 2011). In rice, reduced expression or loss of function of *OsCKX2* (cytokinin oxidase/dehydrogenase) caused by deletions in the 5′-untranslated or exonic regions increase the number of reproductive organs, resulting in enhanced grain yield (Ashikari et al. 2005). Palaisa et al. (2003) proposed that yellow endosperm color in maize (*Zea mays* L.), resulting in increased nutritional value, is caused by the upregulation of the *Y1* phytoene synthase gene due to insertions in its promoter. The loss of function or suppression of the *ds1* protein kinase gene in sorghum (*Sorghum bicolor* (L.) Moench) caused by indels or SNPs leads to leaf spot resistance (Kawahigashi et al. 2011). However, for crops lacking whole genome sequence information, identification of FNPs is laborious and the development of genomic resources including DNA markers used for genetic analysis is still required to identify loci and genes of interest.

As one type of indels, polymorphisms in simple sequence repeat (SSR) or microsatellite regions in the genome have been used for marker development of targeted loci based on variation of the number of repeat motifs (Morgante and Olivieri 1993). At present, SSR markers are one of the most efficient tools for genetic analysis due to their abundance, ubiquitous distribution in plant genomes, and especially high ability to detect polymorphisms as co-dominant multi-allelic loci (Kalia et al. 2011). DNA markers in genic regions enable comparative genomic studies due to orthologous relationships among related species (Gale and Devos 1998). Co-linearity revealed by comparative genomic studies enables design of genic markers at specific loci even in species without available genome sequence information. Expression sequence tag (EST)-SSR markers are often used as genic markers. Sonah et al. (2011) reported the frequency of SSR motifs in the cording DNA sequences ranges from 68.1 to 203.7 SSR/Mb in three monocot and three dicot plant genomes. However, there appear to be a limit to the number of SSR motifs in coding regions because they can cause a disruption of the coding sequence.

Recently, polymorphisms in intron regions including indels (intron length polymorphisms, ILPs) have been established as molecular markers. Originally, markers targeting intron polymorphisms in specific genes were used in population genetics, and called exon-primed intron-crossing (EPIC) markers (Palumbi and Baker 1994). Since the beginning the twenty-first century, several genome projects in plant species such as *Arabidopsis thaliana* (L.) Heynh. (Arabidopsis Genome Initiative 2000) and rice (International Rice Genome Sequencing Project 2005) have revealed the exon-intron structures of genes over complete genomes. Thereafter, the development and application of DNA markers targeting intron regions in the whole genome has been progressing in plant species. This concept of marker development was previously called comparative anchor tagged sequence (CATS) markers (Lyons et al. 1997), intron length polymorphism (ILP) markers (Wang et al. 2005), Intron-flanking EST markers (Wei et al. 2005), PCR-based landmark unique gene (PLUG) markers (Ishikawa et al. 2007), and conserved intron scanning (CIS) markers

(Jayashree et al. 2008). In this review, markers targeting intron polymorphism are called 'intron polymorphism (IP) markers' and especially ones targeting ILPs are called 'ILP markers'.

Polymorphisms in Introns

Introns are non-coding sequences in genes that are transcribed into mRNA but removed by splicing. Although introns may have functions such as control of transcription (Fiume et al. 2004) and support of miRNA production (Ruby et al. 2007), they do not code for a protein sequence. Therefore, introns are expected to be exposed to reduced selection pressure during evolution compared to the exon region, resulting in increased variation of intron versus exon DNA sequences. Several studies reported respective differences in polymorphism rates between exons and introns in plants. Between two *Medicago truncatula* Gaertn. genotypes, the frequency of polymorphisms in intron and exon regions of 47 loci were on average, 1 SNP/142 bp and 1 SNP/509 bp, respectively (Choi et al. 2004). Feltus et al. (2006) reported that on a per locus basis, IPs were 12.1/kb while exon polymorphisms were 3.6/kb on average in 114 loci among eight rice genotypes. Frequencies of inter-species polymorphisms in conserved orthologue sets were reported in *Solanum* (87 loci, *S. melongena* L. vs *S. linnaeanum* Hepper & Jaeger, Wu et al. 2009a) and *Capsicum* (214 loci, *C. annuum* L. vs *C. frutescens* L., Wu et al. 2009b). SNP frequencies in introns were 1 SNP/103 bp and 1 SNP/128 bp, respectively, and those in exons were 1 SNP/180 bp and 1 SNP/182 bp in *Solanum* and *Capsicum*, respectively. Indels were identified in 32 out of the 74 introns but only 1 out of 21 exons in *Solanum*, and in 71 out of 171 introns but only 1 out of 43 exons in *Capsicum*. Sequencing of 45 pairs of amplicons from the temperate forage grass perennial ryegrass (*Lolium perenne* L.) and meadow fescue (*Festuca pratensis* Huds.) revealed that all (126) indel sites were found in intron regions, even though about a quarter of the compared sequences covered exon regions (Tamura et al. 2009). These findings suggest that introns are more polymorphic than coding sequences, and this tendency is even more prevalent for indels.

Higher plant species have five to six introns per gene (Roy and Gilbert 2006). Thus, an important question for development of IP markers is: how many polymorphisms are there in intron regions in the whole genome? For already sequenced rice genomes, Wang et al. (2005) estimated that there are 19,064–23,037 (0.414/gene) ILPs between *japonica* rice cultivar 'Nipponbare' and *indica* rice cultivar '93-11'. They reported that the ILP density fluctuates dramatically in the genome and varies among chromosomes, ranging from 23.04 per 1 Mb (chromosome 12) to 45.31 per 1 Mb (chromosome 2). Arai-Kichise et al. (2011) performed resequencing of the whole genome in *japonica* rice cultivar 'Omachi' using next-generation sequencing and compared polymorphisms with those of the 'Nipponbare' genome. Out of 21,149 SNPs and 5,901 indels in genic regions, 10,181 SNPs and 3,450 indels were found in intron regions. Although polymorphic frequencies depend on the genetic diversity among targeted genotypes, these examples of rice indicate that IPs are expected to be not randomly, but ubiquitously distributed over the genome with remarkable density. Further studies reported that the frequency of ILPs is lower (about 1/3) than that of SNPs in introns (Yang et al. 2007; Arai-Kichise et al. 2011).

The frequency of ILPs in intra- and inter-species comparisons was evaluated. Wang et al. (2005) investigated the frequency of polymorphisms of candidate ILP markers among 10 accessions of *japonica* and *indica* rice using electrophoresis of 6% non-denaturing PAGE or 2% agarose gels. As a result, 123 (71.1%) out of the 173 ILP markers were polymorphic between, but monomorphic within subspecies, and thus, subspecies-specific. High species-specificity of ILPs was also reported in *Lolium-Festuca* comparisons (Tamura et al. 2009). Evaluation of 61 ILP markers by genotyping of four cultivars and accessions (32 individuals) of *L. perenne* and *F. pratensis* respectively, suggested that many ILPs were species-specific for *L. perenne* and *F. pratensis*, including fifteen markers, completely distinguishing between *L. perenne* and *F. pratensis* genomes, while being monomorphic within each species. Thus, ILPs are suitable for detecting inter-species polymorphisms. These are useful, e.g., for construction of inter-species genetic linkage maps and phylogenetic and evolutionary studies across related species. However, the frequency of ILPs depends on the genetic diversity among the targeted genotypes. Resequencing of six maize inbred lines by the paired-end sequencing identified 10,436 indels in intron regions ranging from 1 to 6 bp in length (Lai et al. 2010). Panjabi et al. (2008) evaluated polymorphisms between two *Brassica juncea* (L.) Czern. genotypes of parents of a di-haploid mapping population. Electrophoresis of PCR amplicons by 1,180 intron flanking primer pairs using 1.2–2% agarose gels detected intra-species polymorphisms for 383 (32%) markers. Similarly, Li et al. (2010) identified 28–31% of PCR amplicons, including intron regions, showing intra-species ILPs between *Brassica rapa* L. genotypes used as parents of a mapping population. Thus, depending on the species and genotypes used, there may be ILPs with remarkable frequency present within species. Separation methods of amplicons with high separation resolution would increase the frequency of ILP detection.

Examining the intron sequences of 5,811 candidate ILP loci between *japonica* and *indica* rice revealed that only 3.58% of ILPs were due to SSR variation (Wang et al. 2005). Similarly, SSR variation with mono-, di-, and tri-nucleotide repeats were only 11.9% in 126 inter-species ILPs between *L. perenne* and *F. pratensis* (Tamura et al. 2009). Thus, the overlap between ILPs and SSRs is limited, so that ILPs are a source of genetic markers complementing SSRs.

Development of Intron Polymorphic Markers

In this decade, thousands of IP markers including ILP ones have been developed in many plant species (Table 10.1). A general method of development of IP markers is schematically shown in Fig. 10.1. First, intron positions in genes must be identified.

Table 10.1 List of published plant IP marker studies

Publication	Marker name	EST source	Reference genome	No. of primer pairs	Considering uniqueness in genome	Considering conservation of primer region	Species tested for polymorphisms	Types of markers checked for polymorphisms[a]	Data base in web site
Choi et al. (2004)	–	*Medicago truncatula*	*Arabidopsis thaliana*	60	Yes	Yes	*Medicago truncatula* (intra-species)	indel, CAPS, dCAPS, SNP (sequencing)	–
Wang et al. (2005)	ILP marker	(genomic sequence of *Oryza sativa japonica* and *indica* subspecies)		5,811	Yes	No	*Oryza sativa* (inter-subspecies)	indel	–
Wei et al. (2005)	Intron-flanking EST-PCR marker	*Rhododendron catawbiense*	*Arabidopsis thaliana*	44	Yes	No	*Rhododendron* spp. (inter-species)	indel	–
Feltus et al. (2006)	CIPS marker	*Sorghum* spp.. *Pennisetum ciliae*	*Oryza sativa*	384	Yes	Yes (0–1 mismatch)	Poaceae	(sequencing)	–
Fredslund et al. (2006b)	CATS marker	*Lotus japonicus* *Medicago truncatula* *Triticum aestivum* *Sorghum bicolor*	*Arabidopsis thaliana* *Lotus japonicus* *Medicago truncatula* *Oryza sativa*	148 (*Lotus japonicus*) 311 (*Medicago truncatula*) 1,355 (grass species)	Yes	Yes	–	–	http://cgi-www.daimi.au.dk/cgi-chili/GeneticMarkers/table

(continued)

Table 10.1 (continued)

Publication	Marker name	EST source	Reference genome	No. of primer pairs	Considering uniqueness in genome	Considering conservation of primer region	Species tested for polymorphisms	Types of markers checked for polymorphisms[a]	Data base in web site
Wu et al. (2006)	COSII marker	Euasterid I	*Arabidopsis thaliana*	548	Yes	Yes (complete match at 8 bp in 3′ terminal)	*Solanum lycopersicum*, *S. pennellii* (inter-species)	CAPS	http://solgenomics.net/markers/cosii_markers.pl
Chapman et al. (2007)	COS marker	*Lactuca sativa*, *Helianthus annuus*	*Arabidopsis thaliana*	192	Yes	Yes (allow not more than 4 bp difference between lettus and sunflower sequences)	*Helianthus annuus*	(sequencing)	–
Lohithaswa et al. (2007)	CIPS marker	*Allium, Musa*	*Oryza sativa*	2,286 (*Allium*) 2,582 (*Musa*)	Yes	Yes	–	–	http://www.plantgenome.uga.edu/CISP
Sargent et al. (2007)	–	*Fragaria, Malus*	*Fragaria, Malus*	24	No	No	*Fragaria* spp. (inter-species)	indel	–
Yang et al. (2007)	PIP marker	59 species	*Arabidopsis thaliana*, *Oryza sativa*	57,658	No	No	–	–	http://ibi.zju.edu.cn/pgl/pip/index.html
Panjabi et al. (2008)	IP marker	*Brassica*	*Arabidopsis thaliana*	1,180	Yes	Yes	*Brassica juncea* (intra-species)	indel	–

10 Insertion-Deletion Marker Targeting for Intron Polymorphisms

Reference	Marker	Design species	Test species	Number	Transferable	Applicable to other	Application	Type	URL
Ishikawa et al. (2009)	PLUG marker	Triticum aestivum	Oryza sativa	960	Yes	No	Triticum aestivum (among homoeologous genomes)	indel, CAPS	–
Tamura et al. (2009)	Intron-flanking EST marker	Lolium, Festuca	Oryza sativa	209	Yes	Yes	Lolium, Festuca (inter-species)	indel, CAPS	–
Poczai et al. (2010)	IT marker	Solanum tuberosum	Arabidopsis thaliana Solanum lycopersicum	56	Yes	No	Solanum tuberosum, S. migrum (inter species)	indel	–
Shirasawa et al. (2010a)	TEI marker	Solanum lycopersicum	Arabidopsis thaliana	674	Yes	No	Solanum lycopersicum, S. pennellii (inter species)	(high resolution melting analysis)	http://www.kazusa.or.jp/tomato/
Gupta et al. (2011)	ILP marker	Setaria italica	Oryza sativa	98	Yes	No	Setaria italica	indel	–
Tamura et al. (2012)	COTER marker	Triticum aestivum Festuca arundinacea	Oryza sativa	222 (Triticum aestivum) 52 (Festuca arundinacea)	Yes	Yes (complete match at 8 bp in 3′ terminal)	Lolium, Festuca (intra-species)	indel	–

[a]*indel* insertion-deletion, *CAPS* cleavage amplified polymorphic sequences, *dCAPS* derived cleavage amplified polymorphic sequences, *SNP* single nucleotide polymorphism

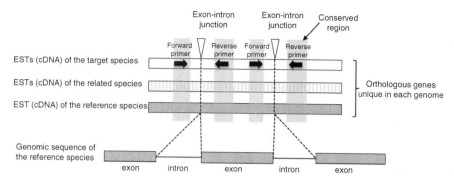

Fig. 10.1 Schematic presentation of the design of conserved primer pairs for intron polymorphism (*IP*) markers. Single copy genes are suitable for developing markers, available as 'anchor' in comparative genomic studies. Copy number of genes can be estimated from the genomic structure of the reference species such as rice and *Arabidopsis*. To amplify intron sequences with a higher polymorphic frequency than in exons, PCR primer pairs are designed in the exon region flanking the exon-intron junction in ESTs (cDNA) of the target species. Exon-intron junctions are conserved among species, and are, thus, estimated from the genomic structure of the orthologous gene in the reference species. If primers are designed in conserved exon regions among related species, they are expected to have a high transferability, useful for comparative studies

Information of genomic sequences, including the exon-intron structure, are limited to few plant species such as rice, but abundant sequence data of expressed genes composed of transcribed exon regions, i.e., cDNAs/ESTs have been collected for many plant species. The exon-intron structures are largely conserved among orthologous genes from different taxa with a large genetic diversity (Fedorov et al. 2002). Therefore, the alignment between cDNA/EST sequences of target species and genomic sequences of the orthologous genes in the related species such as rice and *Arabidopsis* enable deduction of the splice junction sites, which reflects the intron position in genes. In case of single-copy genes, the resulting markers are expected to map to a single locus and they can be used as 'anchors' for linkage maps of related species in comparative genomic and phylogenetic or population genetic studies (Wu et al. 2006; Fredslund et al. 2006a). The gene copy number can be estimated using a reference species with a complete genomic sequence (Fredslund et al. 2006a). For example, Fredslund et al. (2006a) selected 4,062 and 5,644 putative single copy genes from 28,460 to 36,878 gene indices from *Lotus japonicus* (Regel) K. Larsen and *M. truncatula*, respectively, homologous to single or two copy genes in *Arabidopsis* (*Arabidopsis* has undergone a recent whole genome duplication, while legumes have not). Wu et al. (2006) identified 2,869 single copy orthologous genes among Euasterid plants using a combination of bioinformatic and phylogenetic methods.

Second, a pair of PCR primers must be designed to amplify intron regions with potential polymorphisms. Exon regions are highly conserved among individuals and even among species, thus the design of primer pairs from exon sequences increases the chance of successful PCR reactions. Therefore, usually IP primer pairs are designed based on exon sequences, while flanking an intron region. Furthermore, primers derived from conserved exons among a diverse range of species, called conserved intron-scanning primers (CISP), allow to amplify genomes from a wide range of species, which is useful for comparative genomic studies (Feltus et al. 2006). The transferability of CISP sets was evaluated by several researchers. Feltus et al. (2006) designed 384 primer pairs complementary to conserved (completely identical or with 1 mismatch) exon regions flanking introns, using *Sorghum* and *Pennisetum* EST alignments to the rice genome sequence, and confirmed 53–81% transferability of the markers to four Poaceae species with 42 million years of divergence. *Allium–Oryza* and *Musa–Oryza* CISPs showed 30.2–61.5% single-copy amplification success rates in ten monocot species (Lohithaswa et al. 2007). Although conservation in the primer sequence regions is needed for high transferability, a requirement for complete conservation over the primer-annealing region among related species would limit the options for designing primers. Several mismatches in the 5′ end regions of the primer do not necessarily prevent successful amplification (Sommer and Tautz 1989). On wheat (*Triticum aestivum* L.) derived intron-flanking primer sets called PLUG (Ishikawa et al. 2007, 2009), Tamura et al. (2012) investigated the relationship between the homology of primer sequences from wheat to the corresponding rice sequence and the amplification rate of *Lolium* and *Festuca* genomes, which are related to wheat. As a result, identity of the three bases at the 3′- end does not significantly increase the amplification rate over a lower homology, but high complementarity between wheat and rice within the eight bases at the 3′- end of the primer appears to increase the possibility of amplification of *Lolium* and *Festuca* orthologs. Based on these results, conserved 3′- end region (COTER) primers, with complete similarity to reference orthologs for eight bases at the 3′- end of each primer were designed. COTER primer sets developed from both tall fescue and wheat showed high transferability in six temperate grasses (mean amplification rates of 95% for tall fescue primers; 79% for wheat primers). Previously, Wu et al. (2006) developed conserved primer sets for single copy genes in Euasterid I with a similar concept. The single-band amplification rate of ~100 primer sets in six solanaceous species were 40–89% (Wu et al. 2006). Other limiting factors to designing primer pairs include the intron length suitable for PCR amplification and T_m in the primer regions. Although the intron length in orthologous genes varies among species, approximate lengths of introns tend to be conserved. For example, in 145 orthologous genes, wheat intron lengths were correlated significantly with the predicted intron lengths from the rice genome with a small difference of the mean intron length (259.8 bp in wheat and 354.7 bp in rice) (You et al. 2009). Therefore, in addition to intron positions, intron length in orphan crops could be roughly estimated from the reference gene structure of related model species.

Several automated bioinformatics pipelines have been developed for the design of intron-flanking primer sets. GeMprospector (Fredslund et al. 2006b, http://cgi-www.daimi.au.dk/cgi-chili/GeMprospector/main) deals with cross-species intron flanking primer sets in legume and grass families. Of the submitted ESTs, those corresponding to single copy genes containing at least one intron are selected using the genomic sequences of *Lotus japonica*, *M. truncatula* and *Arabidopsis* in the legume application and that of *O. sativa* in the grass application. Primer pairs to amplify intron regions are automatically designed in conserved regions among gene index collections from related species. In a putative intron polymorphism (PIP) database web site (Yang et al. 2007, http://ibi.zju.edu.cn/pgl/pip/index.html), from query ESTs, intron-flanking primer pairs can be developed using the reference genome from rice or *Arabidopsis*. The PIP database in addition contains 57,658 primer pairs from 59 plant species. Although the PIP method does not concern itself with primer conservation between the ESTs and reference genome, corresponding PIP markers with the same or similar primers in other species can be searched, which are useful for comparative genomic research. In the ConservedPrimers 2.0 (You et al. 2009, http://probes.pw.usda.gov/ConservedPrimers/index.html) web application, genome sequences from eight species (*O. sativa*, *S. bicolor*, *Brachypodium distachyon* (L.) P. Beauv., *A. thaliana*, *Glycine max* (L.) Merr., *M. truncatula*, *Vitis vinifera* L., *Populus trichocarpa* Torr. & A. Gray) can be chosen as a reference for an estimation of the uniqueness of the genome, intron position and rough intron length. Primers are designed based on the input ESTs to span intron region(s), but do not consider the conservation with the reference genome. Parameters for primer design can be determined in detail by users.

If the whole genome of the target species is already sequenced, it is possible to develop IP markers with a high frequency of polymorphisms. Wang et al. 2005 developed 5,811 ILP markers based on the ILPs between *japonica* and *indica* rice revealed by the whole genome sequencing. Shu et al. (2010) reported different methods for designing a primer set for ILP markers from the general methods described above. To develop candidate ILP markers, they selected introns with exon-derived insertion such as transposons, using blastn between soybean intron sequences and exon sequences, and designed sequence characterized amplified region (SCAR) primer sets for these loci. Among nine soybean (*G. max*) varieties, 161 of 331 SCAR markers showed polymorphisms.

Detection of Polymorphisms

Two types of polymorphisms are expected in intron regions, namely, ILPs and SNPs. If polymorphisms are unknown among the genotypes of interest, electrophoresis of PCR amplicons including the intron region should be performed, because ILPs can be more easily detected than SNPs. The average length of ILPs between *japonica* and *indica* rice cultivars was 11.42 bp, and 72.6% of ILPs had a length of fewer than 5 bp (Wang et al. 2005). Tamura et al. (2009) reported that 54% of 126 ILPs between *L. perenne* and *F. pratensis* were <5 bp. Agarose gel electrophoresis is an easy method for detecting ILPs but the resolution limited (generally >10% of fragment size), depending on gel concentration and fragment length. Therefore, if detection of small ILPs is required, high-resolution electrophoresis methods should be adapted such as polyacrylamide gel and capillary gel electrophoresis. To obtain a smaller PCR amplicon suitable for detection of small ILPs after amplicon sequencing, design of a PCR primers near the ILP within the intron region is an alternative effective method, which is called Whithin INtron PCR (WIN-PCR) (Wang et al. 2005). Recently, microarray-based methods for genome wide genotyping using indel polymorphisms have been developed in model plants such as *Arabidopsis* (Salathia et al. 2007) and rice (Edwards et al. 2008). A combination of high-throughput genotyping methods and ILPs would facilitate genome wide genotyping in genic regions.

SNPs are more abundant than ILPs in introns as described above. To detect unidentified SNPs, the cleavage amplified polymorphic sequences (CAPS) method using a restriction enzyme with 4 bp recognition is one of the easiest and cheapest methods (Ishikawa et al. 2009; Tamura et al. 2009). Other methods such as high resolution melting analysis (Shirasawa et al. 2010a, b) and single strand conformation polymorphism (SSCP) (Wang et al. 2010) are also applicable to detect undefined SNPs.

Application of Intron Polymorphic Markers

A ubiquitous distribution of IP markers across the whole genome is suitable for genetic map construction and to locate genes and QTL. Due to the low detecting frequency of polymorphisms within species, IP markers have often been used for the construction of interspecific genetic maps, such as in almond (*Prunus dulcis* (Mill.) D.A. Webb) × peach (*P. persica* (L.) Batsch) (Cabrera et al. 2009), related species of pepper (*C. annuum* × *C. frutescens*) (Wu et al. 2009a), eggplant (*S. melongena* × *S. linnaeanum*) (Wu et al. 2009b), tomato (*S. lycopersicum* L. × *S. pennellii* Correll) (Shirasawa et al. 2010a) and wheat × intermediate wheatgrass (*Thinopyrum intermedium* (Host) Barkworth & D.R. Dewey) (Wang et al. 2010). In several crops, low genetic diversity within species due to genetic bottlenecks during domestication and development of modern varieties, makes construction of inter-species linkage maps with related wild relatives necessary, such as in tomato (Shirasawa et al. 2010a). Inter-species linkage maps are also used for determination of species-specific loci for agronomic traits, so that unique chromosome segments can be introduced from donor into agronomic species by inter-species crosses using introgression approaches (Wang et al. 2010). Finally, Ishikawa et al. (2007, 2009) demonstrated the utility of IP markers to distinguish three homoeologous loci in allohexaploid wheat, which aid to identify QTL and associated genes in a specific genome within a complex hexaploid genome structure.

DNA markers, which are specific for species or genera but not polymorphic within these species or genera, such as several ILP markers, are useful for phylogenetic studies on genetic differentiation of related species (Wu et al. 2006; Zhao et al. 2009). Using subspecies-specific ILP markers, Zhao et al. (2009) revealed the *indica-japonica* subspecies differentiation in rice formed in *O. rufipogon* Griff., the progenitor of *O. sativa*, before the beginning of domestication. Species-specific ILP markers are also useful tools for inter-specific hybrid and introgression breeding to detect the constitution of the genomic structure, where multiple genotypes are involved. For example, we successfully used IP markers to distinguish *L. perenne* and *F. pratensis* for the dissection of the genomic structures in the breeding materials of Festulolium hybrids and introgression populations, i.e., *L. perenne* populations with introgressed partial chromosomal fragments from *F. pratensis* (unpublished data).

IP markers have been also applied for constructing intra-species linkage maps such as in *M. truncatula* (Choi et al. 2004), *B. juncea* (Panjabi et al. 2008) and *B. rapa* (Li et al. 2010) based on SNPs in addition to ILPs. IP markers based on specific genes showing intra-species polymorphisms were used for population genetic studies on gene diversity and biodiversity (Ferreira et al. 2009) and identification of cultivars for protecting breeder's right (Shimada et al. 2009).

IP markers with primer sets designed at conserved exons of a single-copy gene could be used as anchors to establish orthologous relationships for comparative genomic studies. Comparison of orthologous loci among different taxa reveals the similarity of genomic structure, namely, synteny at the macro level such as conserved genomic blocks, and at the micro level such as co-linearity (similar order of orthologs). In addition, differences of genomic structure at the macro level can be detected, i.e., chromosome rearrangements mainly caused by paracentric inversions and translocations including respective breaking points, and single gene transpositions at the micro level (Choi et al. 2004; Wu et al. 2006, 2009a, b; Panjabi et al. 2008; Li et al. 2010). These facilitate not only the evolutionary study of plant genomes as basic research, but also applied studies such as genetic analyses of traits of interest contributing to molecular breeding. Syntenic relationships can help in the development of markers in regions of interest, which might support map-based cloning of QTL and may allow educated guesses for candidate genes in the region under investigation elucidated from the reference genome (Fredslund et al. 2006a). For example, to identify the locus related to resistance to the barley yellow dwarf virus (BYDV) on *T. intermedium* 2Ai#2 chromosome, Wang et al. (2010) designed intron-flanking primer pairs from genes on wheat chromosome 2B, homoeologous to *T. intermedium* 2Ai#2. Using wheat-2Ai#2 disomic and substitution lines, the relationships between each marker locus of alien chromosome introductions and the BYDV resistance were analyzed. Burrell et al. (2011) identified a tandem array of three *myb* transcription factors as possible candidate genes underlying differences in sepal color between two *Caulanthus amplexicaulis* S. Wats. varieties, using the syntenic relationship between *C. amplexicaulis* and *Arabidopsis*, revealed by IP markers.

Intron Polymorphisms for Functional Markers

Annotated functional information of genes from which IP markers are derived may help to identify related loci or QTL of interest and may lead to the development of functional and diagnostic markers. Shang et al. (2010) designed 150 intron-flanking primer pairs from wheat resistance gene analogs (RGA) encoding conserved domains such as the nucleotide binding site plus leucine rich repeat (NBS-LRR) using rice homologs as the reference, called RGA-ILP. Twenty-eight RGA-ILP showing ILPs or dominant polymorphisms between mapping parents were mapped in a wheat linkage map and the correspondence was confirmed between the RGA-ILP genetic loci and the physical loci of resistance genes previously reported. Similarly, 1,972 RGA-ILP markers were developed based on 632 RGAs in the maize genome, and 23.2–31.9% of 69 RGA-ILP markers selected randomly showed polymorphisms between any two of the four inbred lines (Liu et al. 2012).

The close proximity of introns to exons makes IP markers well suited for linkage disequilibrium studies to identify the haplotypes associated with traits of interest. Furthermore, polymorphisms in the intron region, especially ILPs, may be directly associated to the variation of traits. Introns are non-coding sequences in genes, but are suggested to have functions such as control of transcription (Fiume et al. 2004). Recently, FNPs in intron regions have been reported in several plant species. In wheat and barley (*Hordeum vulgare* L.) with the winter growth habitat, vernalization, a prolonged exposure to cold during winter, is required for flowering. Up-regulation by low temperature of *VERNALIZATION1* (*VRN1*) coding a MADS-box transcription factor promotes flowering (Yan et al. 2003). Wheat and barley varieties having dominant *Vrn1* alleles with insertions in first intron, *VRN1* are expressed without prior cold treatment, showed the spring growth habitat i.e. flowering without vernalization (Fu et al. 2005; Hemming et al. 2009). Epigenetic regulation through histone modification in the first intron of *VRN1* is suggested to contribute to repression of *VRN1* before winter (Oliver et al. 2009). Using diagnostic markers for *Vrn-A1*, *Vrn-B1* and *Vrn-D1*, in addition to the photoperiod sensitivity gene *Ppd-D1*, Eagles et al. (2009) identified and estimated the effects of combinations of the alleles of the vernalisation and photoperiod genes in Australian wheat. In rice, *GS3* gene for QTL of the grain length have been identified, and SSR polymorphisms in the forth intron is confirmed to be associated with the grain length, in addition to other polymorphisms in exon (Fan et al. 2006; Wang et al. 2011). Iwata et al. (2012) revealed that the continuous flowering during favorable season, an important trait in rose, are caused by the insertion of a retrotransposon in the second intron of the *TERMINAL FLOWER 1* (*TFL1*) homolog. As a result of the insertion mutation, the second intron is not spliced out in the continuous flowering roses and the absence of the matured TFL1 product provokes continuous blooming. Even in orphan crops, functionally annotated information on genes, which is being rapidly collected in species such as rice and other major crop studies, can be advantageously used for the development of functional markers, including those based on FNPs in introns.

Conclusion and Perspectives

IPs, especially ILPs, are more easily detectable and at lower costs than SNPs, with wide length variation. Especially in orphan crops with limited genomic information, due to their cost effectiveness in development, ease of use, functional and locus information from orthologs of reference species, and stability and reliability, ILP markers are useful for genetic analyses of traits of interest as well as comparative genomics, while facilitating molecular breeding. Because of the larger size of polymorphisms compared to SNPs, they are more likely to cause trait variation. Therefore, in some cases, ILPs might be used as FNPs for diagnostic markers directly, and be useful for molecular breeding as described above. Massively parallel 'Next Generation Sequencing' technology is becoming a prevailing genotyping method (Lister et al. 2009). Whole genome sequencing of new plants and crops will reveal detailed genic structures, such as exon-intron junctions, genomic structures including gene orders and copy number variants, and conserved syntenic relationships of genomes among related species and their co-linearity. Resequencing of whole genome will reveal a vast amount of polymorphisms including ILPs (Arai-Kichise et al. 2011; Lai et al. 2010). However, to detect indels especially by short read sequences, technical progress in bioinformatics is required. Deep sequencing of amplicons from specific genes and exome sequencing, a technique focusing on only the protein-coding portion of the genome using sequence capture methods (Albert et al. 2007), could detect a massive amount of indels and SNP variants in genic region and could in some cases be directly diagnostic. These rapidly accumulating information for indels in genic regions might accelerate use of high-throughput genotyping methods (Salathia et al. 2007; Edwards et al. 2008), while especially in orphan crops, single marker systems for indels are still valuable for detailed genetic analyses such as fine mapping of agronomic traits.

References

Albert TJ, Molla MN, Muzny DM, Nazareth L, Wheeler D, Song XZ, Richmond TA, Middle CM, Rodesch MJ, Packard CJ, Weinstock GM, Gibbs RA (2007) Direct selection of human genomic loci by microarray hybridization. Nat Methods 4:903–905

Arabidopsis Genome Initiative (2000) Analysis of the genome sequence of the flowering plant *Arabidopsis thaliana*. Nature 408:796–815

Arai-Kichise Y, Shiwa Y, Nagasaki H, Ebana K, Yoshikawa H, Yano M, Wakasa K (2011) Discovery of genome-wide DNA polymorphisms in a landrace cultivar of *japonica* rice by whole-genome sequencing. Plant Cell Physiol 52:274–282

Ashikari M, Sasaki A, Ueguchi-Tanaka M, Itoh H, Nishimura A, Datta S, Ishiyama K, Saito T, Kobayashi M, Khush GS, Kitano H, Matsuoka M (2002) Loss-of-function of a rice gibberellin biosynthetic gene, *GA20 oxidase* (*GA20ox-2*), led to the rice 'green revolution'. Breed Sci 52:143–150

Ashikari M, Sakakibara H, Lin SY, Yamamoto T, Takashi T, Nishimura A, Angeles ER, Qian Q, Kitano H, Matsuoka M (2005) Cytokinin oxidase regulates rice grain production. Science 309:741–745

Burrell AM, Taylor KG, Williams RJ, Cantrell RT, Menz MA, Pepper AE (2011) A comparative genomic map for *Caulanthus amplexicaulis* and related species (Brassicaceae). Mol Ecol 20:784–798

Cabrera A, Kozik A, Howad W, Arus P, Iezzoni AF, van der Knaap E (2009) Development and bin mapping of a Rosaceae Conserved Ortholog Set (COS) of markers. BMC Genomics 10:562

Chapman MA, Chang J, Weisman D, Kesseli RV, Burke JM (2007) Universal markers for comparative mapping and phylogenetic analysis in the Asteraceae (Compositae). Theor Appl Genet 115:747–755

Choi HK, Kim D, Uhm T, Limpens E, Lim H, Mun JH, Kalo P, Penmetsa RV, Seres A, Kulikova O, Roe BA, Bisseling T, Kiss GB, Cook DR (2004) A sequence-based genetic map of *Medicago truncatula* and comparison of marker colinearity with *M. sativa*. Genetics 166:1463–1502

Eagles HA, Cane K, Vallance N (2009) The flow of alleles of important photoperiod and vernalisation genes through Australian wheat. Crop Pasture Sci 60:646–657

Edwards JD, Janda J, Sweeney MT, Gaikwad AB, Liu B, Leung H, Galbraith DW (2008) Development and evaluation of a high-throughput, low-cost genotyping platform based on oligonucleotide microarrays in rice. Plant Methods 4:13

Fan CH, Xing YZ, Mao HL, Lu TT, Han B, Xu CG, Li XH, Zhang QF (2006) *GS3*, a major QTL for grain length and weight and minor QTL for grain width and thickness in rice, encodes a putative transmembrane protein. Theor Appl Genet 112:1164–1171

Fedorov A, Merican AF, Gilbert W (2002) Large-scale comparison of intron positions among animal, plant, and fungal genes. Proc Natl Acad Sci USA 99:16128–16133

Feltus FA, Singh HP, Lohithaswa HC, Schulze SR, Silva TD, Paterson AH (2006) A comparative genomics strategy for targeted discovery of single-nucleotide polymorphisms and conserved-noncoding sequences in orphan crops. Plant Physiol 140:1183–1191

Ferreira AO, Cardoso HG, Macedo ES, Breviario D, Arnholdt-Schmitt B (2009) Intron polymorphism pattern in *AOX1b* of wild St John's wort (*Hypericum perforatum*) allows discrimination between individual plants. Physiol Plant 137:520–531

Fiume E, Christou P, Giani S, Breviario D (2004) Introns are key regulatory elements of rice tubulin expression. Planta 218:693–703

Fredslund J, Madsen LH, Hougaard BK, Nielsen AM, Bertioli D, Sandal N, Stougaard J, Schauser L (2006a) A general pipeline for the development of anchor markers for comparative genomics in plants. BMC Genomics 7:207

Fredslund J, Madsen LH, Hougaard BK, Sandal N, Stougaard J, Bertioli D, Schauser L (2006b) GeMprospector – online design of cross-species genetic marker candidates in legumes and grasses. Nucleic Acids Res 34:W670–W675

Fu D, Szűcs P, Yan L, Helguera M, Skinner JS, von Zitzewitz J, Hayes PM, Dubcovsky J (2005) Large deletions within the first intron in *VRN-1* are associated with spring growth habit in barley and wheat. Mol Genet Genomics 273:54–65

Gale MD, Devos KM (1998) Comparative genetics in the grasses. Proc Natl Acad Sci USA 95:1971–1974

Guillet-Claude C, Birolleau-Touchard C, Manicacci D, Rogowsky PM, Rigau J, Murigneux A, Martinant JP, Barriere Y (2004) Nucleotide diversity of the *ZmPox3* maize peroxidase gene: relationships between a MITE insertion in exon 2 and variation in forage maize digestibility. BMC Genet 5:19

Gupta S, Kumari K, Das J, Lata C, Puranik S, Prasad M (2011) Development and utilization of novel intron length polymorphic markers in foxtail millet (*Setaria italica* (L.) P. Beauv.). Genome 54:586–602

Hemming MN, Fieg S, Peacock WJ, Dennis ES, Trevaskis B (2009) Regions associated with repression of the barley (*Hordeum vulgare*) *VERNALIZATION1* gene are not required for cold induction. Mol Genet Genomics 282:107–117

International Rice Genome Sequencing Project (2005) The map-based sequence of the rice genome. Nature 436:793–800

Ishikawa G, Yonemaru J, Saito M, Nakamura T (2007) PCR-based landmark unique gene (PLUG) markers effectively assign homoeologous wheat genes to A, B and D genomes. BMC Genomics 8:135

Ishikawa G, Nakamura T, Ashida T, Saito M, Nasuda S, Endo TR, Wu JZ, Matsumoto T (2009) Localization of anchor loci representing five hundred annotated rice genes to wheat chromosomes using PLUG markers. Theor Appl Genet 118:499–514

Iwata H, Gaston A, Remay A, Thouroude T, Jeauffre J, Kawamura K, Oyant LHS, Araki T, Denoyes B, Foucher F (2012) The *TFL1* homologue *KSN* is a regulator of continuous flowering in rose and strawberry. Plant J 69:116–125

Jayashree B, Jagadeesh VT, Hoisington D (2008) CISprimerTOOL: software to implement a comparative genomics strategy for the development of conserved intron scanning (CIS) markers. Mol Ecol Resour 8:575–577

Kalia RK, Rai MK, Kalia S, Singh R, Dhawan AK (2011) Microsatellite markers: an overview of the recent progress in plants. Euphytica 177:309–334

Kawahigashi H, Kasuga S, Ando T, Kanamori H, Wu JZ, Yonemaru J, Sazuka T, Matsumoto T (2011) Positional cloning of *ds1*, the target leaf spot resistance gene against *Bipolaris sorghicola* in sorghum. Theor Appl Genet 123:131–142

Kobayashi S, Goto-Yamamoto N, Hirochika H (2004) Retrotransposon-induced mutations in grape skin color. Science 304:982

Lai JS, Li RQ, Xu X, Jin WW, Xu ML, Zhao HN, Xiang ZK, Song WB, Ying K, Zhang M, Jiao YP, Ni PX, Zhang JG, Li D, Guo XS, Ye KX, Jian M, Wang B, Zheng HS, Liang HQ, Zhang XQ, Wang SC, Chen SJ, Li JS, Fu Y, Springer NM, Yang HM, Wang JA, Dai JR, Schnable PS, Wang J (2010) Genome-wide patterns of genetic variation among elite maize inbred lines. Nat Genet 42:1027–1130

Li X, Ramchiary N, Choi SR, Van Nguyen D, Hossain MJ, Yang HK, Lim YP (2010) Development of a high density integrated reference genetic linkage map for the multinational *Brassica rapa* Genome Sequencing Project. Genome 53:939–947

Lister R, Gregory BD, Ecker JR (2009) Next is now: new technologies for sequencing of genomes, transcriptomes, and beyond. Curr Opin Plant Biol 12:107–118

Liu HL, Lin YA, Chen GB, Shen Y, Liu J, Zhang SZ (2012) Genome-scale identification of resistance gene analogs and the development of their intron length polymorphism markers in maize. Mol Breed 29:437–447

Lohithaswa HC, Feltus FA, Singh HP, Bacon CD, Bailey CD, Paterson AH (2007) Leveraging the rice genome sequence for monocot comparative and translational genomics. Theor Appl Genet 115:237–243

Lyons LA, Laughlin TF, Copeland NG, Jenkins NA, Womack JE, Obrien SJ (1997) Comparative anchor tagged sequences (CATS) for integrative mapping of mammalian genomes. Nat Genet 15:47–56

Miura K, Ashikari M, Matsuoka M (2011) The role of QTLs in the breeding of high-yielding rice. Trends Plant Sci 16:319–326

Morgante M, Olivieri AM (1993) PCR-amplified microsatellites as markers in plant genetics. Plant J 3:175–182

Oliver SN, Finnegan EJ, Dennis ES, Peacock WJ, Trevaskis B (2009) Vernalization-induced flowering in cereals is associated with changes in histone methylation at the *VERNALIZATION1* gene. Proc Natl Acad Sci USA 106:8386–8391

Palaisa KA, Morgante M, Williams M, Rafalski A (2003) Contrasting effects of selection on sequence diversity and linkage disequilibrium at two phytoene synthase loci. Plant Cell 15:1795–1806

Palumbi SR, Baker CS (1994) Contrasting population-structure from nuclear intron sequences and mtDNA of humpback whales. Mol Biol Evol 11:426–435

Panjabi P, Jagannath A, Bisht NC, Padmaja KL, Sharma S, Gupta V, Pradhan AK, Pental D (2008) Comparative mapping of *Brassica juncea* and *Arabidopsis thaliana* using Intron Polymorphism (IP) markers: homoeologous relationships, diversification and evolution of the A, B and C Brassica genomes. BMC Genomics 9:113

Poczai P, Cernák I, Gorji AM, Nagy S, Taller J, Polgár Z (2010) Development of intron targeting (IT) markers for potato and cross-species amplification in *Solanum nigrum* (Solanaceae). Am J Bot 97:e142–e145

Roy SW, Gilbert W (2006) The evolution of spliceosomal introns: patterns, puzzles and progress. Nat Rev Genet 7:211–221

Ruby JG, Jan CH, Bartel DP (2007) Intronic microRNA precursors that bypass Drosha processing. Nature 448:83–86

Salathia N, Lee HN, Sangster TA, Morneau K, Landry CR, Schellenberg K, Behere AS, Gunderson KL, Cavalieri D, Jander G, Queitsch C (2007) Indel arrays: an affordable alternative for genotyping. Plant J 51:727–737

Sargent DJ, Rys A, Nier S, Simpson DW, Tobutt KR (2007) The development and mapping of functional markers in *Fragaria* and their transferability and potential for mapping in other genera. Theor Appl Genet 114:373–384

Selinger DA, Chandler VL (1999) Major recent and independent changes in levels and patterns of expression have occurred at the *b* gene, a regulatory locus in maize. Proc Natl Acad Sci USA 96:15007–15012

Shang W, Zhou R, Jia J, Gao L (2010) RGA-ILP, a new type of functional molecular markers in bread wheat. Euphytica 172:263–273

Shimada N, Nakatsuka T, Nakano Y, Kakizaki Y, Abe Y, Hikage T, Nishihara M (2009) Identification of gentian cultivars using SCAR markers based on intron-length polymorphisms of flavonoid biosynthetic genes. Sci Hortic (Amsterdam) 119:292–296

Shirasawa K, Asamizu E, Fukuoka H, Ohyama A, Sato S, Nakamura Y, Tabata S, Sasamoto S, Wada T, Kishida Y, Tsuruoka H, Fujishiro T, Yamada M, Isobe S (2010a) An interspecific linkage map of SSR and intronic polymorphism markers in tomato. Theor Appl Genet 121:731–739

Shirasawa K, Isobe S, Hirakawa H, Asamizu E, Fukuoka H, Just D, Rothan C, Sasamoto S, Fujishiro T, Kishida Y, Kohara M, Tsuruoka H, Wada T, Nakamura Y, Sato S, Tabata S (2010b) SNP discovery and linkage map construction in cultivated tomato. DNA Res 17:381–391

Shomura A, Izawa T, Ebana K, Ebitani T, Kanegae H, Konishi S, Yano M (2008) Deletion in a gene associated with grain size increased yields during rice domestication. Nat Genet 40:1023–1028

Shu Y, Li Y, Zhu Y, Zhu Z, Lv D, Bai X, Cai H, Ji W, Guo D (2010) Genome-wide identification of intron fragment insertion mutations and their potential use as SCAR molecular markers in the soybean. Theor Appl Genet 121:1–8

Sommer R, Tautz D (1989) Minimal homology requirements for PCR primers. Nucleic Acids Res 17:6749

Sonah H, Deshmukh RK, Sharma A, Singh VP, Gupta DK, Gacche RN, Rana JC, Singh NK, Sharma TR (2011) Genome-wide distribution and organization of microsatellites in plants: an insight into marker development in *Brachypodium*. PLoS One 6:e21298

Tamura K, Yonemaru J, Hisano H, Kanamori H, King J, King IP, Tase K, Sanada Y, Komatsu T, Yamada T (2009) Development of intron-flanking EST markers for the *Lolium/Festuca* complex using rice genomic information. Theor Appl Genet 118:1549–1560

Tamura K, Kiyoshi T, Yonemaru J (2012) The development of highly transferable intron-spanning markers for temperate forage grasses. Mol Breed 30:1–8

Wang X, Zhao X, Zhu J, Wu W (2005) Genome-wide investigation of intron length polymorphisms and their potential as molecular markers in rice (*Oryza sativa* L.). DNA Res 12:417–427

Wang MJ, Zhang Y, Lin ZS, Ye XG, Yuan YP, Ma W, Xin ZY (2010) Development of EST-PCR markers for *Thinopyrum intermedium* chromosome 2Ai#2 and their application in characterization of novel wheat-grass recombinants. Theor Appl Genet 121:1369–1380

Wang C, Chen S, Yu S (2011) Functional markers developed from multiple loci in *GS3* for fine marker-assisted selection of grain length in rice. Theor Appl Genet 122:905–913

Wei H, Fu Y, Arora R (2005) Intron-flanking EST-PCR markers: from genetic marker development to gene structure analysis in *Rhododendron*. Theor Appl Genet 111:1347–1356

Wu F, Mueller LA, Crouzillat D, Petiard V, Tanksley SD (2006) Combining bioinformatics and phylogenetics to identify large sets of single-copy orthologous genes (COSII) for comparative, evolutionary and systematic studies: a test case in the euasterid plant clade. Genetics 174:1407–1420

Wu F, Eannetta NT, Xu Y, Durrett R, Mazourek M, Jahn MM, Tanksley SD (2009a) A COSII genetic map of the pepper genome provides a detailed picture of synteny with tomato and new insights into recent chromosome evolution in the genus *Capsicum*. Theor Appl Genet 118:1279–1293

Wu F, Eannetta NT, Xu Y, Tanksley SD (2009b) A detailed synteny map of the eggplant genome based on conserved ortholog set II (COSII) markers. Theor Appl Genet 118:927–935

Yamamoto T, Yonemaru J, Yano M (2009) Towards the understanding of complex traits in rice: substantially or superficially? DNA Res 16:141–154

Yan L, Loukoianov A, Tranquilli G, Helguera M, Fahima T, Dubcovsky J (2003) Positional cloning of the wheat vernalization gene *VRN1*. Proc Natl Acad Sci USA 100:6263–6268

Yang L, Jin G, Zhao X, Zheng Y, Xu Z, Wu W (2007) PIP: a database of potential intron polymorphism markers. Bioinformatics 23:2174–2177

You FM, Huo N, Gu YQ, Lazo GR, Dvorak J, Anderson OD (2009) ConservedPrimers 2.0: a high-throughput pipeline for comparative genome referenced intron-flanking PCR primer design and its application in wheat SNP discovery. BMC Bioinformatics 10:331

Zhao X, Yang L, Zheng Y, Xu Z, Wu W (2009) Subspecies-specific intron length polymorphism markers reveal clear genetic differentiation in common wild rice (*Oryza rufipogon* L.) in relation to the domestication of cultivated rice (*O. sativa* L.). J Genet Genomics 36:435–442

Zhou Y, Zhu JY, Li ZY, Yi CD, Liu J, Zhang HG, Tang SZ, Gu MH, Liang GH (2009) Deletion in a quantitative trait gene *qPE9-1* associated with panicle erectness improves plant architecture during rice domestication. Genetics 183:315–324

Chapter 11
Evolving Molecular Marker Technologies in Plants: From RFLPs to GBS

Reyazul Rouf Mir, Pavana J. Hiremath, Oscar Riera-Lizarazu, and Rajeev K. Varshney

Introduction

Analysis of DNA-sequence variation (or allelic state) at a specific chromosomal location in an individual/genotype is referred to as genotyping. Variation in the DNA sequence may or may not have functional significance. For example, variation may result either in a synonymous or non-synonymous change in a codon. Such alterations may either cause a favorable change or deleterious mutations (mis-sense or non-sense) in an organism. Genetic variation may be small changes in frame (point-mutations, substitutions) or frame-shifts (insertions or deletions) (Jones et al. 2009). Nevertheless, these variations have been used as molecular markers to understand genome architecture as well as for plant breeding applications. Marker genotyping has various applications including parental genotype selection,

---

R.R. Mir
Centre of Excellence in Genomics (CEG), International Crops Research Institute for the Semi-Arid Tropics (ICRISAT), Patancheru, Hyderabad 502 324, India

Division of Plant Breeding & Genetics, Shere-Kashmir University of Agricultural Sciences & Technology of Jammu (SKUAST-J), Chatha, Jammu 180 009, India

P.J. Hiremath • O. Riera-Lizarazu
Centre of Excellence in Genomics (CEG), International Crops Research Institute for the Semi-Arid Tropics (ICRISAT), Patancheru, Hyderabad 502 324, India

R.K. Varshney (✉)
Centre of Excellence in Genomics (CEG), International Crops Research Institute for the Semi-Arid Tropics (ICRISAT), Patancheru, Hyderabad 502 324, India

CGIAR-Generation Challenge Programme (GCP), c/o CIMMYT, Int APDO Postal 6-641, 06600 Mexico City, DF, Mexico

School of Plant Biology (M084), The University of Western Australia, 35 Stirling Highway, Crawley, WA 6009, Australia
e-mail: r.k.varshney@cgiar.org

screening mapping populations, genome mapping, trait mapping, germplasm diversity assessment, marker-assisted selection, linkage drag elimination in backcrossing and identification of genomic re-arrangements across taxa (Jain et al. 2002).

Variation in germplasm collections has been harnessed at both the morphological as well as molecular level. When morphological traits, including plant height, tillering, photoperiod, seed type, texture, leaf shape, and flower colour, have been used for assessing and utilizing genetic variation, they are referred to as 'morphological markers' (Tanksley 1983; Emami and Sharma 1999). As morphological markers are normally limited in number, the genetics and breeding community found a need to use enzymes and DNA polymorphisms as markers, which are referred to as biochemical and DNA-based 'molecular markers', respectively. Although biochemical markers are also molecular markers, the term is mostly used to refer to DNA-based polymorphisms. Molecular markers can provide genomic information for plant evaluation before entering the next cycle of selection which is critical for success in plant breeding (Bagge and Lűbberstedt 2008) and also help track polymorphisms with no obvious phenotype.

Due to advances in automation coupled with the demand of increasing throughput in a cost-effective manner, molecular marker technology has evolved during the last three decades. Based on their degree of multiplexing capacity /throughput, i.e., number of genetic loci per experiment, available molecular markers can be classified into the following categories: (i) low-throughput (100s of loci on 100s of lines), (ii) medium-throughput (from 100s up to 1,000s of loci on 1,000s of lines), (iii) high-throughput (1,000s of loci on 1,000s of lines), and (iv) ultra-high throughput marker systems (from 1,000s loci up to 50,000 loci on 1,000s of lines) (Fig. 11.1). This article provides a brief overview of the different molecular markers in these categories with a major emphasis on emerging genotyping technologies including genotyping-by-sequencing (GBS). It is anticipated that new marker technologies/genotyping platforms will facilitate development of functional molecular markers (Table 11.1, Fig. 11.1).

Low-Throughput Marker Systems

Restriction Fragment Length Polymorphisms (RFLPs)

RFLPs initialized the era of DNA marker technology during the 1980s in plant genetic studies and are, therefore, referred to as 'First generation molecular markers' (Jones et al. 2009). The polymorphisms detected by RFLPs are due to changes in nucleotide sequences in recognition sites of restriction enzymes or due to insertions or deletions of several nucleotides leading to detectable shift in fragment size (Tanksley et al. 1989). RFLPs have several advantages including high reproducibility, a co-dominant nature, no need of prior sequence information, and high locus-specificity. By using RFLP markers, genetic maps have been developed in several crop species including rice (McCouch et al. 1988), maize (Helentjaris 1987), wheat (Chao et al.

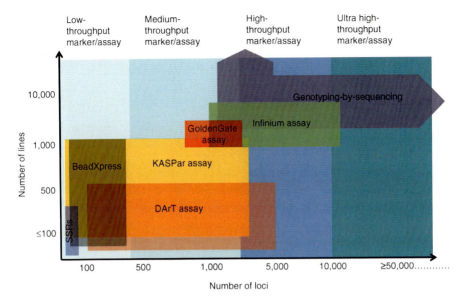

Fig. 11.1 Low to ultra high-throughput cost-effective marker assay platforms for genotyping. *Horizontal axis* indicates number of loci that can be assayed in a single experiment, while the *vertical axis* indicates the number of lines/samples that can be genotyped in high-throughput manner at low cost

1989), soybean (Keim et al. 1990), tomato and potato (Tanksley et al. 1992), barley (Graner et al. 1991), and chickpea (Simon and Muehlbauer 1997). Although these markers have also been used for trait mapping (see Varshney et al. 2005; Gupta et al. 2010), they have not been found to be very useful for plant breeding applications. This can be attributed to the tedious and time consuming procedure involving their use as well as a general inability to automate the procedure.

Medium-Throughput Marker Systems

The revolutionary advent of PCR during the 1980s stimulated development of different molecular marker types. A brief overview over some of these markers is provided below.

Random Amplified Polymorphic DNA's (RAPDs)

RAPDs are probably the first PCR based genetic markers that were easy to use and inexpensive as no prior sequence information is required (Williams et al. 1990). They are used as universal markers for species with little or no genomic resources

Table 11.1 Comparison between different marker systems

Sl. No.	Attribute	Low-throughput markers	Medium-throughput markers			High-throughput markers				Ultra high-throughput markers	
		RFLP	RAPD	AFLP	SSR	KASPar	GoldenGate Beadchip	GoldenGate Beadarray	DArT assays	Infinium assays	GBS
1	DNA amount/reaction	2–10 μg	5–10 ng	~1 μg	10–20 ng	≥5 ng	+	+	50–100 ng	50–100 ng +	100 ng
2	DNA quality	High	Moderate	Moderate	Moderate	Moderate	Moderate	Moderate	High	Moderate	High
3	Cost	High	Less	High	High	Low	High	High	Cheapest	High	Low/moderate
4	Reproducibility	Very high	Low	High	High	High	High	High	High	High	High
5	Radioactivity	Yes	–	Yes	–	–	–	–	–	–	–
6	Markers assayable	<100	<100	>100	>500	As per requirement	48–384	3,072	96–5,000	1,000–5,000	>1,000–
7	Technical procedure	Tedious	Simpler	Tedious	Simpler	Automated	Automated	Automated	Simpler	Automated	Automated
8	Sample size	<50–100	<100	<100	48–384	<100–1,000	>1,000–3,072	48–1,000	100–500	>1,000	>1,000
9	Sequence information	–	–	–	Yes	Yes	Yes	Yes	–	Yes	Yes
10	Multiplexing	Difficult	Difficult	Possible	Possible	Possible	Possible	Possible	No	Possible	Possible

available. RAPD markers have been extensively used in different plant species for fingerprinting, assessment of genetic variation in populations and species, study of phylogenetic relationships among species/subspecies and cultivars, and for many other purposes including gene tagging (see Gupta et al. 1999). However, RAPD markers are dominant and cannot distinguish between homozygous and heterozygous individuals. Furthermore, due to their random nature of amplification and short primer length, they are not a preferred choice for genome mapping. In addition, these markers do not exhibit reliable amplification patterns, are not reproducible, and vary with the experimental conditions (Heun and Helentjaris 1993; Ellsworth et al. 1993).

Simple Sequence Repeats (SSRs)

Simple sequence repeats (SSRs) or microsatellites were developed during 1990s and provided a choice for various studies since they are amenable to low, medium and high-throughput approaches. SSRs are easily assayable by gel electrophoresis for few to hundreds of samples, which could be affordable by laboratories with limited resources. SSRs are often derived from non-coding/anonymous genomic regions, such as genomic survey sequences (GSSs) and bacterial artificial chromosomes (BACs). As a result, development of SSR markers used to be expensive and laborious. In recent years, however, due to the availability of large-scale gene/EST (expressed sequence tag) sequence information for various plant species, SSR markers can easily be developed *in silico*. Such markers have been referred to as genic SSR markers and have been developed in a very cost-effective manner (Varshney et al. 2005). The high degree of polymorphism as compared to RFLPs and RAPDs, their locus specific and co-dominant nature, make them the markers of choice for a variety of purposes including practical plant breeding (Gupta and Varshney 2000). SSR markers dominated genetics research and breeding applications, especially in plants for more than a decade. SSR markers are probably the only class of markers that have been used for almost all aspects of genetics research and breeding in a wide range of plant species (Gupta and Varshney 2000; Varshney et al. 2005).

Amplified Fragment Length Polymorphism's (AFLPs)

Amplified fragment length polymorphism is a multi-locus marker technique that combines the techniques of restriction digestion and selective PCR amplification of restriction fragments and can be applied to DNA of any origin or complexity (Vos et al. 1995). The use of AFLP markers is cost-effective, since it needs moderate amounts of DNA, and a single assay allows simultaneous detection of a large number of co-amplified restriction fragments. Moreover, AFLPs are considered to be a robust and reliable genotyping technique, as stringent primer annealing conditions are used. The high frequency of identifiable AFLP bands coupled with a high reproducibility makes this technology an attractive tool for fingerprinting, constructing genetic maps and saturating genomic regions with low marker density (Gupta et al. 1999). In addition, the property of reliable inheritance and transferability of these markers have encouraged their application in genetic diversity analyses in several crop species like rice (Mackill et al. 1996; Zhu et al. 1998; Maheshwaran et al. 1997), wheat (Huang et al. 2000; Xu and Ban 2004; Barrett et al. 1998; Shan et al. 1999; Soleimani et al. 2002), barley (Faccioli et al. 1999; Shan et al. 1999), and also in legume species like soybean (Maughan et al. 1996; Young et al. 1999) and chickpea (Winter et al. 2000; Nguyen et al. 2003). While AFLPs have also been used for trait mapping in several instances, the conversion of associated AFLP markers into a locus-specific and user-friendly marker such as a sequence tagged site (STS) or sequence characterized amplified region (SCAR) has not always been straightforward. Therefore, use of AFLP markers has not been common for molecular breeding applications (Xu and Ban 2004).

High-Throughput Marker Systems

Molecular breeding in general involves screening of large segregating populations with molecular markers. Therefore, screening of markers in a high-throughput manner can offer cost-effective marker genotyping and enhance adoption of molecular markers in plant breeding applications. In this context, genotyping of SSR markers in a high-throughput manner has been adopted by using ABI capillary sequencing electrophoresis and the Multiplex-Ready^{TM} marker technology (MRT) (Appleby et al. 2009). Despite of those high-throughput SSR platforms, costs are still prohibitive for many breeding programs.

Single Nucleotide Polymorphisms (SNPs)

Single nucleotide polymorphisms (SNPs) are the most abundant sequence variation in nature (frequency varies with each organism/species) (Rafalski 2002). SNPs are mostly bi-allelic and arise either due to substitutions/point mutations (transition and transversion) or due to insertion/deletion of nucleotides and are detectable when similar genomic regions from different genotypes of same or different species are aligned. Their occurrence in coding sequence may be linked to phenotypic changes in an organism. SNPs are not only efficient in terms of reliability, reproducibility and transferability, but are also amenable to automation and high-throughput approaches.

Although initially development of SNP markers was considered expensive as it mainly involved allele-specific sequencing, the advent of NGS or second generation sequencing technologies (454/FLX, Solexa/Illumina, SOLiD/ABI) has brought sequencing cost down (Thudi et al. 2012). Very recently, the third (or future) generation sequencing technologies such as single molecule sequencing (PacBio/Pacific Biosciences, USA; HeliScope/Helicos Biosciences, USA), and Polonator (Dover/Harvard, USA) started to emerge (Thudi et al. 2012). These third generation sequencing technologies are expected to further reduce sequencing costs drastically to levels below $1 per mega base compared to $60, $2, and $1 estimated costs for sequences generated by 454/FLX, SOLiD/ABI, and Solexa/Illumina, respectively. All these sequencing technologies are being used for whole genome *de novo* and re-sequencing studies (syteny.cnr.berkeley.edu/wiki/index.php/Sequenced_plant_genomes), reduced representation sequencing (Hyten et al. 2010a; Davey et al. 2011), targeted genomic sequencing (Delmas et al. 2011; Griffin et al. 2011), paired-end sequencing (Rounsley et al. 2009), meta-genomic sequencing (Ottesen et al. 2011), transcriptome sequencing (Cheung et al. 2006; Hiremath et al. 2011), small RNA sequencing (Gonzalez-Ibeas et al. 2011; Zhou et al. 2009), and chromatin immune-precipitation sequencing (ChIP) (Shendure and Ji 2008; Varshney et al. 2009). As a result, it has become easier and very-cost effective to quickly identify a large number of SNPs in short time in any plant species.

For genotyping SNP markers in low to medium-throughput approaches, more than 30 assays are currently available that can be classified into four reaction principles or chemistries: hybridization with allele-specific oligonucleotide probes, oligonucleotide ligation, single nucleotide primer extension, and enzymatic cleavage (Gupta et al. 2001; Kwok 2001; Syvanen 2005; Steemers et al. 2006). However, very recently additional SNP genotyping platforms from the company Illumina have been developed and discussed below in detail.

GoldenGate Assays

Illumina's GoldenGate assay provides SNP genotyping for genome-wide marker profiling. Thus, one can select any number of SNPs (for each of the samples to be genotyped) and the throughput level best suited for a study. GoldenGate assays may be developed for any crop species using either BeadArray, or VeraCode technology (Thomson et al. 2011). On the basis of level of multiplexing and through put, GoldenGate assays can be classified into: (i) GoldenGate BeadArray allowing simultaneous genotyping of 96-, 192-, 384, 768-, 1,536- and 3,072-SNP loci in a fairly large collection of samples, (ii) GoldenGate VeraCode (*BeadXpress*) allowing genotyping of 48-, 96-, 192-, and 384-plexes, and (iii) GoldenGate Indexing allowing genotyping of 96–384 SNPs simultaneously. Among these GoldenGate Indexing screen up to 16 times more samples per reaction than one can do with the standard GoldenGate assay thereby decreasing costs of the genotyping assay. These assays are used for a variety of applications such as association mapping, linkage mapping, and diversity analyses in crops like rice (see McCouch et al. 2010; Thomson et al. 2011), wheat (Akhunov et al. 2009; Chao et al. 2010), barley

(Rostoks et al. 2006; Close et al. 2009; Druka et al. 2011), maize (Yan et al. 2010; Mammadov et al. 2012), soybean (Hyten et al. 2008; Hyten et al. 2009), common bean (Hyten et al. 2010b), pea (Deulvot et al. 2010) and cowpea (Muchero et al. 2009).

Competitive Allele-Specific PCR (KASPar) Assays

Above mentioned GoldenGate (GG) assays by Illumina seem to be superior for genotyping a large number of SNPs/sample for several samples. This makes KASPar a simple, cost-effective and flexible genotyping system, since the assays can be adjusted with a range of DNA samples. However, some molecular breeding applications such as marker-assisted selection (MAS) or marker-assisted backcrossing (MABC) employ genotyping of large number of lines with only few SNPs. In such cases, new genotyping assays that involve competitive allele-specific PCR for a given SNP, followed by SNP detection via Fluorescence Resonance Energy Transfer (FRET) have been developed (Chen et al. 2010). These assays for the target SNPs are being developed and used for genotyping commercially by Kbioscience UK (http://www.kbioscience.co.uk/) and are referred as KBioScience Allele-Specific Polymorphism (KASP) or KASPar assays. One of the advantages of using KASPar assays is that there is no need of sequencing to identify SNPs for assay development, instead SNP flanking sequences already known while developing different types of genotyping assays (e.g., Illumina) can readily be used for primer design (one common and two allele-specific primers) for KASPar assays (for review see McCouch et al. 2010). Although KASPar genotyping assays have come to the market very recently, they have started to be used for genetic diversity studies (Maughan et al. 2011; Cortes et al. 2011; Hiremath et al. 2012) and genetic mapping (Allen et al. 2011; Hiremath et al. 2012; Saxena et al. 2012).

Diversity Array Technology (DArT)

Diversity array technology (DArT) is a high-throughput microarray hybridization based assay involving genotyping of several hundred polymorphic loci simultaneously spread over the genome without prior sequence information (Jaccoud et al.2001). DArT markers are bi-allelic and behave mostly in a dominant (presence *vs* absence) or sometimes in a co-dominant (2 doses *vs* 1 dose *vs* absent) manner. These markers usually detect polymorphisms due to single base-pair changes (SNPs) within restriction sites recognized by endonucleases, or due to insertion/deletion (InDels) or rearrangements (Jaccoud et al. 2001). The technique is reproducible and cost-effective, and has become available for >70 species of both plants and animals (http://www.diversityarrays.com/genotypingserv.html). In plants, DArTs have been already developed in all major crop species including rice (Jaccoud et al. 2001), wheat (Akbari et al. 2006; Semagn et al. 2006; White et al.

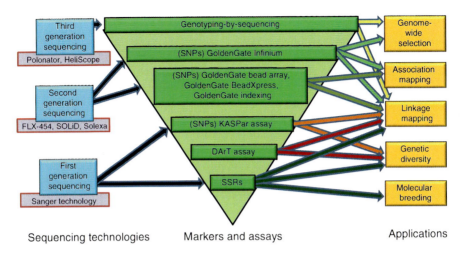

Fig. 11.2 Marker assay platforms for plant genetic analysis. A diagrammatic representation of utilization of different sequencing platforms for marker discovery and their subsequent use in plant genetic analyses

2008; Peleg et al. 2008; Jing et al. 2009), sorghum (Mace et al. 2008; Mace et al. 2009), rye (Bolibok-Brągoszewska et al. 2009), oat (Newell et al. 2011), triticale (Badea et al. 2011) and more than 30 other plant species (Jing et al. 2009). It is important to note that for wheat alone more than 50,000 samples (>95% as service at ~1 cent per marker assay), >350 mapping populations have been processed, which resulted in preparation of >100 genetic maps with ~7,000 markers assigned to chromosomes (A. Kilian, personal communication). DArT markers have been extensively used for diversity studies, genetic mapping, bulked segregant analysis (BSA), QTL interval mapping, and association mapping.

Ultra High-Throughput Marker Systems

Some modern genetics and breeding approaches like genome-wide association studies (GWAS) and genome-wide selection (GWS) or genomic selection (GS) require genotyping of large populations with a large number of markers. Such studies require ultra-high throughput marker systems (Figs. 11.1 and 11.2).

Infinium Assay for Whole-Genome Genotyping

Illumina's Infinium assay based on BeadChip™ technology is a high-density SNP genotyping technology for whole-genome genotyping allowing for genotyping of hundreds of thousands of SNPs simultaneously. One of the advantages of this system is that it allows simultaneous measurement of both signal intensity as well as changes in allelic composition (Gupta et al. 2008; Varshney 2010). This assay involves the use of 12-, 24-, 48-, or 96-sectioned BeadChips simultaneously, where each section of a BeadChip contains 1.1 million beads carrying oligonucleotides with known functions (Syvanen 2005; Gunderson et al. 2005; Steemers and Gunderson 2007). The challenge for the development of infinium assays in plants was the availability of a sufficient number of SNPs. This problem has been solved with the advent of NGS technologies, which allowed discovery of sufficient high density SNPs for infinium assays. Infinium assays have already been developed and used in crop plants. For instance in soybean, the *Illumina Infinium iSelect SoySNP50 chip* containing 44,299 informative SNPs was used to resolve the issue of origin of genomic heterogeneity in William 82 cultivars of soybean (Haun et al. 2011). In maize, a 50 K SNP Infinium chip containing SNPs in approximately two-thirds of all maize genes providing an average marker density of ~1 marker every 40 kb was developed (Ganal et al. 2010). Infinium genotyping assays have been developed in tree species like loblolly pine to study population structure and environmental associations to aridity (Eckert et al. 2010). The commercial availability of these high density SNP platforms will undoubtedly facilitate the application of SNP markers in molecular plant breeding (Mammadov et al. 2012).

Genotyping-By-Sequencing (GBS)

Recent advances in NGS technologies have helped us in providing unmatched discovery and characterization of molecular polymorphisms e.g. SNPs. However, before assaying the identified polymorphisms, there is a need to develop the genotyping platform. Genotyping-by-sequencing (GBS) is an approach that identifies and genotypes the SNPs simultaneously. GBS is a robust, cost-effective, highly multiplexed sequencing approach considered a powerful approach for association studies and also to facilitate the refinement (anchoring and ordering) of the reference genome sequence while providing tools for genomics-assisted breeding (GAB). With the continuous increase in NGS machine output, thereby continuous reduction in cost/sample, GBS will clearly become the marker genotyping platform of choice in coming years. Unlike other SNP discovery and genotyping platforms, GBS overcomes the issue of ascertainment bias of SNPs in a new germplasm. Keeping the cost/sample in view, it is also believed that GBS will provide an attractive option for genomic selection applications in breeding programs where cost per sample is considered a critical factor (Huang et al. 2010; Elshire et al. 2011; Poland et al. 2012).

GBS approach involves the use of restriction enzymes (REs) for reducing the complexity of genomes followed by targeted sequencing of reduced proportions, so that each marker can be sequenced at high coverage across many individuals at low cost and high accuracy. Overall, the process of GBS involves the following

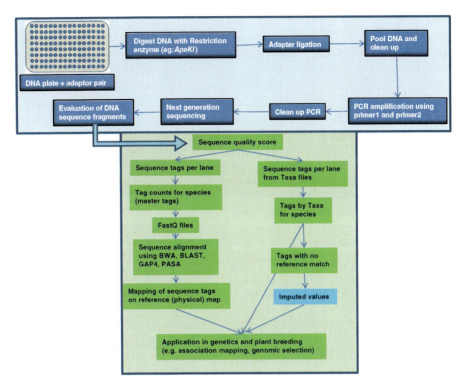

Fig. 11.3 A workflow for genotyping-by-sequencing (*GBS*) approach. A schematic representation of various steps involved in GBS approach (Adapted from Elshire et al. 2011; Poland et al. 2012, has been shown)

sequential steps: (i) isolation of high quality DNA, (ii) selection of a suitable RE and adaptor, (iii) preparation of libraries for NGS, (iv) single-end sequencing of either 48-plex or 96-plex library on NGS platforms like Genome Analyzer II or HiSeq 2000 of Illumina Inc. (www.illumina.com/systems.ilmn), (v) sequence quality assessment/filtering, (vi) sequence reads alignment, (vii) calling of SNPs. The complete procedure of GBS has been described elsewhere (Elshire et al. 2011) and a modified approach has been also developed and tested in wheat and barley recently (Poland et al. 2012). A workflow of GBS has been presented in Fig. 11.3. Comparison of GBS approach with other marker systems has also been presented in Table 11.1.

The choice of an appropriate RE is a critical factor in GBS approach for masking the repetitive regions of the genomes and, thereby, increases the chance of sampling markers from hypo-methylated gene rich regions of the genome. In the original GBS approach used in case of maize and barley, only one RE "*Ape*KI" (methylation-sensitive enzyme) was used to reduce the complexity and to select hypo-methylated regions of genome for sequencing (Elshire et al. 2011). However, recently, two REs (one "rare-cutter" and one "common-cutter")-based

GBS protocol has been developed and used for a species without a reference genome sequence. The two REs approach has advantages of generating suitable and uniform complexity reduction of complex genomes and has been earlier successfully tested in sequencing pools of BAC libraries for construction of physical maps (van Oeveren et al. 2011). Such GBS protocol has recently been used for genotyping bi-parental populations of wheat and barley for developing a genetically anchored reference map of identified SNPs and tags. This approach resulted in identification and mapping of >34,000 SNPs and 240,000 tags onto the Oregon Wolfe Barley reference map, and 20,000 SNPs and 367,000 tags on the Synthetic W97846 X Opata85 (SynOpDH) wheat reference map (Poland et al. 2012). In addition to above, Ion Torrent NGS platform has been also used for GBS in maize (http://www.invitrogen.com/etc/medialib/images/agricultural-biotechnology/pdf.Par.20344.File.dat/Maize-Genotyping-by-Sequencing-on-Ion-Torrent.pdf). This involves a two-step GBS protocol for genotyping of maize inbreds/RILs at up to a few hundred pre-defined SNPs in only two working days. The method in brief involves: (i) amplification (via multiplex PCR) of Genotyping by Multiple Amplicon Sequencing (GBMAS) targets, and (ii) addition of unique barcodes to the PCR products from each individual RIL and pooling of all the PCR products for Ion library construction and sequencing.

In summary, GBS is a highly multiplexed approach that can typically lead to the discovery of thousands of SNPs in one experiment and may be suitable for population studies, germplasm characterization, high-density genetic mapping, genomic selection and other breeding applications in diverse organisms (Huang et al. 2010; Elshire et al. 2011; Poland et al. 2012). The GBS approach can be used even in those plant species that do not have the reference genome available. In such cases, the sequence tags can be treated as dominant markers for kinship analysis. Moreover, availability of the genome sequence in a given species helps in increasing the number of marker loci analyzed through imputation.

Cost Effectiveness of Different High-Throughput Markers

One of the critical requirements of deployment of markers in molecular breeding programs is their cost effectiveness. While comparing different high-throughput markers system, the DArT marker system offers the lowest costs per marker data point. The cost per marker assay in commercial service offered by Triticarte P/L is ~US$ 0.02 (or approximately US $50 per genotype; Mantovani et al. 2008), which may be >6 times lower than the costs of SSR genotyping. A similar comparison with Illumina GoldenGate assays indicate, that DArT assays are only ~3 times cheaper (Yan et al. 2010). However, GoldenGate assay based-SNP genotyping is 100-fold faster than gel-based SSR methods leading to cost savings of ~75% (Yan et al. 2010). GBS available in -48, -96, and -384 array-plexes may further reduce the cost of genotyping and may become the method of choice for future plant genotyping. The continuous reduction in costs of GBS is due to increases in multiplexing, and, thus, lower labor, reagent, and sequencing costs. For instance, the labor cost was decreased from ~$2.00 for 48 to ~$0.50 for 384-plexes, while sequencing costs decreased from ~$33.00 for a 48-plex to ~$9.00 for a 384-plex assay. It is, therefore, obvious that the increase in throughput of markers is coupled to a reduction in their costs. Therefore, advances in NGS technologies will continuously help in reducing the costs of sequencing and, thus, the reduction in the cost of marker development and application (Davey et al. 2011).

Summary and Outlook

As is evident from the discussion above, that varying levels of throughput (low to ultra-high) are available. Thus, an appropriate marker system can be selected based on the need. For instance, Illumina's GoldenGate assays and Infinium assays as well as DArT markers are suitable for the construction of genetic linkage maps and GWAS studies, but these marker systems may not be suitable for molecular breeding applications such as marker-assisted selection (MAS), or marker-assisted backcrossing (MABC). One of the reasons for this is that genotyping costs for all the SNPs present in GoldenGate or Infinium assays is in lieu of only few informative SNP markers that are linked to the traits of interest. Alternatively, the associated markers present in GoldenGate or Infinium assays need to be converted into a user-friendly assay like KASPar or TaqMan assays. KASpar assays have become very cost effective in case of large populations (Fig. 11.2).

SNP markers that are transferable across different genotyping chemistries will serve as flexible selection tools for plant breeders in marker-assisted selection (MAS). However, technical issues may jeopardize the conversion and application of a particular marker for MAS (Mammadov et al. 2012). Recently, a set of 695 putative functional GoldenGate based-SNP assays were identified in maize and converted into Infinium, TaqMan, and KASPar chemistries with a high efficiency ranging from 89% for GG-to-Infinium to 98% from GG-to-KASPar (Mammadov et al. 2012). As a result of this conversion, a set of 162 highly polymorphic, putative functional and versatile SNP assays were identified and will be universally utilized in molecular genetics and breeding projects.

In contrast, low to moderate through put marker systems like SSRs can be deployed for selection of targeted genomic loci in breeding populations without any difficulty. While comparing the value of SSR with DArT markers, it was found that SSR markers can be preselected and may, therefore, represent whole genome coverage, which is not the case of DArT markers (Yan et al. 2010). The other obstacle is that one cannot use a selected DArT marker, identified through QTL interval mapping or association mapping, directly for marker-assisted selection (MAS) procedures. For using an associated DArT marker in MAS, the marker needs to be converted to a user-friendly assay. For instance, five robust SCARs were developed from three non-redundant DArTs, that co-segregated with crown rust resistance gene "*Pc91*" in oat (McCartney et al. 2011). However, the conversion of the associated DArT marker to a PCR-based marker is not always possible especially in cases where sequence data for DArT clones are not available.

For marker genotyping of a large number of marker loci for applications such as genome-wide association studies (GWAS) and genomic selection (GS), the GBS approach seems to be the best approach in terms of costs as well as throughput. With the increasing availability of reference genome sequences in a range of crop species, GBS is going to be the approach of choice in majority of the plant species in the coming years. It is anticipated that availability and routine use of GBS technology may re-orient molecular breeding programmes from MAS to GS, which will allow the realization the full potential of genomics-assisted breeding in crop improvement.

Acknowledgements Authors are thankful to Theme Leader Discretionary Grant of CGIAR-Generation Challenge Programme (GCP) and Centre of Excellence (CoE) grant from Department of Biotechnology (DBT) for funding the research of authors.

References

Akbari M, Wenzl P, Caig V, Carling J, Xia L, Yang S, Uszynski G, Mohler V, Lehmensiek A, Kuchel H, Hayden MJ, Howes N, Sharp P, Vaughan P, Rathmell B, Huttner E, Kilian A (2006) Diversity arrays technology (DArT) for high-throughput profiling of the hexaploid wheat genome. Theor Appl Genet 113:1409–1420

Akhunov E, Nicolet C, Dvorak J (2009) Single nucleotide polymorphism genotyping in polyploidy wheat with the Illumina GoldenGate assay. Theor Appl Genet 119:507–517

Allen AM, Barker GLA, Berry ST, Coghill JA, Gwilliam R, Kirby S, Robinson P, Brenchley RC, D'Amore R, McKenzie N, Waite D, Hall A, Bevan M, Hall N, Edwards KJ (2011) Transcript-specific, single-nucleotide polymorphism discovery and linkage analysis in hexaploid bread wheat (*Triticum aestivum* L.). Plant Biotechnol J 9:1086–1099

Appleby N, Edwards D, Batley J (2009) New technologies for ultra-high throughput genotyping in plants. In: Somers DJ, Langridge P, Gustafson JP (eds) Plant genomics methods & protocols. Humana Press, Hertfordshire, pp 19–39

Badea A, Eudes F, Salmon D, Tuvesson S, Vrolijk A, Larsson CT, Caig V, Huttner E, Kilian A, Laroche A (2011) Development and assessment of DArT markers in triticale. Theor Appl Genet 122:1547–1560

Bagge M, Lűbberstedt T (2008) Functional markers in wheat: technical and economic aspects. Mol Breed 22:319–328

Barrett BA, Kidwell KK, Fox PN (1998) Comparison of AFLP and pedigree-based genetic diversity assessment methods using wheat cultivars from the pacific Northwest. Crop Sci 38:1271–1278

Bolibok-Brągoszewska H, Heller-Uszyńska K, Wenzl P, Uszyński G, Kilian A, Rakoczy-Trojanowska M (2009) DArT markers for the rye genome-genetic diversity and mapping. BMC Genomics 10:578

Chao S, Sharp PJ, Worland AJ, Warham EJ, Koebner RMD, Gale MD (1989) RFLP-based genetic maps of wheat homeologous group7 chromosomes. Theor Appl Genet 78:495–904

Chao S, Dubcovsky J, Dvorak J, Luo MC, Baenziger SP, Matnyazov R, Clark DR, Talbert LE, Anderson JA, Dreisigacker S, Glover K, Chen J, Campbell K, Bruckner PL, Rudd JC, Haley S, Carver BF, Perry S, Sorrells ME, Akhunov ED (2010) Population- and genome-specific patterns of linkage disequilibrium and SNP variation in spring and winter wheat (*Triticum aestivum* L.). BMC Genomics 11:727

Chen W, Mingus J, Mammadov J, Backlund JE, Greene T, Thompson S, Kumpatla S (2010) KASPar: a simple and cost-effective system for SNP genotyping. In: Plant and Animal Genomes XVII conference, San Diego, USA, p 194

Cheung F, Haas BJ, Goldberg SM, May GD, Xiao Y, Town CD (2006) Sequencing *Medicago truncatula* expressed sequenced tags using 454 life sciences technology. BMC Genomics 7:272

Close TJ, Bhat PR, Lonardi S, Wu Y, Rostoks N, Ramsay L, Druka A, Stein N, Svensson TJ, Wanamaker S, Bozdag S, Roose ML, Moscou MJ, Chao S, Varshney RK, Szucs P, Sato K, Hayes PM, Matthews DE, Kleinhofs A, Muehlbauer GJ, Young JD, Marshall DF, Madishetty K, Fenton RD, Condamine P, Graner A, Waugh R (2009) Development and implementation of high-throughput SNP genotyping in barley. BMC Genomics 10:582

Cortes AJ, Chavarro MC, Blair MW (2011) SNP marker diversity in common bean (*Phaseolus vulgaris* L.). Theor Appl Genet 123:827–845

Davey JW, Hohenlohe PA, Etter PD, Bone JQ, Catchen JM, Blaxter ML (2011) Genome-wide genetic marker discovery and genotyping using next-generation sequencing. Nat Rev Genet 12:499–510

Delmas CE, Lhuillier E, Pornon A, Escaravage N (2011) Isolation and characterization of microsatellite loci in *Rhododendron ferrugineum* (Ericaceae) using pyrosequencing technology. Am J Bot 98:e120–e122

Deulvot C, Charrel H, Marty A, Jacquin F, Donnadieu C, Lejeune-Henaut I, Burstin J, Aubert G (2010) Highly-multiplexed SNP genotyping for genetic mapping and germplasm diversity studies in pea. BMC Genomics 11:468

Druka A, Franckowiak J, Lundqvist U, Bonar N, Alexander J, Houston K, Radovic S, Shahinnia F, Vendramin V, Morgante M, Stein N, Waugh R (2011) Genetic dissection of barley morphology and development. Plant Physiol 155:617–627

Eckert AJ, Heerwaarden JV, Wegryzn JL, Nelson CD, Ross-Ibarra J, Gonzalez-Martinez SC, Neaale DB (2010) Patterns of population structure and environmental associations to aridity across the range of loblolly pine (*Pinustaeda* L., Pinaceae). Genetics 185:969–982

Ellsworth DL, Rittenhouse KD, Honeycutt RL (1993) Artifactual variation in randomly amplified polymorphic DNA banding patterns. Biotechniques 14:214–217

Elshire RJ, Glaubitz JC, Sun Q, Poland JA, Kawamoto K, Buckler ES, Mitchell SE (2011) A robust, simple genotyping-by-sequencing (GBS) approach for high diversity species. PLoS One 6:e19379

Emami MK, Sharma B (1999) Linkage between three morphological markers in lentil. Plant Breed 118:579–581

Faccioli P, Pecchioni N, Stanca AM, Terzi V (1999) Amplified fragment length polymorphism (AFLP) markers for barley malt fingerprinting. J Cereal Sci 29:257–260

Ganal MW, Durstewitz G, Polley A, Bérard A, Buckler ES, Charcosset A, Clarke JD, Graner EM, Hansen M, Joets J, Le Paslier M, McMullen MD, Montalent P, Rose M, Schön CS, Sun Q, Walter H, Martin OC, Falque M (2010) A large maize (*Zea mays* L.) SNP genotyping array: Development and germplasm genotyping, and genetic mapping to compare with the B73 reference genome. PLoS ONE 6:e28334

Gonzalez-Ibeas D, Blanca J, Donaire L, Saladie M, Mascarell-Creus A, Delgado A, Garcia-Mas J, Llave C, Aranda MA (2011) Analysis of the melon (*Cucumis melo*) small RNAome by high-throughput pyrosequencing. BMC Genomics 12:393

Graner A, Jahoor A, Schondelmaier J, Siedler H, Pillen K, Fischbeck G, Wenzel G, Herrmann RG (1991) Construction of an RFLP map of barley. Theor Appl Genet 83:250–256

Griffin PC, Robin C, Hoffmann AA (2011) A next-generation sequencing method for overcoming the multiple gene copy problem in polyploidy phylogenetics, applied to Poa grasses. BMC Biol 9:19

Gunderson KL, Steemers FJ, Lee G, Mendoza LG, Chee MS (2005) A genome-wide scalable SNP genotyping assay using microarray technology. Nat Genet 37:549–554

Gupta PK, Varshney RK (2000) The development and use of microsatellite markers for genetic analysis and plant breeding with emphasis on bread wheat. Euphytica 113:163–185

Gupta PK, Varshney RK, Sharma PC, Ramesh B (1999) Molecular markers and their application in wheat breeding. Plant Breed 118:369–390

Gupta PK, Roy JK, Prasad M (2001) Single nucleotide polymorphisms: a new paradigm for molecular marker technology and DNA polymorphism detection with emphasis on their use in plants. Curr Sci 80:524–535

Gupta PK, Rustagi S, Mir RR (2008) Array-based high-throughput DNA markers for crop improvement. Heredity 101:1–14

Gupta PK, Kumar J, Mir RR, Kumar A (2010) Marker-assisted selection as a component of conventional plant breeding. Plant Breed Rev 33:145–217

Haun WJ, Hyten DL, Xu WW, Gerhardt DJ, Albert TJ, Richmond T, Jeddeloh JA, Jia G, Springer NM, Vance CP, Stupar RM (2011) The composition and origins of genomic variation among individuals of the soybean reference cultivar Williams 82. Plant Physiol 155:645–655

Helentjaris T (1987) A genetic linkage map for maize based on RFLPs. Trends Genet 3:217–221

Heun M, Helentjaris T (1993) Inheritance of RAPDs in F_1 hybrids of corn. Theor Appl Genet 85:961–968

Hiremath PJ, Farmer A, Cannon SB, Woodward J, Kudapa H, Tuteja R, Kumar A, Bhanuprakash A, Mulaosmanovic B, Gujaria N, Krishnamurthy L, Gaur PM, KaviKishore PB, Shah T, Srinivasan R, Lohse M, Xiao Y, Town CD, Cook DR, May GD, Varshney RK (2011) Large-scale transcriptomic analysis in chickpea (*Cicer arietinum* L.), an orphan legume crop of the semi-arid tropics of Asia and Africa. Plant Biotechnol J 9:922–931

Hiremath PJ, Kumar A, Penmetsa RV, Farmer A, Schlueter JA, Chamarthi SK, Whaley AM, Carrasquilla-Garcia N, Gaur PM, Upadhyaya HD, KaviKishor PB, Shah TM, Cook DR, Varshney RK (2012) Large-scale development of cost-effective SNP marker assays for diversity assessment and genetic mapping in chickpea and comparative mapping in legumes. Plant Biotechnol J 10:716–732

Huang X, Zeller FJ, Hsam SL, Wenzel G, Mohler V (2000) Chromosomal location of AFLP markers in common wheat utilizing nulli-tetrasomic stocks. Genome 43:298–305

Huang X, Wei X, Tap S, Zhao Q, Feng Q, Zhao Y, Li C, Zhu C, Lu T, Zhang Z, Li M, Fan D, Guo Y, Wang A, Wang L, Deng L, Li W, Lu Y, Weng O, Liu K, Huang T, Zhou T, Jing Y, Li W, Lin Z et al (2010) Genome-wide association studies of 14 agronomic traits in rice landraces. Nat Genet 42:961–967

Hyten DL, Song Q, Choi IY, Yoon MS, Specht JE, Matukumalli LK, Nelson RL, Shoemaker RC, Young ND, Cregan PB (2008) High-throughput genotyping with the GoldenGate assay in the complex genome of soybean. Theor Appl Genet 116:945–952

Hyten DL, Smith JR, Fredrick RD, Tucker ML, Song Q, Cregan PB (2009) Bulked segregant analysis using the GoldenGate assay to locate the Rpp 3 locus that confers resistance to soybean rust in soybean. Crop Sci 49:265–271

Hyten DL, Cannon SB, Song Q, Weeks N, Fickus EW, Shoemaker RC, Specht JE, Farmer AD, May GD, Cregan PB (2010a) High-throughput SNP discovery through deep resequencing of a reduced representation library to anchor and orient scaffolds in the soybean whole genome sequence. BMC Genomics 11:38

Hyten DL, Song Q, Fickus EW, Quigley CV, Lim J, Choi I, Hwang E, Pastor-Corrales M, Cregan PB (2010b) High-throughput SNP discovery and assay development in common bean. BMC Genomics 11:475

Jaccoud D, Peng K, Feinstein D, Kilian A (2001) Diversity arrays: a solid state technology for sequence information independent genotyping. Nucleic Acids Res 29:25–31

Jain SM, Brar DS, Ahloowalia BS (2002) Molecular techniques in crop improvement. Kluwer, Dordrecht, p 601

Jing H-C, Bayon C, Kanyuka K, Berry S, Wenzl P, Huttner E, Kilian A, Hammond-Kosack KE (2009) DArT markers: diversity analyses, genomes comparison, mapping and integration with SSR markers in *Triticum monococcum*. BMC Genomics 10:458

Jones N, Ougham H, Thomas H, Pasakinskiene I (2009) Markers and mapping revisited: finding your gene. New Phytol 183:935–966

Keim P, Diers BW, Olson TC, Shoemaker RC (1990) RFLP mapping in soybean: association between marker loci and variation in quantitative traits. Genetics 126:735–742

Kwok P (2001) Methods for genotyping single nucleotide polymorphisms. Annu Rev Genomics Hum Genet 2:235–258

Mace EM, Xia L, Jordan DR, Halloran K, Parh DK, Huttner E, Wenzl KA (2008) DArT markers: diversity analyses and mapping in *Sorghum bicolor*. BMC Genomics 9:26

Mace ES, Rami JF, Bouchet S, Klein PE, Klein RR, Kilian A, Wenzl P, Xia L, Halloran K, Jordan DR (2009) A consensus genetic map of sorghum that integrates multiple component maps and high-throughput Diversity Array Technology (DArT) markers. BMC Plant Biol 9:13

Mackill DJ, Zhang Z, Redona ED, Colowit PM (1996) Level of polymorphism and genetic mapping of AFLP markers in rice. Genome 39:969–977

Maheshwaran M, Subudhi PK, Nandi S, Xu JC, Parco A, Yang DC, Huang N (1997) Polymorphism, distribution and segregation of AFLP markers in a doubled haploid rice population. Theor Appl Genet 94:39–45

Mammadov J, Chen W, Mingus J, Thompson S, Kumpatla S (2012) Development of versatile gene-based SNP assays in maize (*Zeamays* L.). Mol Breed 29:779–790

Mantovani P, Maccaferri M, Sanguineti MC, Tuberosa R, Kilian A et al (2008) An integrated DArT-SSR linkage map of durum wheat. Mol Breed 22:629–648

Maughan PJ, Maroof MAS, Buss GR, Huetis GM (1996) Amplified fragment length polymorphism (AFLP) in soybean: species diversity, inheritance, and near-isogenic line analysis. Theor Appl Genet 93:392–401

Maughan PJ, Smith SM, Fairbanks DJ, Jellen EN (2011) Development, characterization and linkage mapping of single nucleotide polymorphisms in the grain Amaranths (*Amaranthus* sp.). Plant Genome 4:92–101

McCartney CA, Stonehouse RG, Rossnagel BG, Eckstein PE, Scoles GJ, Zatorski T, Beattie AD, Chong J (2011) Mapping of the oat crown rust resistance gene *Pc91*. Theor Appl Genet 122:317–325

McCouch SR, Kochert G, Yu ZH, Wang ZY, Khush GS, Coffman WR, Tanksley SD (1988) Molecular mapping of rice chromosomes. Theor Appl Genet 76:815–829

McCouch SR, Zhao K, Wright M, Tung CW, Ebana K, Thomson M, Reynolds A, Wang D, DeClerck G, Ali L, McClung A, Eizenga G, Bustamanate C (2010) Development of genome-wide SNP assays for rice. Breed Sci 60:524–535

Muchero W, Diop NN, Bhat PR, Fenton RD, Wanamaker S, Pottorff M, Hearne S, Cisse N, Fatokun C, Ehlers JD, Roberts PA, Close TJ (2009) A consensus genetic map of cowpea [*Vigna ungiculata* (L) Walp.] and synteny based on EST-derived SNPs. Proc Natl Acad Sci USA 106:18159–18164

Newell MA, Cook D, Tinker NA, Jannink JL (2011) Population structure and linkage disequilibrium in oat (*Avena sativa* L.): implications for genome-wide association studies. Theor Appl Genet 122:623–632

Nguyen TT, Taylor PWJ, Redden RJ, Ford R (2003) Genetic diversity estimates in Cicer using AFLP analysis. Plant Breed 123:173–179

Ottesen EA, Marin R, Preston CM, Young CR, Ryan JP, Scholin CA, Delong EF (2011) Metatranscriptomic analysis of autonomously collected and preserved marine bacterioplankton. ISME J 5:1881–1895

Peleg Z, Saranga Y, Suprunova T, Ronin Y, Roder MS, Kilian A, Korol AB, Fahima T (2008) High-density genetic map of durum wheat × wild emmer wheat based on SSR and DArT markers. Theor Appl Genet 117:103–115

Poland JA, Brown PJ, Sorrells ME, Jannink J-L (2012) Development of high-density genetic maps for barley and wheat using a novel two-enzyme genotyping-by-sequencing approach. PLoS One 7:e32253

Rafalski A (2002) Applications of single nucleotide polymorphisms in crop genetics. Curr Opin Plant Biol 5:94–100

Rostoks N, Ramsay L, MacKenzie K, Cardle L, Bhat PR, Roose ML, Svensson JT, Stein N, Varshney RK, Marshall DF, Graner A, Close TJ, Waugh R (2006) Recent history of artificial outcrossing facilitates whole genome association mapping in elite inbred crop varieties. Proc Natl Acad Sci USA 103:18656–18661

Rounsley R, Marri PR, Yu Y, He R, Sisneros N, Goicoechea JL, Lee SJ, Angelova A, Kudrna D, Luo M, Affourtit J, Desany B, Knight J, Niazi F, Egholm M, Wing RA (2009) De novo next generation sequencing of plant genomes. Rice 2:35–43

Saxena RK, Penmetsa RV, Upadhyaya HD, Kumar A, Carrasquilla-Garcia N, Schlueter JA, Farmer A, Whaley AM, Sarma BK, May GD, Cook DR, Varshney RK (2012) Large-scale development of cost-effective single-nucleotide polymorphism marker assays for genetic mapping in pigeonpea and comparative mapping in legumes. DNA Res doi:10.1093/dnares/dss025

Semagn K, Bjornstad A, Skinnes H, Maroy AG, Tarkegne Y, William M (2006) Distribution of DArT, AFLP and SSR markers in a genetic linkage map of a doubled-haploid hexaploid wheat population. Genome 49:545–555

Shan X, Blake TK, Talbert LE (1999) Conversion of AFLP markers to sequence-specific PCR markers in barley and wheat. Theor Appl Genet 98:1072–1078

Shendure J, Ji H (2008) Next-generation DNA sequencing. Nat Biotechnol 26:1135–1145

Simon CJ, Muehlbauer FJ (1997) Construction of a chickpea linkage map and its comparison with maps of pea and lentil. J Hered 88:115–119

Soleimani VD, Baum BR, Johnson DA (2002) AFLP and pedigree-based genetic diversity estimates in modern cultivars of durum wheat [*Triticum turgidum* L. sub sp. durum (Desf.) Husn.]. Theor Appl Genet 104:350–357

Steemers FJ, Gunderson KL (2007) Whole genome genotyping technologies on the BeadArray platform. Biotechnol J 2:41–49

Steemers FJ, Chang W, Lee G, Barker DL, Shen R, Gunderson KL (2006) Whole-genome genotyping with the single-base extension assay. Nat Methods 3:31–33

Syvanen A (2005) Towards genome-wide SNP genotyping. Nat Genet 37:S5–S10

Tanksley SD (1983) Molecular markers in plant breeding. Plant Mol Biol Rep 1:3–8

Tanksley SD, Young ND, Paterson AH, Bonierbale MW (1989) RFLP mapping in plant breeding: new tools for an old science. Nat Biotechnol 7:257–264

Tanksley SD, Ganal MW, Prince JP, deVicente MC, Bonierbale MW, Broun P, Fulton TM, Giovannoni JJ, Grandillo S, Martin GB, Messenger J, Miller C, Miller L, Patreson AH, Pineda O, Roder MS, Wing RA, Wu W, Young ND (1992) High density molecular linkage maps of the tomato and potato genomes. Genetics 132:41141–41160

Thomson MJ, Zhao K, Wright M, McNally KL, Rey J, Tung C-W, Reynolds A, Scheffler B, Eizenga G, McClung A et al (2011) High-throughput single nucleotide polymorphism for breeding applications in rice using the BeadXpress platform. Mol Breed 29:875–886

Thudi M, Li Y, Jackson SA, May GD, Varshney RK (2012) Current state-of-art of sequencing technologies for plant genomics research. Brief Funct Genomics 11:3–11

van Oeveren J, de Ruiter M, Jesse T, van der Poel H, Tang J et al (2011) Sequence-based physical mapping of complex genomes by whole genome profiling. Genome Res 21:618–625

Varshney RK (2010) Gene-based marker systems in plants: high throughput approaches for marker discovery and genotyping. In: Jain SM, Brar DS (eds) Molecular techniques in crop improvement. Springer, Dordrecht, pp 119–142

Varshney RK, Graner A, Sorrells ME (2005) Genic microsatellite markers in plants: features and applications. Trends Biotechnol 23:48–55

Varshney RK, Nayak SN, May GD, Jackson SA (2009) Next generation sequencing technologies and their implications for crop genetics and breeding. Trends Biotechnol 27:522–530

Vos P, Hogers R, Bleeker M, Reijans M, van deLee T, Hornes M, Frijters A, Pot J, Peleman J, Kuiper M, Zabeau M (1995) AFLP: a new technique for DNA fingerprinting. Nucleic Acids Res 23:4407–4414

White J, Law JR, MacKay KJ, Chalmers KJ, Smith JSC, Kilian A, Powell W (2008) The genetic diversity of UK, US and Australian cultivars of *Triticum aestivum* measured by DArT markers and considered by genome. Theor Appl Genet 116:439–453

Williams JGK, Kubelic AR, Livak KJ, Rafalsky JA, Tingey SV (1990) DNA polymorphism amplified by arbitrary primers are useful as genetic markers. Nucleic Acids Res 18:6532–6535

Winter P, Benko-Iseppon AM, Hűttel B, Ratnaparkhe M, Tullu A, Sonnante G, Pfaff T, Tekeoglu M, Santra D, Sant VJ, Rajesh PN, Kahl G, Muehlbauer FJ (2000) A linkage map of the chickpea (*Cicer arietinum* L.) genome based on recombinant inbred lines from a *C .arietinum* × *C. reticulatum* cross: localization of resistance genes for *Fusarium* wilt races 4 and 5. Theor Appl Genet 101:1155–1163

Xu DH, Ban T (2004) Conversion of AFLP markers associated with FHB resistance in wheat into STS markers with an extension-AFLP method. Genome 47:660–665

Yan J, Yang X, Shah T, Villeda HS, Li H, Warburton M, Zhou Y, Crouch JH, Xu Y (2010) High-throughput SNP genotyping with the GoldenGate assay in maize. Mol Breed 25:441–451

Young WP, Schupp JM, Keim P (1999) DNA methylation and AFLP marker distribution in the soybean genome. Theor Appl Genet 99:785–792

Zhou X, Sunkar R, Jin H, Zhu JK, Zhang W (2009) Genome-wide identification and analysis of small RNAs originated from natural antisense transcripts in *Oryza sativa*. Genome Res 19:70–80

Zhu J, Gale MD, Quarrie S, Jackson MT, Bryan GJ (1998) AFLP markers for the study of rice biodiversity. Theor Appl Genet 96:602–611

Part V
Development of Non-DNA Biomarkers

Chapter 12
Methylation-Based Markers

Emidio Albertini and Gianpiero Marconi

Introduction

DNA methylation is a heritable epigenetic enzymatic modification resulting from the addition of a methyl group in the cyclic carbon-5 of cytosine (Tsaftaris et al. 2005). DNA methylation can increase the functional complexity of prokaryotic and eukaryotic genomes by providing additional avenues for the control of cellular processes. Levels of methylation vary greatly between organisms: 0–3% in insects, 2–7% in vertebrates, 10% in fish, more than 30% in some plants (Adams 1996). This depends on several factors, but likely the most important one is addition of a methyl group to cytosine residues occurs at CpG or CpNpG sequences (where N could be any nucleotide) in plants, while methylation in mammalian genomes is generally restricted to symmetric CpG sequences (CG islands). Tsaftaris et al. (2005) hypothesized four main roles of methylation: (i) to provide a heritable epigenetic mark directing the developmental program of organisms (Holliday and Pugh 1975; Regev et al. 1998; Wolffe and Matzke 1999), (ii) to provide defense against the activity of parasitic mobile elements (Yoder et al. 1997), (iii) to reduce background transcriptional noise in organisms that have a large number of genes (Bird 1995), and (iv) to "memorize" patterns of gene activity by stabilizing gene silencing brought about by other mechanisms (Bird 2002). In plants, epimutations (due to DNA methylation) have important phenotypic consequences, since they can be inherited through the transmission of epigenetic alleles (epialleles) over many generations (Kakutani 2002). These heritable epigenetic alleles can be considered as a source of polymorphism, which may produce novel phenotypes and could be a source of variation for selection. For example, epialleles can originate as

E. Albertini (✉) • G. Marconi
Department of Applied Biology, University of Perugia, Borgo XX Giugno 74, 06121 Perugia, Italy
e-mail: emidio.albertini@unipg.it

a genome response to stressful environments to enable plants to tolerate stresses (Tsaftaris et al. 2005). Experimental evidence of this was given by Sano et al. (1990) by treatment of germinated rice seeds with 5-azadeoxycytidine, a chemical that powerfully induces demethylation of DNA *in vivo*. The authors showed that plants exhibited global demethylation and altered phenotypes at maturity, including dwarfism, and that these acquired traits and demethylation patterns were inherited for at least six generations. Steward et al. (2002) reported genome-wide demethylation occurring in maize root tissues, when seedlings were exposed to cold stress. Screening of genomic DNA identified one particular fragment (designated ZmMI1) that was demethylated during chilling. In particular, ZmMI1 was expressed only under cold stress. These results indicate that ZmMI1 provides a selective advantage toward cold stress adaptation and its epialleles could be used in plant breeding of maize.

DNA Methylation Biomarkers

A major challenge is to identify CpG modifications among millions of CpG dinucleotides and thousands of gene-associated CpG islands (Gentil and Maury 2007). Assessing the importance of methylated epialleles in plant breeding requires to determine the extent of variation in methylation and to verify, whether and which of these methylation patterns affect phenotypes. The technical potential exists to assess methylation pattern differences between individuals and thus, estimate levels of methylation-associated epiallelic diversity and its impact on phenotypic diversity (Tsaftaris et al. 2005). Here, we report some of the major approaches which allow to distinguish the 5-methylcytosine (5mC) DNA methylation from unmethylated cytosine (C) and, thus, to establish methylation biomarkers in plants (Fig. 12.1). These approaches are based on three techniques: digestion with methylation-sensitive restriction enzymes, bisulfite conversion, and affinity purification.

Restriction Endonuclease-Based Analysis

Methylation-sensitive restriction endonucleases are one of the most powerful tools in DNA methylation analysis because of the high versatility of many restriction enzymes. Most of these are inhibited by methylation of their recognition site, whereas some, most notably *McrBC*, specifically digest methylated DNA (Zilberman and Henikoff 2007). In the last decade, several variations of restriction enzyme-based methods have been developed and used (Lippman et al. 2005; Rollins et al. 2006; Lister et al. 2009; Edwards et al. 2010). Generally, a comparison is made either between a sample treated with a methylation-sensitive enzyme and a control treated with a methylation-insensitive isoschizomer or between two test samples (i.e., a genotype grown under optimal conditions and the same genotype grown

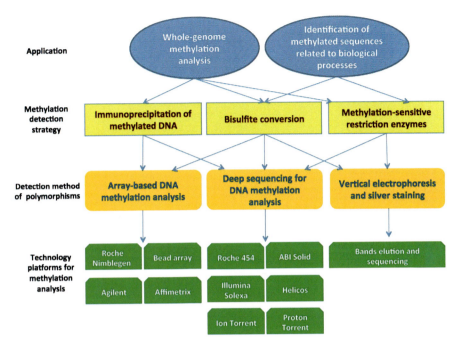

Fig. 12.1 Methods for DNA methylation analysis. DNA immunoprecipitation methods, bisulfite-converted DNA, and DNA restriction generated by methylation-specific restriction enzymes can be used for advanced array-based and high-throughput deep sequencing–based genome-wide DNA methylation analyses, and for the identification of genes activated or inactivated due to methylation or demethylation, induced by biological processes

under stress), both treated with the same enzyme. Probably the best known pair of isoschizomers is *Hpa*II and *Msp*I, both of which recognize the 5′-CCGG-3′ motif. Cleavage by *Hpa*II is inhibited by methylation at any of the two cytosine residues, whereas *Msp*I is blocked only if the external C is methylated. In addition, *Hpa*II (but not *Msp*I) also cleaves if the CCGG motif is hemimethylated (i.e., only one strand is methylated) (Korch and Hagblom 1986; McClelland et al. 1994; Weising et al. 2005). Initially, restriction endonucleases were limited to the study of DNA methylation patterns within individual genomic regions, but in recent years, the approach has been adapted for genome-wide DNA methylation analysis. An example of this application was reported by Zhang et al. (2008), who analyzed the extent of natural genomic variation in cytosine methylation among *Arabidopsis thaliana* (L.) wild accessions Columbia (Col) and Vancouver (Van) by comparing hybridization intensity differences between genomic DNA digested with either *Hpa*II or *Msp*I restriction enzyme. The authors demonstrated that at least 8% of all analyzed CCGG sites were constitutively methylated across the two strains, while about 10% of all analyzed CCGG sites were differentially methylated between them. Moreover, polymorphic methylation occurred much more frequently in gene ends than constitutive methylation. Gene expression analyses in matching tissue samples showed that the magnitude of methylation polymorphisms immediately upstream or downstream of genes was inversely correlated with the degree of expression variation (Zhang et al. 2008). In contrast, methylation polymorphisms within genic regions showed weak positive correlations with expression variation. This demonstrates presence of extensive epigenetic variation between Arabidopsis accessions and suggests a possible relationship between natural CG methylation variation and gene expression variation. In 1997, Reyna-López et al. (1997) applied a modified version of AFLP (Amplified Fragment Length Polymorphisms, Vos et al. 1995), in which AFLPs based on methylation-dependent restriction enzymes were employed to genomic DNA samples, isolated from representatives of three major fungal taxa. Xiong et al. (1999) applied this technique to plants and named it MSAP (Methylation-Sensitive Amplified Polymorphism). In MSAP, template DNA is usually digested with one 6-cutter (such as *Eco*RI), and one pair of isoschizomers (e.g., *Hpa*II-*Msp*I). Xiong and collaborators monitored patterns of cytosine methylation in rice hybrids and their parental lines as well as in different rice tissues. These authors detected an increase of methylation in hybrids compared with their parents, as well as in seedling DNA compared to leaf DNA. In 2001, Ashikawa (2001) applied MSAP to rice and found conserved as well as cultivar-specific methylation at CCGG sites, demonstrating stable Mendelian inheritance of methylation patterns over six generations. In recent years, the MSAP technique has been applied by Yi et al. (2010) and Zhao et al. (2010) on *Jatropha curcas* and *Gossypium hirsutum*, respectively. Yi and collaborators analyzed five populations of *J. curcas* belonging to China, Indonesia, Suriname, Tanzania, and India planted at one farm under the same agronomic practices. Genetic and epigenetic diversity, evaluated using AFLP and MSAP markers, showed a very low level of genetic diversity (polymorphic bands <0.1%) and a significant epigenetic variation (25.3% of polymorphic bands) within and among populations. Zhao et al. (2010) applied the MSAP technique to two different salt-tolerant cotton lines, and inferred a relation between DNA methylation and abiotic stress responses. Based on the MSAP, authors noticed extensive cytosine methylation alterations, including hypermethylation and demethylation in salt-treated cotton lines compared with corresponding controls.

By applying MSAP, Verhoeven et al. (2010) revealed de novo methylation variation between triploid F_1 individuals obtained by crossing diploid × tetraploid parents. Marconi and Albertini (in preparation) compared the extent and pattern of cytosine methylation in one salt-tolerant and one salt-sensitive rapeseed (*Brassica napus* (L.)) cultivar, using the technique of methylation-sensitive amplified polymorphism (MSAP). Under salt stress conditions, the tolerant cultivar showed a lower level of methylation than the sensitive cultivar (38.3% vs. 49.4%). Some methylation-related fragments were recovered and showed high homology with Arabidopsis genes related to stress tolerance. Validation by Real-Time PCR confirmed the involvement of methylation in gene expression.

Bisulfite Conversion of DNA

Methylated and unmethylated Cytosine residues are indistinguishable by traditional Sanger sequencing. Treatment of DNA with sodium bisulfite leads to conversion of unmethylated cytosine to uracil, while leaving methylated cytosine intact (Clark et al. 1994). PCR amplification of converted DNA replaces uracil with thymine, and the following sequencing gives an estimate of the extent of methylation. During the last 15 years, several methodologies relying on the use of bisulfite-converted DNA have been developed and applied (Herman et al. 1996; Eads et al. 2000; Meissner et al. 2008; Lister et al. 2008). As an example, Baubec et al. (2010) applied bisulfite-conversion and Sanger sequencing to Arabidopsis DNA and revealed the presence of epigenetic transcriptional silencing, surprisingly resistant to genetic and chemical interference. Based on their results authors provided evidence that two epigenetic features, namely, symmetric DNA methylation and histone methylation, cooperate to generate a double safeguard system that controls transcriptional suppression. Hence, both modifications have to be unlocked to convert the silent epiallele into an active one. Baek et al. (2011) applied the same methods (bisulfite conversion and Sanger sequencing) to Arabidopsis mutants having a T-DNA insertion in the AtHKT1 promoter and identified several important elements for gene expression and regulation. These authors also found that the AtHKT1 promoter contains a putative small RNA target region. CG methylation of this region in leaves is increased compared to roots. They concluded that this might cause higher levels of expression of AtHKT1 in roots.

Immunoprecipitation or Affinity Purification Based Methods

One of the most recent strategies to enrich methylated DNA is by immunoprecipitation (or affinity purification), which is based on the ability of some proteins to bind methylated CG sites due to their methyl-binding domain (MBD). Several immunoprecipitation-based methods employing affinity columns containing MBD have been developed for DNA methylation analysis (Cross et al. 1994; Weber et al. 2005; Zhang et al. 2006; Koga et al. 2009). A system based on a monoclonal antibodies that recognize methylated cytosine and enable immunoprecipitation of methylated DNA, is commercially available (Reynaud et al. 1992; Keshet et al. 2006; Weber et al. 2007; Zilberman et al. 2007). Plant researchers prefer the MBD method, since it purifies only CG-methylated DNA, while the monoclonal antibody is directed against any methylated DNA. However, as almost all the methylated loci that have been characterized in plants have CG methylation, the results obtained with the two methods should be closely correlated. Moreover, since these methods do not require either digestion of genomic DNA or bisulfite treatment, they generate data that are relatively easy to analyze and interpret (Zilberman and Henikoff 2007). The drawback of immunoprecipitation-based methods is that they do not provide DNA methylation information at single-nucleotide resolution. Due to the ease of these methods over bisulfite conversion–based methods, several studies have used immunoprecipitation-based approaches for genome-wide DNA methylation analysis (Weber et al. 2005; Zilberman et al. 2007; Koga et al. 2009; Gupta et al. 2010). Methylcytosine immunoprecipitation (mCIP) and tiling array hybridization (Zhang et al. 2006) were used in Arabidopsis to produce the first detailed whole-genome map of cytosine methylation.

Array-Based Genome-Wide DNA Methylation Analysis

Several methodologies have been developed, allowing large-scale analysis of DNA methylation and relying on the use of array-based platforms (bead or tiling arrays) to either restricted-digested, immunoprecipitated or bisulfite-converted DNA. To date, both commercial and custom oligonucleotides arrays are available.

Bead Array Technology

The bead array-based analysis of DNA methylation developed by Illumina is one of the most advanced array-based approaches for both custom and large-scale DNA methylation analysis (Bibikova et al. 2006; Fan et al. 2006). This system is based on bisulfite-converted DNA and has been used by several researchers (Bibikova et al. 2006; Suhr et al. 2009; Noushmehr et al. 2010; Liu et al. 2010). Bisulfite-converted DNA is assayed with two primers, each labeled with a different fluorescent dye. One primer is designed to hybridize to a target sequence, when cytosine is methylated (and unconverted), whereas the other primer will only hybridize to a converted target sequence. The two primers are used in PCR reactions with a locus-specific methylation-insensitive primer (Zilberman and Henikoff 2007) and the ratio of PCR products is determined by the Illumina bead array platform. The strength of this technique is, that it provides quantitative evaluation of specific cytosine sites, and that it allows to process many samples in parallel. Therefore, this method is well suited to compare a set of known methylated loci among a large number of cell lines or individuals to evaluate for methylation polymorphisms. The main drawback is, that methylation sites have to be known a priori (to design primers). Therefore, this strategy cannot be applied to de novo investigations of DNA methylations.

Tiling Array Technology

Tiling arrays function based on similar principles compared to traditional microarrays, in that labeled target molecules are hybridized to unlabeled probes fixed on to a solid surface. Depending on probe length and spacing, different degrees of resolution can be achieved.

Short oligonucleotides arrays. Affymetrix provides arrays, used in several methylation studies, which contain millions of probes consisting of 25-mer oligonucleotides. These short probes, guarantee a good specificity. However, those short probes result in decreased sensitivity and increased random signal variation (noise) when compared with longer probes (Kreil et al. 2006; Zilberman and Henikoff 2007). Paired samples (e.g., same genotype grown under stress and normal conditions) are hybridized to separate arrays and the resulting signals are compared. A study employing the Arabidopsis array was carried out by Zhang et al. (2006) to profile methylated DNA enriched by MBD and antibody affinity purification to yield a high-resolution methylation map of the entire Arabidopsis genome.

Long oligonucleotides arrays. NimbleGen and Agilent provide several different arrays including custom arrays, which have been used to analyze DNA methylation on a genome-wide scale (Irizarry et al. 2009; Koga et al. 2009; Ruzov et al. 2009). The main difference between the two arrays is the number of probes (400,000–2.1 million 60-mer oligonucleotides for NimbleGen vs. 250,000–1M 60-mers for Agilent). For both arrays, paired samples are labeled with different fluorescent dyes and hybridized on a single chip. Because of the probe length, the probe density of these arrays is lower compared to short probe-arrays, which results in a better balance between specificity, sensitivity, and noise (Kreil et al. 2006). Therefore, long probe-array data require less statistical manipulation (Zilberman and Henikoff. 2007). In plants, NimbleGen arrays have been used for comparing whole-genome DNA methylation of wild-type Arabidopsis with loss-of-function mutants for DNA demethylase genes by Penterman et al. (2007) and allowed to detect about 200 methylation differences.

Genome-Scale DNA Methylation Analysis Using High-Throughput Sequencing

Over the past few years, there has been a fundamental shift away from the application of automated Sanger sequencing for genome analysis. The automated Sanger method is considered as a 'first-generation' technology, while newer methods of high throughput sequencing are referred to as next-generation sequencing (NGS). These technologies rely on a combination of template preparation, sequencing and imaging, genome alignment and assembly methods. High-throughput, next-generation sequencing (NGS) technologies have revolutionized research in biological sciences allowing a rapid analysis of the entire genome of any organism, and have changed the way we think about scientific approaches in basic and applied research (Metzker 2010). High-throughput sequencing enables to produce a very large amount of sequence information more rapidly, at a lower cost than conventional Sanger sequencing, and without the need for cloning. The number of approaches is increasing and some of them are still under development or improvement (Braslavsky et al. 2003; Meyers et al. 2004; Shendure et al. 2005; Bentley 2006). High-throughput sequencing can be employed as an alternative to oligonucleotide arrays for analyzing DNA methylation. Instead of labeling and hybridizing the test and control samples, as in array experiments, they can be sequenced directly. The frequency of a given sequence will be represented by its abundance in the sample. With sufficient sequences, information density comparable to microarray data can be achieved (Zilberman and Henikoff 2007). The choice of the best NGS method for methylation detection depends on the size of genome to be analyzed. For small, low repetitive genomes, direct sequencing of bisulfite-converted DNA is optimal. Restriction enzyme- and affinity-based methods would also be suitable for the analysis of small genomes, even if the data resolution level is not comparable with that of bisulfite analysis. For large genomes with a high repeat content, direct bisulfite sequencing or affinity-based purification of methylated DNA would be challenging, as those genomes are highly methylated.

High-Throughput Deep Sequencing of Bisulfite-Converted DNA

To date the most promising method for studying covalent cytosine modification is bisulfite conversion followed by high-throughput DNA sequencing. Several modified versions of the standard bisulfite sequencing method have been developed for the major next generation sequencing platforms (Roche/454, SoLid and Illumina). After bisulfite sequencing, the genome is largely composed of only three nucleotides (A, G, and T). Therefore, the downstream analysis requires two reference genomes, of which one represents an in silico matrix of the bisulfite conversion. Thus, data analysis is very cumbersome and requires highly trained bioinformaticians and specialized analysis software. In plants, a successful combination of deep sequencing and DNA bisulfite-conversion has been reported by Lister et al. (2008). These authors examined DNA methylation changes in floral tissue of *Arabidopsis thaliana* and produced an integrated map of the genomic distribution of methylcytosines, smRNAs, and transcripts at unprecedented resolution. Through the simultaneous analysis of these three interrelated phenomena in wild-type plants and in informative mutant backgrounds, they were able to study interactions between methylation and smRNA at a genome-wide scale, as well as their impact on transcriptional regulation.

Deep Sequencing of Methylation-Specific Restriction Enzyme Digested or Methylation-Specific Affinity Purified DNA

NGS in combination with either methylation-specific restriction endonucleases or affinity-based methylated DNA enrichment for methylation analysis is very challenging, and has so far mainly been applied to human genomes (Oda et al. 2009; Edwards, et al. 2010; Ruike et al. 2010). Recently, Yan et al. (2010) employed an immunoprecipitation strategy combined with Illumina sequencing for genome-wide mapping of cytosine methylation in rice. The pattern of methylated DNA distribution in rice chromosomes was similar to that of heterochromatin distribution. Moreover, DNA methylation patterns of rice and Arabidopsis genes were shown to be very similar (Yan et al. 2010).

Single Molecule Real Time Sequencing Technology

The next technology method to hit the commercial sector is likely to be real-time sequencing, and Pacific Biosciences is currently leading this effort (Metzker 2010). This company developed a method for Single Molecule Real Time (SMRT), which involves imaging the continuous incorporation of dye-labeled nucleotides during DNA synthesis (Eid et al. 2009). With the Pacific Biosciences platform, single DNA polymerase molecules are attached to the bottom surface of individual zero-mode waveguide detectors (Levene et al. 2003) that can obtain sequence information, while phospholinked nucleotides are incorporated into the growing primer strand.

SMRT sequencing is capable of detecting DNA methylation without the need for bisulfite conversion, since it utilizes arrival times and duration of the resulting fluorescence pulse to generate information about polymerase kinetics, which allows direct detection of modified nucleotides in the DNA template. Flusberg et al. (2010) applied SMRT to directly detect DNA methylation by sequencing both methylated and control DNA. The long read lengths of SMRT sequencing permit methylation profiling in highly repetitive genomic regions, in which a substantial fraction of mCytosine residues reside.

Applications in Plant Breeding

DNA methylation was shown to be relevant for several important traits and phenomena related to plant breeding, and is receiving attention to complement classical genetics. DNA methylation affects important parameters of conventional plant breeding programs including creation of favorable genetic variation that will form the basis for subsequent selection schemes, selection of superior genotypes through their phenotypes, single plant heritability, hybrid vigor, plant-environment interactions, stress tolerance and preservation and stability or even further improvement of pure line cultivars (Tsaftaris and Polidoros 2000; Tsaftaris et al. 2005). DNA methylation was also shown to be important for transgenic technology because of the interference with the phenotypic stability in new transgenic cultivars (Brandle et al. 1995; Matzke and Matzke 1996; Vaucheret et al. 1998). Until recently, the general assumption was that heritable variation involves only changes in DNA sequences. However, epigenetic information is meanwhile an accepted and important source of variation (Tsaftaris and Polidoros 2000).

Biomarkers based on DNA methylation are complementary to classical markers. In fact, they represent different DNA variation, which can be used to develop better varieties and to better understand mechanisms underlying plant development, diseases resistance, and plant reproduction and evolution. Recently, a Methylation Polymorphisms (MPs) have been suggested for application in plant breeding. For example, Nimmakayala et al. (2011) investigated the dynamics of genetic diversity among American watermelon (*Citrullus lanatus*) cultivars at the methylation level. Their aim was to compare methylation-specific with DNA-based genetic diversity for the same set of American watermelon heirloom cultivars using DNA markers (ISSR and AFLP) (Levi et al. 2004). In their study, epigenetic diversity was noted to be 16–43% among cultivated watermelons in contrast to 3.2–19.8% of genetic diversity estimated using conventional DNA markers by Levi et al. (2004) (Nimmakayala et al. 2011). Therefore, the study revealed that diversity at the methylation level was three times higher than the genetic diversity revealed by DNA markers on the same set of genotypes. A similar result was obtained by Sae-Eung et al. (2012). These authors investigated DNA MPs as well as nucleotide polymorphisms in Cycas species localized in Thailand, using MSAP in order to elucidate the role of epigenetics for genetic diversity of these plants and found that the percentage of DNA methylation was different among the studied cycads ranging from 36.1 to 57.4%. These data together with those reported in Arabidopsis (Cervera et al. 2002; Riddle and Richards 2002), rice (Ashikawa 2001; Wang et al. 2004; Sakthivel et al. 2010), Pisum (Knox and Ellis 2001), and cotton (Keyte et al. 2006) suggest that MPs are widespread among plants and can serve as epigenetic markers for use in plant breeding.

References

Adams RLP (1996) DNA methylation. In: Bittar EE (ed) Principles of medical biology, vol 5. JAI Press Inc, New York, pp 33–66

Ashikawa I (2001) Surveying CpG methylation at 5-CCGG in the genomes of rice cultivars. Plant Mol Biol 45:31–39

Baek D, Jiang J, Chung JS, Wang B, Chen J, Xin Z, Shi H (2011) Regulated AtHKT1 gene expression by a distal enhancer element and DNA methylation in the promoter plays an important role in salt tolerance. Plant Cell Physiol 52:149–161

Baubec T, Dinh HQ, Pecinka A, Rakic B, Rozhon W, Wohlrab B, Haeseler A, Scheida OM (2010) Cooperation of multiple chromatin modifications can generate unanticipated stability of epigenetic states in Arabidopsis. Plant Cell 22:34–47

Bentley DR (2006) Whole-genome re-sequencing. Curr Opin Genet Dev 16:545–552

Bibikova M, Lin Z, Zhou L, Chudin E, Garcia EW, Wu B, Doucet D, Thomas NJ, Wang Y, Vollmer E, Goldmann T, Seifart C, Jiang W, Barker DL, Chee MS, Floros J, Fan J (2006) High-throughput DNA methylation profiling using universal bead arrays. Genome Res 16:383–393

Bird AP (1995) Gene number, noise reduction and biological complexity. Trends Genet 11:94–100

Bird AP (2002) DNA methylation patterns and epigenetic memory. Genes Dev 16:6–21

Brandle JE, McHugh SG, James L, Labbe H, Miki BL (1995) Instability of transgene expression in field grown tobacco carrying the csr1-1 gene for sulfonylurea herbicide resistance. Biotechnology 13:994–998

Braslavsky I, Hebert B, Kartalov E, Quake SR (2003) Sequence information can be obtained from single DNA molecules. Proc Natl Acad Sci USA 100:3960–3964

Cervera MT, Ruiz-Garcia L, Martinez-Zapater JM (2002) Analysis of DNA methylation in *Arabidopsis thaliana* based on methylation-sensitive AFLP markers. Mol Genet Genomics 268:543–552

Clark SJ, Harrison J, Paul CL, Frommer M (1994) High sensitivity mapping of methylated cytosines. Nucleic Acids Res 22:2990–2997

Cross SH, Charlton JA, Nan X, Bird AP (1994) Purification of CpG islands using a methylated DNA binding column. Nat Genet 6:236–244

Eads CA, Danenberg KD, Kawakami K, Saltz LB, Blake C, Shibata D, Danenberg PV, Laird PW (2000) MethyLight: a high-throughput assay to measure DNA methylation. Nucleic Acids Res 28:E32

Edwards JR, O'Donnell AH, Rollins RA, Peckham HE, Lee C, Milekic MH, Chanrion B, Fu Y, Su T, Hibshoosh H, Gingrich JA, Haghighi F, Nutter R, Bestor TH (2010) Chromatin and sequence features that define the fine and gross structure of genomic methylation patterns. Genome Res 20:972–980

Eid J, Fehr A, Gray J, Luong K, Lyle J, Otto G, Peluso P, Rank D, Baybayan P, Bettman B, Bibillo A, Bjornson K, Chaudhuri B, Christians F, Cicero R, Clark S, Dalal R, Dewinter A, Dixon J, Foquet M, Gaertner A, Hardenbol P, Heiner C, Hester K, Holden D, Kearns G, Kong X, Kuse R, Lacroix Y, Lin S, Lundquist P, Ma C, Marks P, Maxham M, Murphy D, Park I, Pham T, Phillips M, Roy J, Sebra R, Shen G, Sorenson J, Tomaney A, Travers K, Trulson M, Vieceli J, Wegener J, Wu D, Yang A, Zaccarin D, Zhao P, Zhong F, Korlach J, Turner S (2009) Real-time DNA sequencing from single polymerase molecules. Science 323:133–138

Fan JB, Gunderson KL, Bibikova M, Yeakley JM, Chen J, Wickham Garcia E, Lebruska LL, Laurent M, Shen R, Barker D (2006) Illumina universal bead arrays. Methods Enzymol 410:57–73

Flusberg BA, Webster DR, Lee JH, Travers KJ, Olivares EC, Clark TA, Korlach J, Turner SW (2010) Direct detection of DNA methylation during single-molecule, real-time sequencing. Nat Methods 7:461–465

Gentil M, Maury S (2007) Characterization of epigenetic biomarkers using new molecular approaches. In: Varshney RK, Tuberosa R (eds) Genomics-assisted crop improvement. Vol 1: genomics approaches and platforms. Springer, Dordrecht, pp 351–370

Gupta R, Nagarajan A, Wajapeyee N (2010) Advances in genome-wide DNA methylation analysis. Biotechniques 49:iii–xi

Herman JG, Graff JR, Myohanen S, Nelkin BD, Baylin SB (1996) Methylation-specific PCR: a novel PCR assay for methylation status of CpG islands. Proc Natl Acad Sci USA 93:9821–9826

Holliday R, Pugh JE (1975) DNA modification mechanisms and gene activity during development. Science 187:226–232

Irizarry RA, Ladd-Acosta C, Wen B, Wu Z, Montano C, Onyango P, Cui H, Gabo K, Rongione M, Webster M, Ji H, Potash JB, Sabunciyan S, Feinberg AP (2009) The human colon cancer methylome shows similar hypo- and hypermethylation at conserved tissue-specific CpG island shores. Nat Genet 41:178–186

Kakutani T (2002) Epi-alleles in plants: inheritance of epigenetic information over generations. Plant Cell Physiol 43:1106–1111

Keshet I, Schlesinger Y, Farkash S, Rand E, Hecht M, Segal E, Pikarski E, Young RA, Niveleau A, Cedar H, Simon I (2006) Evidence for an instructive mechanism of de novo methylation in cancer cells. Nat Genet 38:149–153

Keyte A, Percifield R, Liu B, Wendel JF (2006) Infraspecific DNA methylation polymorphism in cotton (*Gossypium hirsutum* L.). J Hered 97:445–446

Knox MR, Ellis THN (2001) Stability and inheritance of methylation states at *Pst*I sites in *Pisum*. Mol Genet Genomics 265:497–507

Koga Y, Pelizzola M, Cheng E, Krauthammer M, Sznol M, Ariyan S, Narayan D, Molinaro AM, Halaban R, Weissman SM (2009) Genome-wide screen of promoter methylation identifies novel markers in melanoma. Genome Res 19:1462–1470

Korch C, Hagblom P (1986) In-vivo-modified gonococcal plasmid pJD1. A model system for analysis of restriction enzyme sensitivity to DNA modifications. Eur J Biochem 161:519–524

Kreil DP, Russell RR, Russell S (2006) Microarray oligonucleotide probes. Methods Enzymol 410:73–98

Levene MJ, Korlach J, Turner SW, Foquet M, Craighead HG, Webb WW (2003) Zero-mode waveguides for single-molecule analysis at high concentrations. Science 299:682–686

Levi A, Thomas CE, Newman M, Reddy UK, Zhang X, Xu Y (2004) ISSR and AFLP markers differ among American watermelon cultivars with limited genetic diversity. J Am Soc Hortic Sci 129:553–558

Lippman Z, Gendrel AV, Colot V, Martienssen R (2005) Profiling DNA methylation patterns using genomic tiling microarrays. Nat Methods 2:219–224

Lister R, O'Malley RC, Tonti-Filippini J, Gregory BD, Berry CC, Millar AH, Ecker JR (2008) Highly integrated single-base resolution maps of the epigenome in Arabidopsis. Cell 133:523–536

Lister R, Pelizzola M, Dowen RH, Hawkins RD, Hon G, Tonti-Filippini J, Nery JR, Lee L, Ye Z, Ngo QM, Edsall L, Antosiewicz-Bourget J, Stewart R, Ruotti V, Millar AH, Thomson JA, Ren B, Ecker JR (2009) Human DNA methylomes at base resolution show widespread epigenomic differences. Nature 462:315–322

Liu J, Morgan M, Hutchison K, Calhoun VD (2010) A study of the influence of sex on genome wide methylation. PLoS One 5:e10028

Matzke MA, Matzke AJM (1996) Stable epigenetic states in differentiated plant cells: implications for somaclonal variation and gene silencing in transgenic plants. In: Russo VEA, Martienssen RA, Riggs AD (eds) Epigenetic mechanisms of gene regulation. Cold Spring Harbor Laboratory Press, New York, pp 377–392

McClelland M, Nelson M, Raschke E (1994) Effect of site-specific modifi- cation on restriction endonucleases and DNA modification methyltransferases. Nucleic Acids Res 22:3640–3659

Meissner A, Mikkelsen TS, Gu H, Wernig M, Hanna J, Sivachenko A, Zhang X, Bernstein BE, Nusbaum C, Jaffe DB, Gnirke A, Jaenisch R, Lander ES (2008) Genome-scale DNA methylation maps of pluripotent and differentiated cells. Nature 454:766–770

Metzker ML (2010) Sequencing technologies-the next generation. Nat Rev Genet 11:31–46

Meyers BC, Vu TH, Tej SS, Ghazal H, Matvienko M, Agrawal V, Ning J, Haudenschild CD (2004) Analysis of the transcriptional complexity of *Arabidopsis thaliana* by massively parallel signature sequencing. Nat Biotechnol 22:1006–1011

Nimmakayala P, Vajja G, Gist RA, Tomason YR, Levi A, Reddy UK (2011) Effect of DNA methylation on molecular diversity of watermelon heirlooms and stability of methylation specific polymorphisms across the genealogies. Euphytica 177:79–89

Noushmehr H, Weisenberger DJ, Diefes K, Phillips HS, Pujara K, Berman BP, Pan F, Pelloski CE, Sulman EP, Bhat KP, Verhaak RG, Hoadley KA, Hayes DN, Perou CM, Schmidt HK, Ding L, Wilson RK, Van Den Berg D, Shen H, Bengtsson H, Neuvial P, Cope LM, Buckley J, Herman JG, Baylin SB, Laird PW, Aldape K (2010) Identification of a CpG island methylator phenotype that defines a distinct subgroup of glioma. Cancer Cell 17:510–522

Oda M, Glass JL, Thompson RF, Mo Y, Olivier EN, Figueroa ME, Selzer RR, Richmond TA, Zhang X, Dannenberg L, Green RD, Melnick A, Hatchwell E, Bouhassira EE, Verma A, Suzuki M, Greally JM (2009) High-resolution genome-wide cytosine methylation profiling with simultaneous copy number analysis and optimization for limited cell numbers. Nucleic Acids Res 37:3829–3839

Penterman J, Zilberman D, Huh JH, Ballinger T, Henikoff S, Fischer RL (2007) DNA demethylation in the Arabidopsis genome. Proc Natl Acad Sci USA 104:6752–6757

Regev A, Lamb MJ, Jablonka E (1998) The role of DNA methylation in invertebrates: developmental regulation or genome defense? Mol Biol Evol 15:880–891

Reyna-Lopez GE, Simpson J, Ruiz-Herrera J (1997) Differences in DNA methylation patterns are detectable during the dimorphic transition of fungi by amplification of restriction polymorphisms. Mol Gen Genet 253:703–710

Reynaud C, Bruno C, Boullanger P, Grange J, Barbesti S, Niveleau A (1992) Monitoring of urinary excretion of modified nucleosides in cancer patients using a set of six monoclonal antibodies. Cancer Lett 61:255–262

Riddle NC, Richards EJ (2002) The control of natural variation in cytosine methylation in Arabidopsis. Genetics 162:355–363

Rollins RA, Haghighi F, Edwards JR, Das R, Zhang MQ, Ju J, Bestor TH (2006) Large-scale structure of genomic methylation patterns. Genome Res 16:157–163

Ruike Y, Imanaka Y, Sato F, Shimizu K, Tsujimoto G (2010) Genome- wide analysis of aberrant methylation in human breast cancer cells using methyl-DNA immunoprecipitation combined with high-throughput sequencing. BMC Genomics 11:137

Ruzov A, Savitskaya E, Hackett JA, Reddington JP, Prokhortchouk A, Madej MJ, Chekanov N, Li M, Dunican DS, Prokhortchouk E, Pennings S, Meehan RR (2009) The non-methylated DNA-binding function of Kaiso is not required in early *Xenopus laevis* development. Development 136:729–738

Sae-Eung C, Kanchanaketu T, Sangduen N, Hongtrakul V (2012) DNA methylation and genetic diversity analysis of genus Cycas in Thailand. Afr J Biot 11:743–751

Sakthivel K, Girishkumar K, Ramkumar G, Shenoy VV, Kajjidoni T, Salimath PM (2010) Alterations in inheritance pattern and level of cytosine DNA methylation, and their relationship with heterosis in rice. Euphytica 175:303–314

Sano H, Kamada I, Youssefian S, Katsumi M, Wabilko H (1990) A single treatment of rice seedlings with 5-azacytidine induces heritable dwarfism and undermethylation of genomic DNA. Mol Gen Genet 220:441–447

Shendure J, Porreca GJ, Reppas NB, Lin X, McCutcheon JP, Rosenbaum AM, Wang MD, Zhang K, Mitra RD, Church GM (2005) Accurate multiplex polony sequencing of an evolved bacterial genome. Science 309:1728–1732

Steward N, Ito M, Yamachuchi Y, Koizumi N, Sano H (2002) Periodic DNA methylation in maize nucleosomes and demethylation by environmental stress. J Biol Chem 277:37741–37746

Suhr ST, Chang EA, Rodriguez RM, Wang K, Ross PJ, Beyhan Z, Murthy S, Cibelli JB (2009) Telomere dynamics in human cells reprogrammed to pluripotency. PLoS One 4:e8124

Tsaftaris AS, Polidoros AN (2000) DNA methylation and plant breeding. Plant Breed Rev 18:87–176

Tsaftaris AS, Polidoros AN, Koumproglou R, Tani E, Kovacevic N, Abatzidou E (2005) Epigenetic mechanisms in plants and their implications in plant breeding. In: Tuberosa R, Phillips RL, Gale M (eds) Proceedings of the international congress "in the wake of the double helix: from the green revolution to the gene revolution", Bologna, Italy, 27–31 May 2003. Avenue Media, Bologna, pp 157–171

Vaucheret H, Beclin C, Elmayan T, Feuerbach F, Godon C, Morel JB, Mourrain P, Palauqui JC, Vernhettes S (1998) Transgene-induced gene silencing in plants. Plant J 16:651–659

Verhoeven KJF, van Dijk PJ, Biere A (2010) Changes in genomic methylation patterns during the formation of triploid asexual dandelion lineages. Mol Ecol 19:315–324

Vos P, Hogers R, Bleeker M, Reijans M, van de Lee T, Hornes M, Fijters A, Pot J, Peleman J, Kuiper M, Zabeau M (1995) AFLP: a new technique for DNA fingerprinting. Nucleic Acids Res 23:4407–4414

Wang JL, Tian A, Madlung H, Lee S, Chen M (2004) Sto- chastic and epigenetic changes of gene expression in Arabidopsis polyploids. Genetics 167:1961–1973

Weber M, Davies JJ, Wittig D, Oakeley EJ, Haase M, Lam WL, Schubeler D (2005) Chromosome-wide and promoter-specific analyses identify sites of differential DNA methylation in normal and transformed human cells. Nat Genet 37:853–862

Weber M, Hellmann I, Stadler MB, Ramos L, Pääbo S, Rebhan M, Schübeler D (2007) Distribution, silencing potential and evolutionary impact of promoter DNA methylation in the human genome. Nat Genet 39:457–466

Weising K, Nybom H, Wolf K, Kahl G (2005) DNA Fingerprinting in plants, 2nd edn. CRC Press, Taylor & Francis, Boca Raton, p 444

Wolffe AP, Matzke MA (1999) Epigenetics: regulation through repression. Science 286:481–486

Xiong LZ, Xu CG, Saghai Maroof MA, Zhang Q (1999) Patterns of cytosin methylation in an elite rice hybrid and its parental lines, detected by a methylation- sensitive amplification polymorphism technique. Mol Gen Genet 261:439–446

Yan H, Kikuchi S, Neumann P, Zhang W, Wu Y, Chen F, Jiang J (2010) Genome-wide mapping of cytosine methylation revealed dynamic DNA methylation patterns associated with genes and centromeres in rice. Plant J 63:353–365

Yi C, Zhang S, Liu X, Bui HTN, Hong Y (2010) Does epigenetic polymorphism contribute to phenotypic variances in *Jatropha curcas* L.? BMC Plant Biol 10:259

Yoder JA, Walsh CP, Bestor TH (1997) Cytosine methylation and the ecology of intragenomic parasites. Trends Genet 13:335–340

Zhang X, Yazaki J, Sundaresan A, Cokus S, Chan SW-L, Chen H, Henderson IR, Shinn P, Pellegrini M, Jacobsen SE, Ecker JR (2006) Genome-wide high-resolution mapping and functional analysis of DNA methylation in Arabidopsis. Cell 126:1189–1201

Zhang X, Shiu S, Cal A, Borevitz JO (2008) Global analysis of genetic, epigenetic and transcriptional polymorphisms in *Arabidopsis thaliana* using whole genome tiling arrays. PLoS Genet 4:3

Zhao Y, Yu S, Ye W, Wang H, Wang J, Fang B (2010) Study on DNA cytosine methylation of cotton (*Gossypium hirsutum* L.) genome and its implication for salt tolerance. Agric Sci China 9:783–791

Zilberman D, Henikoff S (2007) Genome-wide analysis of DNA methylation patterns. Development 134:3959–3965

Zilberman D, Gehring M, Tran RK, Ballinger T, Henikoff S (2007) Genome-wide analysis of *Arabidopsis thaliana* DNA methylation uncovers an interdependence between methylation and transcription. Nat Genet 39:61–69

Chapter 13
Transcriptome-Based Prediction of Heterosis and Hybrid Performance

Stefan Scholten and Alexander Thiemann

Introduction

Heterosis is defined as the ability of hybrids to outperform their parents with respect to various characteristics and agronomical important traits (Shull 1948). Various plant traits and particularly pronounced yield display heterosis. Hybrid breeding is based upon the phenomenon and prediction of hybrid performance and heterosis are important applications to increase the efficiency of hybrid breeding programs. Traditional phenotypic evaluation is still the common methodology to estimate general combining abilities for the predictions of hybrid traits (Choudhary et al. 2008). Molecular markers like RFLPs (restriction fragment length polymorphism), RAPDs (randomly amplified polymorphic DNA) or AFLPs (amplified fragment length polymorphism) can be employed to improve the prediction of complex traits like yield. An example is the application of AFLP-Markers in a linear regression approach to predict hybrid performance and SCA in intergroup-crosses of maize (Vuylsteke et al. 2000). For a long time molecular marker-based genetic distances has not proven to be a reliable approach for the prediction of hybrid characteristics in crops (Melchinger 1999) and might have limitations for prediction, as e.g. in *Arabidopsis* the correlation between heterosis and genetic distance was not significant (Meyer et al. 2004; Stokes et al. 2007). However, the improvement of DNA marker based approaches and its application to factorial crosses in maize led to a prediction accuracy that is comparable to phenotype based estimates (Schrag et al. 2006, 2007). To further increase prediction abilities for hybrid breeding, various molecular compounds of the plant, like DNA methylation states, transcripts, proteins and metabolites, are currently considered and tested as new markers with

---

S. Scholten (✉) • A. Thiemann
Biocenter Klein Flottbek and Botanical Garden, Developmental Biology and Biotechnology, University of Hamburg, Ohnhorststraße 18, 22609 Hamburg, Germany
e-mail: s.scholten@botanik.uni-hamburg.de

diagnostic and predictive potential. Current results related to specific compounds are discussed in the various chapters of this section. Here we address the development and applicability of RNA expression data for the prediction of heterosis and hybrid performance.

Heterosis Associated Gene Expression Pattern Exhibit Predictive Characteristics

The molecular basis of heterosis is complex. Genetic mechanisms like dominance, overdominance and epistasis are thought to underlie the phenomenon (Birchler et al. 2003; Schön et al. 2010). These genetic mechanisms are suggested to be generic features of gene regulatory networks and might be, at least partially, explained by ectopic or temporarily altered mRNA expression levels (Omholt et al. 2000).

Variation in gene expression is a major basis for phenotypic variation. Therefore, fundamental questions regarding the molecular mechanism of heterosis are how the two different alleles brought together in the hybrid are expressed and how is the relationship between yield heterosis and gene expression in the hybrids. As a matter of principle two modes of expression variation in hybrids with two different parental alleles can be distinguished. On one hand an additive allelic expression with the average of both parents' expression levels may occur. On the other hand gene expression in hybrids might differ from the mid-parental value resulting in a non-additive level (Birchler et al. 2003). Non-additive gene expression can arise when the combination of diverse alleles leads to gene regulatory interactions in the hybrid, whereas additive expression is expected if solely *cis*-regulatory differences are responsible for expression level regulation in hybrids (Wittkopp et al. 2004). The latter mode of expression regulation in hybrids is of special interest with respect to RNA based prediction approaches, because *cis*-regulatory variation results from control elements physically linked to the genes, like promoters or enhancers, determined by DNA sequence and thus is directly inherited from the parents.

Extensive gene expression level changes in hybrids relative to their parents have been documented in the past by a large number of studies (reviewed by Hochholdinger and Hoecker 2007). These studies reported on extensive transcriptome remodelling in various tissues and developmental stages of various plant species' heterotic hybrids. (Xiong et al. 1998; Sun et al. 1999; Ni et al. 2000; Wu et al. 2003; Song and Messing 2003; Guo et al. 2004). It has also been hypothesized that differential gene expression in inbreds and hybrids contribute to heterosis (Song and Messing 2003; Guo et al. 2004).

The exploration of non-additive gene expression in hybrids of *Arabidopsis* (Vuylsteke et al. 2005) and in diploid and triploid maize hybrids (Auger et al. 2005) indicated gene regulatory interactions among the parental alleles to be largely responsible for the expression variation in hybrids. Indeed, up to date all heterosis related gene expression studies revealed dominant and overdominant expression pattern to a certain extend, indicative of new interactions between the combined genomes in the hybrid state. The possibility to predict heterosis based on parental gene expression profiles would be questionable, if these expression patterns are the main cause underlying heterosis. However, in maize for example, little consistency regarding the relative occurrence of specific modes of gene action between parental inbred lines and the corresponding hybrids in various tissues could be found. Global expression analysis using microarray hybridizations detected all possible modes of gene action but predominant additive expression variation in the above ground tissue of 14 day-old seedlings (Swanson-Wagner et al. 2006). Expression profiles of meristems of 21 day-old plants from different maize inbred combinations also revealed all possible modes of gene action, although the proportion of genes in expression behaviour classes differed with predominant non-additive expression (Uzarowska et al. 2007). In early embryo and endosperm development 6 days after pollination and immature ear tissue as well mainly additive but also dominant and overdominant expression patterns were found (Guo et al. 2006; Meyer et al. 2007; Jahnke et al. 2010).

Together, data on differential gene expression between inbred lines and hybrids support the idea that gene regulatory networks at the level of transcription are involved in the control of hybrid vigour in plants. In most studies the majority of genes showed additive expression, which is expected if solely allelic *cis*-regulatory differences are responsible for expression level regulation in hybrids (Wittkopp et al. 2004). Supportive to this hypothesis, allele-specific expression analyses in maize revealed that *cis*-regulatory differences between inbred lines can explain the expression profiles of their hybrids to a significant extend (Stupar and Springer 2006). Furthermore, extensive sequence variation between rice inbred parents affecting *cis*-regulatory elements were found in putative promoter regions of orthologous genes with predominant additive expression in hybrids (Zhang et al. 2008).

If transcriptome regulatory networks provide a basis for heterosis and the expression variation is largely explainable by heritable *cis*-regulatory factors, the transcriptome characteristics of inbred lines resulting from their individual genomic constitution should be useful as markers to predict the performance of the hybrids generated by crosses of these inbred lines. To reliably establish relations between transcriptome characteristics and heterosis or hybrid performance systematic approaches are required with the numbers of RNA species measured and genotypes analyzed sufficiently high to identify significant correlations.

Two studies provide examples for more systematic approaches to establish a relation of the transcriptome and heterosis in maize. Guo et al. (2006) analyzed an extensive series of 16 maize hybrids that vary in the degree of yield heterosis and the parental inbred lines by GeneCalling mRNA profiling technology. GeneCalling is an open-ended, gel-based technology that allows profiling of mRNA abundance for both known and novel genes in an unbiased way. It is well suited for comprehensive gene expression pattern analyses, since it detects 80–90% of RNA species in a given tissue (Shimkets et al. 1999). The experimental maize population structure used one common female parent, which was crossed with a series of male inbreds. Yield data of all genotypes were generated in 2 years, four locations and two replicates per location and year. The gene expression data were generated from immature ears before pollination. These expression profiles revealed the percentage of mid-parental, additive expression positively correlated with heterosis and a positive correlation between the percentage of inter-parental differentially expressed genes and yield heterosis as well. Both correlations were strong and highly significant (Guo et al. 2006). Especially the strong relation of inter-parental expression pattern to heterosis of yield indicates the potential of parental expression profiles for prediction approaches.

Stupar et al. (2008) studied expression profiles of inbreds and hybrids of six different maize genotype combinations with Affymetrix GeneChip maize arrays, which represent 13,339 genes, to determine whether the degree of non-additive gene expression vary in hybrids with different level of heterosis. The frequencies of expression patterns were found to be highly similar between hybrids of diverse heterosis responses. The majority of differentially expressed genes in each of the six different hybrids exhibited additive expression and expression levels outside the parental range where found to be with 1% extremely rare (Stupar et al. 2008). Interestingly, a high correlation between transcriptional variation and genetic diversity, as estimated by SNP based sequence analyses (Hamblin et al. 2007), was established. Although a significant correlation of genetic diversity with heterosis was found for seedling biomass, as one of five traits, only, these data add further indication that heterosis may be related to transcriptional variation between parental inbred lines (Stupar et al. 2008).

Advantages of RNA Expression Measures for the Prediction of Heterosis and Hybrid Performance

A large number of small genetic effects are thought to underlie quantitative traits (Buckler et al. 2009) and their heterosis (McMullen et al. 2009). Therefore, capturing as much of the genomic constitution of the parental inbred lines as possible is necessary for successful modelling and prediction of heterosis. RNA expression profiles may have major advantages over DNA markers to accomplish this task, because transcript abundance results from the integration of variegated genetic information. The influence ranges from *trans*-acting factors, like transcriptions factors or regulators of transcript stability, to *cis*-acting factors, like promoter sequences. Additionally, variations of the chromatin states with differences in DNA methylation levels and histone modification states integrate to the specific expression profile of an inbred line. As a result, the expression values measured reflect the specific combination of alleles and epialleles of a given genotype.

Transcriptome data are yet likely to contain additional information to DNA marker data, irrespective how dense the genome is covered by DNA markers and even when whole genome sequences are used. A relationship between transcriptional variation and genetic distance was shown in maize (Stupar et al. 2008). The higher information content was indicated by the comparison of transcriptome based distance measures of 21 inbred lines with DNA marker based genetic distances, based on QTL associated amplified fragment length polymorphism (AFLP) marker (Frisch et al. 2010; Schrag et al. 2006). The range of transcriptome based distances of inter-pool crosses was much larger than the range of inter-pool genetic distances (Frisch et al. 2010). These results are indication that transcriptome data are likely to be superior to marker data for the characterization of inbred lines belonging to the different heterotic pools that are usually used to generate high yielding hybrids. These presumptions are supported by moderate but highly significant correlations of heterosis and hybrid performance with the transcriptome based distances in contrast to DNA marker based distances, which showed no correlation at all (Frisch et al. 2010).

Another advantage of transcriptome data with respect to DNA markers is that no assumptions on linkage disequilibrium are required for prediction. Founder effects, selection, and random genetic drift can cause differences in the linkage disequilibrium between marker alleles and functional alleles of different heterotic pools (Boppenmaier et al. 1993; Charcosset and Essioux 1994). These effects are expected to reduce severely the precision of DNA marker based predictions.

Development of Transcriptome-Based Prediction Methods

The development of RNA markers and methods to predict hybrid performance based on data generated from inbred lines is in its infancy. Two first attempts have been made to demonstrate the predictive power of transcriptome data (Frisch et al. 2010; Stokes et al. 2010). Both studies commonly considered the transcript abundance levels of parental lines as quantitative variables to characterize a genotype and to predict characteristics of new hybrids, but differ fundamentally with respect to the population structures and mathematical models used.

Stokes et al. (2010) studied *Arabidopsis* and maize to established relationships between transcript abundance of specific genes and the value of heterosis and yield. For both species they used experimental hybrid populations that were produced by using one common accession or inbred line as maternal parent and genetically diverse accessions or inbred lines as paternal parents. For *Arabidopsis* and maize 12 and 20 paternal lines were used, respectively. Affimetrix GeneChips, either Arabidopsis ATH1 or Maize Genome Arrays were used for expression profiling representing either 24,000 or 13,339 genes, respectively. The relationship between transcript levels and the magnitude of heterosis or yield were established by linear regression. Only few genes in *Arabidopsis* and several hundred genes in maize exhibiting correlation were identified. Of these, a highly restrictive significance threshold setting retained 2 and 185 genes for the prediction approach in *Arabidopsis* and maize, respectively. For any given transcript within the defined significance threshold heterosis or yield was calculated by using the parameters from the regression analyses. The mean of the predicted values across all genes used provided the overall prediction.

The prediction ability of the gene expression levels were estimated by calculating the correlation of predicted versus measured heterosis or yield of the training set and additional test set hybrid genotypes. For both, *Arabidopsis* seedling biomass heterosis and yield of maize hybrids a considerably high correlation of about 0.69 was found (Stokes et al. 2010). These results demonstrate the principal relationship of inbred line transcriptome data and heterosis and their suitability for prediction.

In the second study to demonstrate the predictive power of transcriptome data, by Frisch et al. (2010), 21 maize inbred lines of 2 heterotic pools, 7 flint and 14 dent lines, were used in a 7×14 factorial mating design to produce 98 hybrids. Heterosis and hybrid performance for yield of all genotypes were evaluated in field trials. The transcript levels of whole seedlings were measured by hybridizations of a 46 k oligonucleotide microarray (www.maizearray.org, University of Arizona, USA) with 43,381 gene-oriented probes. Five biological replicates were pooled for each genotype and subjected to an interwoven loop design resulting in sampling each dent line five and each flint line eight times for direct comparisons. The expression profiling revealed a set of 10,810 genes with differential expression (≥ 1.3 fold change at 1% FDR) between at least one pair of inbred lines. Of these, trait associated genes were selected by a binominal probability test and used to construct transcriptome-based distances analogous to DNA marker based genetic distances.

To assess the actual prediction efficiency of transcriptome-based distances a cross validation procedure with 100 runs was applied to determine the correlation coefficient of observed with predicted values for heterosis and hybrid performance. In each run the lines for estimation of the prediction parameters were randomly chosen, the corresponding trait associated genes were determined and used to construct the distances and estimate regression parameters with the corresponding hybrids. To predict the heterosis or hybrid performance of new hybrids the regression parameters were applied to distances constructed with the previously determined trait associated gene's expression values of the corresponding inbred lines. To roughly meet the reality of maize breeding programs with respect to the amount of field data available to develop prediction parameters, less than half of the lines were used to estimate the prediction parameters and more than half of the lines were used for validation. The prediction efficiency was estimated with a correlation coefficient of 0.8 (Frisch et al. 2010).

With this approach previous results of hybrid performance prediction efficiency of the same experimental setting but predictions based on phenotypic data to estimate the general combining ability (GCA) or QTL associated genetic AFLP marker (Schrag et al. 2007) were excelled with approximately 0.1 higher correlation values. Additionally, the transcriptome-based predictions had much lower variability (Frisch et al. 2010), which is an important parameter for the applicability in breeding programs. These results clearly demonstrate the high predictive power of transcript data, however, because in each validation run the set of trait associated genes were newly determined, thus specific predictive genes or RNA markers are not yet available. The overlapping genes between the cross validation runs are interesting in this respect.

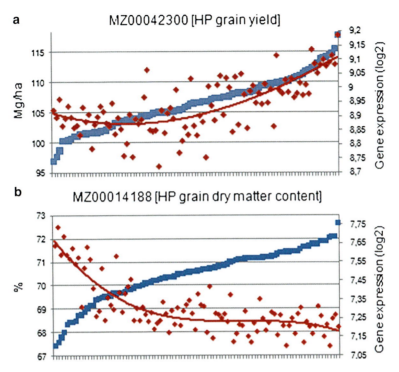

Fig. 13.1 Correlation of transcript abundance in parental lines with hybrid performance for two traits. The genes with the highest correlation for each trait from Thiemann et al. (2010) are shown. The calculated mid-parental transcript abundance (log2, *rhombs*) of both inbred lines was used for Pearson correlation with the performance of the respective hybrids (HP) for (**a**) grain yield (Mg/ha, *squares*) or (**b**) grain dry matter content (%, *squares*). The *continuous line* represents a polynomial trend line for the mid-parental expression values

Based on the same factorial cross experiment and data set Thiemann et al. (2010) correlated the inbred line expression values with hybrid characteristics to identify trait associated genes for hybrid performance of grain yield and grain dry matter content as well as for heterosis of grain yield. The correlations were based on calculated mid-parental values, corresponding to the prevalent additive expression in hybrids (e.g. Guo et al. 2006; Stupar et al. 2008). This approach led to the identification of highly correlated genes for each trait. The whole gene set was published (Thiemann et al. 2010) and revealed functional properties of the traits (Fu et al. 2010; Thiemann et al. 2010). The genes exhibiting the highest correlations for hybrid performance of grain yield and grain dry matter content are shown in Fig. 13.1. They demonstrate the tight relation of the calculated mid-parental expression levels to the traits. The high correlations of some 2,000 genes for each trait suggest that the genes identified are useful as RNA markers for prediction approaches.

Applicability in Breeding

Both studies on transcriptome-based prediction in maize demonstrated high predictive power of transcriptome data of young seedling material to predict end-of-season yield data (Frisch et al. 2010; Stokes et al. 2010). Analyses of young seedlings provide valuable advantages with respect to the applicability of inbred line gene expression data in breeding programs. Seedlings have low space requirements and the growth period is limited to a few days. Thus, the required data can be generated quickly with low input of resources and importantly the growth conditions can be exactly controlled and replicated. The high predictive power of the seedling expression data for yield justifies the assumption that transcript levels in seedlings are highly related to agronomic performance. Supportive to this view are the functional characterizations of end-of-season traits, which were based on correlations with seedling expression profiles and revealed comprehensible and conclusive results with respect to known mechanism involved in trait formation (Fu et al. 2010; Thiemann et al. 2010). In addition, the co-localization of genomic fragments that are enriched for yield heterosis and hybrid performance associated genes with QTL for the same traits (Thiemann et al., unpublished) support a close relation. This may be due to prevalently additive expression in hybrids that influence the late hybrid traits indirectly by continuously supporting conditions for growth and development throughout plant life, thereby establishing favourable conditions for later appearing traits. The neutrality of transcript abundance in evolution that seems to be characteristic for plant transcriptomes (Broadley et al. 2008), is likely to contribute to the close relation as well. Neutral evolution of transcript abundance would lead to the prevalence of genotype specific expression levels with functional significance for specific processes throughout the plant and all developmental stages as long as they do not exhibit selective disadvantage. Together the available data to date indicate that it is justified to make use of the advantages of seedlings for transcriptome based prediction approaches.

The number of RNA measurements required to perform reliable predictions of hybrid performance is an important parameter in view of practical applicability. In any case a large number of samples need to be analysed to go along and support a hybrid breeding program with transcriptome based predictions. For low numbers up to several hundred genes real-time quantitative polymerase chain reaction can be applied. For the analysis of up to several thousand genes custom oligonucleotide or synthesized microarrays are most suitable and cost effective. If the exploration of whole transcriptomes is required for all samples to achieve robust predictions large genome wide microarrays can be used. Alternatively next generation deep sequencing technologies, like Illumina/Solexa (Illumina) or SOLiD (AppliedBiosciences, Life Technologies) might be applied. Deep sequencing is well suited for quantitative transcriptome analyses and the costs per sequence read for these relative new technologies are dropping rapidly.

The possibility to develop few universal RNA markers to reliably predict the performance of new genotypes was indicated by the approach of Stokes et al. (2010).

The expression values of two genes only were sufficient to predict fairly precise heterosis of biomass in *Arabidopsis* based on linear regressions. For a comparable precise prediction of maize yield the expression values of 163 genes were used (Stokes et al. 2010). The actual influence of number of genes on prediction accuracy was not systematically analysed in this study.

Prediction methods based on transcriptome-based distances indicate that transcription values of more genes are required to precisely predict heterosis of hybrid performance for yield in maize. With fixed numbers of trait-associated genes for each cross validation run the construction of the distances for prediction 1,000–1,500 genes were determined to be optimal with respect to median prediction accuracy and its variability. Reducing the number of genes down to a minimum number of 50 had only a small influence of the average prediction accuracy whereas its variability strongly increases (Frisch et al. 2010). The determination of overlapping genes between the validation runs might lead to a core set of highly predictive genes. If fixed sets of genes improve the prediction efficiency and how many genes are required for prediction approaches sufficiently precise to be applied in breeding programs remains to be determined.

Fixed gene sets open as well the question how large the genetic distance of inbred lines can be while still allowing useful prediction accuracy. Is the prediction only possible with hybrids highly related to the inbred lines used to determine the genes and to train the algorithm or is it possible to define a fixed gene set which can be used in several breeding cycles and for less related genetic backgrounds? The initial results concerning transcriptome-based prediction are promising in this respect. To roughly meet the reality of maize breeding programs with respect to the amount of field data available to develop prediction parameters, less than half of the lines were used to estimate the prediction parameters and more than half of the lines were used for cross validation in the approach of Frisch et al. (2010). This led to the circumstance that in each validation run hybrids without common parental lines to the ones used to estimate the parameters were predicted. Although the high prediction accuracy was the mean result of all validation runs and trait-associated genes were selected in each run, the relative low variability indicate transferability between genotypes to some extent. Stokes et al. (2010) successfully predicted new hybrids' performance by relative small fixed gene sets. Whereas the use of genetically highly diverse inbred lines in this approach supports the view of transferability of marker genes, the shared pedigree of all hybrids due to the use of one common paternal line weaken this notion.

It is conceivable that the complementation of genetic data with the transcript data of few genes is an effective strategy. The combination of two different markers was already realized in *Arabidopsis* by combining metabolic markers with genetic SNP markers (Gärtner et al. 2009; Steinfath et al. 2010) and could improve heterosis prediction to some extent. However, the questions which and how many genes need to be analysed to provide sufficient predictive power need to be addressed in further experiments. In respect thereof, monitoring an actual breeding program over several breeding cycles would be a promising approach.

Applications of RNA Marker in Plant Functional Genetics

Several studies used differential gene expression between parental inbred lines and hybrids to identify molecular mechanism underlying heterosis (e.g. Huang et al. 2006; Meyer et al. 2007; Uzarowska et al. 2007; Hoecker et al. 2008; Wei et al. 2009; Jahnke et al. 2010). Functional aspects of non-additively expressed genes were the focus in most of these studies. Although the different genotype combinations analysed were considerably high in some studies, the experimental design does not allow to reliably connect transcriptome characteristics with heterosis since the observed expression need not to be the real cause of heterosis, but could also be caused by heterosis.

The approaches developed for the prediction of heterosis and hybrid characteristics involve large inbred line and hybrid populations and effectively associates quantitative RNA measures with quantitative traits (Stokes et al. 2010; Frisch et al. 2010). The predictive power of the parental inbred line expression profiles of certain genes for heterosis, even for unrelated hybrids, strongly strengthen the assumption that those genes are involved in heterosis formation. The approach of Stokes et al. (2010) revealed both genes with the strongest correlation to heterosis in *Arabidopsis* and maize to be involved in gene regulation. These results are consistent with the hypothesis that transcriptional regulatory networks are a main molecular basis of heterosis. Although the study of Thiemann et al. (2010), which was based on the same parental expression profiles used for prediction by Frisch et al. (2010), revealed a gene involved in energy metabolism as the highest heterosis correlated gene (Fig. 13.1), the analysis of overrepresented gene functions among heterosis correlated genes revealed, amongst others, gene regulation and signal transduction to be involved in heterosis formation as well. Interestingly, the genes identified are not randomly distributed throughout the genome. Some genomic fragments are enriched for these heterosis-correlated genes and co-localize with QTLs for yield (Thiemann et al., unpublished). This might indicate a possibility to identify QTL underlying genes by this transcriptome based approach.

Future Directions

All approaches towards transcriptome-based prediction to date considered exclusively polyadenylated, messenger transcripts (mRNA), which are mainly transcribed from protein coding genes. An additional measure with a high potential to represent the genomic constitution of complex genomes are small, non-coding RNAs (smRNAs). smRNAs act as important negative regulators of genes and other nucleotide sequences. MicroRNAs (miRNAs) have been implicated in the regulation of genes involved in development and homeostasis. Small interfering RNAs (siRNAs) are important suppressors of transposons and viruses, but are also implicated in processes of homeostasis as well as in the maintenance of epigenetic states (Mallory and Vaucheret 2006). Importantly, smRNAs are largely transcribed from non-gene coding regions and therefore are likely to be highly complementary to mRNA transcriptional profiles with respect to genome representation.

The effect of smRNAs might be especially important for the regulation of large genomes. In maize, smRNA populations were shown to be extremely complex by the identification of a new group of 22-nucleotide siRNAs (Nobuta et al. 2008; Wang et al. 2009), which preferentially target gene-coding regions and a group of miRNA-like smRNAs potentially targeting genes in *trans* (Wang et al. 2009). This high complexity of smRNA species and function correspond well to the extremely large and complex maize genome. Additionally, transposable elements play a key role in the genome-wide distribution of epigenetic marks of large genomes with influence on the regulation of neighboring genes (Weil and Martienssen 2008).

Based on the uniparental expression and the population complexity of smRNAs in *Arabidopsis* seed development, it was proposed that small RNAs have an effect on expression variation in hybrids and could contribute to heterotic phenotypes (Mosher et al. 2009). smRNAs are involved in the regulation of epigenetic states and several lines of evidence indicate that epigenetic differences in the parents may lead to heterosis (reviewed by Groszmann et al. 2011a). Both rice and *Arabidopsis* hybrids exhibit locus specific variation in siRNA accumulation (He et al. 2010; Groszmann et al. 2011b) and in *Arabidopsis* intra-species hybrids dramatically reduced levels of siRNAs occur at loci with large differences in siRNA levels between the parental alleles (Groszmann et al. 2011b). These correlations of siRNA levels in hybrids with differences in siRNA levels between parents strongly support the assumption that small non-coding RNAs have a prediction potential. Additional support indicates the findings in *Arabidopsis* that epigenetic states with respect to DNA methylation patterns as well as histone marks are mainly additively inherited to hybrids (Moghaddam et al. 2010, 2011).

Another aspect for future investigations is the evaluation of different statistical methods to find the most powerful approach for transcriptome based prediction. Currently, two methods were applied to relate transcript levels with quantitative traits for prediction: Linear regressions techniques (Stokes et al. 2010) and transcriptome based distance measures which were related to hybrid performance by linear regressions (Frisch et al. 2010), both described in more detail above. Alternative prediction methods might be developed on the basis of partial least square regression (PLS), which were applied in studies with metabolites as markers (Gärtner et al. 2009; Steinfath et al. 2010). Additionally, support vector machine regression (SVM) was suggested for prediction approaches with genetic markers (Maenhout et al. 2007). It is important to carefully evaluate these methods with the same dataset to ensure comparability. Critical criteria are the overall prediction accuracy and accuracy in relation to the amount of data available. The comparison of the performances for prediction of untested hybrids when parental line data are available and of fully unrelated hybrids of which no testcross data are available might give an estimate about the data amount required to effectively support a breeding program.

Which statistical method is best suitable to integrate data of different markers to achieve higher prediction accuracy than with one marker type alone is an additional aspect for future investigations.

Conclusions

Transcriptome-based prediction approaches are in its infancy and initial approaches are very promising. Clearly, a close relation of mRNA levels of certain genes and trait performance could be established. The amount of measurements required to effectively support hybrid breeding with respect to the number of genes and the number of inbred lines remains to be determined and will largely decide whether such an approach can be applied economically in short term.

References

Auger DL, Gray AD, Ream TS, Kato A, Coe EH, Birchler JA (2005) Nonadditive gene expression in diploid and triploid hybrids of maize. Genetics 169:389–397
Birchler JA, Auger DL, Riddle NC (2003) In search of the molecular basis of heterosis. Plant Cell 15:2236–2239
Boppenmaier J, Melchinger AE, Seitz G, Geiger HH, Herrmann RG (1993) Genetic diversity for RFLPs in European maize inbreds.3. Performance of crosses within versus between heterotic groups for grain traits. Plant Breeding 111:217–226
Broadley MR, White PJ, Hammond JP, Graham NS, Bowen HC, Emmerson ZF, Fray RG, Iannetta PP, McNicol JW, May ST (2008) Evidence of neutral transcriptome evolution in plants. New Phytol 180(3):587–593
Buckler ES, Holland JB, Bradbury PJ, Acharya CB, Brown PJ, Browne C, Ersoz E, Flint-Garcia S, Garcia A, Glaubitz JC et al (2009) The genetic architecture of maize flowering time. Science 325:714–718
Charcosset A, Essioux L (1994) The effect of population-structure on the relationship between heterosis and heterozygosity at marker loci. Theor Appl Genet 89:336–343
Choudhary K, Choudhary OP, Shekhawat NS (2008) Marker assisted selection: a novel approach for crop improvement. American-Eurasian J Agron 1:26–30
Frisch M, Thiemann A, Fu JJ, Schrag TA, Scholten S, Melchinger AE (2010) Transcriptome-based distance measures for grouping of germplasm and prediction of hybrid performance in maize. Theor Appl Genet 120:441–450
Fu J, Thiemann A, Schrag TA, Melchinger AE, Scholten S, Frisch M (2010) Dissecting grain yield pathways and their interactions with grain dry matter content by a two-step correlation approach with maize seedling transcriptome. BMC Plant Biol 10:63
Gärtner T, Steinfath M, Andorf S, Lisec J, Meyer RC, Altmann T, Willmitzer L, Selbig J (2009) Improved heterosis prediction by combining information on DNA- and metabolic markers. PLoS One 4(4):e5220
Groszmann M, Greaves IK, Albert N, Fujimoto R, Helliwell CA, Dennis ES, Peacock WJ (2011a) Epigenetics in plants-vernalisation and hybrid vigour. Biochim Biophys Acta 1809(8):427–437
Groszmann M, Greaves IK, Albertyn ZI, Scofield GN, Peacock WJ, Dennis ES (2011b) Changes in 24-nt siRNA levels in Arabidopsis hybrids suggest an epigenetic contribution to hybrid vigor. Proc Natl Acad Sci USA 108:2617–2622

Guo M, Rupe MA, Zinselmeier C, Habben J, Bowen BA, Smith OS (2004) Allelic variation of gene expression in maize hybrids. Plant Cell 16:1707–1716

Guo M, Rupe MA, Yang XF, Crasta O, Zinselmeier C, Smith OS, Bowen B (2006) Genome-wide transcript analysis of maize hybrids: allelic additive gene expression and yield heterosis. Theor Appl Genet 113:831–845

Hamblin MT, Warburton ML, Buckler ES (2007) Empirical comparison of simple sequence repeats and single nucleotide polymorphisms in assessment of maize diversity and relatedness. PLoS One 2:e1367

He G, Zhu X, Elling AA, Chen L, Wang X, Guo L, Liang M, He H, Zhang H, Chen F, Qi Y, Chen R, Deng XW (2010) Global epigenetic and transcriptional trends among two rice subspecies and their reciprocal hybrids. Plant Cell 22:17–33

Hochholdinger F, Hoecker N (2007) Towards the molecular basis of heterosis. Trends Plant Sci 12:427–432

Hoecker N, Keller B, Muthreich N, Chollet D, Descombes P, Piepho HP, Hochholdinger F (2008) Comparison of maize (*Zea mays* L.) F1-hybrid and parental inbred line primary root transcriptomes suggests organ-specific patterns of nonadditive gene expression and conserved expression trends. Genetics 179:1275–1283

Huang Y, Zhang L, Zhang J, Yuan D, Xu C, Li X, Zhou D, Wang S, Zhang Q (2006) Heterosis and polymorphisms of gene expression in an elite rice hybrid as revealed by a microarray analysis of 9198 unique ESTs. Plant Mol Biol 62:579–591

Jahnke S, Sarholz B, Thiemann A, Kühr V, Gutiérrez-Marcos JF, Geiger HH, Piepho HP, Scholten S (2010) Heterosis in early seed development: a comparative study of F1 embryo and endosperm tissues 6 days after fertilization. Theor Appl Genet 120:389–400

Maenhout S, De Baets G, Haesaert G, Bockstaele EV (2007) Support vector machine regression for the prediction of maize hybrid performance. Theor Appl Genet 115:1003–1013

Mallory AC, Vaucheret H (2006) Functions of microRNAs and related small RNAs in plants. Nat Genet 38:S31–S36

McMullen MD, Kresovich S, Villeda HS, Bradbury P, Li HH, Sun Q et al (2009) Genetic properties of the maize nested association mapping population. Science 325:737–740

Melchinger AE (1999) Genetic diversity and heterosis. In: Coors JG, Pandey S (eds) The genetics and exploitation of heterosis in crops. ASA-CSSA, Madison, pp 99–118

Meyer RC, Torjek O, Becher M, Altmann T (2004) Heterosis of biomass production in Arabidopsis. Establishment during early development. Plant Physiol 134:1813–1823

Meyer S, Pospisil H, Scholten S (2007) Heterosis associated gene expression in maize embryos 6 days after fertilization exhibits additive, dominant and overdominant pattern. Plant Mol Biol 63:381–391

Moghaddam AM, Fuchs J, Czauderna T, Houben A, Mette MF (2010) Intraspecific hybrids of *Arabidopsis thaliana* revealed no gross alterations in endopolyploidy, DNA methylation, histone modifications and transcript levels. Theor Appl Genet 120:215–226

Moghaddam AM, Roudier F, Seifert M, Bérard C, Magniette ML, Ashtiyani RK, Houben A, Colot V, Mette MF (2011) Additive inheritance of histone modifications in *Arabidopsis thaliana* intraspecific hybrids. Plant J. doi:10.1111/j.1365-313X.2011.04628.x

Mosher RA, Melnyk CW, Kelly KA, Dunn RM, Studholme DJ, Baulcombe DC (2009) Uniparental expression of PolIV-dependent siRNAs in developing endosperm of Arabidopsis. Nature 460:283–286

Ni Z, Sun Q, Liu Z, Wu L, Wang X (2000) Identification of a hybrid-specific expressed gene encoding novel RNA-binding protein in wheat seedling leaves using differential display of mRNA. Mol Gen Genet 263:934–938

Nobuta K, Lu C, Shrivastava R, Pillay M, De Paoli E, Accerbi M, Arteaga-Vazquez M, Sidorenko L, Jeong DH, Yen Y, Green PJ, Chandler VL, Meyers BC (2008) Distinct size distribution of endogenous siRNAs in maize: evidence from deep sequencing in the mop1-1 mutant. Proc Natl Acad Sci USA 105:14958–14963

Omholt SW, Plahte E, Øyehaug L, Xiang K (2000) Gene regulatory networks generating the phenomena of additivity, dominance and epistasis. Genetics 155:969–981

Schön C, Dhillon B, Utz H, Melchinger A (2010) High congruency of QTL positions for heterosis of grain yield in three crosses of maize. Theor Appl Genet 120:321–332

Schrag TA, Melchinger AE, Sorensen AP, Frisch M (2006) Prediction of single-cross hybrid performance for grain yield and grain dry matter content in maize using AFLP markers associated with QTL. Theor Appl Genet 113:1037–1047

Schrag TA, Maurer HP, Melchinger AE, Piepho HP, Peleman J, Frisch M (2007) Prediction of single-cross hybrid performance in maize using haplotype blocks associated with QTL for grain yield. Theor Appl Genet 114:1345–1355

Shimkets RA, Lowe DG, Tai JT-N, Sehl P, Jin H, Yang R, Predki PF, Rothberg BEG, Murtha MT, Roth ME, Shenoy SG, Windemuth A, Simpson JW, Simons JF, Daley MP, Gold SA, McKenna MP, Hillan K, Went GT, Rothberg JM (1999) Gene expression analysis by transcript profiling coupled to a gene database query. Nat Biotechnol 17:798–803

Shull GH (1948) What is 'Heterosis'? Genetics 33:439–446

Song RT, Messing J (2003) Gene expression of a gene family in maize based on noncollinear haplotypes. Proc Natl Acad Sci USA 100:9055–9060

Steinfath M, Gärtner T, Lisec J, Meyer RC, Altmann T, Willmitzer L, Selbig J (2010) Prediction of hybrid biomass in *Arabidopsis thaliana* by selected parental SNP and metabolic markers. Theor Appl Genet 120:239–247

Stokes D, Morgan C, O'Neill C, Bancroft I (2007) Evaluating the utility of *Arabidopsis thaliana* as a model for understanding heterosis in hybrid crops. Euphytica 156:157–171

Stokes D, Fraser F, Morgan C, O'Neill CM, Dreos R, Magusin A et al (2010) An association transcriptomics approach to the prediction of hybrid performance. Mol Breed 26:91–106

Stupar RM, Springer NM (2006) Cis-transcriptional variation in maize inbred lines B73 and Mo17 leads to additive expression patterns in the F-1 hybrid. Genetics 173:2199–2210

Stupar RM, Gardiner JM, Oldre AG, Haun WJ, Chandler VL, Springer NM (2008) Gene expression analyses in maize inbreds and hybrids with varying levels of heterosis. BMC Plant Biol 8:33

Sun QX, Ni ZF, Liu ZY (1999) Differential gene expression between wheat hybrids and their parental inbreds in seedling leaves. Euphytica 106:117–123

Swanson-Wagner RA, Jia Y, DeCook R, Borsuk LA, Nettleton D, Schnable PS (2006) All possible modes of gene action are observed in a global comparison of gene expression in a maize F-1 hybrid and its inbred parents. Proc Natl Acad Sci USA 103:6805–6810

Thiemann A, Fu J, Schrag TA, Melchinger AE, Frisch M, Scholten S (2010) Correlation between parental transcriptome and field data for the characterization of heterosis in *Zea mays* L. Theor Appl Genet 120:401–413

Uzarowska A, Keller B, Piepho HP, Schwarz G, Ingvardsen C, Wenzel G, Lübberstedt T (2007) Comparative expression profiling in meristems of inbred-hybrid triplets of maize based on morphological investigations of heterosis for plant height. Plant Mol Biol 63:21–34

Vuylsteke M, Kuiper M, Stam P (2000) Chromosomal regions involved in hybrid performance and heterosis: their AFLP-based identification and practical use in prediction models. Heredity 85:208–218

Vuylsteke M, van Eeuwijk F, Van Hummelen P, Kuiper M, Zabeau M (2005) Genetic analysis of variation in gene expression in *Arabidopsis thaliana*. Genetics 171:1267–1275

Wang X, Elling AA, Li X, Li N, Peng Z, He G, Sun H, Qi Y, Liu XS, Deng XW (2009) Genome-wide and organ-specific landscapes of epigenetic modifications and their relationships to mRNA and small RNA transcriptomes in maize. Plant Cell 21:1053–1069

Wei G, Tao Y, Liu G, Chen C, Luo R, Xia H, Gan Q, Zeng H, Lu Z, Han Y, Li X, Song G, Zhai H, Peng Y, Li D, Xu H, Wei X, Cao M, Deng H, Xin Y, Fu X, Yuan L, Yu J, Zhu Z, Zhu L (2009) A transcriptomic analysis of superhybrid rice LYP9 and its parents. Proc Natl Acad Sci USA 106:7695–7701

Weil C, Martienssen R (2008) Epigenetic interactions between transposons and genes: lessons from plants. Curr Opin Genet Dev 18:188–192

Wittkopp PJ, Haerum BK, Clark AG (2004) Evolutionary changes in cis and trans gene regulation. Nature 430:85–88

Wu LM, Ni ZF, Meng FR, Lin Z, Sun QX (2003) Cloning and characterization of leaf cDNAs that are differentially expressed between wheat hybrids and their parents. Mol Genet Genomics 270:281–286

Xiong LZ, Yang GP, Xu CG, Zhang QF, Maroof MAS (1998) Relationships of differential gene expression in leaves with heterosis and heterozygosity in a rice diallel cross. Mol Breed 4:129–136

Zhang H, He H, Chen L, Li L, Liang M, Wang X, Liu X, He G, Chen R, Ma L, Deng XW (2008) A genome-wide transcription analysis reveals a close correlation of promoter INDEL polymorphism and heterotic gene expression in rice hybrids. Mol Plant 1:720–731

Chapter 14
Metabolite-Based Biomarkers for Plant Genetics and Breeding

Olga A. Zabotina

Metabolites, Metabolomics, and Biomarkers

Harnessing the vast genetic potential that exists in wild exotic species and modern crop elite varieties for plant breeding requires the establishment of rapid, predictive tools and concepts to understand the mechanistic basis for traits and to associate traits with genomic or other diagnostic information. This first step enables subsequent crop improvement by breeding and selection, using the diagnostic information to guide plant breeding to combine key traits in improved varieties. In large-scale germplasm – enhancement programs working to develop techniques to associate markers with phenotypes impacting crop quality, phenotyping is the rate-limiting step (Zamir 2001). Commonly, phenotyping of plants requires growing a set of plants and assaying the organs of interest in a time- and cost- intensive process. Current technology links these phenotypes to genetic markers to allow marker-assisted selection. However, recently, metabolomics has emerged as a highly promising approach for prediction of a variety of agronomically important phenotypes of crop plants grown in different environments, and particularly for discovering signature metabolites or biomarkers for traits of interest (Sumner et al. 2003). Biomarkers are used to predict phenotypic properties before these features become apparent and, therefore, are valuable tools for both fundamental and applied research. Diagnostic biomarkers were discovered in medicine many decades ago and are now broadly applied in clinical studies. Although routine in medicine, this approach has only recently received attention in plant biology, specifically in breeding (Steinfath et al. 2010a, b). Such metabolite biomarkers can assist in developing fast, targeted and

---

O.A. Zabotina (✉)
Department of Biochemistry, Biophysics and Molecular Biology,
Iowa State University, Ames, IA 50011, USA
e-mail: zabotina@iastate.edu

low-cost diagnostic assays that will facilitate crop breeding programs and quality control by increasing prediction power. At the early selection stages in breeding, when the number of independent lines is high and the number of individual plants per line is limited, the predictive ability of biomarkers can significantly increase the efficiency of selection. In addition, metabolite-based biomarkers can improve marker-assisted approaches, particularly when the prediction power of other molecular markers is limited.

The major requirements for successful implementation of biomarker selection are: the predictive power of the biomarkers should not be affected by environmental variation, and the biomarkers should be applicable to broader plant populations different from their populations of origin (Steinfath et al. 2010a, b). Conventional genetic or molecular markers commonly used for marker-assisted selection in modern plant breeding have been very successful in providing a powerful tool for the identification of specific lines that carry positive traits, but only for diploid crops. Application of genetic markers is significantly limited in species with complex polyploid genomes, which is common in agriculturally important crops such as wheat, cotton, and sugarcane. Another limitation of molecular markers is that they cannot be directly applied for polygenic traits and to predict epistatic or environmental effects (Steinfath et al. 2010a, b). These limitations of genetic markers can be overcome by introduction of metabolite-based biomarkers, which can predict phenotype independent of available genomic information and environmental variation. Metabolite biomarkers can enable the development of targeted diagnostic assays for breeding programs and can guide investigation of the biochemical mechanisms that determine the trait phenotype. Metabolic screening can also be used to predict quantitative phenotypic properties that become apparent much later after the metabolites were sampled, and can be used to find biomarkers for traits with uncharacterized biochemical mechanisms.

The immediate task in biomarker discovery is to establish techniques for untargeted, high-throughput, comprehensive screens for plant metabolites. The metabolic profiles of a large set of informative plant lines could then be analyzed with a predictive modeling machine learning approach. Such a combination of metabolomics and bioinformatics would allow identification, from the thousands of detected compounds, of the small set of specific metabolites that can serve as biomarkers for prediction of various complex traits, such as yield, disease resistance and stress tolerance, nutritional value, etc. Predictive biomarkers can then be used to develop low-cost, high-throughput, targeted diagnostic assays for the pre-selection of segregating crosses in various crops, including those for which the biochemical mechanisms underlying the traits of selection are yet unknown. Thus, metabolite-assisted breeding offers new opportunities for crops with limited or even unavailable genomic information or with highly complex genetic mechanisms underlying the traits of interest (Steinfath et al. 2010a, b). Hence, the increasing amount of available information in genomics and proteomics and the decreasing costs of techniques for metabolic screening build a strong argument for the integration of metabolomics platforms as a valuable component in plant breeding programs.

Metabolomic Platforms and Technologies

Plants have developed complex metabolic combinations; some combinations are common but others are found only in certain species or even found only in particular organs or specialized tissues (Sumner et al. 2003). The total number of metabolites produced in all plants, including primary and secondary metabolites, is estimated to be between 100,000 and 200,000 (Oksman-Caldentey and Inze 2004). The elucidation of plant metabolic composition would allow an effective use of this knowledge for diagnostics and predictive modeling.

Metabolomics is an emerging core scientific discipline that complements genome, transcriptome, and proteome information in systems biology studies. Metabolomics is direct, not dependent on genotyping, and addresses the features that are directly relevant to biological function and thus to plant phenotype and agronomic traits (Stitt and Fernie 2003). Metabolites are low molecular weight (compared to proteins and nucleic acids) organic and inorganic compounds that are the substrates, intermediates, or products of enzyme-mediated biochemical reactions (Dunn et al. 2011). Metabolites can be classified by polarity, molecular size, structure, or reaction similarity, and their compositional/structural diversity reflects the broad variation in metabolite physicochemical properties in living organisms. Quantitative and qualitative studies of metabolites directly monitor the biochemical status of an organism and can be used to link phenotypic differences to the genetic differences that cause them (Sumner et al. 2003; Maloney 2004). Metabolites can be the products of synthesis (anabolism) or degradation (catabolism) and are also involved in many biochemical processes not directly related to their formation or consumption, such as metabolic regulation. Moreover, metabolites are related either to primary or secondary metabolism. Central or primary metabolism includes reactions that are required for growth, development, and reproduction and are conserved across many species; secondary metabolites are not directly involved in these life-supporting processes and are found in more limited sets of species. Metabolomics has been defined as the technology developed to acquire the broadest, generally untargeted insight into the complex population of small constituents present in living organisms (Hall 2006). There are several specific terms and definitions emerging in the field of metabolomics (Table 14.1).

Many significant advances have been already made in the development of technology for metabolomic analysis; however, the final goal – to obtain a comprehensive overview of the entire metabolic composition of a plant in a single analysis – is currently inconceivable (Hall 2006) because of the high complexity of the metabolic arsenal developed by living organisms and in particular by plants. Thus, the number of human metabolites is estimated to be about 10,000, but in plants this number is roughly predicted to be 200,000–300,000. This metabolic complexity, which is largely driven by major variations in secondary metabolism, is further enhanced by the broad dynamic range of various metabolites present in different concentrations and at particular times. In addition, large groups of structurally related compounds, such as 6,000 different flavonoids (Schijlen et al. 2004) or

Table 14.1 Definition of terms used in metabolomics

Term	Definition
Metabolome	The comprehensive combination of small molecules present in an organism.
Metabolomics	The technology that provides unbiased, comprehensive quantitative and qualitative insight into metabolic composition of an organism.
Metabolic fingerprinting	Global snapshot screening (rapid and simple) of metabolic composition with the primary aim of comparison and discrimination analysis, generally without identification of the metabolites.
Metabolic profiling	Identification and quantification of the metabolites in valid and robust manner. Currently, this is feasible only for limited number of metabolites that are either available through databases or can be artificially synthesized.
Targeted metabolic analyses	The quantitative study of a small number of metabolites, frequently related by similarities in structure or properties. This technology follows broad-scale metabolomics analysis, or requires previous knowledge of biochemical pathway, and requires optimized extraction and dedicated separation/detection.

Adopted from Hall (2006)

12,000 alkaloids (Facchini et al. 2004), have been already discovered and represent another challenge to separation and detection technology. Chemical complexity, heterogeneity, dynamics and ease of extraction make all current techniques, irrespective of their sophistication, intrinsically biased towards certain metabolites (Hall 2006). However, complementary biochemical data can be rapidly built up using carefully selected combinations of extraction, separation, and detection platforms.

Extraction and Separation

Because various metabolites have distinctly different chemical properties, for example polar and non-polar, volatile and non-volatile, etc., all current extraction protocols have been developed to enrich specific groups of metabolites, for example polar metabolites extracted with aqueous solutions and less polar metabolites with non-polar organic solvents. Therefore, in order to obtain the most comprehensive list of metabolites present in particular plant samples, the combination of multi-parallel platforms with diverse extraction and separation capabilities is required. Selection of extraction protocols also closely depends on the subsequent separation and detection techniques that will be applied. One general requirement for any chosen extraction protocol is immediate inactivation of metabolism because of extremely rapid metabolite turnover (Lisec et al. 2006); this inactivation is usually achieved by rapid freezing at −80°C. Two major separation techniques dominate metabolite profiling strategies: gas and liquid chromatography (Hall 2006; Dunn et al. 2011); both have high throughput capacity although the latter usually requires a longer time for each run. Lately, capillary electrophoresis as an alternative separation technique is becoming more and more popular because of its additional selectivity and sensitivity (Soga et al. 2003; Sato et al. 2004) together with its high throughput capacity, which is similar to gas chromatography.

Gas chromatography (GC) is currently accepted as the most robust and well developed global analysis method (Hall 2006; Jenkins et al. 2007; Dunn et al. 2011). Because separation occurs between the gas and liquid phases, gas chromatography is primarily applied for metabolites that are naturally volatile at temperatures up to 250°C, such as alcohols, esters, monoterpenes, etc., however, thermolabile metabolites can be missed. For comprehensive metabolic analysis, additional extraction of volatile metabolites using either solvents (pentane) or solid phase micro-extraction (SPME) (using different fibers) is desirable (Tikunov et al. 2005). Comparisons with standard compounds and commercially available compound databases can be used to confirm metabolite identity. Gas chromatography is also broadly used for nonvolatile, polar compounds, particularly primary metabolites such as amino acids, sugars, organic acids, and to a lesser extent for secondary metabolites such as alkaloids, terpenoids and glycosides. These nonvolatile metabolites require chemical derivatization to convert them into volatile thermostable compounds before analysis by GC; hence the comprehensive profile of most key primary metabolites can be obtained in a single run (Desbrosses et al. 2005). There is an extensive list of chemicals available for GC derivatization, including methylating, alkylating, acylating and silylating reagents (Knapp 1979), among which, thrimethylsilylation is the most frequent choice (Birkemeyer et al. 2003). The recent introduction of comprehensive GC × GC technology that increases analyte peak capacity and resolving power by diverting each peak obtained after separation on the first GC column to a separation on a second column, provides separation in two dimensions, and thus, further improves the GC separation platform (Dunn et al. 2011).

Liquid chromatography (LC) is a particularly important and more versatile technology for analysis of secondary metabolites (Verhoeven et al. 2006). Development of ultra-performance liquid chromatography (UPLC) together with the diversification of column chemistry has significantly improved the separation capabilities of this platform, increasing its analytical precision (Wilson et al. 2005). The main advantage of LC technology is direct application of the metabolite mixture without derivatisation, which reduces the time for sample preparation and the risk of missing underivatized or unstable compounds. However, separation of complex mixtures usually requires a longer running time on the column compared to GC. Lately, multi-dimensional LC has gained in popularity for proteomics analyses (America et al. 2006). Somewhat similarly to GC × GC, this technology is based on orthogonal separation mechanisms, e.g. the combination of a first separation on a strong cation-exchange column and a second separation of individual peaks (fractions) on a reverse-phase column. Considering the broad range of chemical properties of plant metabolites, this strategy could be valuable for metabolomics in the near future (Hall 2006).

Capillary electrophoresis (CE) is another separation technology that offers extra selectivity (Soga et al. 2003; Sato et al. 2004). Electrically charged species are separated by electro-osmotic flow in an electrically conductive liquid phase under an externally applied electrical field (Dunn et al. 2011). Separation efficiency of CE is generally similar to or even better than GC or UPLC, and smaller sample volumes are required, but the analysis of non-polar metabolites is technically limited, requiring additional derivatisation. For water soluble extracts, high-resolution chromatographic separation makes capillary electrophoresis suitable for the analysis of a diverse range of primary and secondary metabolites (Sato et al. 2004), giving significant coverage of the metabolome in a single analysis.

Detection

Two detection techniques dominate metabolome platforms: mass spectrometry (MS) and nuclear magnetic resonance (NMR). Other diverse techniques are also used in different applications with more specific goals, for example UV/VIS spectroscopy, photo diode array (PDA), electrochemical detection, etc.; depending on the nature of the metabolites of interest, these techniques can give higher detection sensitivity and selectivity.

Mass spectrometry (MS) is currently the most powerful detection method that is used in a broad range of formats. MS can be used as standalone or, most frequently, be coupled with the various separation techniques described above. Gas-chromatography-mass-spectrometry (GC-MS), gas-chromatography- time-of-flight-mass-spectrometry (GC-TOF-MS) and liquid- chromatography-mass-spectrometry (LC-MS) are currently the most popular standard mass-spectrometry methods for metabolite analyses (Roessner 2007; Dunn et al. 2011). In addition, capillary electrophoresis-mass spectrometry (CE-MS), Fourier transformion cyclotron resonance-mass spectrometry (FTMS) and flow- or direct-injection/infusion-mass spectrometry (FI/DI-MS) have been used in various applications, although they are more rarely used for plant metabolomic analysis.

Mass spectrometry is based on the formation, separation, and detection of positively or negatively charged ions formed from metabolites of interest. Separation and detection is carried out under vacuum to limit ion-ion or ion-molecule collisions and improve mass resolution and sensitivity of measurements. Ionisation, in turn, can be performed either under vacuum (MALDI or electron impact) or at atmospheric pressure (Electro Spray Ionisation, ESI or Atmospheric Pressure Chemical Ionisation, APCI). Currently, most metabolomics applications use time-of-flight (TOF), quadruple (Q), Fourier transform (FT) and hybrid (Q-TOF, ion trap-Orbitrap, QQQ) instruments, which have their own advantages and weaknesses. The most common advantages include high sensitivity, fast scan and signal acquisition, high mass resolution and accuracy. Electron impact ionization is usually used in GC-MS instruments, while ESI is more commonly used in LC-MS and CE-MS. Other ion sources are applied less frequently, for example chemical ionization in GC-MS and APCI in LC-MS. When combined with chromatography, mass spectrometry can detect thousands of metabolites in a single sample and make them easier to identify. However, MS has some disadvantages; for example, molecules physically interact with the instrument causing changes in response over time. The degree and timing of the signal attenuation is not consistent across various samples; hence, periodic application of quality control (QC) samples is highly recommended (Dunn et al. 2011). The chosen QC samples should be identical for the entire experiment. Later, during data processing and before statistical analysis, the QC responses can be used to assess the quality of the data, to remove poorly reproducible peaks and to correct signal attenuation (van de Greef et al. 2007; Zelena et al. 2009). Another disadvantage is that response factors depend on sample composition, which can vary in different samples, causing variation in the measured responses for the same metabolite concentrations. To compensate for this irreproducibility, the application of a chemical analogue of the metabolites under analysis (an internal standard) is recommended. However, this is not feasible for untargeted metabolomic profiling where metabolites of interest are unknown.

GC-MS and GC-TOF-MS (Table 14.2) are the most frequently utilized techniques in metabolic analysis, and they provide high chromatographic resolution together with highly sensitive detection. Complete ionization of molecules in the complex sample can be easily achieved in the gas phase coming out of the column, minimizing interference between molecules in the MS detector, i.e., reducing convolution. Availability of extensive databases of fragmentation patterns obtained by GC-MS, which greatly facilitates identification of compounds present in the sample, also makes these techniques highly popular.

LC-MS and UPLC-MS (Table 14.2) are restricted to molecules that can be ionized, either as positively or negatively charged ions, and some machines can switch continuously between positive and negative modes within a single run giving broader coverage of molecules that either gain or lose a proton. UPLC-MS can detect thousands of ions in a given sample, and generally does not require any derivatisation although the latter can be applied to improve selectivity for more targeted metabolomic analysis (Barry et al. 2003). Unlike GC-MS, few mass spectral libraries are available for LC-MS, and this is a key topic being given considerable attention at present (Verhoeven et al. 2006).

Direct infusion mass spectrometry (DIMS) is applied with ESI spectrometers, and samples are directly injected into the mass spectrometer using an automated flow injection mode. To analyze highly complex samples, instruments with high mass resolution and mass accuracy are required, such as TOF (Verhoeven et al. 2006) and FTMS (Aharoni et al. 2002; Hirai et al. 2004). DIMS provides a high-throughput analytical platform, albeit with reduced identification capability. In complex metabolome samples, ionization suppression is usually observed due to the competition among multiple species for the available charge. Some recent innovations have improved the mass accuracy and number of detected metabolites. For example, Single Ion Monitoring (SIM)-stitching experiments applying multiple SIM windows in FT-MS instruments reduce space-charging effects observed in trap-based instruments, thereby, reducing the noise of the signal (Southam et al. 2007; Payne et al. 2009).

Table 14.2 Main characteristics of standard techniques used in the metabolomics platforms

Technique	Application	Comprehensiveness	Throughput	Sensitivity	Mass range
GC-MS GC × GC-MS	Polar or lipophilic metabolites	High	High	High (10–12 M)	≤350 Da
SPME GC-MS	Volatile metabolites	Medium-High	High	High	≤350 Da
CE-MS	Polar metabolites	High	Medium	High (10–13 M)	≤1,000 Da
LC-MS	Mainly secondary metabolites	High	High	High (10–15 M)	≤1,500 Da
LC-EC-MS	Mainly secondary metabolites	High	High	High	≤1,500 Da
LC-NMR	Identification of unknowns	High	Low	Low (10–6 M)	≤50 Da
LC-UV	Secondary metabolites	Very low	High	Medium-High	≤1,500 Da
FTICR-MS	Identification of unknowns	High	High	High	≤1,500 Da
NMR	Nondestructive analyses in a sample	High	Low	Low (10–6 M)	≤50 Da
Direct-injection MS	Fingerprinting of metabolic contents	High	High	High	≤1,500 Da
IR	Fingerprinting of metabolic contents	Low	High	Low	≤1,500 Da

This information was taken from Fernie and Schauer (2009), Sumner et al. 2003, and Weckwerth and Morgenthal (2005)

EC electrochemical detection, *FTICR* Fourier transform ion cyclotron resonance, *IR* infrared spectroscopy

Nuclear magnetic resonance spectroscopy (NMR) has become an invaluable tool for many applications. NMR is a quantitative, highly reproducible, and non-selective technology, the sensitivity of which does not depend on the hydrophobicity or pK_a of the species to be analyzed (Ratcliffe and Shachar-Hill 2001, 2005). Furthermore, this technology provides broad structural information enabling the identification of individual compounds within a complex sample (Table 14.2). In addition, NMR is non-destructive, which allows several analyses to be performed using the same sample. The main disadvantage of NMR is its relatively low sensitivity, particularly in comparison with MS; therefore, only metabolites present in high concentrations can be reliably detected (Kaddurah-Daouk et al. 2004; Defernez et al. 2004), and other important compounds can be missed if they are present in plant extracts at levels below NMR detection thresholds. Using LC to selectively concentrate metabolites during chromatographic runs, followed by NMR analysis has been shown to increase the sensitivity of analysis (Bailey et al. 2000; Simpson et al. 2004). In addition, NMR can be used non-invasively (on living cells), providing subcellular information, and it is easier to derive atomic information for flux modeling from NMR than from MS-based approaches (Fernie and Schauer 2009).

A major limitation of untargeted metabolomic platforms is the vast amount of detected chemical features that are not structurally characterized or putatively classified (Weckwerth 2011). Identification of unknown metabolites can be difficult and expensive. Furthermore, the frequent fragmentation that takes place during MS-based analysis leads to the loss of ions that represent the intact chemical structure, thereby complicating identification of unknown metabolites. The fact that analytical procedures also produce many artifacts has to be considered as well. For example, Giavalisco and colleagues (2009) examined plants fully labeled with ^{13}C, which allowed the authors to distinguish between labeled and non-labeled chemical structures, and demonstrated thousands of analytical artifacts, chemical or electronic noise. Utilization of synthetic precursor compounds is the one way to overcome these problems. Using softer types of ionization, such as chemical ionization, can help to obtain the correct mass of the structure of interest. There are many databases created for mass spectra, but only a few plant related metabolites are present in these databases. Therefore, there is a strong need to further extend and combine existing libraries such as the NIST (Stein 1999), GMD (Kopka et al. 2005), the FiehnLib (Kind et al. 2009), and PM (Bais et al. 2010). In general, all the developments in analytical procedures for metabolomic profiling should be accompanied by artificial synthesis of authentic compounds that would extend existing libraries and facilitate identification of unknown metabolites.

In addition to the determination of unknowns, another important issue in metabolomic platforms developed to-date is the obtaining accurate quantitative data. Mass spectrometry used as a detection method in most of analytical techniques rarely produces quantitative measurements due to fluctuations in the ionization efficiency of analytes (Harada et al. 2006). Hence, the stable dilution method is usually used to perform accurate quantitative measurements in mass spectrometry (Matuszewski et al. 2003). Two strategies are applied for stable isotope dilution: (i) post-extraction derivatization using the isotope coded affinity tag (ICAT) commercially available for proteomics analysis; (ii) *in vivo* labeling by up taking isotopes from culture media during plant growth (Harada et al. 2006 and references therein). Another approach used for normalization of metabolomics data is introduction of multiple internal (added to sample prior to extraction) and external (added to sample after extraction) standards (Sysi-Aho et al. 2007). Specific standards can be assigned to metabolite peaks based on similarity in specific chemical property such as retention time and/or mass-to-charge ratio. An important component of any metabolomics studies, and in particular for obtaining exact quantitative data, is the performance of the daily quality control as it was mentioned already above. Usually, it involves method blanks (reagents and equipment used to control for lab contaminations) and several calibration curve samples including pure reference compounds (Fiehn et al. 2008). It is also to ensure that the protocol used for metabolic profiling yields reproducible data sets, independently of the individual who carried out the sample preparation.

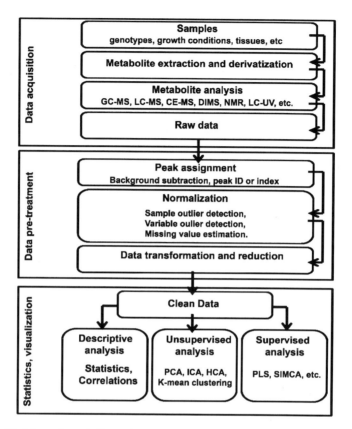

Fig. 14.1 Workflow of metabolic profiling including generation and processing of the samples, metabolite analysis to collect the raw data, pre-processing of raw data, computational analysis, and visualization of final data

Data Processing and Analysis

The fundamental goal of any metabolomic analysis is to extract biological meaning from raw data (Dunn et al. 2011). To do this, effective tools for handling raw data are currently under development. Powerful analytical software is necessary to carry out raw data processing through several key steps (Fig. 14.1): (1) collection and preprocessing the raw data to allow comparison of outputs from various metabolomics platforms; (2) processing the data to mine out the components of interest; (3) visualization of the data in understandable forms; and (4) effective storage of the final data files for easier access and further applications.

Raw data acquired on any analytical instrument described in the previous section are usually exported in different computer-readable formats and need to be converted into a specific format for pre-processing. The pre-processing step has two main aims: first to reduce the file size through a reduction of data complexity and to convert data into a format suitable for application of software packages, and second, to align data, ensuring that specific compounds are identified as the same metabolites in all analyzed samples (Hall 2006). Usually, drift in detected parameters (retention time or response factors in MS, and chemical shifts in NMR) are observed during analysis. Raw data pre-processing typically converts continuous data to segmented data. For example in mass spectrometry, files are typically converted from the manufacturer's data format to a text-based file format known as NetCDF. There are several other XML-based data formats available, such as mzXML, mxData, etc. For chromatography-MS and CE-MS, alignment of retention times must be performed, for which a collection of software packages is available (BinBase, MSFACTs, XCMS, Metalign, MZmine, MathDAMP). This process, called "peak-deconvolution", provides alignment of retention times and accurate masses (Steinfath et al. 2008; Dunn et al. 2011). For example to assess underexploited biodiversity in fungi (Smedsgaard and Nielsen 2005) and plants (Hendriks et al. 2005), various Correlation Optimized Warping (COW) tools for chromatographic matching have been employed (Hall 2006). COW aligns two chromatographic profiles by piece-wise linear stretching and compression of the time axis of one of the profiles (Nielson et al. 1998; Maloney 2004) using two input parameters, section lengths and flexibility, which can be estimated from the peak width. This method does not require determination of all peaks and knowledge about compounds present in the sample.

Pre-processed or "clean" data need to be further pre-treated before statistical analysis, i.e., normalized, scaled, missing values replaced, and outliers detected and removed (Steinfath et al. 2008; Dunn et al. 2011). It is also helpful to subject the data file to some size reduction or clustering (algorithm creation), particularly in untargeted metabolomic profiling when highly dimensional raw data are too complex. These algorithms are called "unsupervised learning methods", and their basic principal is translation of extremely high-dimensional raw data (M) into a lower dimensional function (P) while preserving the maximal amount of experimental information. Either unsupervised approaches for discriminatory analysis such as Principal Component Analysis (PCA), hierarchical clustering (HCA), K-means clustering, or supervised approaches (machine-learning methods) such as partial least squares (PLS) and SIMCA are most frequently used during this step of data processing (Stitt et al. 2010).

PCA is the one of the most popular multivariate techniques, where raw data are orthogonally transformed into a set of principal components, which are uncorrelated variables, where usually $P \leq M$. PCA is a linear additive model, where each principal component accounts for a portion of the total variance of the dataset. Plotting the data in the area defined by the two to three largest principal components is a fast way to visualize similarities or differences, perhaps allowing better discrimination of samples (Sumner et al. 2003; Steinfath et al. 2008).

Another common pre-treatment approach is called "signal correction", which uses QC samples periodically analyzed through the experiment to reduce the effects of known or unknown bias in the data set. When the cause of bias is unknown, then Orthogonal Signal Correction (OSC) can be introduced. In this case, algorithms correct for any multivariate effects that are completely uncorrelated with the experimental conditions (Dunn et al. 2011).

Hierarchical cluster analysis (HCA) groups samples in a dataset according to similarity and performs a progressive pair-wise grouping of samples by distance. Several distance measures can be used in HCA, such as Euclidean distance, or Manhattan distance, and the results are visualized as a dendogram or a tree. Another method, K-means clustering, groups samples using a fixed number (K) of groups. HCA and K-means clustering require the definition of a distance metric to govern the clustering, but the way of grouping used in these two approaches is different (Sumner et al. 2003). Self-organizing maps (SOMs) are similar to K-means clustering because they both require the number of groups for data classification to be predefined (number must be a power of two) (Toronen et al. 1999). Clustering is very useful for classifying samples into groups and can be applied after transformation of data with PCA, thereby becoming a means of identifying groups in the reduced dimension data space (Sumner et al. 2003).

All techniques mentioned above are called "unsupervised" because they require nothing more than the original datasets. By contrast, "supervised" techniques discriminate samples using a "training" dataset, i.e. a set of characteristics that have been independently classified. Although supervised methods can be used only if a set of known examples is provided, usually they are more powerful than unsupervised methods. There are several reports where supervised methods have been applied to metabolomic data (Sumner et al. 2003 and references therein).

Metabolomics, like all "omic" approaches, relies heavily on bioinformatics to process, store, retrieve, and analyze the huge resultant datasets. Changes in metabolite levels may be dramatic or subtle. Dramatic changes can be easily recognized, but subtle changes require careful experimental design and extensive statistical processing to determine their significance. Generally, statistical tests (ANOVA, Student's t-test, etc.) have to be performed to exclude erroneous data, following by calculation of means and standard deviations and determination if the differences are significant at a chosen confidence level (Sumner et al. 2003; Dunn et al. 2011).

Data visualization tools are critical to allow simple and easy comprehension of the multidimensional complexities of metabolomic data and for extraction of effective signature metabolites that can serve as biomarkers. Currently, several tools are available, such as KEGG (Kanehisa et al. 2002), AraCyc (Mueller et al. 2003), MetNet (Wurtele et al. 2003) BioPathAtMAPS combined with BioPathAtDB (Lange and Ghassemian 2005), MapMan (Usadel et al. 2005), and VANTED (Klukas and Schreiber 2010), however these tools are limited to metabolites from confirmed biochemical pathways (Hall 2006).

Receiver Operating Characteristic (ROC) is a visualization tool for classification, which is increasingly used in machine learning and data mining research. ROC curves visualize true positive rates versus false positive rates; this type of analysis provides the tool to define optimal thresholds and rule sets for classification and to select effective for prediction models. For example in human disease biomarker discovery, the ROC curves have been successfully used to illustrate the biomarker utility for metabolite based predictive models (Dunn et al. 2011); however, this approach is limited to two-state experimental designs, e.g., case-control.

Metabolomic data can be used to construct computer models, such as Classification models (Partial least squares discriminant analysis is a frequently used classification method) (Westerhuis et al. 2008), Structural models (Sweetlove and Ratcliffe 2011), or Kinetic models (Mendes 2001; Rohwer 2012), which can summarize large amounts of disparate data and confirm their consistency. Thus, the objective of computational biochemistry is to construct models of metabolism directly from "omic" data. Modeling can help to confirm the applicability of selected biomarkers and investigate their prediction power. For example, when a kinetic model for the specific physiological process is developed, it can be used to predict pathway behavior with a specific outcome in mind and to identify key control points that are most important for determining the value of certain fluxes or metabolite concentrations during this process, and ultimately can reveal biomarkers for the particular pathways (Rohwer 2012).

The main difficulty is confirming whether the selected biomarker is valid. For example in clinical studies, numerous reports point out the potentially misleading aspects of profiling techniques such as metabolomics, proteomics, transcriptomics and genomics, in biomarker discovery (references from Dunn et al. 2011). Therefore, the vigorous cross-validation of constructed models is always required, which sometimes can be far from straightforward (Westerhuis et al. 2008). The most robust method accepted for validation of biomarkers is to repeat the experiment with an independent sample set (Dunn et al. 2011) called the "hold-out set", which was not used in the generation of the model for discovery of signature biomarkers. The set of data used for model generation is called the "training set".

Therefore in general, three separate studies should be performed (Fig. 14.2): (1) a discovery study, where a smaller set of samples are used as a training set; (2) validation using a similarly small set of samples as a hold-out set; and finally (3) cohort validation where a significantly broader set of independent samples is used. The key important requirement is that discovery study experiments have to be rigorous enough (untargeted analytical platforms with high comprehensiveness) so that the resulting biomarker metabolites are robust and validated by the hold-out data (step 2). The independent validation studies (step 3) define the utility of discovered biomarkers in the target population; therefore, targeted metabolic profiling can be used. Since this study includes significantly larger set of samples, the applied analytical techniques must have high throughput capacity.

Application of Metabolomics for Plant Genetics and Breeding

Advances in technology promote the introduction of a broader diversity of alleles for trait genes into breeding programs, either via biotechnology or by application of molecular markers in combination with broader crosses (Kopka et al. 2004).

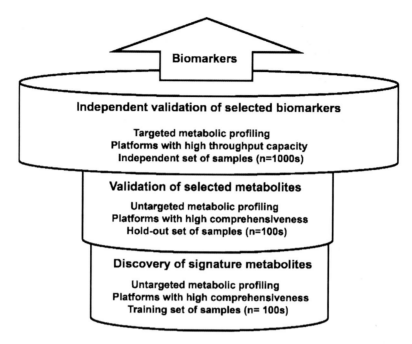

Fig. 14.2 Schematic representation of the multiple-step studies in biomarker discovery including discovery and validation studies using two independent populations of relatively low number of samples and final cohort validation studies using a larger set of independent samples. Modified from Dunn et al. 2011

Metabolic profiling can help with phenotypic characterization of this diversity, i.e. determination of metabolic composition, providing a powerful tool for guiding the breeding process by identifying promising or detrimental traits at early stages of selection. The power of this technology can be vastly increased if it is combined with a systematic survey of the metabolomic composition of the parental lines and exotic (wild) plant species, thereby providing a baseline or rational framework for prediction assessments. Metabolomic profiling can also be applied for assessment of genotypic variation, without requirement for prior development of molecular tools for a particular species (Fiehn et al. 2000; Roessner et al. 2002).

Application of metabolite-based biomarkers or metabolomic platforms in plant breeding programs has not yet been reported, and metabolite biomarkers have not been commercially utilized. The main challenge of using metabolic biomarkers is the major dependence of metabolite composition on environmental and experimental variation, which makes experimental design and sample preparation critical. By contrast, genetic markers are stable under different environmental conditions; therefore traits in breeding are currently monitored using molecular genetic markers such as single nucleotide polymorphisms (SNPs). However, polygenic traits, polyploidy, and epigenetically or environmentally controlled traits are difficult to assess using gene-based technologies. Moreover, metabolite composition is more directly linked to the phenotype than genes, and metabolomic analysis is independent of the plant sample or variety and availability of genomic information. Recently reported metabolomic studies (discussed below) demonstrate the potential of metabolic profiling in revealing signature metabolites with predictive power, which may be successfully used in diagnostic plant breeding in the near future.

In the past, traditional crop improvement methods were based on a single or, rarely, a handful of metabolic traits that were valuable for industrial applications or for human nutrition. Examples of such targeted approaches include carotenoid content in tomato, protein content in maize and starch content in potato and rice. One example of a long-term program for improvement of crop composition is the Illinois long-term selection experiment for protein and oil content in maize (http://www.ideals.uiuc.edu/handle/2142/3524), which began in 1896. Created populations contain the known phenotypic extremes for maize kernel composition and are still used in breeding programs as a favorable source of alleles associated with oil, protein, and starch content. The combination of metabolomics and association mapping approaches showed associations between genomic regions of maize and kernel composition, starch content in potato, pigment content in tomato and pro-vitamin A in maize (Fernie and Schauer 2009 and references therein). These targeted metabolic platforms demonstrated that metabolomics could in turn benefit from multi-parallel strategies because of contribution of a higher mapping resolution, a greater allele number, and faster turnover in establishing associations in comparison with linkage analysis (Yu and Buckler 2006).

Later, pathway-based approaches were applied to identify the genetic determinants of crop compositional qualities. These approaches led to a better understanding of glucosinolate biosynthesis (Kliebenstein et al. 2001), seed oil synthesis (Hobbs et al. 2004) and oligosaccharide metabolism (Bentsink et al. 2000) in Arabidopsis, and flavonoid biosynthesis in Arabidopsis (Tohge et al. 2005), tomato (Spencer et al. 2005), and poplar (Morreel et al. 2006).

The advantage of untargeted metabolic screens is the broad and simultaneous search for several metabolite-based markers, which increases the reliability of selected biomarkers and the possibility of discovering novel components of the biochemical mechanisms underlying the trait. The ability to screen a broad range of metabolites simultaneously can often help to predict any unexpected consequences of metabolic engineering on plant yield or on the levels of other metabolites, which usually are unwanted traits (Fernie and Schauer 2009). Such untargeted screens can also enable the greater understanding of complex metabolic network interactions resulting in development of new phenotypes. Metabolomic data can be complemented by proteome, transcriptome and phenotype or environmental data, thereby resulting in the identification of multiple physiological metabolite biomarkers embedded in correlative molecular networks that are not assessable by targeted studies (Weckwerth and Morgenthal 2005). From this perspective, for example in medical research, data integration was identified as being a bottleneck for future research in drug discovery (Searls 2005), which is also true for biomarker research in plants.

The shift from single metabolite measurements to metabolomic platforms has already led to the development of better models to describe the links between different biochemical pathways and between metabolism and yield-associated traits. Thus, metabolomic studies in tomato, wheat, rice, sesame, broccoli, mustard and Arabidopsis have facilitated the identification of important alleles to be used in metabolic engineering (Fernie and Schauer 2009 and references therein). Results from these studies will aid in the future selection of breeding lines because genetic engineering alone cannot substitute for plant breeding in the generation of better varieties, particularly when a flux increase in complex pathways is essential (Morandini and Salamini 2003). Plant genetic engineering can provide genomic resources that are unavailable using conventional breeding tools; these resources will produce varieties with "specific" phenotypes via expression of one or a few genes, thereby creating an additional resource for plant breeding programs.

Breeding using hybrids represents a significant innovation in the history of plant breeding and agriculture (Gartner et al. 2009). Hybrid breeding exploits the phenomenon of heterosis or hybrid vigor, the superiority of hybrid lines to their parental inbred lines in multiple parameters, most importantly yield. In addition, hybrid lines show such superiority not only in comparison with their parental lines but also with lines derived from classical breeding. Crops that strongly rely on hybrid breeding include maize (Duvick 2001), rye (Miedaner et al. 2002), sugar beet (Panella and Lewellen 2007), rice (Virmani 1994) and oilseed rape (Ofori and Becker 2008). However, knowledge about the molecular basis of the heterosis is limited. The main observation is that heterosis strongly depends on the parental combinations; this represents a big challenge for breeding programs because the best combinations can be confirmed only through multiple trials with numerous testcrosses. To reduce the level of uncertainty in the search for suitable parental combinations and to increase predictability, approaches based on genetic markers in the two parents have been used (Gartner et al. 2009 and references therein). However, genetic markers are of limited utility for complex phenomena involving many genes, such as heterosis. Therefore significant efforts have recently been made to investigate parental metabolomes to determine their predictive power for heterosis in valuable traits and for selection of biomarkers for these traits. Thus, Gartner and colleagues (2009) searched for predictive biomarkers for biomass heterosis in Arabidopsis and demonstrated that metabolic analysis can significantly improve the predictive power of genetic data, which suggests the complex mechanism underlying heterosis. To search for predictive markers in different cross-combinations, the number of variables in the respective models had to be reduced to a minimal set of variables whose predictive power does not differ significantly from the optimal predictive power. To do this, Gartner and colleagues developed a marker identification via Minimum-Description-Length (MDL) based strategy (Gartner et al. 2009) which led to a substantial reduction in the number of variables in comparison with all measured variables and identification of the most important ones with respect to prediction of the trait of interest. When comparing the distribution levels of different metabolic markers, it became apparent that the highly predictive biomarkers tend to deviate from normal distribution, displaying bimodal distribution (Gartner et al. 2009). When they analyzed the selected set of metabolic and genetic markers needed for heterosis prediction in two different Arabidopsis testcrosses, they could indentify three classes of markers: a group of highly predictive markers for both testcrosses, a group of markers that were specific for only one of the testcross combinations, and markers that were negligible in any model.

Another example of metabolic profiling for biomass heterosis is the work of Meyer and colleagues (2007), where they used efficient GC-MS technology to identify a metabolic signature strongly linked to the integrative trait biomass in Arabidopsis recombinant inbred line (RIL) and introgression line (IL) populations. The authors did not obtain strong correlations between any primary metabolites and plant biomass; instead, they indentified a metabolite signature that was composed of contributions from several different metabolites, and their linear combination correlated with the biomass trait. A median correlation (.58) was determined for the biomass of previously unknown Arabidopsis cultivars, and a highly significant canonical correlation (0.73) was observed between biomass and a specific combination of metabolites in the training set. Based on these findings, Lisec and colleagues (2008, 2009) screened homozygous mapping populations of 422 RILs and 97 ILs and observed a canonical correlation between a simple derivation of the metabolites of two parental lines (RIL and P1 or P2) and the biomass heterosis value of the corresponding cross. This correlation was lower than previous results but still significant and indicates that parental metabolite profiles carry significant information about the biomass heterosis displayed by the cross.

The objective of the subsequent work of Steinfath and colleagues (2010a, b) was to improve methods for the prediction of properties of the hybrid based on the properties of the parents, and thereby to facilitate quality assessment and selection in breeding programs. These authors used three types of potential predictors (macroscopic phenotypes, genetic markers, and metabolites) and presented a proof of concept for a new approach that uses metabolite profiling in addition to SNP markers and introduces a new feature selection procedure. The biomarkers found by the new feature selection procedure were confirmed to be robust to small changes in the data set. Additional investigation of a relationship between plant biomass and metabolite profiling in Arabidopsis ecotypes (Sulpice et al. 2009) demonstrated that metabolite levels change reciprocally to biomass across diverse genotypes. In a subsequent study (Sulpice et al. 2010), the authors additionally measured enzymatic activities and end products of these activities and revealed a link between starch and biomass. They found that plants having the most efficient utilization of the starch as a carbon source can turn this energy into biomass accumulation; this conclusion hints at the potential value of metabolic profiling for plant breeding and biotechnology.

Another metabolic profiling study used an integrative approach based on independent component analysis (ICA) and unbiased identification of hundreds of individual compounds in Arabidopsis (wild type and mutants). This enabled the identification of characteristic biomarker metabolites within metabolite-protein correlation networks (Morgenthal et al. 2005). The authors demonstrated that multivariate data analysis of the integrative metabolite-protein data matrix allows the visualization of inherent time-dependent biological characteristics and thereby, the identification of the most discriminatory metabolites and proteins within a dynamic network of correlations, which can be promising approach for the diagnostic technology and biomarker discovery.

Using natural variation in mapping populations (F_2 populations, RILs, near-isogenic lines NILs, etc.) makes it possible to locate genetic factors that determine observed variation and thereby, offers the possibility of discovering the biomarker metabolites reflecting these genetic-phenotypic correlations. Keurentjes and colleagues (2007, 2008) conducted integrative multi-parallel analyses of gene expression, enzyme activity and metabolite accumulation in Arabidopsis and demonstrated that the natural variations in plant primary metabolism can be attributed to allelic differences in structural genes of catalytic enzymes, and to the presence of regulatory loci. The authors presented evidence for metabolic signaling that modulates metabolic routes. Their findings define the need for integrative studies of the complex regulation of plant metabolism; the results of these studies can be used for classical breeding and for metabolic engineering of agronomically important crops.

The most extensive studies on natural metabolomic variations have been carried out using Arabidopsis, but more recently, the focus has shifted to crop species. So far, most crop studies have been done using material from a single harvest (Kusano et al. 2007; Laurentin et al. 2008; Rochfort et al. 2008; Fraser et al. 2007), which makes it impossible to distinguish between genotype and environmental effects. However, a few studies have used material from multi-harvest crops. For example, a wide range of compositional traits including protein and oil contents, fatty acid, amino acid and organic acid content were analyzed in three maize hybrids grown at three separate locations (Harrigan et al. 2007). These studies revealed a major influence of environmental conditions on metabolic profiles.

A broad profiling of tomato fruit volatiles, which are extremely important flavor components, in a population of 74 *Solanum lycopersicum* × *S. pennellii* ILs yielded 100 QTL that were conserved across harvests (Tieman et al. 2006). In this study, similar to the study in maize, a strong effect of environment on metabolites was observed. Another series of studies in tomato using GC-MS based metabolite profiling of a set of introgression lines obtained through commercial × exotic crosses revealed a strong correlation between harvest index and metabolite content (Schauer et al. 2005, 2006; Schauer and Fernie 2006). Metabolite QTL from tomato fruits and leaves were correlated to phenotypic traits: yield, harvest index, seed number, and total soluble solids content. Three independent field harvests of introgression lines were profiled and known correlations between sugars, organic acids, and total soluble solids were confirmed, thus demonstrating that an applied approach is suitable for identifying novel relationships. These studies demonstrated that the mean hereditability of the metabolite QTL was at an intermediate rate, as was also found in Arabidopsis (Rowe et al. 2008), but a handful of the traits displayed reasonable heritability. Also, most metabolite QTL were dominantly inherited with either an additive or recessive mode of action. Very few QTL displayed overdominant inheritance characteristics. Different classes of metabolites, e.g. sugars and acids, showed quantitatively different patterns of inheritance. Importantly for breeding strategies, the association between morphological and metabolic traits was less prominent in the next generation plants that were heterozygous for the ILs in comparison with ILs themselves (Fernie and Schauer 2009).

Steinfath and colleagues (2010a, b) established a method that enabled the selection of a subset of metabolites correlated with quality parameters to predict browning and chip color traits in commercial potato cultivars. They performed metabolite profiling for several potato cultivars that showed a wide range of variation in black spot bruising and chip coloring during production, and then integrated metabolite data with quality analysis data. They were looking for a specific biomarker that corresponded to a single metabolite with the highest (close to 100%) prediction power, but they found that the biomarker comprised a range of between 2 and 100 compounds with prediction power less than 100%. Their mathematical model was validated in a following season using harvest collected from different areas, and the predictive power of their selected biomarkers was confirmed. Additional validation of these biomarkers was performed using a set of segregating populations, where biomarkers again strongly correlated with chip coloring and black spot bruising.

In another type of experiments, targeted profiling of glycoalkaloids extracted from cultivated, wild, and somatic hybrids of potato demonstrated that the glycoalkaloid pattern of the somatic hybrids represented the sum of their parents' profiles along with additional new glycoalkaloid compounds (Savarese et al. 2009). These results provided evidence that glycoalkaloids in potato can be used as additional biomarkers for detection of somatic hybrids and their parents.

Lui and colleagues (2005), using GC-MS based profiling, identified several volatile metabolite markers that helped to discriminate among five pathogenic diseases that affect storage of potato tubers and produce significant crop losses. The authors identified 13 metabolites that differed between infected and non-infected tubers. Since infections occur not only in the field but also at harvest as the result of tuber wounding by harvesting equipment, it was proposed that the developed method could be used to assess the quality of potatoes either before harvest, at harvest, or during storage. In addition, using the metabolite markers in combination can enable disease discrimination with fewer samples. In the future, if the biochemical pathways underlying the formation of these marker metabolites are revealed, then these metabolites might be useful for selection of resistant potato lines.

Untargeted comprehensive metabolic profiling using GC-TOF-MS and UPLC-FT-MS of a number of diverse herbs demonstrated that metabolomic techniques are a valuable diagnostic tool to assess the variation in individual metabolic phenotypes both among diverse species and across different environmental conditions. Such a tool is helpful in investigation of epigenetic adaptation by species to their environment, allowing discovery of marker metabolites specific for these processes (Scherling et al. 2010). In the past, numerous studies have shown the usefulness of natural biodiversity for the elucidation of agronomically important traits, and pleiotropic loci have been identified that control different traits simultaneously (Koorneef et al. 2004). Understanding the mechanisms that explain natural variation in metabolite profiles and how this correlates with phenotype is a primary challenge to defining natural biodiversity and maximizing its use through directed plant breeding. The parallel genetic analysis of physiological, transcriptional and biochemical profiling can greatly enhance our understanding of the metabolic regulatory circuitry and its relationship with phenotypic traits segregating in the same population. Genome-wide genetic correlative metabolic analysis can be successfully used for any set of metabolites analyzed in any mapping population.

Metabolite biomarkers representing the developmental stages in rice tillering were discovered in a study by Tarpley and colleagues (2005). These authors performed PCA analysis of GC-MS comprehensive metabolomic profiling data followed by K-mean clustering of metabolites representing variation determined by loading on the first three PC. This allowed them to select the biomarkers with predictive power for metabolite changes in response to changes in environment, and later reconfirm those predictive metabolites in an independent rice metabolomic study. From this work, the authors concluded that, in general, the set of metabolite biomarkers can be used in comparative screening of metabolite patterns of various developmental, genotypic or environmental sets of samples and thus, provide a complementary tool to diagnostic biomarker approaches if: (1) the metabolites represent much of the metabolic variance observed during an experiment that improves crop quality, (2) these metabolites are relatively independent of each other, and (3) they are common and found in any typical plant sample. While a particular biomarker set cannot be optimal for every experimental condition, it should be fairly robust in capturing physiologically real differences among various conditions and be responsive to eventual metabolic interpretation (Tarpley et al. 2005). Since metabolites influence plant development, particularly morphology during early stages of development (Alba et al. 2005; Lumba and McCourt 2005), quantitative measurements of different classes of compounds such as proteins and metabolites, and various processes, such as gene expression in combination with other phenotypic analyses would significantly contribute to understanding of biological functions (Keurentjes et al. 2006).

The discovery of the phenomenon of allelopathy, when plant development is depressed by biochemicals produced by other plant species, prompted intensive efforts to use it for weed management. Several crop species including rice, wheat, barley, and sorghum possess potent allelopathic interference mediated by root exudation of such active biochemicals called allelochemicals (Belz 2007). Biosynthesis and exudation of allelopathic metabolites follow a distinct temporal pattern and can be induced by biotic and abiotic stresses. Creation of weed-suppressive cultivars with improved allelopathic interference is still a challenge, but traditional breeding and biotechnology can contribute significantly to this effort. Currently, scientists are still far from completely understanding crop/weed interactions; therefore, an untargeted screen for biomarkers could both assist breeding efficiency and reveal the mechanisms underlying this interaction (Belz 2007).

Continuing efforts to investigate the metabolic responses to biotic and abiotic factors suggest that metabolomics-assisted breeding can be used in the development of stress resistant crops (Morandini and Salamini 2003). The application of metabolomic and other post-genomic technologies can improve and shorten the process of elite line selection. Thus, metabolomics-assisted breeding can be applied to crop species in a manner similar to that which has already proven successful in breeding programs to increase disease resistance and herbicide or salinity tolerance (McCouch 2004; Zamir 2001; Takeda and Matsuoka 2008), and this is certainly a viable option for crop improvement.

Chemical composition traits in plants have also become urgently important for increasing nutritional and market values of crops. Currently, understanding of these traits lags behind that of yield or stress resistance traits and breeders continue to face problems in the identification of potentially valuable lines with improved characteristics, particularly for market quality. The nutritional status of agricultural crops is ultimately a reflection of their metabolic composition; hence, the compositional quality of crops is highly important for human health (Fernie and Schauer 2009). For example, such post-harvest traits as flavor or coloring are highly complex and difficult to track on a genomics level because of the major contributions of post-translational modifications and the regulation of enzymatic activities involved in the control of such quality traits. In plant breeding, combining molecular composition with nutrition and health is a fast-developing approach. Using metabolite profiling provides a snapshot of a transient quality state and permits determination of the relationship between trait and metabolome. This approach for the creation of new lines can be powerful if combined with systematic profiling of metabolite composition of the plant products already on the market compared to the metabolite composition of the novel product. The latter would also provide important input into the public debate about the acceptability of changes in food-production chains (Kopka et al. 2004).

Current approaches for biomarker selection face some problems as well. For example, if more factors are responsible for the trait of interest in addition to the selected biomarkers then the biomarkers will not have adequate predictive power and their applicability will be not be confirmed during validation. This can be solved by increasing the number and diversity of the samples selected for creation of the predictive model, i.e. by broadening of the number of cultivars, crosses and environmental conditions (Steinfath et al. 2010a, b).

Metabolite profiling opens the possibility of broader trait-oriented high-throughput phenotyping of natural and generated genetic diversity (Zamir 2001; Wang and Larkins 2001). In addition, metabolomic profiling combined with data-mining tools can provide a platform to assess genotype variation without acquiring molecular information for a particular species (Fiehn et al. 2000; Roessner et al. 2002). Moreover, once broad metabolite databases are created by profiling a wide variety of species, cultivars and conditions, it will be possible to perform various screens to discover new phenotypes. Natural diversity provides a spectrum of functionally important changes displayed in polymorphisms (Stitt et al. 2010), information on which will soon be available as a result of massive genome sequencing efforts (Clark et al. 2007; Weigel and Mott 2009; Dooner and He 2008). This will demand the generation of metabolomic data matrices, which can be combined with comprehensive genetic information to become an important interface between basic research and breeding programs allowing broader efforts in biomarker discovery. Therefore, biomarker discovery and application in plant breeding is the next valuable technology that will be empowered in the nearest future (Steinfath et al. 2010a, b; Fernie and Schauer 2009).

Future Prospects

Scientists are only beginning to understand how a genotype confers a particular set of properties to a living organism. The subtleties of phenotypic plasticity resulting from environmental changes and the levels of genetic redundancy that determine biological systems represent an "iceberg of unknowns". It is clear that enhancing the potential for genetic diversity is one previously ignored direction for future breeding programs (McCouch 2004). Currently, breeders face problems in the selection of lines with improved traits because the complex and environmentally influenced traits or post-harvest traits that determine market quality are difficult to track using gene-based markers. However, metabolite profiling, by providing a snapshot of the plant's quality state, allows identification of the relationship between metabolites and quality traits, thereby making it possible to select biomarkers that have stronger ties with the ultimate phenotypic characteristics. In addition, metabolite-assisted breeding can allow breeding programs to use exotic varieties for which genetic information is unavailable or the genetic mechanisms that determine the trait of interest are too complex. Thus, comprehensive metabolomic profiling of a wide variety of genotypes, including breeding populations and cultivars, will help to build up databases that will allow creation of predictive models and will reveal novel phenotypes. Hence, metabolite profiling will provide an invaluable technology for biomarker discovery for plant breeding in the near future.

Currently, metabolomic profiling and therefore biomarker discovery is still an expensive technology which requires costly analytical instrumentation and the participation of different experts in chemistry, bioinformatics, mathematics, and biology. In addition, a large sample set must be profiled to increase the prediction power of selected biomarkers, and to confirm their applicability. However, rapid advances in instrumentation and software development stimulate establishment of more automated and high-throughput platforms that allow significant reductions in profiling costs and substantial extension of metabolome coverage. Initial trials, none-comprehensively presented in this chapter, have demonstrated that prediction models can be successfully used for a broad variety of plants. So far, only a few metabolites with predictive power as biomarkers have been revealed. Considering the number of metabolites predicted to be present in plant genera, this small group of signature metabolites can be seen as only the very beginning of the enormous potential of the metabolite-based biomarker discovery field. Taking this into account, metabolomics urgently needs tools for metabolite identification and accurate quantification, which will lead to more comprehensive coverage of metabolic networks. This requires broader involvement of experts in synthetic and combinatorial chemistry into creation of diverse metabolite libraries which would be available to the scientific community.

Another future prospect for metabolomics is broadening the range of molecules included in metabolic profiles. In general, compounds with relatively low molecular weights (MW $\leq$ 1,500) are considered to represent the plant metabolome; however, some larger molecules can also participate in metabolic networks, particularly in their regulation. Thus, oligopeptides (for example, systemins, ENDO40, phytosulfokines) accumulate in different plant species and participate as signalling molecules in the regulation of plant growth or stress responses (Schaller 1999). Another example of larger molecules participating in metabolic networks is oligosaccharides (MW about 2 kD), the products of cell wall polysaccharide catabolism, which transiently accumulate in plant cells and initiate particular metabolic processes involved in plant stress responses or organogenesis (Zabotina and Zabotina 2011). The endogenous concentration of these compounds is extremely low, so it is difficult to detect them with currently available techniques. But in the future when analytical platforms will be more advanced, these molecules might become valuable as potential biomarkers for stress related traits.

The field of metabolomics progresses at a tremendous pace, keeping up with and largely due to emerging technologies and bioinformatics. In plant biology, metabolomics has a key role as a fundamental tool in systems biology research, but also has great potential for predictions and diagnostics for plant breeding and biotechnology. In coming years, further intensive development of comprehensive databases that will accumulate and combine detailed information about metabolic networks and genotype-phenotype correlations will provide rich sources for the search for new valuable phenotypic traits and their metabolite markers.

References

Aharoni A, de Vos CHR, Verhoeven HA, Maliepaard CA, Kruppa G, Bino RJ, Goodenowe DB (2002) Nontargeted metabolome analysis by use of Fourier Transform Ion Cyclotron Mass Spectrometry. OMICS 6:217–234

Alba R, Payton P, Fei ZJ, McQuinn R, Debbie P, Martin GB, Tanksley SD, Giovannoni JJ (2005) Transcriptome and selected metabolite analyses reveal multiple points of ethylene control during tomato fruit development. Plant Cell 17:2954–2965

America AHP, Cordewender JNG, van Geffe A, Lommen A, Vissers JPC, Bino RJ, Hall RD (2006) Alignment and statistical difference analysis of complex peptide data sets generated by multidimensional LC-MS. Proteomics 6:641–653

Bailey NJC, Stanley PD, Hadfield ST, Lindon JC, Nicholson JK (2000) Mass spectrometrically detected directly coupled high performance liquid chromatography/nuclear magnetic resonance spectroscopy/mass spectrometry for the identification of xenobiotic metabolites in maize plants. Rapid Commun Mass Spectrom 14:679–684

Bais P, Moon SM, He K, Leitao R, Dreher K, Walk T, Sucaet Y, Barkan L, Wohlgenuth G, Roth MR et al (2010) PlantMetabolomics.org: a web portal for plant metabolomics experiments. Plant Physiol 152:1807–1816

Barry SJ, Carr RM, Lane SJ, Leavens WJ, Monte S, Waterhouse I (2003) Derivatisation for liquid chromatography/electrospray mass spectrometry: synthesis of pyridinium compounds and their amine and carboxylic acid derivatives. Rapid Commun Mass Spectrom 17:603–620

Belz RG (2007) Allelopathy in crop/weed interactions – an update. Pest Manag Sci 63:308–326

Bentsink L, Alonso-Blanco C, Vreugdenhil D, Tesnier K, Groot SPC, Koornneef M (2000) Genetic analysis of seed-soluble oligosaccharides in relation to seed storability of Arabidopsis. Plant Physiol 124:1595–1604

Birkemeyer C, Kolasa A, Kopka J (2003) Comprehensive chemical derivatization for gas chromatography-mass spectrometry-based multi-targeted profiling of the major phytohormones. J Chromatogr A 993:89–102

Clark RM, Schweikert G, Toomajian C, Ossowski S, Zeller G, Shinn P, Wartmann N, Hu TT, Fu G, Hinds DA et al (2007) Common sequence polymorphisms shaping genetic diversity in *Arabidopsis thaliana*. Science 317:338–342

Defernez M, Gunning YM, Parr AJ, Shepherd LVT, Davies HV, Coloquhoun IJ (2004) NMR and HPLC-UV profiling of potatoes with genetic modifications to metabolic pathways. J Agric Food Chem 52:6075–6085

Desbrosses GG, Kopka J, Udvardi MK (2005) Losus japonicas metabolic profiling: development of gas chromatography-mass spectrometry resources for the study of plant-microbe interactions. Plant Physiol 137:1302–1318

Dooner HK, He L (2008) Maize genome structure variation: interplay between retrotransposon polymorphisms and genic recombination. Plant Cell 20:249–258

Dunn WB, Broadhurst DI, Atherton HJ, Geedacre R, Griffin JL (2011) Systems level studies of mammalian metabolomes: the roles of mass spectrometry and nuclear magnetic resonance spectroscopy. Chem Soc Rev 40:387–426

Duvick DN (2001) Biotechnology in the 1930s: the development of hybrid maize. Nat Rev Genet 2:69–74

Facchini PJ, Bird DA, St-Pierre B (2004) Can Arabidopsis make complex alkaloids? Trends Plant Sci 9:116–122

Fernie AR, Schauer N (2009) Metabolomics – assisted breeding: a viable option for crop improvement? Trends Genet 25:39–48

Fiehn O, Kopka J, Dormann P, Altmann T, Trethewey RN, Willmitzer L (2000) Metabolite profiling for plant functional genomics. Nat Biotechnol 18:1157–1161

Fiehn O, Wohlgemuth G, Scholz M, Kind T, Lee DY, Lu Y, Moon S, Nikolau B (2008) Quality control for plant metabolomics: reporting MSI-compliant studies. Plant J 53:691–704

Fraser PD, Enfissi EMA, Goodfellow M, Eguchi T, Bramley PM (2007) Metabolite profiling of plant carotenoids using the matrix-assisted laser desorption ionization time-of-flight mass spectrometry. Plant J 49:552–564

Gartner T, Steinfath M, Andorf S, Lisec J, Meyer RC, Altmann T, Willmitzer L, Selbig J (2009) Improved heterosis prediction by combining information on DNA- and metabolic markers. PLoS One 4:e5220

Giavalisco P, Kohl K, Hummel J, Seiwert B, Willmitzer L (2009) C-13 isotope-labeled metabolomes allowing for improved compound annotation and relative quantification in liquid chromatography-mass spectrometry-based metabolomic research. Anal Chem 81:546–6551

Hall RD (2006) Plant metabolomics: from holistic hope, to hype, to hot topic. New Phytol 169:453–468

Harada K, Fukusaki E, Bamba T, Kobayashi A (2006) *In vivo*^{15}N-enrichment of metabolites in *Arabidopsis* cultured cell T87 and its application to metabolomics. In: Nikolau BJ, Wurtele ES (eds) Concepts in plant metabolomics. Springer, Dordrecht, pp 287–297

Harrigan GG, Stork LG, Riordan SG, Reynolds TL, Ridley WP, Masucci JD, MacIsaac S, Halls SC, Orth R, Smith RG et al (2007) Impact of genetics and environment on nutritional and metabolite components of maize grain. J Agric Food Chem 55:6177–6185

Hendriks MMWB, Cruz-Juarez L, De Bont D, Hall RD (2005) Preprocessing and exploratory analysis of chromatographic profiles of plant extracts. Analytika Chemica Acta 545:53–64

Hirai MY, Yano M, Goodenowe DB, Kanays S, Kimura T, Awazuhara Mjiwara T, Saito K (2004) Integration of transcriptomics and metabolomics for understanding of global responses to nutritional stresses in *Arabidopsis thaliana*. Proc Natl Acad Sci USA 101:10205–10210

Hobbs DH, Flintham JE, Hills MJ (2004) Genetic control of storage oil synthesis in seeds of Arabidopsis. Plant Physiol 136:3341–3349

Jenkins H, Beckmann M, Draper J, Hardy N (2007) GC-MS peak labeling under ArMet. In: Nikolau BJ, Wurtele ES (eds) Concepts in plant metabolomics. Springer, Dordrecht, pp 19–28

Kaddurah-Daouk R, Beecher C, Kristal BS, Matson WR, Bogdanov M, Asa DJ (2004) Bioanalytical advances for metabolomics and metabolic profiling. Pharma Genomics 4:46–52

Kanehisa M, Goto S, Kawashima S, Nakaya A (2002) The KEGG databases at GenomeNet. Nucleic Acids Res 30:42–46

Keurentjes JJB, Fu J, de Vos CHR, Lommen A, Hall RD, Bino RJ, van der Plas LHW, Jansen RC, Vreugdenhil D, Koornneef M (2006) The genetics of plant metabolism. Nat Genet 38:842–849

Keurentjes JJB, Fu JY, Terpstra IR, Garcia JM, van den Ackerveken G, Snoek LB, Peeters AJM, Vreugdenhil D, Koornneef M, Jansen RC (2007) Regulatory network construction in Arabidopsis by using genome-wide gene expression quantitative trait loci. Proc Natl Acad Sci USA 104:1708–1713

Keurentjes JJB, Sulpice R, Gibon Y, Steinhauser M-C, Fu J, Koornneef M, Stitt M, Vreugdenhil D (2008) Integrative analyses of genetic variation in enzyme activities of primary carbohydrate metabolism reveal distinct modes of regulation in *Arabidopsis thaliana*. Genome Biol 9:R129

Kind T, Wohlgemuth G, Lee DY, Lu Y, Palazoglu M, Shahbaz S, Fiehn O (2009) FiehnLib: mass spectral and retention index libraries for metabolomics based on quadrupole and time-of-flight gas chromatography/mass spectrometry. Anal Chem 81:10038–10048

Kliebenstein DJ, Gershenzon J, Mitchell-Olds T (2001) Comparitive quantitative trait loci mapping of aliphatic, indolic and benzylic glucosinolate production in *Arabidopsis thaliana* leaves and seeds. Genetics 159:359–370

Klukas C, Schreiber F (2010) Integration of –omics data and networks for biomedical research with VANTED. J Integr Bioinform 7(2):112

Knapp DR (1979) Handbook of analytical derivatisation reactions. Wiley, New York

Koorneef M, Alonso-Blanco C, Vreugdenhil D (2004) Naturally occurring genetic variation in *Arabidopsis thaliana*. Annu Rev Plant Biol 55:141–172

Kopka J, Fernie A, Weckwerth W, Gibon Y, Stitt M (2004) Metabolite profiling in plant biology: platforms and destinations. Genome Biol 5:1465–1469

Kopka J, Schauer N, Krueger S, Birkemeyer C, Usadel B, Bergmuller E, Dormann P, Weckwerth D (2005) GMD@CSB.DB: the Golm metabolome database. Bioinformatics 21:1635–1638

Kusano M, Fukushima A, Kobayashi M, Hayashi N, Jonsson P, Moritz T, Ebana K, Saito K (2007) Application of a metabolomics method combining one-dimensional and two-dimensional gas chromatography-time-of-flight/mass spectrometry to metabolic phenotyping of natural variants in rice. J Chromatogr B Analyt Technol Biomed Life Sci 855:71–79

Lange BM, Ghassemian M (2005) Comprehensive post-genomic data analysis approaches integrating biochemical pathway maps. Phytochemistry 66:413–451

Laurentin H, Ratzinger A, Karlovsky P (2008) Relationship between metabolic and genomic diversity in sesame (*Sesamum indicum* L). BMC Genomics 9:250

Lisec J, Schauer N, Kopka J, Willmitzer L, Fernie A (2006) Gas chromatography mass spectrometry – based metabolite profiling in plants. Nat Protoc 1:387–396

Lisec J, Meyer R, Steinfath M, Redestig H, Becher M, Witucka-Wall H, Fiehn O, Torjek O, Selbig J, Altmann T, Willmitzer L (2008) Identification of metabolic and biomass QTL in Arabidopsis thaliana in a parallel analysis of RIL and IL populations. Plant J 53:960–972

Lisec J, Steinfath M, Meyer RC, Selbig J, Melchinger AE, Willmitzer L, Altmann T (2009) Identification of heterotic metabolite QTL in Arabidopsis thaliana RIL and IL populations. Plant J 59:777–788

Lui LH, Vikram A, Abu-Nada Y, Kushalappa AC, Raghavan GSV, Al-Mughrabi K (2005) Volatile metabolic profiling for discrimination of potato tubers inoculated with dry and soft rot pathogens. Am J Potato Res 82:1–8

Lumba S, McCourt P (2005) Preventing leaf identity theft with hormones. Curr Opin Plant Biol 8:501–505

Maloney V (2004) Plant metabolomics. BioTeach J 2:92–99

Matuszewski MK, Constanzer ML, Chavez-Eng CM (2003) Strategies for the assessment of matrix effect in quantitative bioanalytical methods based on HPLC-MS/MS. Anal Chem 75:3019–3030

McCouch S (2004) Diversifying selection in plant breeding. PLoS Biol 2:e347

Mendes P (2001) Modeling large scale biological systems from functional genomics data: parameter estimation. In: Kitano H (ed) Foundation of systems biology. MIT Press, Cambridge, MA, pp 163–186

Meyer RC, Steinfath M, Lisec J, Beceher M, Witucka-Wall H, Torjek O, Fiehn O, Eckardt A, Willmitzer L, Selbig J, Altmann T (2007) The metabolic signature related to high plant growth rate in *Arabidopsis thaliana*. Proc Natl Acad Sci USA 104:4759–4764

Miedaner T, Gey A-K, Sperling U, Geiger HH (2002) Quantitative-genetic analysis of leaf-rust resistance in seedling and adult-plant stage of inbred lines and their testcrosses in winter rye. Plant Breed 121:475–479

Morandini P, Salamini F (2003) Plant biotechnology and breeding: allied for years to come. Trends Plant Sci 8:70–75

Morgenthal K, Wienkoop S, Scholz M, Selbig J, Weckwerth W (2005) Correlative GC-TOF-MS – based metabolite profiling and LC-MS – based protein profiling reveal time – related systemic regulation of metabolite – protein networks and improve pattern recognition for multiple biomarker selection. Metabolomics 1:109–121

Morreel K, Goeminne G, Storme V, Sterck L, Ralph J, Coppieters W, Breyne P, Steenackers M, Georges M, Messens E, Boerjan W (2006) Genetic metabolomics of flavonoid biosynthesis in Populus: a case study. Plant J 47:224–237

Mueller LA, Zhang P, Phee SY (2003) AracCyc: a biochemical pathway database for Arabidopsis. Plant Physiol 135:453–460

Nielson NPV, Carstensen JM, Smedsgaard J (1998) Aligning of single and multiple wavelength chromatographic profiles for chemometric data analysis using correlation optimized warping. J Chromatogr 805:17–35

Ofori A, Becker H (2008) Breeding of *Brassica rapa* for biogas production: heterosis and combining ability of biomass yield. BioEnergy Res 1:98–104

Oksman-Caldentey K-M, Inze D (2004) Plant cell factories in the post-genomic era: new ways to produce designer secondary metabolites. Trend Plant Sci 9:433–440

Panella L, Lewellen R (2007) Bradening the genetic base of sugar beet: introgression from wild relative. Euphytica 154:383–400

Payne TG, Southam AD, Arvanitis TN, Viant MR (2009) A signal filtering method for improved quantification and noise discrimination in fourier transform Ion cyclotron resonance mass spectrometry-based metabolomics data. J Am Soc Mass Spectrom 20:1087–1095

Ratcliffe RG, Shachar-Hill Y (2001) Probing plant metabolism with NMR. Ann Rev Plant Physiol Plant Mol Biol 52:499–526

Ratcliffe RG, Shachar-Hill Y (2005) Revealing metabolic phenotypes in plants: inputs from NMR analysis. Biol Rev 80:27–43

Rochfort SJ, Trenerry VC, Imsic M, Panozzo J, Jones R (2008) Class targeted metabolomics: ESI ion trap screening methods for glucosinolates based on MSn fragmentation. Phytochemistry 69:1671–1679

Roessner U (2007) Uncovering the plant metabolome: current and future challenges. In: Nikolau BJ, Wurtele ES (eds) Concepts in plant metabolomics. Springer, Dordrecht, pp 71–85

Roessner U, Willmitzer L, Fernie AR (2002) Metabolic profiling and biochemical phenotyping of plant systems. Plant Cell Rep 31:189–196

Rohwer JM (2012) Kinetic modeling of plant metabolic pathways. J Exp Bot 63:2275–2292

Rowe HC, Hansen BG, Halkier BA, Kliebenstein DJ (2008) Biochemical networks and epistasis shape the *Arabidopsis thaliana* metabolome. Plant Cell 20:1199–1216

Sato S, Soga T, Nishioka T, Tomita M (2004) Simultaneous determination of the main metabolites in rice leaves using capillary electrophoresis mass spectrometry and capillary electrophoresis diode array detection. Plant J 40:151–163

Savarese S, Andolfi A, Cimmino A, Carputo D, Frusciante L, Evidente A (2009) Glycoalkaloids as biomarkers for recognition of cultivated, wild, and somatic hybrids of potato. Chem Biodivers 6:437–446

Schaller A (1999) Oligopeptide signalling and the action of systemin. Plant Mol Biol 40:763–769

Schauer N, Fernie AR (2006) Plant metabolomics: towards biological function and mechanism. Trends Plant Sci 11:508–516

Schauer N, Zamir D, Fernie AR (2005) Metabolic profiling of leaves and fruit of wild species tomato: a survey of the *Solanum lycopersicum* complex. J Exp Bot 56:297–307

Schauer N, Semel Y, Roessner U, Gur A, Balbo I, Carrari F, Pleban T, Perez-Melis A, Bruedigam C, Kopka J, Willmitzer L, Zamir D, Fernie AR (2006) Comprehensive metabolic profiling and phenotyping of interspecific introgression lines for tomato improvement. Nat Biotechnol 24:447–454

Scherling C, Roscher C, Giavalisco P, Schulze E-D, Weckwerth W (2010) Metabolomics unravel contrasting effects of biodiversity on the performance of individual plant species. PLoS One 5:e12569

Schijlen EGWM, de Vos CHR, van Tunen AJ, Bovy AG (2004) Modification of flavonoid biosynthesis in crop plants. Phytochemistry 65:2631–2648

Searls DB (2005) Data integration: challenges for drug discovery. Nat Rev Drug Discov 4:45–58

Simpson AJ, Tseng LH, Simpson MJ, Spraul M, Braumann U, Kingery WL, Kelleher BP, Hayes MHB (2004) The application of LC-NMR and LC-SPE-NMR to compositional studies of natural organic matter. Analyst 129:1216–1222

Smedsgaard J, Nielsen J (2005) Metabolite profiling of fungi and yeast: from phenotype to metabolome by MS and informatics. J Exp Bot 56:273–286

Soga T, Ohashi Y, Ueno Y, Naraoka H, Tomita M, Nishioka T (2003) Quantitative metabolome analysis using capillary electrophoresis mass spectrometry. J Proteome Res 2:488–494

Southam AD, Payne TG, Cooper HJ, Arvanitis TN, Viant MR (2007) Dynamic range and mass accuracy of wide-scan direct infusion nanoelectrospray fourier transform ion cyclotron resonance mass spectrometry-based metabolomics increased by the spectral stitching method. Anal Chem 79:4595–4602

Spencer JP, Kuhnle GG, Hajirezaei M, Mock HP, Sonnewald U, Rice-Evans C (2005) The genotypic variation of the antioxidant potential of different tomato varieties. Free Radic Res 39:1005–1016

Stein SE (1999) An integrated method for spectrum extraction and compound identification from gas chromatography/mass spectrometry data. J Am Soc Mass Spectrom 10:770–781

Steinfath M, Froth D, Lisec J, Selbig J (2008) Metabolite profile analysis: from raw data to regression and classification. Physiol Plant 132:150–161

Steinfath M, Gartner T, Lisec J, Meyer RC, Altmann T, Willmitzer L, Selbig J (2010a) Prediction of hybrid biomass in *Arabidopsis thaliana* by selected parent SNP and metabolic markers. Theor Appl Genet 120:239–247

Steinfath M, Strehmel N, Peters R, Schauer N, Groth D, Hummel J, Steup M, Selbig J, Kopka J, Geigenberger P, van Gongen JT (2010b) Discovering plant metabolic biomarkers for phenotype prediction using an untargeted approach. Plant Biotechnol J 8:900–911

Stitt M, Fernie AR (2003) From measurements of metabolites to metabolomics: an 'on the fly' perspective illustrated by recent studies of carbon-nitrogen interactions. Curr Opin Biotechnol 14:136–144

Stitt M, Sulpice R, Keurentjes J (2010) Metabolic networks: how to identify key components in the regulation of metabolism and growth. Plant Physiol 152:428–444

Sulpice R, Pyl E-T, Ishibara H, Trenkamp S, Steinfath M, Witucka-Wall H, Gibon Y, Usadel B, Poree F, Piques MC, Von Korff M, Steinhauser MC, Keurentjes JJB, Guenther M, Hoehne M, Selbig J, Fernie AR, Altmann T, Stitt M (2009) Starch as a major integrator in the regulation of plant growth. Proc Natl Acad Sci USA 106:10348–10353

Sulpice R, Trenkamp S, Steinfath M, Usadel B, Gibon Y, Witucka-Wall H, Pyl E-T, Tschoep H, Steinhauser MC, Guenther M, Hoehne M, Rohwer JM, Altmann T, Fernie AR, Stitt M (2010) Network analysis of enzyme activities and metabolite levels and their relationship to biomass in a large panel of Arabidopsis accessions. Plant Cell 22:2872–2893

Sumner LW, Mendez P, Dixon RA (2003) Plant metabolomics: large – scale phytochemistry in the functional genomics era. Phytochemistry 62:817–836

Sweetlove LJ, Ratcliffe RG (2011) Flux-balance modeling of plant metabolism. Front Plant Sci 2:38

Sysi-Aho M, Katajamaa M, Yetukuri L, Oresic M (2007) Normalization method for metabolomics data using optimal selection of multiple internal standards. BMC Bioinformatics 8:93

Takeda S, Matsuoka M (2008) Genetic approaches to crop improvement: responding to environmental and population changes. Nat Rev Genet 9:444–457

Tarpley L, Duran AL, Kebrom TH, Sumner LW (2005) Biomarker metabolites capturing the metabolite variance present in a rice plant developmental period. BMC Plant Biol 5:8

Tieman DM, Zeigler M, Schmelz EA, Taylor MG, Bliss P, Kirst M, Klee HJ (2006) Identification of loci affecting flavor volatile emissions in tomato fruits. J Exp Bot 57:887–896

Tikunov Y, Lommen A, de Vos CHR, Verhoeven HA, Bino RJ, Hall RD, Bovy AG (2005) A novel approach for nontargeted data analysis for metabolomics. Large-scale profiling of tomato fruit volatiles. Plant Physiol 139:1125–1137

Tohge T, Nishiyama Y, Hirai MY, Yano M, Nakajima J, Awazuhara M, Inoue E, Takahashi H, Geedenowe DB, Kitayama M (2005) Functional genomics by integrated analysis of metabolome and transcriptome of Arabidopsis plants over-expressing and MYB transcription factor. Plant J 42:218–235

Toronen P, Kolehmainen M, Wong G, Castren E (1999) Analysis of gene expression data using self-organizing maps. FEBS Lett 451:142–146

Usadel B, Nagel A, Thimm O, Redestig H, Blaesing OE, Palacios-Rojos N, Selbig J, Hennemann J, Conceicao Piques M, Steinhauser D, Scheible W-R, Gibon Y, Morcuende R, Weicht D, Meyer S, Stitt M (2005) Extension of the visualization tool MapMan to allow statistical analysis of arrays, display of corresponding genes, and comparison with known responses. Plant Physiol 138:1195–1204

van de Greef J, Martin S, Juhasz P, Adourian A, Plasterer T, Verheij ER, McBurney RN (2007) The art and practice of systems biology in medicine: mapping patterns of relationships. J Proteome Res 6:1540–1559

Verhoeven HA, de Vos CHR, Bino RJ, Hall RD (2006) Plant metabolomics strategies based upon Quadruple Time of Flight Mass Spectrometry (QTOF-MS). In: Saito K, Dixon R, Willmitzer L (eds) Biotechnology in agriculture and forestry. Plant metabolomics. V57. Springer, Berlin/Heidelberg, pp 33–48

Virmani SS (1994) Heterosis and hybrid rice breeding. Monogr Theor Appl Genet 22:142–154

Wang XL, Larkins BA (2001) Genetic analysis of amino acid accumulation in opaque-2 maize endosperm. Plant Physiol 125:1766–1777

Weckwerth W (2011) Unpredictability of metabolism – the key role of metabolomics science in combination with next – generation genome sequencing. Anal Bioanal Chem 400: 1967–1978

Weckwerth W, Morgenthal K (2005) Metabolomics: from pattern recognition to biological interpretation. Drug Discov Today 10:1551–1558

Weigel D, Mott R (2009) The 1001 genomes project for *Arabidopsis thaliana*. Genome Biol 10:107

Westerhuis JA, Hoefsloot HCJ, Smit S, Vis DJ, Smidle AK, van Velzen EJJ, van Duijnhoven JPM, van Dorsten FA (2008) Assessment of PLSDA cross validation. Metabolomics 4:81–89

Wilson ID, Nicholson JK, Castro-Perez J, Granger JH, Jonson KA, Smith BW, Plumb RS (2005) High resolution "Ultra performance" liquid chromatography coupled to a TOF mass spectrometry as a tool for differential metabolic pathway profiling in functional genomic studies. Proteome Res 4:591–598

Wurtele ES, Li J, Diao L, Zhang H, Foster CM, Fatland D, Dickerson J, Brown A, Cox Z, Cook D, Lee E-K, Hofmann H (2003) MetNet: software to build and model the biogenetic lattice of Arabidopsis. Comp Funct Genomics 4:239–245

Yu J, Buckler ES (2006) Genetic association mapping and genome organization of maize. Curr Opin Biotechnol 17:155–160

Zabotina OA, Zabotina AI (2011) Biologically active oligosaccharide functions in plant cell: updates and prospects. In: Gordon NS (ed) Oligosaccharides: sources, properties and applications. Nova Science Publishers Inc, New York, pp 209–243

Zamir D (2001) Improving plant breeding with exotic genetic libraries. Nat Rev 2:983–989

Zelena E, Dunn WB, Broadhurst D, Francis-McIntyre S, Carroll KM, Begley P, O'Hagan S, Knowles JD, Halsall A, Wilson ID, Kell DB (2009) Development of a robust and repeatable UPLC-MS method for the long-term metabolomic study of human serum. Anal Chem 81:1357–1364

Part VI
Deposition of Diagnostic Marker Information

Chapter 15
Plant Databases and Data Analysis Tools

Mary L. Schaeffer, Jack M. Gardiner, and Carolyn J. Lawrence

Introduction

Completion of various plant genome sequencing projects has created a wealth of both opportunities and challenges for plant biologists. Opportunities for understanding the biochemical and genetic processes that have driven and continue to drive plant productivity are substantial, but so, too, are the challenges that are presented in understanding the avalanche of data resulting from a complete genome sequence. Further increasing the challenges of data management in the past several years is the availability of very inexpensive DNA sequencing technologies, termed "NextGen sequencing". Additional inexpensive "omics" technologies that have made their way into the research laboratory include both protein (proteomic) and small molecule (metabolomic) profiling, which provide complementary information to more established gene expression (functional genomic) profiling technologies such as microarrays and more recently, RNA sequencing (RNA-Seq). The evolution of these technologies has resulted in DNA, RNA, and small molecule "markers".

M.L. Schaeffer
Plant Genetics Research Unit, USDA-ARS, Columbia, MO, USA

Division of Plant Sciences, University of Missouri, Columbia, MO, USA
e-mail: mary.schaeffer@ars.usda.gov

J.M. Gardiner
Department of Genetics, Development, and Cell Biology, Iowa State University, Ames, IA, USA

School of Plant Sciences, University of Arizona, Tucson, AZ, USA
e-mail: jmgardin@iastate.edu

C.J. Lawrence (✉)
Department of Genetics, Development, and Cell Biology, Iowa State University, Ames, IA, USA

Crop and Insect Genetics, Genomics, and Informatics Research Unit, USDA-ARS, Ames, IA, USA
e-mail: carolyn.lawrence@ars.usda.gov

These molecular markers expand the toolset available for plant breeding beyond what may be observed by eye, and greatly aid in fine mapping of candidate genes for phenotypes important to the plant breeder (Ingvardsen et al. 2010; Ku et al. 2011; Salvi et al. 2002). Here we limit our discussions of markers to molecular markers: DNA/RNA, and small molecules.

Markers

DNA-based markers include RFLPs, AFLPs, CAPs, INDEL, SSR, SNP, etc. and are used as signposts where a particular marker variant is associated with a desired phenotypic trait. RNA markers, quantified by microarray and RNA-Seq experiments, are used to determine various states of stress, developmental transitions, etc. by assessing the expression of genes known to be associated with particular states. Metabolite markers are small molecules that can be measured by gas chromatography/mass spectrometry (GC-MS) methods.

Marker-Assisted Selection (MAS) is an indirect selection process that uses DNA/RNA-based methods to choose plants carrying desirable traits to carry forward in the breeding program. The recent availability of whole genome sequences coupled with diversity data placed on reference (sequenced) genomes have improved the MAS process in that it enables researchers to develop markers that are closely tied to (or synonymous with) genes directly responsible for traits of interest.

DNA-based MAS is a well-proven and useful tool to breed improved plants, especially for diploid organisms. However, for crops with more complex genomes, these DNA-based marker analyses can be confusing. Non-DNA based markers that may be used include protein electrophoretic variations or isozymes (Edwards et al. 1992), and protein markers associated with QTL (Burstin et al. 1994; Consoli et al. 2002). Metabolic markers/profiling methods may also be used to characterize agronomic traits, and aid in candidate gene assessment for a region (Riedelsheimer et al. 2012).

Online Resources

There are various online resources available for accessing and analyzing available marker data. Resources discussed here are listed in Table 15.1. Most make available information for analysis via some sort of Web interface connected to a relational database that stores or warehouses some of the data. Data stores also can be made available for others to access directly and serve via a separate website. In this case, information viewed at a particular website actually resides elsewhere. Tools and technologies that enable remote representation and analysis of off-site data are collectively referred to as Web services. The World Wide Web Consortium (W3C; http://www.w3.org/) defines Web services as, "a software system designed

Table 15.1 Online resources and available tools outlined in this chapter that are available to access and analyze marker data

Resource Name	Link
BRENDA	http://www.brenda-enzymes.info/
CottonDB	http://www.cottondb.org
CoGe	http://genomevolution.org/CoGe/
DDBJ	http://www.ddbj.nig.ac.jp/
EMBL	http://www.ebi.ac.uk/embl/
GenBank	http://www.ncbi.nlm.nih.gov/genbank/
Gene Ontology	http://www.geneontology.org/
GDR	http://www.rosaceae.org/
GrainGenes	http://wheat.pw.usda.gov
GRASSIUS	http://grassius.org/
GRIN	http://www.ars-grin.gov/
Gramene	http://www.gramene.org/
KEGG	http://www.genome.jp/kegg/
MaizeGDB	http://www.maizegdb.org/
Panzea	http://www.panzea.org/
Plant Ontology	http://www.plantontology.org/
POPcorn	http://popcorn.maizegdb.org
SGN	http://solgenomics.net/
SoyBase	http://soybase.org/
TAIR	http://www.arabidopsis.org/
UniProtKB/TrEMBL	http://www.ebi.ac.uk/uniprot/
VPhenoDBS	http://vphenodbs.rnet.missouri.edu/
Breeders Toolboxes	**Link**
CottonDB	http://www.cottondb.org/wwwroot/toolbox.php
GDR	http://www.rosaceae.org/breeders_toolbox
SGN	http://solgenomics.net/breeders
SoyBase	http://soybase.org/
Specific Tools	**Link**
BioCyc	http://biocyc.org/
eFP Browser	http://bar.utoronto.ca
FlapJack	http://bioinf.scri.ac.uk/flapjack/
GBrowse	http://gmod.org/wiki/GBrowse
Locus Lookup	http://www.maizegdb.org/cgi-bin/locus_lookup.cgi

to support interoperable machine-to-machine interaction over a network." In effect, the data sit off-site but are made available to users from the website in context-appropriate locations. Some resources also offer a hybrid approach with some of the data stored on-site (warehoused) and other data brought in from off-site repositories (using Web services).

Databases that are limited in scope to one or a few species can be classified as Model Organism Databases (MODs), which cater to a single research species like maize (MaizeGDB; http://www.maizegdb.org; Sen et al. 2009) or soybean (SoyBase; http://www.soybase.org; Grant et al. 2010) or Clade-Oriented Databases

(CODs), which serve information on a phylogenetically related group like the Solanaceae (SGN; http://solgenomics.net/; Bombarely et al. 2011) or Leguminosae (LIS; http://www.comparative-legumes.org/; Gonzales et al. 2005). The more general repositories that serve information limited to a particular datatype include the sequence repositories DDBJ (http://www.ddbj.nig.ac.jp/; Kaminuma et al. 2011), EMBL (http://www.ebi.ac.uk/embl/; Kulikova et al. 2007), and GenBank (http://www.ncbi.nlm.nih.gov/genbank/; Benson et al. 2011) as well as automatic annotation groups like PlantGDB (http://www.plantgdb.org; Duvick et al. 2008), UniProtKB/TrEMBL (http://www.ebi.ac.uk/uniprot/index.html; The Uniprot Consortium 2011) and CoGe (the "Place to Compare Genomes"; http://synteny.cnr.berkeley.edu/CoGe; Lyons and Freeling 2008; Lyons et al. 2008) where sequence information is processed and annotated for ease of use by biologists in their analyses.

There also are various databases that specialize, but cover multiple species, including animals and fungi. They are generally heavily curated, with links to scientific literature and external resources. Examples include GRIN (US breeding germplasm; http://www.ars-grin.gov/), BRENDA (enzymes for all species, with literature; http://www.brenda-enzymes.org/; Chang et al. 2009), KEGG (general metabolism; http://www.genome.jp/kegg/pathway.html; Kanehisa et al. 2004); Gene Ontology (gene product functional annotations with controlled vocabulary, and evidence codes, typically supplied by collaborating databases; http://www.geneontology.org/; Ashburner et al. 2000).

A number of data resources are for ongoing projects that may be expected to share data with a central database at some point. These include The Maize Diversity Project (which specializes in SNPs for maize and its relatives; http://www.panzea.org; Canaran et al. 2008), GRASSIUS (http://www.grassius.org; Yilmaz et al. 2009), which specializes in transcription factors and promoter sequences for grasses, mainly rice, maize, and *Brachypodium* at this time.

Notable datatypes that support the elucidation of candidate gene function in plants include sequence-indexed mutant collections in Arabidopsis, rice, maize, etc. where various database genome browsers provide links to stocks resources (e.g., TAIR [Lamesch et al. 2012], Gramene [Youens-Clark et al. 2011], and MaizeGDB); tissue-specific gene expression datasets (e.g., Sekhon et al. 2011); and epigenetics information (e.g., Gendler et al. 2008).

As evident in the discussion above, a large number of resources are available, but with access provided by various servers, and database teams. However, the average user would prefer to have a single, one-stop-shopping location or interface. A number of strategies are in place to facilitate access to multiple data resources, but most require extensive curation and file-sharing by participants, and may need updating as new resources become available.

One strategy practiced early on, is to have reciprocal links among records in various databases. Another is to relate information to a controlled vocabulary of terms shared among many data repositories to enable cross-referencing among participating websites and resources. When terms are hierarchically related, they are called ontologies. Across all forms of life, the Gene Ontology (GO) is used to indicate molecular functions, biological processes, and cellular components associate with the function of individual genes (Harris et al. 2004). Most biological databases make use of GO for functional annotation. The Plant Ontologies (PO; Avraham et al. 2008) are structured controlled vocabularies that allow data sharing across species, and include terms that are related to plant anatomy as well as growth and developmental stages. Thus the term 'kernel' in maize, can be translated 'behind the scenes' to more generic term 'fruit'. At the central repository, (http://www.plantontology.org), one could find, for example, a listing of all maize stocks and tomato germplasm that have associated traits or phenotype affecting 'fruit/kernel'. The VPhenoDBS "Query by Text Annotation" function (http://vphenodbs.rnet.missouri.edu/QBTA.php) is an interesting application that calls up phenotype images in maize with some reliance on closely related terms and synonyms in various ontologies (Green et al. 2011). A third mechanism to relate data across resources is the use of Web services, a mechanism (discussed above) by which information may be passed directly from one resource to another for representation. POPcorn, the PrOject Portal for corn, uses Web services heavily to search sequence and related data stored at various repositories (http://popcorn.maizegdb.org). Most repositories rely upon some sharing of information and many use all types of strategies described here to create useful resources.

Tools

Bioinformatic tools allow researchers to query and interrogate data rather than just browsing through information. They exist online at websites as well as via stand-alone applications that reside on a personal computer. The online tools exist both independently via websites crafted to accomplish a particular task and as components of more complex websites, especially MODs and CODs, as well as other sites that serve diverse data.

Tools for DNA Markers

The information researchers use to develop molecular markers can be searched via customized tools like the one shown in Fig. 15.1 from TAIR, or via software that has been developed to support visualization of marker data within its genomic context. Because so many crops now have fully-sequenced reference genomes, the availability of the latter has increased over the past few years with the genome browser software GBrowse (Stein et al. 2002) dominating plant databases including GrainGenes (O'Sullivan 2007), GDR (Jung et al. 2008), SGN, SoyBase, and others. In Fig. 15.2, molecular markers are shown relative to the tomato genome at SGN via their instance of GBrowse, along with other items useful for MAS including restriction sites and known genes.

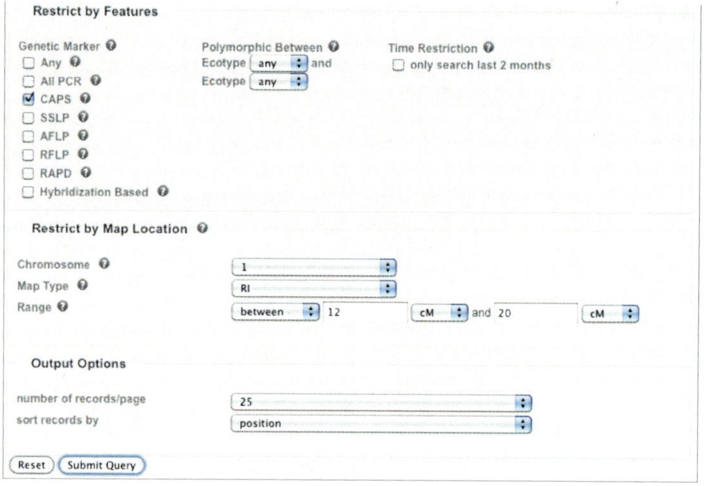

Fig. 15.1 The TAIR Markers Search enables researchers to locate markers based upon various criteria. Here the results are limited to CAPS markers on chromosome 1 using the RI map in the 12–20 cM range

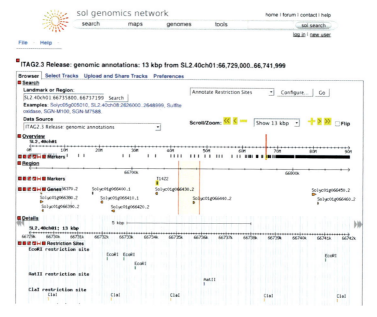

Fig. 15.2 The SGN tomato GBrowse instance. The track labeled "Overview" shows the genomic context of the region shown in detail by bounding the detail region by *red vertical lines*. Beneath that, the same red lines are shown as a reference point. For MAS, items like markers and restriction sites are shown and can be used to develop and make use of markers in a specific region with known distance to nearby genes of interest

15 Plant Databases and Data Analysis Tools

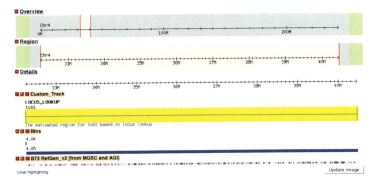

Fig. 15.3 MaizeGDB's Locus Lookup result for *tcb1*, the *teosinte crossing barrier1* gene. This region was defined using method 3 described in the text: i.e., genetically mapped probes that are nearest the input locus were identified, the tool checked whether those probes have known genomic coordinates (working outward until appropriate probes are identified) and finally the region of the genome contained by the identified probes was reported. From *top* to *bottom*, horizontal sections show the following. "Overview" shows the region in the assembly that is the focus of sections below. "Region" and "Details" display zoomed views of the portion of the pseudomolecule from "Overview". "Custom Track" LOCUS_LOOKUP defines the region returned by the Locus Lookup tool. "Bins" reports the sections of the maize genome defined as bins. Each bin is approximately 10 cM in length. "B73 RefGen_v2" shows pieces of BACs that were assembled to form the pseudomolecule. From the output, it is evident that *tcb1* at the end of bin 4.04 or within bin 4.05. Note that each purple mark within the "B73 RefGen_v2" track shows regions of a BAC. Without more work toward fine-mapping *tcb1*, it would not be possible to define dependable molecular markers for this gene

Another tool that springboards off GBrowse has been developed at MaizeGDB. This tool, called Locus Lookup (http://www.maizegdb.org/cgi-bin/locus_lookup.cgi?id=IBM2&locus=; Andorf et al. 2010), enables researchers to locate the region of the genome that should contain their locus of interest, even if the sequence of that locus has not been identified. As stated on all outputs of the tool, Locus Lookup works by:

> (1) checking physical map coordinates to find out whether the locus is already placed. If so, your physically mapped locus is highlighted in red in the region returned. If not, the tool (2) checks the locus record at MaizeGDB to find out if any BACs are known to detect the locus and that BAC is returned within its genomic context. If not, (3) genetically mapped probes that are nearest the input locus are identified, the tool checks whether those probes have known genomic coordinates (working outward until appropriate probes are identified) and finally the region of the genome contained by the identified probes is reported with bounding probes shown in red.

In outputs, the method that produces the result returned is highlighted (i.e., items 1, 2, or 3 in the list above are highlighted to indicate which method determined the region for the locus of interest) and a snapshot of that region of the genome is returned. Clicking on that snapshot drops the user into the MaizeGDB Genome Browser (a GBrowse instance) with the region determined shown as a new track labeled "Custom_Track" (see Fig. 15.3). Expanding on this idea is the Locus Pair Lookup, which can be used along with the loci bounding a QTL of interest to define

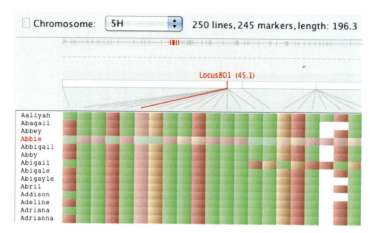

Fig. 15.4 A portion of FlapJack SNP data visualization. Regions of chromosome 5H displayed are *highlighted red*. It is apparent that *Abbie* has a genotype much different that other lines shown in this display

a QTL region. For the Locus Pair Lookup, two loci are entered as input and the region contained by both results (i.e., the left marker's region plus the right marker's region plus the region bounded by the markers) is returned.

Graphical visualization of genotypes is useful, especially for understanding and making use of available diversity. Milne et al. created the FlapJack tool to enable the visualization of diversity data across thousands of lines and datatypes (http://bioinf.scri.ac.uk/flapjack/; Milne et al. 2010). FlapJack is an example of a tool that can be downloaded to a local machine to analyze datasets generated in a particular lab or downloaded from websites rather than residing at a MOD or COD to interact with only public datasets. FlapJack is unique in its ability to show multiple scales of information simultaneously: QTL, heat maps of trait data, and the underlying SNP genotypes for many lines are all shown on the same screen. In Fig. 15.4, the barley line Abbie shows SNPs in a region that among other lines is fairly invariant.

In addition to these specific tools, some MODs and CODs have created "Breeders' Toolboxes" comprised of the set of analysis tools most requested and accessed by researchers. These Breeders' Toolboxes generally limit the set of analysis tools to genetic map interrogation searches, mechanisms to locate DNA-based molecular markers, etc. Examples of websites serving a Breeders' Toolbox include SoyBase, GDR (http://www.rosaceae.org/; Jung et al. 2008), CottonDB (http://cottondb.org/), and SGN.

Metabolic Markers/Functional Genomics

Operationally defined, functional genomics attempts to understand the 'who, what, when, where, and why' of a gene or a group of genes involved in a biological process or pathway. While collecting large quantities of high quality information in the laboratory has become relatively easy in recent years, organizing the resulting information into biologically meaningful representations is still a substantial undertaking and clearly lags behind data generation. Functional genomics software tools enable researchers to organize and analyze large biological datasets in a variety of ways. In the past, these tools have focused on gene expression, which describes what genes are turned on or off. This is changing as additional "omic" technologies have become available that profile both small molecules and proteins. Fortunately, most cellular processes are highly conserved between plants and animals and many are even conserved across all living organisms. This means that publicly available software tools that have been developed for other biological systems can be adapted to suit needs of plant biologists. Examples of tool resources for functional genomics that are adapted to plants include BioCyc (Krummenacker et al. 2005), MapMan (Thimm et al. 2004), and the eFP browser (Winter et al. 2007).

The MapMan (http://mapman.gabipd.org/) software suite allows visualization of a variety of functional genomics datasets (gene expression, protein, enzyme, and metabolite levels) in the context of ~60 well characterized biochemical processes and metabolic pathways. Thus far, MapMan has been used predominately to visualize gene expression data in Arabidopsis with a few limited applications in barley and maize which were due mainly to the limited availability of large data sets for these two important crop species. As noted previously, this limitation has changed dramatically for maize in the past year. A key strength of MapMan is its focus on plant specific processes and it capabilities for cross-species (e.g. maize vs. rice, etc.) comparisons.

BioCyc (http://biocyc.org/) is a collection of over 1,100 Pathway/Genome databases that each describes a single organism and its associated biochemical pathways. Like MapMan, BioCyc allows researchers to display metabolic pathways and visualize a variety of "omics" datasets by projecting gene names or gene models onto curated biochemical pathways. BioCyc has been utilized for various plants including Arabidopsis (AraCyc; http://www.arabidopsis.org/biocyc/), rice (http://pathway.gramene.org/RICE/), soybean (http://www.soybase.org), maize (http://pathway.gramene.org/MAIZE and http://maizecyc.maizegdb.org), and various species in the Solanaceae (http://solgenomics.net/tools/solcyc/index.pl) with other "Cycs" coming online over time. As such, is a good candidate for cross species comparisons.

To enable visualization at the whole plant, organ, or tissue level, the Electronic Fluorescent Pictograph (eFP, http://esc4037-shemp.csb.utoronto.ca/welcome.htm/) browser projects gene expression data onto a series of pictures (pictographs) representing the original plant tissues from which the expression data was derived (Fig. 15.5). Each pictograph is colored according to the level of expression for the gene on of interest. The eFP browser can analyze a single gene across a variety of plant tissues or single tissue undergoing a series of stress treatments (heat, drought, insect and pathogen infection, etc.). An eFP browser can used to display any large-scale data set (gene expression, protein, and metabolite levels) and have been used to display expression data from Arabidopsis, rice, barley, and soybean.

Small molecule datasets are made available at ChEBI (Chemical Entities of Biological Interest; http://www.ebi.ac.uk/chebi/; Degtyarenko et al. 2009), NCBI's

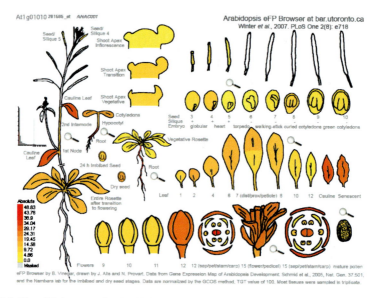

Fig. 15.5 The eFP browser for Arabidopsis. The expression of gene model At1g01010 across various tissues is shown. Tissues with high expression are colored *red*, lower expression is shown as progressively more *yellow*. No expression is *white*

PubChem (http://pubchem.ncbi.nlm.nih.gov/; Wheeler et al. 2006), KEGG (the Kyoto Encyclopedia of Genes and Genomes; http://www.genome.jp/kegg/; Aoki and Kanehisa 2005), the National Cancer Institute's Compound Library (http://cactus.nci.nih.gov/download/nci/), the National Institute of Standards and Technology's MS (http://chemdata.nist.gov/mass-spc/ms-search/), and others. Currently these data resources are more similar to libraries in that they store and allow access to information like MS profiles to which new profiles could be compared rather than focusing on providing analysis tools for small molecules. As the use of metabolic markers for phenotyping increases, it is anticipated that additional tools will become available.

Current Challenges

The various plant reference genome sequences contain high-quality assembled genic regions, but for some species (notably maize), many regions of the genome exist in pieces with some, but not all, positional annotation order and orientation information. At the same time, genomic diversity data are being obtained by the production of whole-genome skim sequencing. These datasets will enable researchers to answer basic research questions as well as to identify genetic variants that can serve as markers to improve crop production. In addition, many researchers are making use of a "Genotype by Sequencing" (GBS) approach that involves sequencing restriction enzyme fragments for reduced genome representations combined with high multiplex. However, these sequencing projects require large-scale data analysis support for two types of discoveries: (1) confirmation of previously identified variation and (2) identification of rare allele variants using modified pipelines such as those developed in the Ware and Buckler laboratories as part of the Maize HapMap project (Gore et al. 2009). GBS analysis pipelines serve as the foundation for downstream population analyses and marker assisted breeding pipelines, which is why CIMMYT (Centro Internacional de Mejoramiento de Maíz y Trigo; International Maize and Wheat Improvement Center) in Mexico plans to use these techniques to evaluate their entire collection over the course of the next few years (http://www.maizegdb.org/cgi-bin/displayrefrecord.cgi?id=1280136). Mechanisms to store GBS data in a way that allows fast querying and downstream visual analytics currently are lacking, and standard methods for storing and interacting with the data are needed.

Another challenge data providers face is longevity of their various resources. This problem is exacerbated by the current funding rates, which are diminished relative to the more recent past. The US sequence repository NCBI (the National Center for Biotechnology Information) is supported by Congressionally mandated funds. For plants, although some resources including MaizeGDB, GrainGenes, SoyBase, and GRIN have Congressionally-mandated base funding provided by the USDA-ARS, most others (e.g., TAIR, GDR, and SGN) do not. Ideas for how to address this problem remain under active discussion, but no clear solutions have been identified or acted upon in a broad manner.

Acknowledgments This work was supported by the USDA-ARS (CJL and MLS) and a grant to the MaizeGDB project by the National Corn Growers Association (JG).

References

Andorf CM, Lawrence CJ, Harper LC, Schaeffer ML, Campbell DA, Sen TZ (2010) The Locus Lookup tool at MaizeGDB: identification of genomic regions in maize by integrating sequence information with physical and genetic maps. Bioinformatics 26(3):434–436

Aoki KF, Kanehisa M (2005) Using the KEGG database resource. Curr Protoc Bioinformatics Chapter 1:Unit 1 12

Ashburner M, Ball CA, Blake JA, Botstein D, Butler H, Cherry JM, Davis AP, Dolinski K, Dwight SS, Eppig JT, Harris MA, Hill DP, Issel-Tarver L, Kasarskis A, Lewis S, Matese JC, Richardson JE, Ringwald M, Rubin GM, Sherlock G (2000) Gene ontology: tool for the unification of biology. The Gene Ontology Consortium. Nat Genet 25(1):25–29

Avraham S, Tung CW, Ilic K, Jaiswal P, Kellogg EA, McCouch S, Pujar A, Reiser L, Rhee SY, Sachs MM, Schaeffer M, Stein L, Stevens P, Vincent L, Zapata F, Ware D (2008) The Plant Ontology Database: a community resource for plant structure and developmental stages controlled vocabulary and annotations. Nucleic Acids Res 36(Database issue):D449–D454

Benson DA, Karsch-Mizrachi I, Lipman DJ, Ostell J, Sayers EW (2011) GenBank. Nucleic Acids Res 39(Database issue):D32–D37

Bombarely A, Menda N, Tecle IY, Buels RM, Strickler S, Fischer-York T, Pujar A, Leto J, Gosselin J, Mueller LA (2011) The Sol Genomics Network (solgenomics.net): growing tomatoes using Perl. Nucleic Acids Res 39(Database issue):D1149–D1155

Burstin J, de Vienne D, Dubreuil P, Damerval C (1994) Molecular markers and protein quantities as genetic descriptors in maize. I. Geneti diversity abmont 21 inbred lines. Theor Appl Genet 89:943–950

Canaran P, Buckler ES, Glaubitz JC, Stein L, Sun Q, Zhao W, Ware D (2008) Panzea: an update on new content and features. Nucleic Acids Res 36(Database issue):D1041–D1043

Chang A, Scheer M, Grote A, Schomburg I, Schomburg D (2009) BRENDA, AMENDA and FRENDA the enzyme information system: new content and tools in 2009. Nucleic Acids Res 37(Database issue):D588–D592

Consoli L, Lefevre A, Zivy M, de Vienne D, Damerval C (2002) QTL analysis of proteome and transcriptome variations for dissecting the genetic architecture of complex traits in maize. Plant Mol Biol 48(5–6):575–581

Degtyarenko K, Hastings J, de Matos P, Ennis M (2009) ChEBI: an open bioinformatics and cheminformatics resource. Curr Protoc Bioinformatics Chapter 14:Unit 14 19

Duvick J, Fu A, Muppirala U, Sabharwal M, Wilkerson MD, Lawrence CJ, Lushbough C, Brendel V (2008) PlantGDB: a resource for comparative plant genomics. Nucleic Acids Res 36(Database issue):D959–D965

Edwards MD, Helentjaris T, Wright S, Stuber CW (1992) Molecular-marker-facilitated investigations of quantitative trait loci in maize. Theor Appl Genet 83(6):765–774

Gendler K, Paulsen T, Napoli C (2008) ChromDB: the chromatin database. Nucleic Acids Res 36(Database issue):D298–D302

Gonzales MD, Archuleta E, Farmer A, Gajendran K, Grant D, Shoemaker R, Beavis WD, Waugh ME (2005) The Legume Information System (LIS): an integrated information resource for comparative legume biology. Nucleic Acids Res 33(Database issue):D660–D665

Gore MA, Chia JM, Elshire RJ, Sun Q, Ersoz ES, Hurwitz BL, Peiffer JA, McMullen MD, Grills GS, Ross-Ibarra J, Ware DH, Buckler ES (2009) A first-generation haplotype map of maize. Science 326(5956):1115–1117

Grant D, Nelson RT, Cannon SB, Shoemaker RC (2010) SoyBase, the USDA-ARS soybean genetics and genomics database. Nucleic Acids Res 38(Database issue):D843–D846

Green JM, Harnsomburana J, Schaeffer ML, Lawrence CJ, Shyu CR (2011) Multi-source and ontology-based retrieval engine for maize mutant phenotypes. Database (Oxford) 2011:bar012

Harris MA, Clark J, Ireland A, Lomax J, Ashburner M, Foulger R, Eilbeck K, Lewis S, Marshall B, Mungall C, Richter J, Rubin GM, Blake JA, Bult C, Dolan M, Drabkin H, Eppig JT, Hill DP, Ni L, Ringwald M, Balakrishnan R, Cherry JM, Christie KR, Costanzo MC, Dwight SS, Engel S, Fisk DG, Hirschman JE, Hong EL, Nash RS, Sethuraman A, Theesfeld CL, Botstein D, Dolinski K, Feierbach B, Berardini T, Mundodi S, Rhee SY, Apweiler R, Barrell D, Camon E, Dimmer E, Lee V, Chisholm R, Gaudet P, Kibbe W, Kishore R, Schwarz EM, Sternberg P, Gwinn M, Hannick L, Wortman J, Berriman M, Wood V, de la Cruz N, Tonellato P, Jaiswal P, Seigfried T, White R (2004) The Gene Ontology (GO) database and informatics resource. Nucleic Acids Res 32(Database issue):D258–D261

Ingvardsen CR, Xing Y, Frei UK, Lubberstedt T (2010) Genetic and physical fine mapping of Scmv2, a potyvirus resistance gene in maize. Theor Appl Genet 120(8):1621–1634

Jung S, Staton M, Lee T, Blenda A, Svancara R, Abbott A, Main D (2008) GDR (Genome Database for Rosaceae): integrated web-database for Rosaceae genomics and genetics data. Nucleic Acids Res 36(Database issue):D1034–D1040

Kaminuma E, Kosuge T, Kodama Y, Aono H, Mashima J, Gojobori T, Sugawara H, Ogasawara O, Takagi T, Okubo K, Nakamura Y (2011) DDBJ progress report. Nucleic Acids Res 39(Database issue):D22–D27

Kanehisa M, Goto S, Kawashima S, Okuno Y, Hattori M (2004) The KEGG resource for deciphering the genome. Nucleic Acids Res 32(Database issue):D277–D280

Krummenacker M, Paley S, Mueller L, Yan T, Karp PD (2005) Querying and computing with BioCyc databases. Bioinformatics 21(16):3454–3455

Ku L, Wei X, Zhang S, Zhang J, Guo S, Chen Y (2011) Cloning and characterization of a putative TAC1 ortholog associated with leaf angle in maize (*Zea mays* L.). PLoS One 6(6):e20621

Kulikova T, Akhtar R, Aldebert P, Althorpe N, Andersson M, Baldwin A, Bates K, Bhattacharyya S, Bower L, Browne P, Castro M, Cochrane G, Duggan K, Eberhardt R, Faruque N, Hoad G, Kanz C, Lee C, Leinonen R, Lin Q, Lombard V, Lopez R, Lorenc D, McWilliam H, Mukherjee G, Nardone F, Pastor MP, Plaister S, Sobhany S, Stoehr P, Vaughan R, Wu D, Zhu W, Apweiler R (2007) EMBL Nucleotide Sequence Database in 2006. Nucleic Acids Res 35(Database issue):D16–D20

Lamesch P, Berardini TZ, Li D, Swarbreck D, Wilks C, Sasidharan R, Muller R, Dreher K, Alexander DL, Garcia-Hernandez M, Karthikeyan AS, Lee CH, Nelson WD, Ploetz L, Singh S, Wensel A, Huala E (2012) The Arabidopsis Information Resource (TAIR): improved gene annotation and new tools. Nucleic Acids Res 40(Database issue):D1202–D1210

Lyons E, Freeling M (2008) How to usefully compare homologous plant genes and chromosomes as DNA sequences. Plant J 53(4):661–673

Lyons E, Pedersen B, Kane J, Alam M, Ming R, Tang H, Wang X, Bowers J, Paterson A, Lisch D, Freeling M (2008) Finding and comparing syntenic regions among Arabidopsis and the outgroups papaya, poplar, and grape: CoGe with rosids. Plant Physiol 148(4):1772–1781

Milne I, Shaw P, Stephen G, Bayer M, Cardle L, Thomas WT, Flavell AJ, Marshall D (2010) Flapjack–graphical genotype visualization. Bioinformatics 26(24):3133–3134

O'Sullivan H (2007) GrainGenes. Methods Mol Biol 406:301–314

Riedelsheimer C, Czedik-Eysenberg A, Grieder C, Lisec J, Technow F, Sulpice R, Altmann T, Stitt M, Willmitzer L, Melchinger AE (2012) Genomic and metabolic prediction of complex heterotic traits in hybrid maize. Nat Genet 44(2):217–220

Salvi S, Tuberosa R, Chiapparino E, Maccaferri M, Veillet S, van Beuningen L, Isaac P, Edwards K, Phillips RL (2002) Toward positional cloning of Vgt1, a QTL controlling the transition from the vegetative to the reproductive phase in maize. Plant Mol Biol 48(5–6):601–613

Sekhon RS, Lin H, Childs KL, Hansey CN, Buell CR, de Leon N, Kaeppler SM (2011) Genome-wide atlas of transcription during maize development. Plant J 66(4):553–563

Sen TZ, Andorf CM, Schaeffer ML, Harper LC, Sparks ME, Duvick J, Brendel VP, Cannon E, Campbell DA, Lawrence CJ (2009) MaizeGDB becomes 'sequence-centric'. Database (Oxford) 2009:bap020

Stein LD, Mungall C, Shu S, Caudy M, Mangone M, Day A, Nickerson E, Stajich JE, Harris TW, Arva A, Lewis S (2002) The generic genome browser: a building block for a model organism system database. Genome Res 12(10):1599–1610

The Uniprot Consortium (2011) Ongoing and future developments at the Universal Protein Resource. Nucleic Acids Res 39(Database issue):D214–D219

Thimm O, Blasing O, Gibon Y, Nagel A, Meyer S, Kruger P, Selbig J, Muller LA, Rhee SY, Stitt M (2004) MAPMAN: a user-driven tool to display genomics data sets onto diagrams of metabolic pathways and other biological processes. Plant J 37(6):914–939

Wheeler DL, Barrett T, Benson DA, Bryant SH, Canese K, Chetvernin V, Church DM, DiCuccio M, Edgar R, Federhen S, Geer LY, Helmberg W, Kapustin Y, Kenton DL, Khovayko O, Lipman DJ, Madden TL, Maglott DR, Ostell J, Pruitt KD, Schuler GD, Schriml LM, Sequeira E, Sherry ST, Sirotkin K, Souvorov A, Starchenko G, Suzek TO, Tatusov R, Tatusova TA, Wagner L, Yaschenko E (2006) Database resources of the National Center for Biotechnology Information. Nucleic Acids Res 34(Database issue):D173–D180

Winter D, Vinegar B, Nahal H, Ammar R, Wilson GV, Provart NJ (2007) An "Electronic Fluorescent Pictograph" browser for exploring and analyzing large-scale biological data sets. PLoS One 2(8):e718

Yilmaz A, Nishiyama MY Jr, Fuentes BG, Souza GM, Janies D, Gray J, Grotewold E (2009) GRASSIUS: a platform for comparative regulatory genomics across the grasses. Plant Physiol 149(1):171–180

Youens-Clark K, Buckler E, Casstevens T, Chen C, Declerck G, Derwent P, Dharmawardhana P, Jaiswal P, Kersey P, Karthikeyan AS, Lu J, McCouch SR, Ren L, Spooner W, Stein JC, Thomason J, Wei S, Ware D (2011) Gramene database in 2010: updates and extensions. Nucleic Acids Res 39(Database issue):D1085–D1094

Part VII
Statistical Considerations

Chapter 16
Prospects and Limitations for Development and Application of Functional Markers in Plants

Everton A. Brenner, William D. Beavis, Jeppe R. Andersen, and Thomas Lübberstedt

Introduction

As genotyping becomes more accessible with faster and cheaper DNA sequencing technologies and single nucleotide polymorphism (SNP) platforms, an ever-increasing number of sequence polymorphisms are revealed in various plant species. In maize, sequence comparison of six recently sequenced inbred lines revealed more than 1,000,000 SNPs and 30,000 insertions/deletions (Indels) in the maize genome (Lai et al. 2010). Knowledge about the effect of polymorphisms on trait variation has also been increasing with results from association and nested association mapping studies. Respective quantitative trait polymorphisms (QTP), sequence polymorphisms associated with phenotypic trait variation, can be converted into Functional Markers (FMs) (Andersen and Lübberstedt 2003). Contrary to linked markers, FMs are derived from polymorphisms causing phenotypic variation.

The development of FMs, as described by Andersen and Lübberstedt (2003), requires the knowledge of functionally characterized loci. Once polymorphic sites are identified within those functional loci, statistical models can be used to test for genotype-phenotype associations. Such association studies provide inferential, i.e., statistical, evidence for correlations, not necessarily reflecting biological causality. Validation of trait-associated polymorphisms, through gene introgression provides biological evidence of functionality. Importantly, functional polymorphisms can be converted into technical assays using, e.g., any of the SNP or insertion/deletion (Indel) detection technologies (Appleby et al. 2009; Gupta et al. 2008).

E.A. Brenner (✉) • W.D. Beavis • T. Lübberstedt
Department of Agronomy, Iowa State University, Agronomy Hall, Ames, IA 50011, USA
e-mail: thomasl@iastate.edu

J.R. Andersen
Department of Genetics and Biotechnology, Research Center Flakkebjerg, University of Aarhus, Aarhus, Denmark

Examples of QTP discovery through association studies include starch biosynthesis (Wilson et al. 2004), cell wall digestibility (Andersen et al. 2008; Brenner et al. 2010; Guillet-Claude et al. 2004), flowering time (Salvi et al. 2007; Thornsberry et al. 2001), carotinoid biosynthesis (Harjes et al. 2008; Palaisa et al. 2003), inflorescence architecture (Bortiri et al. 2006), kernel properties (He et al. 2008; Shi et al. 2008), resistance to bacterial blight (Iyer-Pascuzzi and McCouch 2007), and fruit quality (Costa et al. 2008; Ogundiwin et al. 2008). Thornsberry et al. (2001) pioneered association mapping in plants by developing linkage disequilibrium (LD) mapping and employing this to identify associations between polymorphisms within the Dwarf8 (D8) gene affecting flowering time and plant height. The study was based on 92 diverse maize inbred lines from four populations: Stiff Stalk, non-Stiff Stalks, tropical, and semi-tropical. The association analysis, which was correcting for population structure, identified nine polymorphisms significantly associated with flowering time. Andersen et al. (2005) investigated the applicability of these nine polymorphisms as FMs in an independent set of 71 elite European inbred lines. Ignoring population structure, six of the nine polymorphisms were significantly associated with flowering time, and none with plant height. However, when population structure was considered, only one association between a 2-bp Indel in the promoter region and plant height remained significant, while no association was observed for flowering time. Camus-Kulandaivelu et al. (2006) evaluated a 6 bp Indel identified by Thornsberry et al. (2001) in a larger population consisting of 375 inbred lines and 275 landraces from United States and Europe. This QTP was confirmed to be associated with flowering time under long-day conditions, with different estimated allelic effects for inbreds and landraces.

This example illustrates that availability of qualified candidate genes can facilitate development of informative molecular markers by means of association studies. QTPs may, however, not be consistent across genetic backgrounds and environments. In this review, the challenges in development, estimation of genetic effects for, and application of FMs in plants are discussed.

Power, Precision and Accuracy in QTP Detection

The goal of genetic mapping studies is to identify genomic regions associated with observed phenotypic variation. In plants, linkage mapping started as a great promise to reveal chromosome fragments with higher-than-expected associations with phenotypic variation observed in segregating bi-parental populations. Today, thousands of linkage mapping experiments have been reported (Behn et al. 2004; Blanc et al. 2006; Byrne et al. 1998; Buerstmayr et al. 2003; Pinson et al. 2005; Tang et al. 2000). Identified QTLs and estimated QTL effects, however, have been rarely consistent across and even within populations, and only a minority has been used for cultivar improvement (Bernardo 2008). The reasons for lack of repeatability and application of QTL identified from linkage mapping experiments have been extensively discussed (Beavis 1994; Bernardo 2008; Scho et al. 2004). Most linkage mapping experiments have resulted in inconsistent QTL with overestimated effects mainly due to small population sizes, stringent significance levels, and interactions with different genetic backgrounds and environments (Beavis 1994; Bernardo 2008; Xu 2003a). In addition, the limited number of recombination events accumulated in populations commonly used in linkage mapping experiments (i.e., F2 and backcross) makes it difficult to narrow the associated regions to fewer than several megabases. As a consequence, the identification of causative genes usually requires development of further recombinants of at least 500 individuals for adequate power (Beavis 1994; Lee et al. 2002).

Compared to linkage mapping in families, association mapping in populations can potentially reveal the genetic basis of phenotypic variation with much greater genetic resolution and even identify QTPs. Contrary to linkage mapping, association mapping does not rely on a controlled bi-parental segregating population, but on a collection of lines not necessarily sharing a pedigree, and therefore takes advantage of historical recombination events accumulated among lines. Smaller LD blocks due to accumulated recombination events allow greater genetic resolution, and require, as a consequence, a much higher marker density as compared to linkage mapping within families. In some plant species, like maize, reduced LD combined with large genomes may require hundreds of thousands of molecular markers to adequately cover the genome (Brown et al. 2004; Ching et al. 2002; Flint-Garcia et al. 2005; Hyten et al. 2007; Yu et al. 2008). Evaluating a massive number of markers requires multiple testing corrections to control for false positives, thus decreasing the power of identifying markers associated with phenotypic variation. Reduced power is even more problematic for quantitative traits governed by multiple genes with modest or small phenotypic effects, or alleles with strong phenotypic effect but at low frequencies in the association panel.

A third approach based on the concept of combining LD with linkage mapping has been referred to as Nested Association Mapping (NAM) (Yu et al. 2008). Several NAM populations have been developed in plant species (Guo et al. 2010). Typically these consist of families of Recombinant Inbred Lines (RILs) derived from a sample of inbred lines crossed to a reference inbred line. The relationships among progeny within families are inbred full sibs, while relationships among progeny from different families are half sibs (Bernardo 2002). NAM populations consisting of doubled haploid lines or RILs are "immortalized", meaning that homozygous lines within each family can be evaluated in numerous locations and years without confounding effects of genetic segregation (Nordborg and Weigel 2010).

Yu et al. (2008) developed and released a maize NAM population consisting of 25 families with 200 RILs for each family. Simulations from Guo et al. (2010) suggests that NAM populations similar to the one developed from Yu et al. (2008) have adequate power to accurately and precisely identify additive polymorphisms contributing at least 5% of the variation in the phenotype. Guo et al. (2010) also observed that the resolution and power to detect QTP is maintained even if non-functional alleles are in LD with the causal variant. Two recent studies in the maize NAM populations identified alleles with small effects in association with southern leaf blight resistance and leaf architecture in maize (Kump et al. 2011; Tian et al. 2011). These results demonstrate the potential of nested designs to identify QTPs.

Challenges in Functional Marker Development: LD, Epistasis, Environmental and GxE Effects

Depending upon the sample size, LD, and genetic architecture, the mapping approaches discussed above usually identify genomic intervals associated with phenotypic variation, and FM development requires the identification of the functional variants (SNPs/Indels) within these intervals. The discrimination of functional vs. non-functional variants is often complicated by LD within candidate loci, where non-functional alleles may be associated with phenotypic variation when in LD with functional ones. Varying levels of LD have previously been observed between genes of the phenylpropanoid pathway, decaying within few hundred bps for *CCoAOMT2* and *COMT* (Guillet-Claude et al. 2004; Zein et al. 2007) while spanning more than 3.5 kb at the PAL locus (Andersen et al. 2007). Even in populations with substantial intragenic decay of LD, adjacent polymorphic sites might still be in high or complete LD, leading to an overestimation of SNPs/Indels associated with the investigated phenotype. The identification of causal genetic polymorphisms is a difficult task, and statistical evidences and the biological nature of candidate variants may have to be analyzed mutually in order to discriminate QTPs from closely associated non-causal polymorphisms. SNPs located in coding regions causing non-synonymous non-conservative amino acid changes are more likely to be functional than non-synonymous conservative and synonymous amino acid substitution (Risch 2000). Although SNPs have received more attention in mapping studies, Indels involve larger segments of DNA, and when disrupting or causing frame shifts in coding sequences, are more likely to cause phenotypic variation (i.e. loss of function mutants). Such extreme phenotypes are more likely eliminated or fixed by (natural) selection, and as a result, Indels are usually less frequent in populations as compared to SNPs in genic sequences (Clark et al. 2007; Jones et al. 2009).

Polymorphisms in non-coding regulatory regions are potentially major sources of phenotypic variation when regulating gene expression, while variants in intronic regions may create or delete a splicing site (Talerico and Berget 1990). Salvi et al. (2007) identified a non-coding cis-acting regulatory element located 70 kb upstream of an Ap2-like transcription factor which is involved in flowering time. Clark et al. (2004) and Camus-Kulandaivelu et al. (2008) also identified cis-acting regulatory in regions 60 and 100 kb upstream of the Tb1 and D8 genes, respectively. In effect, the search for QTPs should not be limited to exonic regions, but ideally should also encompass regulatory and intronic regions with potential impact on the investigated trait (Polidoros et al. 2009).

Besides reducing the resolution of association mapping, another LD-related issue is the identification or development of optimal QTP haplotypes when several polymorphisms within the target locus affect the trait of interest. This is a concern especially when favorable QTP alleles for one trait are closely linked to QTP alleles with unfavorable effect on other traits (Chen et al. 2010). If not available in the characterized population, development of optimal QTP allele combinations based on intragenic recombination events might be difficult to achieve, even by use of large populations and intragenic markers. Alternatively, exotic germplasm might provide a source for novel intragenic combinations of QTP alleles. More recently, the use of Zinc finger nucleases (ZFNs) has been proposed as a promising technology to replace alleles by homologous recombination (Shukla et al. 2009). The induction of recombination in defined genomic intervals is, therefore, a promising approach to develop optimal QTP haplotypes even within large LD blocks.

Even after true QTPs have been identified, their transferability might be affected by the composition of populations in different studies, both with regard to allele frequencies at the target locus, and structure of the respective populations. D8 is the only example in plants so far, where the same locus has been studied independently in different experimental populations of inbred lines (Andersen et al. 2005; Camus-Kulandaivelu et al. 2006; Thornsberry et al. 2001). When correcting for population structure, the QTPs identified by Thornsberry et al. (2001) were not significantly associated with flowering time in the study of Andersen et al. (2005) as haplotypes were confounded with population structure in the latter study. Other factors, apart from population structure, with potential impact on the detection of QTPs are epistasis, dominance (so far, association studies in maize were conducted at line per se level), as well as environment and genotype by environment effects.

If the effects of an allele depend on a second allele, either in the same or different loci, the power to detect associations and the accuracy of estimated allelic effects are reduced. Dominance effects cause deviations from additive effects of alleles belonging to the same loci, and simple additive models not accounting for dominance would lead to biased estimation of allelic effects. The relevance of dominance bias for any given trait is directly dependent on the ratio between dominance and additive variances, and it might be reasonably neglected if dominance effects are weak (Hill et al. 2008). In some crop species, the use of RILs or DH lines gives the opportunity to estimate allelic effects free from dominance deviations, permitting more accurate phenotypic predictions of the progeny. In crops evaluated as hybrids, additive effects are still likely the major source of genetic variance among hybrids, but non-estimated dominance effects will probably contribute to phenotypic variation, causing deviations from predicted additive values.

Similarly, epistasis, i.e., the non-additive interaction among alleles at different loci can bias estimates of allelic effects (Cheverud and Routman 1995). Epistasis estimates are often limited by the number of loci included in respective models (Carlborg and Haley 2004). If interacting alleles are not considered or are unknown, it is not possible to model epistatic effects and their consequences in association analysis and FM development. For this reason, if a candidate gene is suspected to interact with other genes, e.g., those belonging to a common genetic network, associations identified for a single gene might be inaccurate and misleading. Numerous mapping studies have detected QTL × QTL epistasis as a statistical feature causing deviation from expected additive effects (Juenger et al. 2005; Yang et al. 2010; Zhang et al. 2008), but only a few studies have investigated gene × gene interaction affecting the phenotypic variation in plant association mapping populations (Li et al. 2010; Manicacci et al. 2009; Stracke et al. 2009).

Mapping experiments often require large population sizes for adequate power to identify QTL and accurately estimate their effects. Collecting phenotypic data across multiple environments, years, and replications is costly and challenging, and accommodating large populations in multiple environments require more efficient experimental designs involving incomplete blocks, e.g., augmented or alpha-lattice designs. Inadequate experimental designs not controlling environmental noise lead to inaccuracy in phenotypic estimation and subsequent identification of QTL and estimation of their effects, even if population sizes are adequate. Control of environmental variation within (with number of plants/plot) and among experimental rows (replications/location) are essential for estimating environment variance within locations. Experiments in multiple locations also account for Genotype x Environment interactions (GxE). Using marginal means across locations might lead to inaccurate associations and estimations of allelic effects, if GxE is significant. In case of weak genetic correlations across environments, association analyses should be conducted on an individual location basis. Clustering environments according to their genetic correlations for all pairwise comparisons across environments (Cooper and DeLacy 1994) is an alternative to classify environments into a smaller number of mega-environments based on their influence on GxE.

In conclusion, the genetic effects of QTPs are background, population, and environment dependent (Fig. 16.1). We propose to employ the term "potential" to describe the presence of a beneficial QTP allele, since this term reflects a certain potential of trait expression and is analogous to the risk concept in human genetic diseases, depending on the genetic effect and penetrance of the respective allele.

In humans, the relative risk of an individual developing a complex disease is estimated by taking into account genetic and non-genetic (i.e.: sex, age, diet, ethnicity, and others) variables. The genetic component of risk assessment is based on odds ratio: the odds of a disease occurring in individuals with a certain allele versus the odds of this disease occurring in individuals without this allele. When more than one gene (marker) is considered, the genetic risk of an individual corresponds to the product of odds ratios of individual alleles (Risch 1990; Wray et al. 2007). The same principle may be applicable in plants. Once lines are genotyped for a FM, breeding values for each genotypic class of this FM can be estimated across lines, environments and years, leading to a normal distribution of breeding values for each genotypic class. These distributions can be further characterize for their "displacement" (Risch 2000), which is defined as the number of standard deviations of the average effect of one homozygous genotypic class in relation to the other. Mendelian alleles with strong phenotypic effects are likely to have larger displacement, while alleles from genes affecting complex inherited traits are likely to have smaller displacements (Fig. 16.2).

Even though the estimation of displacement shows the average effect of one allele in relation to the other, it does not directly measure the likelihood of a genotype to contribute to a desirable phenotype. The "potential" of an allele contributing to a phenotype of interest requires establishment of a threshold separating undesirable from desirable phenotypes (Fig. 16.3). In plant breeding, the threshold may be

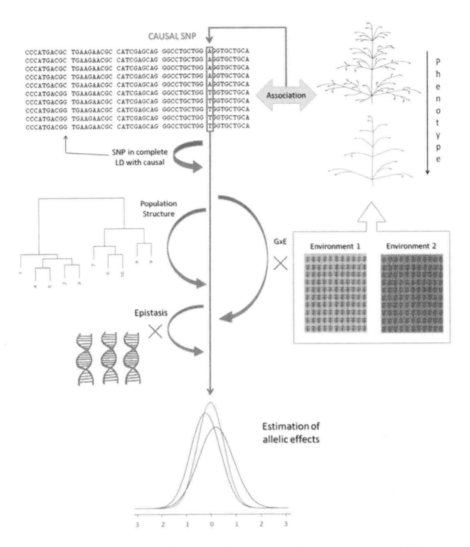

Fig. 16.1 Association between phenotype and genotype, and key components potentially impacting the identification of FMs and the estimation of their effects: *LD, population structure, GxE, epistasis*

defined as a value above the mean phenotype of the best commercial lines (normally used as checks in breeding experimental designs). The estimation of the potential of an allele would be defined as odds of lines passing the threshold with a certain FM genotype versus the odds of lines passing the threshold without this FM genotype.

Fig. 16.2 Potential (*P*) of genotypes from a FM to pass the threshold of a fictional trait. Probabilities vary according to means and distributions. Mendelian traits (*A*) usually display larger displacements and larger differences in probabilities, while differences are subtle in quantitative traits (*C*) (Based on Risch 2000)

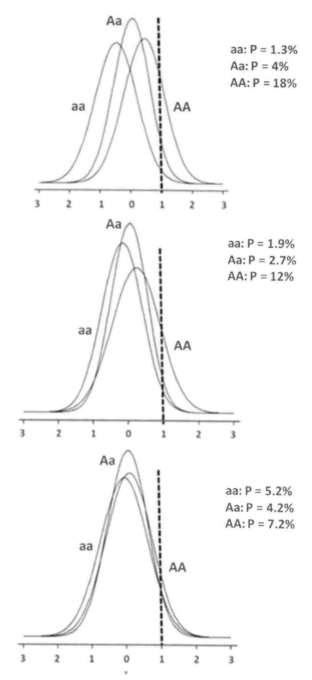

Fig. 16.3 Using the genotype-phenotype relationship to implement selection models. The level of genetic characterization of this relationship may vary from none, such as in genomic selection (*GS*), to highly characterized, such as in functional markers (*FMs*)

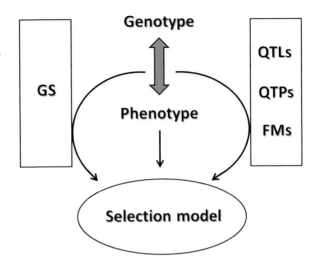

Systematic Collection of Genotypic and Phenotypic Information

Marker and phenotypic data accumulate as mapping experiments designed to investigate genotype-phenotype associations and/or assist breeding decisions are performed. Combining information from different mapping experiments via meta-analysis is a promising approach to enhance statistical power, reduce type 1 errors, and evaluate effects of QTL/QTP in a broader set of genetic backgrounds and environments (Heo et al. 2001). Combining data, however, is not straightforward. The definition of a phenotype and how it is measured is seldom consistent across research groups. Although standard phenotyping techniques are a common practice in the private sector, it would require dialogue among researches in public institutions to reach a consensus. Additionally, detailed description of the experiment including information on germplasm (i.e. maturity), locations, number of replications, check lines, and statistical design, would be required for any researcher to access if an experiment should be considered for meta-analysis or not. Locations and years are not only relevant for estimation of interactions between FM and environments, but the detailed description of an environment (such as maximum/minimum daily temperatures and precipitation) might be important for specific research goals. In drought tolerance studies, for example, temperatures, and amount/distribution of precipitation during different plant development stages is essential information to map drought tolerance genes, given that maize responds differently to water stress in different developmental stages (Barker et al. 2005). With this knowledge, breeders would be able to cluster environments according to relevant climate parameters, and evaluate FM potential in different lines (backgrounds) growing in environments with stress occurring in specific developmental stages.

Besides meta-analysis, comparing mapping outcomes across independent studies is a valuable approach for accessing QTL consistency, refining estimations of QTLs/QTPs effects and narrowing QTL intervals. Pooling results from different mapping experiments is a popular practice in human genetics, where different research groups combine and compared outcomes from large genome wide association studies (GWAS) for common complex diseases, such as type 2 diabetes, coronary disease and breast cancer (McPherson et al. 2007; Scott et al. 2007; Stacey et al. 2007).

In plants, most mapping experiments have consisted of single experiments designed for QTL detection, while less attention has been given to meta and post-hoc analysis. In SoyBase, the USDA-ARS soybean genetics and genomics database, does not routinely archive raw experimental data from QTL experiments (David Grant, pers. comm.). Although the availability of such data could be used to improve QTL mapping, the soybean community has not traditionally done these analyses and so has not made the raw data available. As a consequence the genetic maps in SoyBase are constructed post-hoc by placing the published QTL positions onto a reference genetic map framework using linear scaling between this framework and the reported results. The interpretation of this composite genetic map is complicated by the facts that (1) many of the reported QTL were identified only by analysis of variance (ANOVA) based on a subset of the markers, and (2) the methods and nomenclature used for phenotypic measurements in different experiments are inconsistent. In addition, the choice of QTL mapping procedures has several important ramifications. First, QTL controlled by the same underlying gene can often show different positions due to variation in marker numbers and locations across experiments. Second, the position of the underlying gene cannot be determined relative to the reported QTL. And third, it is not possible to determine the effect of a QTL since the effect and QTL position are confounded if no composite interval mapping is used.

MaizeGDB, the USDA-ARS genetics and genomics database, contains archives for a subset of the raw data for QTL mapping experiments (Carolyn Lawrence, pers. comm.). However, because there is no community agreement on the necessity for submitting such data, it is not possible to do any comprehensive re-analysis of the data due to its incompleteness. As is the case for SoyBase, inconsistencies in trait measurement methodologies and nomenclature along with often imprecise QTL positions impair the ability to compare results between studies.

The current constraints on cross-population comparisons are being addressed by both databases. MIQAS (Minimum Information for QTL and Association Studies, http://miqas.sourceforge.net/) will be adopted to ensure that all QTL studies report a critical minimum of information about a given QTL. In particular, researchers will be encouraged to use interval mapping to identify and position QTL rather than simple ANOVA. Standard ontologies for traits and, where possible, accepted methods used to measure them are being developed.

The Buckler lab has developed standardized phenotyping tools in maize (http://www.maizegenetics.net/phenotyping-tools) which could develop into community standards. Also, all phenotypic data from the NAM population will be made publicly available. This together with the NAM GWAS (http://cbsuapps.tc.cornell.edu/namgwas.aspx) will facilitate unprecedented in silico mapping opportunities. Together these improvements to the public databases will facilitate the re-analysis of combined trait and mapping data from multiple populations. This should produce refined genetic positions for QTL which are needed to identify candidate genes.

Application of Functional Markers

Resulting from the rapid progress in sequencing technology, the genomic sequence of additional maize inbreds beyond B73 is already reality (Lai et al. 2010). Projects like the NAM community approach (Kump et al. 2011; Tian et al. 2011; Yu et al. 2008) will lead to accumulation of further characterized genes and QTPs of agronomic relevance. Thus, the number of functionally characterized polymorphisms in maize as prerequisite for functional marker development will substantially increase over the next decade. FMs might be useful for various steps along the process of cultivar development. These include (1) identification of novel or better alleles (QTPs haplotypes) for characterized genes in exotic germplasm collections, (2) identification of complementary parents for development of new inbreds, (3) description of the "genetic potential" of new inbreds, and (4) variety registration and description. FMs will also be essential to test for negative pleiotropic side-effects. This will in addition lead to a better understanding of the nature of trait correlations, or "pleiotropic" effects described for major genes (Chen and Lübberstedt 2010). Various studies found close genetic correlations between plant height and flowering time. Interestingly, flowering time associated polymorphisms in D8, a gene initially identified by its mutant allele leading to dwarfing, had no effects on plant height (Thornsberry et al. 2001). Similarly, mutant alleles of brown midrib genes in maize were found to affect other agronomic characters, including plant height and biomass yield (Pedersen et al. 2005). However, none of the polymorphisms within the Bm3 gene affecting forage quality affected any of these agronomic traits (Chen et al. 2010). In conclusion, for composition of optimal haplotypes for genes shown to affect one or more traits of interest, multiple traits need to be considered.

It remains to be seen, how FMs will contribute to marker-assisted (recurrent) selection, in particular as compared to genomic selection (GS) procedures based on low cost markers without requirements on their functional characterization (Bernardo and Yu 2007). Although most empirical studies of GS are still limited, accurate estimates of breeding values combined with the possibility for selection of kernels before planting (by seed-chipping) and selection in off-season winter nurseries makes GS very promising for maximizing genetic gain in breeding programs, especially when compared to marker assisted selection based only on markers with statistically significant trait associations (Bernardo and Yu 2007; Heffner et al. 2009; Mayor and Bernardo 2009). GS has been described as brute-force and black box procedure to increase genetic gain (Bernardo and Yu 2007), as selection is based on a large number of markers without prior knowledge of QTL positions or genetic mechanisms involved in phenotypic variation. Markers in LD with favorable QTL receive a large estimated breeding value, even if the QTL is unknown. In GS, lines are selected based on the sum of estimated breeding values of markers across the whole genome, rather than site specific introgression of significant QTL.

Current research on GS is focused on developing statistical methods that incrementally improve the accuracy, i.e., the correlation between predicted and observed breeding values of individuals in a breeding population (de los Campos et al. 2009; Gianola et al. 2006; Habier et al. 2007; Heffner et al. 2009; Kizilkaya et al. 2010; Xu 2003b; Zhong et al. 2009). Alternatively, as functional genomic knowledge increases it seems reasonable to hypothesize that the concept of gene pyramiding could be extended to genome assembly (GA) for polygenic traits. To our knowledge GS has not been compared with gene pyramiding, much less GA. The question is: what criteria should be used to make such a comparison? While genetic gain, or its accuracy component, is a simple criterion, it is not realistic. Actual breeding decisions are based upon multiple breeding objectives, such as maximizing genetic gain, while maintaining genetic diversity throughout the genomes of the breeding population.

Xu et al. (2011) used an operations research approach to address the challenges imposed by varying degrees of LD among favorable functional alleles to assemble a desired phenotype in minimal time while avoiding loss of genetic diversity for other loci in a population. Importantly, using an optimization approach changes the framework for evaluation from a simple criterion of accuracy to the more realistic situation of meeting multiple breeding objectives simultaneously. Hypothetically, GA, based on knowledge of FMs, LD, and genomic diversity should outperform GS for realistic breeding objectives. The likely outcome will be conditional, i.e., depending upon the structure of the breeding population, genetic architecture of the trait, and genome structure we will likely find Pareto Frontiers describing when the hypothesis is true and when it is false.

The question remaining is how GS would benefit from an increasing number of characterized functional genes. Calus et al. (2008) showed that haplotype versus random marker-based GS is more efficient to predict breeding values. It therefore appears likely that marker-multiplexes employed in GS procedures based on previously characterized QTPs are at least superior to random markers in populations with low LD. For populations with high LD, where markers are more likely to be in LD with favorable QTL, prior knowledge of FM might not improve genetic gain in GS (Ødegård et al. 2009). The contribution of FM to the genetic gain in this case, however, will come from an increasing knowledge of allele effects, distributions, and environment/genetic interactions.

Perspective: Future Opportunities

New sequencing platforms have motivated genome sequencing projects in larger populations in different species. In humans, the 1000 Genome project was launched in 2008 as a consortium involving more than 75 universities and companies worldwide. The goal is to sequence genomes and reveal sequence polymorphisms in more than a thousand individuals from different ethnic groups. Another large sequencing initiative is the Genome 10K Project, aiming to sequence the genomes of 10,000 vertebrate species by 2015 (http://genome10k.soe.ucsc.edu/). In plants, the 1001 Genomes Project was initiated in 2008, with the objective of revealing whole-genome sequence variants in 1,001 accessions of *Arabidopsis thaliana*. In maize, seven inbred lines have been resequenced by public institutions in United States and China (Lai et al. 2010; Schnable et al. 2009).

The challenge will be translating this huge amount of genomic information into QTL, QTPs and FMs for crop improvement. In plant breeding, the importance of a marker normally depends on how it predicts the phenotype, and accurate predictions depend on accurate estimation of marker effects based on phenotypic evaluations. Although phenotyping has become more efficient over the years with larger and automated field machinery and hand held computers, field characterization of breeding lines normally requires large allocation of land and labor work. As genotyping costs reduce, phenotyping becomes the major bottleneck in marker assisted breeding. More recently "phenomics", which is using instruments arrays that allows high through-put screening of thousands of lines consistently in short periods of time, has been suggested as the approach that will make phenotyping "catch up" with genomics (Finkel 2009). The use of phenomics, however, will not surrogate field experimentation, and allocation of land and phenotyping labor will still be necessary for major plant breeding traits.

Another challenge associated with FM development is the biological validation of statistically inferred QTPs. Transgenic constructions require time consuming regulations for field evaluations, and are usually vulnerable to position effects, which substantially affect the expression of genes depending on the (random) introgression site in the genome. Backcrossing has been a traditional approach for introgression of moderate number of alleles, but it has the drawback of introgressing unwanted genome from the donor parent by linkage drag. The magnitude of linkage drag can be minimized by selection of recurrent markers flanking the target region. This approach, however, requires larger populations as flanking markers are closer to the target region (Hospital 2001). Recently, ZFN was introduced as a promising technology to assist allele introgression without some of the drawbacks from transgenic and backcross approaches. ZFN promotes recombination in defined chromosome segments, permitting allele introgression without linkage drag with smaller population sizes (Shukla et al. 2009).

Even though phenotyping, validation, and introgression of favorable QTPs are still major drawbacks, identification of candidate QTPs and subsequent FM development are increasingly reported. A number of FMs have already been developed in different plant species (Fan et al. 2009; Ji et al. 2010; Iyer-Pascuzzi and McCouch 2007; Shi et al. 2008; Su et al. 2010; Tommasini et al. 2006). Developing optimal strategies to integrate this increasing knowledge of functionality of genomic regions, and combining this information with phenotypic and GS will be essential to maximize genetic gain. Most likely, FMs will have to be evaluated on a case by cases basis, where their significance to the genetic gain will depend on the populations and environments of individual breeding programs.

References

Andersen JR, Lübberstedt T (2003) Functional markers in plants. Trends Plant Sci 8:554–560

Andersen JR, Schrag T, Melchinger AE, Imad Z, Lübberstedt T (2005) Validation of Dwarf8 polymorphisms associated with flowering time in elite European inbred lines of maize (*Zea mays* L.). Theor Appl Genet 111:206–217

Andersen JR, Zein I, Wenzel G, Krützfeldt B, Eder J, Ouzunova M, Lübberstedt T (2007) High levels of linkage disequilibrium and associations with forage quality at a phenylalanine ammonia-lyase locus in European maize (*Zea mays* L.) inbreds. Theor Appl Genet 114: 307–319

Andersen JR, Zein I, Wenzel G, Darnhofer B, Eder J, Ouzunova M, Lübberstedt T (2008) Characterization of phenylpropanoid pathway genes within European maize (*Zea mays* L.) inbreds. BMC Plant Biol 8:2

Appleby N, Edwards D, Batley J (2009) New technologies for ultra-high throughput genotyping in plants. In: Somers D, Langridge P, Gustafson J (eds) Plant genomics methods and protocols. Humana Press, New York, pp 19–39

Barker T, Campos H, Cooper M, Dolan D, Edmeades GO, Habben J, Schussler J, Wright D, Zinselmeier C (2005) Improving drought tolerance in maize. Plant Breed Rev 25:173–253

Beavis WD (1994) The power and deceit of QTL experiments: lessons from comparative QTL studies. 49th annual corn and sorghum industry research conference, ASTA, Washington, DC, pp 250–266

Behn A, Hartl L, Schweizer G, Wenzel G, Baumer M (2004) QTL mapping for resistance against non-parasitic leaf spots in a spring barley doubled haploid population. Theor Appl Genet 108:1229–1235

Bernardo R (2002) Breeding for quantitative traits in plants. Stemma Press, Woodbury

Bernardo R (2008) Molecular markers and selection for complex traits in plants: learning from the last 20 years. Crop Sci 48:1649–1664

Bernardo R, Yu J (2007) Prospects for genomewide selection for quantitative traits in maize. Crop Sci 47:1082–1090

Blanc G, Charcosset A, Mangin B, Gallais A, Moreau L (2006) Connected populations for detecting quantitative trait loci and testing for epistasis: an application in maize. Theor Appl Genet 113:206–224

Bortiri E, Chuck G, Vollbrecht E, Rocheford T, Martienssen R, Hake S (2006) *ramosa2* encodes a lateral organ boundary domain protein that determines the fate of stem cells in branch meristems of maize. Plant Cell 18:574–585

Brenner EA, Zein I, Chen Y, Andersen JR, Wenzel G, Ouzunova M, Eder J, Darnhofer B, Frei U, Barrière Y, Lübberstedt T (2010) Polymorphisms in O-methyltransferase genes are associated with stover cell wall digestibility in European maize (*Zea mays* L.). BMC Plant Biol 10:27

Brown GR, Gill GP, Kuntz RJ, Langley CH, Neale DB (2004) Nucleotide diversity and linkage disequilibrium in loblolly pine. Proc Nat Acad Sci USA 101:15255–15260

Buerstmayr H, Steiner B, Hartl L, Griesser M, Angerer N, Lengauer D, Miedaner T, Schneider B, Lemmens M (2003) Molecular mapping of QTLs for fusarium head blight resistance in spring wheat. II. Resistance to fungal penetration and spread. Theor Appl Genet 107:503–508

Byrne PF, McMullen MD, Wiseman BR, Snook ME, Musket TA, Theuri JM, Widstrom NW, Coe EH (1998) Maize silk maysin concentration and corn earworm antibiosis: QTLs and genetic mechanisms. Crop Sci 38:461–471

Calus MPL, Meuwissen THE, Roos APW, Veerkamp RF (2008) Accuracy of genomic selection using different methods to define haplotypes. Genetics 178:553–561

Camus-Kulandaivelu L, Veyrieras JB, Madur D, Combes V, Fourmann M, Barraud S, Dubreuil P, Gouesnard B, Manicacci D, Charcosset A (2006) Maize adaptation to temperate climate: relationship between population structure and polymorphism in the Dwarf8 gene. Genetics 172:2449–2463

Camus-Kulandaivelu L, Chevin LM, Tollon-Cordet C, Charcosset A, Manicacci D, Tenaillon MI (2008) Patterns of molecular evolution associated with two selective sweeps in the Tb1-Dwarf8 region in maize. Genetics 180:1107–1121

Carlborg Ö, Haley CS (2004) Epistasis: too often neglected in complex trait studies? Genetics 5:618–625

Chen Y, Lübberstedt T (2010) Molecular basis of trait correlations. Trends Plant Sci 15:454–461

Chen Y, Zein I, Brenner EA, Andersen JR, Landbeck M, Ouzunova M, Lübberstedt T (2010) Polymorphisms in monolignol biosynthetic genes are associated with biomass yield and agronomic traits in European maize (*Zea mays* L.). BMC Plant Biol 10(12)

Cheverud JM, Routman EJ (1995) Epistasis and its contribution to genetic variance components. Genetics 139:1455–1461

Ching A, Caldwell KS, Jung M, Dolan MSO, Smith H, Tingey S, Morgante M, Rafalski AJ (2002) SNP frequency, haplotype structure and linkage disequilibrium in elite maize inbred lines. BMC Genet 14:1–14

Clark RM, Linton E, Messing J, Doebley JF (2004) Pattern of diversity in the genomic region near the maize domestication gene tb1. Proc Nat Acad Sci 101:700–707

Clark TG, Andrew T, Cooper GM, Margulies EH, Mullikin JC, Balding DJ (2007) Functional constraint and small insertions and deletions in the ENCODE regions of the human genome. Genome Biol 8:180

Cooper M, DeLacy IH (1994) Relationships among analytical methods used to study genotypic variation and genotype-by-environment interaction in plant breeding multi-environment experiments. Theor Appl Genet 88:561–572

Costa F, Weg WE, Stella S, Dondini L, Pratesi D, Musacchi S, Sansavini S (2008) Map position and functional allelic diversity of Md-Exp7, a new putative expansin gene associated with fruit softening in apple (Malus × domestica Borkh.) and pear (Pyrus communis). Tree Genet Genomes 4:575–586

de los Campos G, Naya H, Gianola D, Crossa J, Legarra A, Manfredi E, Weigel K, Cotes JM (2009) Predicting quantitative traits with regression models for dense molecular markers and pedigree. Genetics 182:375–385

Fan C, Yu S, Wang C, Xing Y (2009) A causal C-A mutation in the second exon of GS3 highly associated with rice grain length and validated as a functional marker. Theor Appl Genet 118:465–472

Finkel E (2009) With "phenomics" plant scientists hope to shift breeding into overdrive. Science 325:380–381

Flint-Garcia SA, Thuilet AC, Yu J, Pressoir G, Romero SM, Mitchell SE, Doebley J, Kresovich S, Goodman MM, Buckler ES (2005) Maize association population: a high-resolution platform for quantitative trait locus dissection. Plant J 44:1054–1064

Gianola D, Fernando RL, Stella A (2006) Genomic-assisted prediction of genetic value with semiparametric procedures. Genetics 173:1761–1776

Guillet-Claude C, Birolleau-Touchard C, Manicacci D, Fourmann M, Barraud S, Carret V, Martinant JP, Barrière Y (2004) Genetic diversity associated with variation in silage corn digestibility for three O-methyltransferase genes involved in lignin biosynthesis. Theor Appl Genet 110:126–135

Guo B, Sleper DA, Beavis WD (2010) Nested association mapping for identification of functional markers. Genetics 186:373–383

Gupta PK, Rustgi S, Mir RR (2008) Array-based high-throughput DNA markers for crop improvement. Heredity 101:5–18

Habier D, Fernando RL, Dekkers JCM (2007) The impact of genetic relationship information on genome-assisted breeding values. Genetics 177:2389–2397

Harjes CE, Rocheford TR, Bai L, Brutnell TP, Kandianis CB, Sowinski SG, Stapleton AE, Vallabhaneni R, Williams M, Wrutzel ET, Yan J, Buckler ES (2008) Natural genetic variation in lycopene epsilon cyclase tapped for maize biofortification. Science 319:330–333

He XY, Zhang YL, He ZH, Wu YP, XiaoYG MCX, Xia XC (2008) Characterization of phytoene synthase 1 gene (Psy1) located on common wheat chromosome 7A and development of a functional marker. Theor Appl Genet 116:213–221

Heffner EL, Sorrells ME, Jannink J (2009) Genomic selection for crop improvement. Crop Sci 49:1–12

Heo M, Leibel RL, Boyer BB et al (2001) Pooling analysis of genetic data: the association of leptin receptor (LEPR) polymorphisms with variables related to human adiposity. Genetics 159:1163–1178

Hill WG, Goddard ME, Visscher PM (2008) Data and theory point to mainly additive genetic variance for complex traits. PLoS Genet 4:2

Hospital F (2001) Size of donor chromosome segments around introgressed loci and reduction of linkage drag in marker-assisted backcross programs. Genetics 158:1363–1379

Hyten DL, Choi IY, Song Q, Shoemaker RC, Nelson RL, Costa JM, Specht JE, Cregan PB (2007) Highly variable patterns of linkage disequilibrium in multiple soybean populations. Genetics 175:1937–1944

Iyer-Pascuzzi AS, McCouch SR (2007) Functional markers for xa5-mediated resistance in rice (*Oryza sativa*, L.). Mol Breed 19:291–296

Ji Q, Lu J, Chao Q, Zhang Y, Zhang M, Gu M, Xu M (2010) Two sequence alterations, a 136 bp InDel and an A/C polymorphic site, in the S5 locus are associated with spikelet fertility of indica-japonica hybrid in rice. J Genet Genomics 37:57–68

Jones E, Chu WC, Ayele M, Ho J, Bruggeman E, Yourstone K, Rafalski R, Smith OS, McMullen MD, Bezawada C, Warren J, Babayev J, Basu S, Smith S (2009) Development of single nucleotide polymorphism (SNP) markers for use in commercial maize (*Zea mays* L.) germplasm. Mol Breed 24:165–176

Juenger TE, Sen S, Stowe KA, Simms EL (2005) Epistasis and genotype-environment interaction for quantitative trait loci affecting flowering time in *Arabidopsis thaliana*. Genetica 123:87–105

Kizilkaya K, Fernando RL, Garrick DJ (2010) Genomic prediction of simulated multibreed and purebred performance using observed fifty thousand single nucleotide polymorphism genotypes. J Anim Sci 88:544–551

Kump KL, Bradbury PJ, Wisser RJ, Buckler ES, Belcher AR, Oropeza-Rosas MA, Zwonitzer JC, Kresovich S, McMullen MD, Ware D, Balint-Kurti PJ, Holland JB (2011) Genome-wide association study of quantitative resistance to southern leaf blight in the maize nested association mapping population. Nat Genet 43:2

Lai J, Li R, Xu X et al (2010) Genome-wide patterns of genetic variation among elite maize inbred lines. Nature 42:1027–1030

Lee M, Sharopova N, Beavis WD, Grant D, Katt M, Blair D, Hallauer A (2002) Expanding the genetic map of maize with the intermated B73 x Mo17 (IBM) population. Plant Mol Biol 48:453–461

Li L, Paulo MJ, Van Eeuwijk F, Gebhardt C (2010) Statistical epistasis between candidate gene alleles for complex tuber traits in an association mapping population of tetraploid potato. Theor Appl Gen 121:1303–1310

Manicacci D, Camus-Kulandaivelu L, Fourmann M et al (2009) Epistatic interactions between *Opaque2* transcriptional activator and its target gene CyPPDK1 control kernel trait variation in maize. Plant Physiol 150:506–520

Mayor PJ, Bernardo R (2009) Genomewide selection and marker-assisted recurrent selection in doubled haploid versus F2 populations. Crop Sci 49:1719–1725

McPherson R, Persemlidis A, Kavaslar N et al (2007) A common allele on chromosome 9 associated with coronary heart disease. Science 316:1488–1491

Nordborg M, Weigel D (2010) Next-generation genetics in plants. Nature 456:10–13

Ødegård J, Sonesson AK, Yazdi MH, Meuwissen THE (2009) Introgression of a major QTL from an inferior into a superior population using genomic selection. Genet Sel Evol 41:38

Ogundiwin EA, Peace CP, Nicolet CM, Rashbrook VK, Gradziel TM, Bliss FA, Parfitt D, Crisosto CH (2008) Leucoanthocyanidin dioxygenase gene (PpLDOX): a potential functional marker for cold storage browning in peach. Tree Genet Genomes 4:543–554

Palaisa KA, Morgante M, Williams M, Rafalski A (2003) Contrasting effects of selection on sequence diversity and linkage disequilibrium at two phytoene synthase loci. Plant Cell 15:1795–1806

Pedersen JF, Vogel KP, Funnell DL (2005) Impact of reduced lignin on plant fitness. Crop Sci 45:812–819

Pinson SRM, Capdevielle FM, Oard JH (2005) Blight resistance in rice using recombinant inbred lines. Crop Sci 45:503–510

Polidoros AN, Mylona PV, Arnholdt-Schmitt B (2009) Aox gene structure, transcript variation and expression in plants. Physiol Plant 137:342–353

Risch N (1990) Linkage strategies for genetically complex traits. II. The power of affected relative pairs. Am J Hum Genet 46:229–241

Risch N (2000) Searching for genetic determinants in the new millennium. Nature 405:847–856

Salvi S, Sponza G, Morgante M et al (2007) Conserved noncoding genomic sequences associated with a flowering-time quantitative trait locus in maize. Proc Nat Acad Sci 104:11376–11381

Schnable PS, Ware D, Fulton RS et al (2009) The B73 maize genome: complexity, diversity, and dynamics. Science 326:1112–1115

Scho CC, Utz HF, Groh S, Truberg B, Openshaw S, Melchinger AE (2004) Quantitative trait locus mapping based on resampling in a vast maize testcross experiment and its relevance to quantitative genetics for complex traits. Genetics 167:485–498

Scott LJ, Mohlke KL, Bonnycastle LL et al (2007) A genome-wide association study of type 2 diabetes in Finns detects multiple susceptibility variants. Science 316:1341–1345

Shi WYY, Chen S, Xu M (2008) Discovery of a new fragrance allele and the development of functional markers for the breeding of fragrant rice varieties. Mol Breed 22:185–192

Shukla VK, Doyon Y, Miller JC et al (2009) Precise genome modification in the crop species *Zea mays* using zinc-finger nucleases. Nature 459:437–441

Stacey SN, Manolescu A, Sulem P et al (2007) Common variants on chromosomes 2q35 and 16q12 confer susceptibility to estrogen receptor-positive breast cancer. Nat Genet 39:865–869

Stracke S, Haseneyer G, Veyrieras JB, Geiger HH, Sauer S, Graner A, Piepho HP (2009) Association mapping reveals gene action and interactions in the determination of flowering time in barley. Theor Appl Genet 118:259–273

Su Z, Hao C, Wang L, Dong Y, Zhang X (2010) Identification and development of a functional marker of TaGW2 associated with grain weight in bread wheat (*Triticum aestivum* L.). Theor Appl Genet 122:211–223

Talerico M, Berget SM (1990) Effect of 5' splice site mutations on splicing of the preceding intron. Mol Cel Biol 10:6299–6305

Tang D, Wu W, Li W, Lu H, Worland AJ (2000) Mapping of QTLs conferring resistance to bacterial leaf streak in rice. Theor Appl Genet 101:286–291

Thornsberry JM, Goodman MM, Doebley J, Kresovich S, Nielsen D, Buckler ES (2001) Dwarf8 polymorphisms associate with variation in flowering time. Nat Genet 28:286–289

Tian F, Dradbury PJ, Brown PJ et al (2011) Genome-wide association study of leaf architecture in the maize nested association mapping population. Nat Genet 43:159–162

Tommasini L, Yahiaoui N, Srichumpa P, Keller B (2006) Development of functional markers specific for seven *Pm3* resistance alleles and their validation in the bread wheat gene pool. Theor Appl Gen 114:165–175

Wilson LM, Whitt SR, Ibanez AM, Rocheford TR, Goodman MM, Buckler ES (2004) Dissection of maize kernel composition and starch production by candidate gene association. Plant Cell 16:2719–2733

Wray NR, Goddard ME, Visscher PM (2007) Prediction of individual genetic risk to disease from genome-wide association studies. Genome Res 17:1520–1528

Xu S (2003a) Estimating polygenic effects using markers of the entire genome. Genetics 163:789–801

Xu S (2003b) Theoretical basis of the Beavis effect. Genetics 165:2259–2268

Xu P, Wang L, Beavis WD (2011) An optimization approach to gene stacking. Eur J Oper Res 214(1):168–178

Yang X, Guo Y, Yan J, Zhang J, Song T, Rocheford T, Li JS (2010) Major and minor QTL and epistasis contribute to fatty acid compositions and oil concentration in high-oil maize. Theor Appl Genet 120:665–678

Yu J, Holland JB, McMullen MD, Buckler ES (2008) Genetic design and statistical power of nested association mapping in maize. Genetics 178:539–551

Zein I, Wenzel G, Andersen JR, Lübberstedt T (2007) Low level of linkage disequilibrium at the *COMT* (caffeic acid O-methyl transferase) locus in European maize (*Zea mays* L.). Genet Res Crop Evol 54:139–148

Zhang K, Tian J, Zhao L, Wang S (2008) Mapping QTLs with epistatic effects and QTL x environment interactions for plant height using a doubled haploid population in cultivated wheat. J Genet Genomics 35:119–127

Zhong S, Dekkers JCM, Fernando RL, Jannink JL (2009) Factors affecting accuracy from genomic selection in populations derived from multiple inbred lines: a barley case study. Genetics 182:355–364

Part VIII
Applications in Plant Breeding

Chapter 17
Parent Selection – Usefulness and Prediction of Hybrid Performance

Adel H. Abdel-Ghani and T. Lübberstedt

Introduction

Plant breeding programs undergo three main phases including (Schnell 1982; Ceccarelli 2009): (i) generation of genetic variability, (ii) development and selection of the parents of varieties (such as inbreds in case of hybrid breeding), and (iii) testing experimental varieties in multi-season and -location experiments. The most critical challenge to plant breeders is phase I: how to find the best combination of two (or more) parental genotypes to maximize variance within respective breeding populations, and consequently the chance of finding superior transgressive segregants in the segregating population. Schnell and Utz (1975) developed the usefulness concept for line variety development from segregating populations. Usefulness (U_m) for a certain cross is the sum of the population mean (μ_m) of homozygous lines derived from a cross m and the selection response (ΔG_m). The latter is a function of genetic variance among homozygous lines ($\Delta \sigma_{g(m)}$) of a breeding population, heritability within this population (h_m^2), and the standardized selection intensity (i) (Falconer and Mackay 1996). Alternative criteria closely related to U_m are the varietal ability (Wright 1974; Gallais 1979) and the probability of having transgressive segregants among homozygous lines (Jinks and Pooni 1976). Large usefulness of a population will be attractive for plant breeders to increase the chance to find superior recombinants. According to Zhong and Jannink (2007), rather than concentrating on genetic gain within a cross, they suggested focusing on crosses that would generate progenies with higher genotypic value. By giving

---

A.H. Abdel-Ghani (✉)
Department of Plant Production, Faculty of Agriculture, Mu'tah University,
P.O. Box 7, Karak, Jordan
e-mail: abdelghani@mutah.edu.jo

T. Lübberstedt
Department of Agronomy, Iowa State University, 1211 Agronomy Hall, Ames, IA, USA
e-mail: thomasl@iastate.edu

the focus on genotypic value of the cross, they ignored the h_m^2 to obtain what they call a superior progeny value, $S_m = \mu_m + i\Delta\sigma_{g(m)}$. S_m is similar in value to U_m at $h_m^2 = 1$.

Several steps are involved in hybrid seed production, including creation of genetic variability, production of inbred lines by continuous selfing for several generations, testing lines for their combining ability and crossing the best inbred lines to create hybrids. There are two drawbacks facing the selection of the promising line combinations (Souza and Sorrells 1991; Moser and Lee 1994; Melchinger et al. 1998). Selecting the best breeding population is similar to the above mentioned usefulness problem in line breeding programs. The majority of the base populations are usually discarded after preliminary evaluation for *per se* and performance in an "early testing" programs. As inbred lines are typically produced in two opposite heterotic groups, the main challenge in hybrid breeding ultimately is, to identify the best inbred line combination among those two heterotic groups. Presence of 100 inbred lines in each of two heterotic groups would potentially enable production of 10,000 hybrids. Thus, prediction of hybrid performance (HP) and heterosis without having to assess thousands of single-cross hybrids in field trials would reduce the time and efforts required to identify promising inbred combinations substantially.

The ability to predict optimal genotype combinations for different purposes in plant breeding based on molecular-based genetic data would greatly enhance the efficiency of plant breeding programs. Therefore, the objectives of this chapter are to: (i) survey literature aimed at studying the possible relationships between various predictors based on parental information such as genetic distance ($\widehat{GD}$) and midparent value and performance characteristics of their progenies after hybridization such as μ_m and $\Delta\sigma_{g(m)}$, (ii) survey studies aimed at investigating the relationship between $\widehat{GD}$ and HP, and (iii) review the most recent studies using expression profiling techniques and genomic selection based on high-density genetic markers derived from genome sequencing projects in the context of usefulness and HP prediction.

Usefulness of Parent Combinations

The strategies for selection of parent combinations to maximize the chance of finding superior varieties or parents of varieties in respective progeny falls into two categories: methods that assess the value of parents estimated from progeny performance or methods based on parental performance *per se* (Ceccarelli 2009). The first approach is based on progeny test performance. For this purpose, a segregating population is developed (i.e., requiring at least two generations, if starting from inbred parents) to determine the genetic variation for the target trait(s) of the offspring of a particular cross. The usefulness of a parental combination increases with the genetic variation in its offspring. The second category includes selection based on midparental values ($\mu_{\bar{p}}$) and phenotypic distance or $\widehat{GD}$ estimates calculated from morphological or molecular marker data of parental lines. These methods have the advantage that the value of a parental combination can be predicted based parental genotype information, without loss of time for developing and evaluating offspring. To date, prediction of the genetic variance based on parental lines information is still an unsolved problem (Table 17.1). Examples from allogamous crops rather than autogamous crops, because F_1 hybrids in crops such as wheat, oat and barley is rare.

The ability to predict genetic variance in progeny from any cross would be of great benefit to plant breeders in selecting parents and maximizing the usefulness of their offspring. The efficiency of plant breeding programs could be increased if the $\Delta\sigma_{g(m)}$ and h_m^2 of biparental populations could be predicted from genetic or phenotypic parental differences. GD (s) and phenotypic differences between parents have been proposed as a tool for predicting genetic variances of their progenies in order to identify promising parental materials (Souza and Sorrells 1991; Sarawat et al. 1994; Melchinger et al. 1998; Gumber et al. 1999; Utz et al. 2001; Kuczyńska et al. 2007). The number of possible crosses that could be exploited as base populations is excessively large, even if only a moderate number of parent lines is available. Thus, if the potential of crosses could be predicted without producing and testing the crosses or their progenies, plant breeders could concentrate their efforts and resources in testing the most promising populations and developing lines therein. In addition, the availability of a criterion to identify parental combinations that produce larger progeny genetic variances would be useful for quantitative trait locus mapping programs, so that resources could be devoted to crosses with high genetic variance (Bohn et al. 1999).

Prediction of Segregating Population Means and Variance from Mid-Parent Values

It is critical to predict the amount of $\Delta\sigma_{g(m)}$ within crosses, as those parent combinations combining high variances with high μ_m are most promising. However, studies conducted in several crop species including wheat (Bohn et al. 1999; Utz et al. 2001; Kotzamanidis et al. 2008), oat (Souza and Sorrells 1991; Moser and Lee 1994), soybean (Burkhamer 1998), faba bean (Gumber et al. 1999), and maize (Melchinger et al. 1998), showed that the $\mu_{\bar{p}}$ values were good predictors of μ_m for most quantitatively inherited agronomic traits, but $\Delta\sigma_{g(m)}$ for individual crosses cannot reliably be estimated by quantitative-genetic predictors of parental line information (Table 17.1). For quantitative traits with mostly additive inheritance, the $\mu_{\bar{p}}$ is a perfect predictor of μ_m (i.e., population means were not significantly differed from mid-parent means). In a study aimed at predicting μ_m and $\Delta\sigma_{g(m)}$ of winter wheat crosses from parental information by Utz et al. (2001), 30 crosses derived from two sets of five and six winter wheat cultivars were used to predict

Table 17.1 Relationships of various predictors with population mean (μ_m) and genetic variance of segregating population ($\Delta\sigma^2_{g(m)}$)

Crop	Population parameters	Prediction parameters[b]								References
		$\mu_{\bar{p}}$	$|P1-P2|$	$\bar{F}_1$	$\bar{F}_2$	$\sigma^2_{F_2}$	PE	GD	f	
Spring wheat	μ_m	*[a]	—	—	—	—	—	—	—	Busch et al. (1971)
	$\sigma^2_{g(m)}$	—	—	—	—	—	—	—	—	
Winter wheat	μ_m	*	—	—	*	—	—	ns	—	Bohn et al. (1999) and Utz et al. (2001)
	$\sigma^2_{g(m)}$	—	—	—	—	*/ns	ns	ns	ns	
Durum wheat	μ_m	*	—	*	*	—	—	ns	—	Kotzamanidis et al. (2008)
	$\sigma^2_{g(m)}$	*	—	*	*	*	—	ns	—	
Hard red spring wheat	μ_m	—	—	—	—	—	—	—	—	Burkhamer (1998)
	$\sigma^2_{g(m)}$	—	—	—	—	—	—	*/ns	*/ns	
Triticale	μ_m	*	—	—	—	—	—	—	—	Miedaner et al. (2006)
	$\sigma^2_{g(m)}$	—	—	—	—	—	—	—	—	
Oat	μ_m	*	—	—	—	—	—	—	—	Souza and Sorrells (1991) and Moser and Lee (1994)
	$\sigma^2_{g(m)}$	—	—	—	—	—	ns	*/ns	—	
Soybean	μ_m	—	—	—	—	—	ns	*/ns	*/ns	Kisha et al. (1997)
	$\sigma^2_{g(m)}$	—	—	—	—	—	*/ns	—	—	
Faba bean	μ_m	*	*	—	—	—	ns	ns	—	Gumber et al. (1999)
	$\sigma^2_{g(m)}$	*	ns	—	—	—	ns	ns	—	
Maize	μ_m	*	*	*	—	—	—	ns	ns	Melchinger et al. (1998)
	$\sigma^2_{g(m)}$	ns	ns	—	—	—	—	ns	ns	

[a] *, ns, */ns the relation with predictor is existed, not existed and not consistently existed respectively

[b] $\mu_{\bar{p}}$ = midparent means; $|P1-P2|$ = absolute phenotypic difference between two parents; ($\bar{F}_1$) and ($\bar{F}_2$) are means of F1 and F2 generations, respectively; $\sigma^2_{F_2}$ = the genetic variance in F$_2$ generation, PE, $\widehat{GD}$ and f are the euclidean phenotypic distance, genetic distance and coefficient of parentage, respectively

the usefulness of the crosses based on parental information. 44 random F_2-derived F_4 lines ($F_{2:4}$) were tested at two locations in Southern Germany. Recombinant inbred lines (RILs) derived from these crosses were evaluated for heading date, plant height, lodging, thousand kernel weight, grain yield, sedimentation, and grain protein concentration. $\mu_{\bar{p}}$, calculated as the mean yield of the two parents of a cross were highly correlated with the cross mean for all recorded traits except for protein concentration. $\mu_{\bar{p}}$, was generally closely and positively associated with cross means of $F_{2:4}$ families for days to heading, plant height, lodging, kernel weight, and sedimentation (r = 0.90**, 0.90**, 0.76**, 0.79**, 0.74** and 0.71**, respectively). In another study conducted by Miedaner et al. (2006), the $\mu_{\bar{p}}$ was a good predictor of the μ_m for *Fusarium* head blight (FHB) resistance with only two triticale crosses showing significant differences. μ_m and $\mu_{\bar{p}}$ was quite similar in most crosses. In various studies (Souza and Sorrells 1991; Moser and Lee 1994; Burkhamer 1998; Melchinger et al. 1998; Bohn et al. 1999; Gumber et al. 1999; Utz et al. 2001; Kotzamanidis et al. 2008), the usefulness of a cross was mainly influenced by the $\mu_{\bar{p}}$. Therefore, $\mu_{\bar{p}}$ was a fairly a good predictor of the mean performance of crosses μ_m, being in accordance with additive gene effect model (Kearsey and Pooni 1996).

Absolute Difference Between Two Parents and Usefulness

In faba bean, Melchinger (1999) investigated the usefulness of the absolute difference between the phenotypic means of parents (|P1 − P2|) to predict the genetic variance of their $\Delta\sigma^2_{g(m)}$. They found no consistent relationship between |P1 − P2| and $\Delta\sigma^2_{g(m)}$. However, they observed significant association with $\Delta\sigma^2_{g(m)}$ for seed yield and days to heading in one of Mediterranean environments. Similar results were obtained by Melchinger et al. (1998) in maize for the regression analysis between |P1 − P2| and $\Delta\sigma^2_{g(m)}$. The coefficients of determination (R^2) were lower than 0.20 for all traits. According to their findings, the associations between |P1 − P2| and $\Delta\sigma^2_{g(m)}$ were too small to be of predictive value for breeders.

Genetic Similarity ($\widehat{GS}$) and Genetic Distance ($\widehat{GD}$) and Prediction of Progeny Performance

Genetic similarity ($\widehat{GS}$) and $\widehat{GD}$ estimates calculated from molecular marker data were frequently used in plant breeding to categorize individuals, lines and families to establish phylogenetic relationships (Reif et al. 2005). $\widehat{GS}$ and $\widehat{GD}$ measurements were also used to predict μ_m and $\Delta\sigma^2_{g(m)}$ (Bohn et al. 1999; Utz et al. 2001). Estimates for $\widehat{GS}$ among lines/genotypes using DNA markers can be obtained by the formula given by Nei and Li (1979): $G\widehat{S}_{ij} = 2N_{ij}/(N_i + N_j)$, where $G\widehat{S}_{ij}$ is the genetic similarity between two lines/genotypes i and j, N_{ij} is the number of common band in genotype i and j and N_i and N_j are the total number of bands observed for genotypes i and j, respectively. Thus, $G\widehat{S}_{ij}$ is reflecting the percentage of bands in common between two parental lines and may range from 0 (no common bands) to 1 (identical DNA profile of two lines). The Nei genetic distance ($G\widehat{D}_{ij}$) between two genotypes can be calculated as: $G\widehat{D}_{ij} = 1 - G\widehat{S}_{ij}$, which can range from 0, when all bands in two lines identical to 1, when there are no bands in common between two parents. Another similarity coefficient to calculate the genetic relationship among parental lines is the coefficient of parentage (f). f between two genotypes, as defined by Malécot (1948) is the probability that a randomly selected allele at a particular locus of one genotype is the probability that a random allele at a random locus in one genotype is identical by descent to a random allele at the same locus in the other genotype (Cox et al. 1985). For phenotypic data, parental phenotypic differences are usually calculated using Euclidean phenotypic distances ($P\widehat{E}_{ij}$). For calculation of $P\widehat{E}_{ij}$ (Sneath and Sokal 1973), observations for each trait have to be standardized by dividing with the phenotypic standard deviation of the particular trait. The squared difference ($P\widehat{D}$) between parents i and j has to be calculated based on standardized observations for each pairwise comparison of genotypes, applying the formula: $P\widehat{D} = |X_{is} - X_{js}|^2$; where X_{is} and X_{js} are the standardized phenotypic values of parents i and j, respectively, for trait s. Then, the multivariate $P\widehat{E}_{ij}$ between two parents i and j is estimated using the following formula: $\widehat{E}_{ij} = \sqrt{\sum_{s=1}^{t} \tilde{X}_{is} \neq \tilde{X}_{js}}$, where $\tilde{X}_{is}$ and $\tilde{X}_{is}$ are the standardized phenotypic mean values of parents i and j for traits s, and t is the number of traits included in multivariate analysis. In some crop species such as soybean (Manjarrez-Sandoval et al. 1997) and oat (Souza and Sorrells 1989 and 1991), it has been concluded that the best predictor of population variance is f. However, other studies showed positive correlations between $G\widehat{D}$ and f with no association with $\sigma^2_{g(m)}$ (Melchinger et al. 1998; Bohn et al. 1999).

Theoretically, parental phenotypic and genetic differences may be related to $\Delta\sigma^2_{g(m)}$ and h^2_m (Martinez et al. 1983). Burkhamer (1998) consistently found positive correlations between $G\widehat{D}$ determined from Sequence-tagged site (STS) markers and Amplified Fragment Length Polymorphisms (AFLPs) with $\Delta\sigma^2_{g(m)}$ in crosses of hard red spring wheat cultivars. In this study, single trait $\Delta\sigma^2_{g(m)}$ not generally related to $G\widehat{D}$ and f measures (r > 0.5, significant at $P = 0.1$). However, total genetic variance (sum of single trait genetic variance components from standardized data) were significantly correlated with $G\widehat{D}$ and f (r ranged from 0.38 to 0.64, significant at $P = 0.05$ for high correlation coefficient values). Moreover, they concluded that $G\widehat{D}$ based on STS and AFLP markers is not a strong predictor of $\Delta\sigma^2_{g(m)}$ or number of transgressive segregants for a quantitative trait. Utz et al. (2001) investigated the prospects to predict the $\Delta\sigma^2_{g(m)}$ within 30 winter wheat crosses from the differences between the $\mu_{\bar{p}}$ and phenotypic distances based on multivariate analysis of quantitative traits. No significant associations were found between $\Delta\sigma^2_{g(m)}$ and both $P\widehat{D}_{ij}$ and $P\widehat{E}_{ij}$. In a companion study, Bohn et al. (1999) studied the correlations between $\widehat{GD}$ determined from simple sequence repeats (SSRs) and AFLPs with $\Delta\sigma^2_{g(m)}$ values from 30 winter wheat populations. They found consistently poor correlations between $\widehat{GD}_{ij}$ determined from SSR markers and AFLPs with $\Delta\sigma^2_{g(m)}$ in crosses of hard red spring wheat cultivars. Souza and Sorrels (1991) investigated the relation between $\Delta\sigma^2_{g(m)}$ based on quantitatively inherited traits and discreetly inherited morphological and biochemical characters and f in oat. For plant biomass, $\Delta\sigma^2_{g(m)}$ was positively correlated with f. Conversely, for other traits such as grain yield and phenological parameters, crosses between closely related parents produced larger $\Delta\sigma^2_{g(m)}$ than crosses between more distantly parents. These unexpected results were explained by the poor adaptation of one or some parents and their progenies. Moreover, Moser and Lee (1994) found low correlations between f based on RFLP data and $\Delta\sigma^2_{g(m)}$ which they explained by absence of linkage between markers and loci that control the studied morphological characters. In a recent study, Kotzamanidis et al. (2008) examined the effectiveness of six criteria for the prediction of the most promising F_3 populations in durum wheat (*Triticum durum* L.): $\mu_{\bar{p}}$, the F_1, the F_2, the arithmetic mean of F_1 and F_2, and the $\widehat{GD}$ among the parents based on the SSR and RAPD molecular systems. It was concluded that parental pairs with high $\mu_{\bar{p}}$ value and high combined yield ($F_1 + F_2/2$) obtained after evaluation of their F_1 and F_2 at low plant density was the most effective way to predict promising F_3 populations. The poor correlation between $\widehat{GD}$ and $\Delta\sigma^2_{g(m)}$ in selfing populations may indicate that the magnitude of $\widehat{GD}$ (s) of parental lines based on DNA marker information does not provide the required information needed for prediction of segregation variance and consequently parent selection. This could lead to the conclusion that the genetic variability measured at the molecular level may not reflect the genetic variability at the level of agronomic characters.

All in all, several reasons might explain the lack of correlation between parental divergence and $\Delta\sigma^2_{g(m)}$: (i) population size; μ_m and $\Delta\sigma^2_{g(m)}$ tend to be lower in value with small population sizes, (ii) environmental conditions; to increase the repeatability of $\Delta\sigma^2_{g(m)}$, it would be desirable to evaluate more lines per cross in multiple environments (Bohn et al. 1999), and (iii) low LD between the quantitative trait loci (QTL) for quantitative trait and marker loci used to estimate parental divergence. A significant correlation could be expected, if substantial LD exists between the QTL contributing to $\Delta\sigma^2_{g(m)}$ and marker loci used to estimate $\widehat{GS}$. Therefore, markers unlinked to QTL reduce the correlation between $\widehat{GS}$ and $\Delta\sigma^2_{g(m)}$ (Utz et al. 2001; Bohn et al. 1999).

In a simulation study conducted by Zhong and Jannink (2007), $\Delta\sigma_{g(m)}$ prediction was compared under various scenarios such as different values of h^2, marker densities, and number of QTL affecting the traits of interest. While earlier studies (e.g., Moser and Lee 1994; Melchinger et al. 1998; Bohn et al. 1999) emphasized the importance of LD between QTL contributing to $\Delta\sigma_{g(m)}$ and marker loci used to estimate $\widehat{GD}$, absence of LD between QTL and marker loci lead to inaccurate estimation of $\widehat{GD}$ and consequently prediction of $\Delta\sigma^2_{g(m)}$. Zhong and Jannink (2007) came to the following conclusions: (i) the inclusion of all effective QTL markers spaced every 10 cM rather than sparse QTL markers (i.e., markers linked to the trait of interest) every 20 cM in a genome of 10 chromosomes with 100 cM each gave much better estimates of superior progeny values since using sparse markers increase the error in the estimates of marker effects and consequently lead to a poor prediction of $\Delta\sigma_{g(m)}$, (ii) sparse marker spacing and low h^2 (0.1–0.4) reduced the accuracy of $\Delta\sigma_{g(m)}$ estimation due to the reduction of accuracy of marker effects and position estimation. However, including all effective QTL markers, rather than only few markers with significant QTL for traits with low h^2 could lead to improvement of superior progeny values and minimize prediction error in the estimates of the marker effects, and (iii) increasing the population size has a positive effect on the accuracy of a cross's $\Delta\sigma_{g(m)}$ prediction.

The rationale of using high density DNA markers to estimate the $\Delta\sigma^2_{g(m)}$ in a population is, that all genetic effects will be captured if all QTL are in strong LD with the markers, and consequently contribute to the accuracy of $\Delta\sigma^2_{g(m)}$ estimation and lead to an accurate prediction of the usefulness of parental combinations (Zhong and Jannink 2007). The traditional method of marker assisted selection based on QTL mapping is a two steps procedure, localizing the QTL on the chromosomes and estimating their effect using stringent significance thresholds which could lead to the losing of QTL with small effects (Heffner et al. 2009; Jannink et al. 2010; Lorenz et al. 2011). Therefore, a limited fraction of the genetic variation is explained by identified QTL. The idea of "genomic selection, (GS)" proposed by Meuwissen et al. (2001) was to omit significance testing and to use estimates of genetic marker effects. In consequence, dense marker coverage is needed to maximize the number of QTL in LD with the trait of interest, thereby also maximizing the number of QTL whose effects will be captured by markers. According to Meuwissen et al. (2001), GS is developed to accurately predict the breeding value of genotypes with genome-wide marker data as an alternative to marker assisted selection (MAS). Thus, GS attempts to capture the total additive genetic variance with genome-wide marker coverage and effect estimates, while MAS strategies utilize a small number of significant markers for prediction and selection. GS is a three step procedure: (i) marker effects are estimated in a training population by analyzing their phenotypes with high-density marker scores, (ii) breeding values for any genotyped individual in the population are predicted using the marker-effect estimates, and (iii) selection is based on these predictions. In simulation and empirical studies, GS was compared with traditional MAS in agricultural crops. These studies support the superiority of GS for prediction of $\Delta\sigma^2_{g(m)}$ and consequently may lead to a much greater rate of genetic gain (Wong and Bernardo 2008; Zhong et al. 2009; Lorenzana and Bernardo 2009; Heffner et al. 2010). In a simulated maize breeding program,

Bernardo and Yu (2007) found that GS can lead to much higher genetic gains than conventional MAS procedures based on QTL, especially for traits with low h_m^2 and phenotypes that are difficult to record. Discovery of thousands of single nucleotide polymorphisms (SNPs) in genome sequencing projects have provided new opportunities to find genome-wide markers in LD with QTL and to use them for GS in economically important agricultural crops such as soybean, maize, barley, and wheat (reviewed by Lorenz et al. 2011). Thus, genotyping lines based on SNP-based high-density genetic markers with next-generation sequencing methods would greatly improve selection before phenotypic information from the plant or its progeny is available. Therefore, it could be fairly concluded that GS using high-density marker scores should be a promising approach to address deficiencies of traditional MAS, and in consequence could lead to a more accurate prediction of the usefulness of parental combinations to maximize genetic gain from selection in plant breeding programs.

Prediction of Hybrid Performance from Genetic Distances

Field trials to assess HP are laborious, time consuming and expensive. Testing all possible combinations for a large number of inbred lines to select the best inbred combinations is not feasible in a breeding program. Thus, prediction of HP and heterosis based on inbred line information is of great interest for plant breeders to evaluate only a small fraction of available inbred lines in the field. Assessment of $\widehat{GD}$ between inbred lines based on molecular information may be helpful in selecting parental combinations giving the highest HP in plant breeding programs. Heterosis, or hybrid vigor, refers to the phenomenon that the hybrid of two inbred lines shows superior performance to either parent, and is a widely documented phenomenon in diploid organisms that undergo sexual reproduction (Fehr 1993). Heterosis can either be expressed as mid-parent heterosis (MPH), or best-parent heterosis (BPH). MPH applies, if the F_1 is significantly better than the parental mean, while BPH represents cases where the F_1 outperforms the best parent in the cross. Three traditional hypotheses try to explain heterosis: the dominance, overdominance, and epistasis hypothesis (Fehr 1993). According to the dominance hypothesis, superiority of hybrids is caused by complete and partial dominance, due to masking of undesirable recessive alleles from one inbred parent by dominant alleles from the other inbred parent. According to the overdominance hypothesis, hybrid vigor is caused by superior performance of heterozygotes due to overdominance at loci contributing to the trait of interest. The interaction of favorable alleles at different loci (i.e. epistasis) is another classical explanation of hybrid vigor. Frascaroli et al. (2007) mapped QTL involved in the control of heterosis in maize using recombinant inbred lines derived from the heterotic single cross between B73 (Stiff Stalk Synthetic) and H99 (opposite heterotic group Lancaster). For grain yield, 21 QTL were detected and 16 of them showed a marked effect on the expression of heterosis, with the ratio between dominance and additive effects being superior to 1 (heterotic QTL).

The advantage of growing hybrid seed compared to inbred lines comes from heterosis. To obtain hybrids with high HP, elite inbred lines with well documented and consistent phenotypes are crossed. Another factor that is important in hybrid seed production is the combining ability of the parental lines. Although hybrids may produce higher yield than their respective parental lines, it does not necessarily mean that crossing these inbreds will result in the highest yielding hybrids. Combining ability is the way to describe the HP of parental line combinations. Higher combining ability between the parents results in increased HP in the resulting hybrid seed. Hybrids are bred to improve the characteristics of the resulting plants, such as better yield grain yield, plant biomass, increased height and plant vigor (Quinby 1963). One scenario to understand the basis of HP is to analyze the inter-specific variation at a few interacting loci. Variation between closely related species mainly involves either loci with small quantitative effects, or loci conferring no detectable phenotypic effect, known as cryptic variation (Gibson and Dworkin 2004). Gene expression studies have revealed extensive differences between species in both *cis*- and *tran*-regulation across the genome (Fu and Dooner 2002; Hochholdinger and Hoecker 2007). However, the relationship between such variation in gene expression and phenotype has not been extensively explored.

Hybrid seed is produced by cross-pollinating two unrelated male and female genotypes. Several steps are involved in hybrid seed production including production of inbred lines, testing lines for their combining ability and crossing of inbred lines to create hybrids. Inbred parental lines for hybrid seed production in cross pollinated crops are created by either repeated selfing of F_2 plants for six generations or more or by doubled-haploid methods (Fehr 1993). Once inbred lines are developed, breeders have to select a limited number of lines to be used as parents in hybrid seed production. The strategies for selection of inbreds to maximize HP fall into two categories: testing lines for their combining ability or by estimating the $\widehat{GD}$ (s) among inbred lines using DNA markers. In agricultural crops, heterotic groups were established to attain high HP by choosing promising parental combinations. Inter-group hybrids usually have greater parental $\widehat{GD}$ and MPH when compared with intra-group hybrids avoiding testcrosses (Melchinger 1999). The best known examples of heterotic groups in maize are Iowa Stiff Stalk vesus Non-Stiff Stalk in the US Cornbelt and Flint versus Dent in Europe (Duvick et al. 2004). Traditionally, hybrid breeding programs involve development of lines from two heterogenous populations, the examining the performance of a line *per se* and their evaluation for combining ability to select desirable parental combinations for commercial hybrid seed production. Inbred line combinations can be identified and selected mainly based on their predicted General Combining Ability (GCA) and specific combining ability (SCA) effects. As a first step, breeders usually evaluate GCA against one or few testers of the opposite pool. Moreover, testing for GCA enables the breeders to reduce the number of lines that need to be crossed with inbreds to evaluate SCA. The mathematical model for diallel analysis proposed by

Griffing (1956), partitions the total genetic variation into GCA and SCA varicance assigned to parents in hybrids. According to Rojas and Sprague (1952), variance for GCA includes additive variance and additive × additive interaction variance, while that of SCA includes nonadditive portion of total variance arising mainly from dominance variance, additive × additive variance, additive × dominance variance and dominance × dominance variance components. Several researchers have noted that additive variance as estimated by GCA is the primary contributor to total genetic variance in corn (Robinson and Harvey 1955; Lonnquist 1967; Garay et al. 1996; Malik et al. 2004). The relationship between GD and HP has been intensively studied in several agricultural crops and predominantly allogamous crops (Table 17.2). Prediction of HP without having to produce and assess hundreds of single-cross hybrids would reduce the time and effort required to identify promising combinations.

DNA markers have been found to be useful for description or establishment of heterotic groups in various crops, and to assign inbred lines to those groups, including maize, rice sunflower, sorghum, wheat, triticale, and oat (Table 17.2). Subsequently, crosses can be restricted to combinations among divergent groups to maximize HP. The relationship between GD, based on various marker systems, and phenotypic performance of hybrids was not consistent (Table 17.2). Many authors found extremely low non-significant associations between GD and HP (Dhliwayo et al. 2010; Qi et al. 2010b), making the use of these markers unfeasible for prediction of best inbred line combinations. Isozymes were the first molecular marker used in genotype fingerprinting and have been applied to predict HP. Frei et al. (1986) first reported the limitations of GD based on isozyme markers in the search for promising heterotic patterns and groups. They concluded that the correlations between GD and MPH were positive when the parents were closely related, and the level of heterosis declined with increasing GD. The usefulness of isozyme markers was limited due to the small number of isozyme markers available. Other studies based on higher numbers of markers such as restriction fragment length polymorphisms (RFLPs), random amplified polymorphic DNA (RAPDs), AFLPs, inter simple sequence repeats (ISSR), and SSRs, came to the same conclusion (eg., Melchinger et al. 1990; Burstin et al. 1995; Benchimol et al. 2000; Lee et al. 2007; Qi et al. 2010a). With increasing GD, the prediction of HP became more unreliable between line combinations from different heterotic groups. However, results from other studies revealed significant but low to moderate correlations between GD and HP for grain yield in tropical maize (e.g., Reif et al. 2003a, b, c; Betrán et al. 2003, Marsan et al. 1998), rice (Xiao et al. 1996; Tao et al. 2010) and oat (Souza and Sorrells 1991). In conclusion, the success of predicting HP based on GD of parents was at best inconsistent.

The poor prediction of HP from GD could be explained by: (i) low LD between the genes controlling the trait(s) of interest and markers, probably due to a too low genome coverage with the effective markers (Bernardo 1992; Melchinger 1993, 1999; Jordan et al. 2003; Riday et al. 2003), (ii) the limited role of dominant gene action and complementary allele frequencies between the heterotic groups used

Table 17.2 Relationship between $\hat{GD}$ based on molecular markers and General Combining Ability (*GCA*), Specific Combining Ability (*SCA*) using different marker systems in some selected agricultural crops

Crop	$\hat{GD}$	Grain yield GCA	Grain yield SCA	Other agronomic traits GCA	Other agronomic traits SCA	References
Maize	Isozyme	–	ns[a]			Frei et al. (1986)
	RFLPs	–	ns	–	*/ns	Boppenmaier et al. (1992), Boppenmaier et al. (1993), Ajmone et al. (1998), Benchimol et al. (2000) and Betrán et al. (2003)
	AFLPs	*/ns	*/ns	–	*/ns	Ajmone et al. (1998), Legesse et al. (2008), Qi et al. (2010a), Schrag et al. (2010) and Xin et al. (2010)
	RAPD	ns	ns	*/ns	*/ns	Bernardo (1992) and Munhoz et al. (2009)
	SSR	*/ns	*/ns	–	–	Reif et al. (2003a, b, c), Lee et al. (2007) Zheng et al. (2008) and Schrag et al. (2010)
	Expression profiling	–	*	–	–	Frisch et al. (2010)
Rice	RFLPs	–	*/ns	–	*/ns	Saghai-Maroof et al. (1997)
	RAPD	–	ns	–	*/ns	Zhang et al. (2000), Joshi et al. (2001) and Selvaraj et al. (2010)
	SSR	–	*/ns	–	*/ns	Saghaï-Maroof et al. (1997), Ni et al. (2009), Ni et al. (2009) and Tao et al. (2010)
Other crops (Sunflower, sorghum, wheat, triticale and oat)	AFLPs	–	*/ns	–	*/ns	Cheres et al. (2000) and Tames et al. (2006)
	RFLPs	*/ns	–	–	–	Souza and Sorrels (1991), Barbosa-Neto et al. (1996) and Jordan et al. (2003)

[a]*, ns, */ns the relation with predictor is existed, not existed and not consistently existed respectively

in the crosses, which lead to non-significant SCA effects; weak complementing actions of superior dominant alleles from both parental inbred lines at multiple loci over the corresponding favorable alleles might have lead to non-significant SCA effects (Bernardo 1992; Melchinger 1999), (iii) low heritability values of the target traits (as is commonly the case for grain yield) (Lanza et al. 1997; Sun et al. 2004), and (iv) low adaptation of the parent population to the target environment (s) could lead to the low association between $\widehat{GD}$ and HP(Link et al. 1996; Melchinger and Gumber 1998; Reif et al. 2003a, b, c; Dreisigacker et al. 2005). However, in one recent study conducted by Schrag et al. (2010), the usefulness of pedigree information in combination with the covariance between GCA and *per se* performance of parental lines for HP prediction was studied using Best Linear Unbiased Prediction (BLUP) approach proposed by Bernardo (1994). Moreover, marker-based prediction methods using RFLPs and SSR markers were used by building either multiple linear regression (MLR) or 'total effects of associated markers' ('TEAM'), both approaches being described by Schrag et al. (2007). BLUP of GCA and SCA resulted in the highest efficiencies for HP prediction for grain yield and grain dry matter content ($R^2 = 0.6$–0.9), if pedigree and line *per se* data were used. If no pedigree-based data is available, Schrag et al. (2010) concluded that HP for grain yield was more efficiently predicted using molecular markers. Therefore, marker data may substitute pedigree data for the estimation of the genotypic covariance matrix and could be used as reliable predictor of HP.

Transcriptome-Based Distance as a Promising Method to Predict Hybrid Performance

Microarray techniques and Serial Analysis of Gene Expression (SAGE) became a useful tool for exploring gene expression profiling in response to external stimuli such as hormones (De Paepe et al. 2004; Yang et al. 2004) and environmental stresses (Kim and Arnim 2006; Lian et al. 2006), and more recently, to analyze the transcriptome to identify candidate genes related to heterosis (Song et al. 2007; Thiemann et al. 2010). Recent studies on maize using DNA microarrays showed that there was differential expression of <5% and < 10% between maize inbred lines A619 and W23 at different growth stages and in maize inbred lines B73 and Mo17 during seedling stage, respectively (reviewed by Springer and Stupar 2011).

According to Lee et al. (2007), the presence of nonadditive gene action and substantial genome-wide heterozygosity are not required for the expression of heterosis. Gene expression underlying substantial heterosis may not relate to genomic differences based on molecular markers for the trait under consideration (Springer and Stupar 2007). However, although heterosis has been used extensively by breeders to increase the performance of crop plants, our understanding of its molecular basis is still rudimentary (Hochholdinger and Hoecker 2007). The most prominent hypothesis that explain heterosis at the molecular level are violation of genetic colinearity and allele-specific analysis of gene expression in hybrids with *cis* and *trans* influences on gene expression (Hochholdinger and Hoecker 2007). Recently, several studies showed that maize lines significant violation of linearity in maize. Fu and Dooner (2002) found that the *bz* regions of two maize lines (B73 and McC) differ in make-up and location retrotransposon (comprise most of the repetitive DNA in maize) relative to the genes in the *bz* region. In another study aimed at studying the overall genome diversity of among six elite inbred lines (Zheng 58, 5003, 478, 178, Change7-2 and Mol7) by Lai et al. (2010), they identified several hundreds of complete genes that display presence/absence variation among investigated lines indicating a very high level of noncolinearity. Loss of gene colinearity in inbred lines genomes might have only minor quantitative effect on plant performance because these genes might functionally compensated by duplicate copies elsewhere in the genome (Hochholdinger and Hoecker 2007). However, hemizygous complementation of many genes with minor quantitative effects in hybrids might lead to superior performance of F_1 hybrid plants over their parental inbred lines. The presence of hemizygous genes with minor effect could also explain the inbreeding depression after many generations of selfing due to the loss of hemizygous genes (Fu and Dooner 2002; Lia et al. 2010). The combination of *cis*- and *trans*–regulation in allele specific gene expression might lead to significant increase in the HP over the parental lines. However, a gene that is exclusively subjected to *trans*-regulation is expected to provide an equal expression of both alleles in the hybrid, whereas genes exposed to *cis*-regulation will exhibit unequal expression of the two alleles in the hybrid (Hochholdinger and Hoecker 2007).

Recently, the potential utility of the transcriptome-based prediction of HP in hybrid breeding programs has been evaluated and was shown to be promising (Frisch et al. 2010; Thiemann et al. 2010). Changes in RNA expression patterns for seedlings were exploited to assess the genetic relationship between genotypes and consequently to predict HP based on transcriptome-based distances. While, poor correlations between $G\widehat{D}$ based on DNA markers and HP were reported in several agricultural crops, although significant (Melchinger et al. 1990; Burstin et al. 1995; Benchimol et al. 2000; Lee et al. 2007; Qi et al. 2010a). Frisch et al. (2010) found a strong positive correlation between the transcriptome-based distance and HP. Briefly; they analyzed transcription profiles from seedlings of the 21 days parental maize lines of a 7×14 factorial with a 46-k oligonucleotide array to predict the performance 98 hybrid combinations based on the transcriptome-based distances. Five seedlings per entry were pooled for RNA extraction. The maize 46-k array from the maize oligonucleotide array project (http://www.maizearray.org, University of Arizona, USA) that contains 43,381 oligonucleotides (in total 46,128 features) printed on a glass-slide was used for hybridization analyses. The HP of the 98 hybrids was assessed in the field. Multivariate analyses for germplasm grouping showed that the transcriptome-based distances were powerful as other DNA based markers to separate flint from dent inbred lines. Moreover, they compared the efficiency of transcriptome-based distances with $G\widehat{D}$ (s) based on AFLP markers and GCA based on the field data. Prediction of HP with transcriptome-based distances was more precise than those based on DNA markers or GCA. Predictions with the transcriptome-based distances showed greater correlations (r ≈ 0.80) to the HP observed values than predictions with selected AFLP markers (r ranged from 0.06 to 0.13). The close positive significant correlations between the transcriptome-based distances with HP and heterosis may be explained by: (i) the high density of transcriptome loci, which was as a consequence of a high number of differentially expressed genes, indicating good coverage of the genes underlying grain yield, (ii) RNA expression profiling investigates directly the genes, and does not rely on LD between marker alleles and trait of interest, therefore, it is not affected by different linkage phases in different heterotic pools and directly quantifies functional genes between two lines, and (iii) the contribution of additive–additive interactions, which may increase the proportion of phenotypic variance explained by the transcriptome-based distances (Frisch et al. 2010). Moreover and as mentioned earlier, genetic microcolinearity can be violated among inbred lines with high HP which could lead to hemizygous complementation of many genes with minor quantitative effects in hybrids (Fu and Dooner 2002; Lia et al. 2010). This might be also the main reason of the superior performance of hybrids compared to their homozygous parental inbred lines. According to Frisch et al. (2010), transcriptome-based selection is a promising procedure to predict HP in the future. Two main advantages could be attained from RNA expression profiling: (i) enhancing the efficiency of the hybrid breeding program by selecting seedlings directly after inbred line production rather than testing inbred line combinations for many seasons and/or analyzing specific tissues, and (ii) with the reduction in the transcriptome analysis cost in the future, pre-selection at the seedling stage can improve the cost efficiency of hybrid plant breeding programs. It view of high correlations between transcriptome-based distances and HP (r ≈ 0.80), it could be concluded that indirect selection based on transcriptome-based distances has the same efficiency as that of direct selection under field conditions (Frisch et al. 2010).

Conclusions

The ability to predict HP and $\sigma^2_{g(m)}$ based on DNA or non-DNA marker information of respective parents would be highly desirable for plant breeding programs. The earlier DNA marker – based studies were not encouraging and at best inconsistent. However, more recent studies with high marker densities in conjunction with progress in sequencing and marker technologies, based on principles of genomic selection (with different weights of markers), and expression-based biomarkers might open up possibilities, that seemed not viable a decade ago.

Acknowledgements This book chapter was prepared while Dr. Adel Abdel-Ghani was a visiting Fulbright Postdoctoral Fellow and during the sabbatical leave granted to Dr. Adel Abdel-Ghani from Mu'tah University, Jordan during the academic year 2011–2012 at Iowa State University, Ames, USA.

References

Ajmone MP, Castiglioni P, Fusari F, Kuiper M, Motto M (1998) Genetic diversity and its relationship to hybrid performance in maize as revealed by RFLP and AFLP markers. Theor Appl Genet 96:219–227

Barbosa-Neto JF, Sorrells ME, Cisar G (1996) Prediction of heterosis in wheat usingcoefficient of parentage and RFLP-basedestimates of genetic relationship. Genome 39:1142–1149

Benchimol LL, De Souza JR, Garcia AAF, Kono PMS, Mangolin CA, Barbosa AMM, Coelho ASG, De Souza AP (2000) Genetic diversity in tropical maize inbred lines: heterotic group assignment and hybrid performance determined by RFLP markers. Plant Breed 119:491–496

Bernardo R (1992) Relationship between single-cross performance and molecular marker heterozygosity. Theor Appl Genet 83:628–634

Bernardo R (1994) Prediction of maize single-cross performance using RFLPs and information from related hybrids. Crop Sci 34:20–25

Bernardo R, Yu J (2007) Prospects for genome-wide selection for quantitative traits in maize. Crop Sci 47:1082–1090

Betrán FJ, Ribaut JM, Beck D, Gonzalez de Leon D (2003) Genetic diversity, specific combining ability, and heterosis in tropical maize under stress and nonstress environments. Crop Sci 43:797–806

Bohn M, Utz HF, Melchinger AE (1999) Genetic diversity among winter wheat cultivars determined on the basis of RFLPs, AFLPs, and SSRs and their use for predicting progeny variance. Crop Sci 39:228–237

Boppenmaier J, Melchinger AE, Brunklaus-Jung E, Geiger HH, Herrmann RG (1992) Genetic diversity for RFLPs in European maize inbreds: I. Relation to performance of Flint âIJŢ Dent croses for forage traits. Crop Sci 32:895–902

Boppenmaier J, Melchinger AE, Seiltz G, Geiger HH, Herrmann RG (1993) Genetic diversity for RFLPs in European maize inbreds: III. Performance of crosses within versus between heterotic groups for grain traits. Plant Breed 11:217–226

Burkhamer RL (1998) Predicting progeny variance from parental divergence in hard red spring wheat. Crop Sci 38:243–248

Busch RH, Lucken KA, Frohberg RC (1971) F1 hybrids versus random F5 line performance and estimates of genetic effects in spring wheat. 11:357–316

Burstin J, Charcosset A, Barrière Y, Hébert Y, Devienne D, Damerval C (1995) Molecular markers and protein quantities as genetic descriptors in maize. II. Prediction of performance of hybrids for forage traits. Plant Breed 114:427–433

Ceccarelli S (2009) Main stages of a plant breeding program. In: Ceccarelli S, Guimarães EP, Weltzien E (eds) Plant breeding and farmer participation. Food and Agriculture Organization (FAO), Rome, pp 63–74

Cheres MT, Miller JF, Crane JM, Knapp SJ (2000) Genetic distance as a predictor of heterosis and hybrid performance within and between heterotic groups in sunflower. Theor Appl Genet 100:889–894

Cox TS, Lookhart GL, Walker DE, Harrell LG, Albers LD, Rogers DM (1985) Genetic relationships among hard red winter wheat cultivars as evaluated by pedigree analysis and gliadin polyacrylamide gel-electrophoretic patterns. Crop Sci 25:1058–1063

De Paepe A, Vuylsteke M, Van Hummelen P, Zabeau M, Van Der Straeten D (2004) Transcriptional profiling by cDNA AFLP and microarray analysis reveals novel insights into the early response to ethylene in Arabidopsis. Plant J 39:537–559

Dhliwayo T, Pixley K, Menkir A, Warburton M (2010) Combining ability, genetic distances, and heterosis among elite CIMMYT and IITA tropical maize inbred lines. Crop Sci 49:1201–1210

Dreisigacker S, Melchinger AE, Zhang P, Ammar K, Flachenecker C, Hoisington D, Warburton ML (2005) Hybrid performance and heterosis in spring bread wheat, and their relations to SSR-based genetic distances and coefficient of parentage. Euphytica 144:51–59

Duvick DN, Smith JSC, Cooper M (2004) Long-term selection in a commercial hybrid maize breeding program. In: Janick J (ed) Wiley, Engelwood Cliffs. Plant Breed Rev 24:109–51

Falconer DS, Mackay TFC (1996) Introduction to quantitative genetics, 4th edn. Longmans Green, Harlow

Fehr WR (1993) Principles of cultivar development, 1st edn. Macmillian Publishing Compant, New York

Frascaroli E, Cane MA, Landi P, Pea G, Gianfranceschi L, Villa M, Morgante M, Pe ME (2007) Classical genetic and quantitative trait loci analyses of heterosis in a maize hybrid between two elite inbred lines. Genetics 176:625–644

Frei OM, Stuber CW, Goodman MM (1986) Use of allozymes as genetic markers for predicting performance in maize single cross hybrids. Crop Sci 26:37–42

Frisch M, Thiemann A, Tobias JF, Schrag A, Scholten S, Melchinger AE (2010) Transcriptome-based distance measures for grouping of germplasm and prediction of hybrid performance in maize. Theor Appl Genet 120:441–450

Fu H, Dooner HK (2002) Intraspecific violation of genetic colinearity and its implications in maize. Proc Natl Acad Sci U S A 99:9573–9578

Gallais A (1979) The concept of varietal ability in plant breeding. Euphytica 28:811–823

Garay G, Igartua E, Alvarez A (1996) Response to S1 selection in flint and dent synthetic maize populations. Crop Sci 36:1129–1134

Gibson G, Dworkin I (2004) Uncovering cryptic genetic variation. Nat Rev Genet 5:681–690

Griffing B (1956) Concept of general combining ability and specific combining ability to diallele cross system. Aust J Bio Sci 9:463–493

Gumber RK, Schill B, Link W, Kittlitz EV, Melchinger AE (1999) Mean, genetic variance, and usefulness of selfing progenies from intra- and inter-pool crosses in faba beans (*Vicia faba* L.) and their prediction from parental parameters. Theor Appl Genet 98:569–580

Heffner EL, Lorenz AJ, Jannink JL, Sorrells ME (2010) Plant breeding with genomic selection: gain per unit time and cost. Crop Sci 50:1681–1690

Heffner EL, Sorrells ME, Jannink J (2009) Genomic selection for crop improvement. Crop Sci 49:1–12

Hochholdinger F, Hoecker N (2007) Towards the molecular basis of heterosis. Trends Plant Sci 12:427–432

Jannink J, Lorenz AJ, Iwata H (2010) Genomic selection in plant breeding: from theory to practice. Brief Funct Genomics 9:166–177

Jinks JL, Pooni HS (1976) Predicting the properties of recombinant inbred lines derived by single seed descent. Heredity 36:253–266

Jordan DR, Tao Y, Godwin ID, Henzell RG, Cooper M, McIntyre CL (2003) Prediction of hybrid performance in grain sorghum using RFLP markers. Theor Appl Genet 106:559–567

Joshi SP, Bhave SG, Chowdari KV, Apte GS, Dhonukshe BL, Lalitha K, Ranjekar PK, Gupta VS (2001) Use of DNA markers in prediction of hybrid performance and heterosis for a three-line hybrid system in rice. Biochem Genet 39:179–200

Kearsey M, Pooni HS (1996) The genetical analysis of quantitative traits. Chapman & Hall, London

Kim BH, Arnim AG (2006) The early dark-response in Arabidopsis thaliana revealed by cDNA microarray analysis. Plant Mol Biol 60:321–342

Kisha TJ, Sneller CH, Diers BW (1997) Relationship between genetic distance among parents and genetic variance in populations of soybean. Crop Sci 37:1317–1325

Kotzamanidis ST, Lithourgidisb AS, Mavromatisc AG, Chasiotic DI, Roupakias DG (2008) Prediction criteria of promising F$_3$ populations in durum wheat: a comparative study. Field Crop Res 107:257–264

Kuczyńska A, Surma M, Kaczmarek Z, Adamski T (2007) Relationship between phenotypic and genetic diversity of parental genotypes and the frequency of transgression effects in barley (*Hordeum vulgare* L.). Plant Breed 126:361–368

Lai J, Li R, Xu X, Jin W, Xu M (2010) Genome-wide patterns of genetic variation among elite maize inbred lines. Nat Genet 42:1027–1030

Lanza LLB, Souza Júnior CL, Ottoboni LMM, Vieira MLC, de Souza AP (1997) Genetic distance of inbred lines and prediction of maize single-cross performance using RAPD markers. Theor Appl Genet 94:1023–1030

Lee EA, Ash MJ, Good B (2007) Re-examining the relationship between degree of relatedness, genetic effects, and heterosis in maize. Crop Sci 47:629–635

Legesse BW, Myburg AA, Pixley KV, Twumasi-Afriyie S, Botha AM (2008) Relationship between hybrid performance and AFLP based genetic distance in highland maize inbred lines. Euphytica 162:313–323

Lian X, Wang S, Zhang J, Feng Q, Zhang L, Fan D, Li X, Yuan D, Han B, Zhang Q (2006) Expression profiles of 10,422 genes at early stage of low nitrogen stress in rice assayed using a cDNA microarray. Plant Mol Biol 60:617–631

Link W, Schill B, Barbera AC, Cubero JI, Filippetti A, Stringi L, Kittlitz EV, Melchinger AE (1996) Comparison of intra- and inter-pool crosses in fababean (*Viciafaba*L.): I. Hybrid performance and heterosis of crosses in Mediterranean and German environments. Plant Breed 115:352–360

Lonnquist JH (1967) Genetic variability in maize and indicated procedures for its maximum procedures for its maximum utilization. Sciencia y Cultura 19:135–144

Lorenz AJ, Chao S, Asoro FG, Heffner EL, Hayashi T, Iwata H, Smith KP, Sorrells ME, Jannink J (2011) Genomic selection in plant breeding: knowledge and prospects. Adv Agron 110:78–109

Lorenzana RE, Bernardo R (2009) Accuracy of genotypic value predictions for marker-based selection in biparental plant populations. Theor Appl Genet 120:151–161

Malécot G (1948) Les mathématiques de l'hérédité. Masson et Cie, Paris

Malik SI, Malik HN, Minhas NM, Munir M (2004) General and specific combining ability studies in maize diallelcrosses. Int J AgrBiol 6:856–859

Manjarrez-Sandoval P, Carter TE Jr, Webb DM, Burton JW (1997) RFLP genetic similarity estimates and coefficient of parentage as genetic variance predictors for soybean yield. Crop Sci 37:698–703

Marsan AP, Castiglioni P, Fusari F, Kuiper M, Motto M (1998) Genetic diversity and its relationship to hybrid performance in maize, as revealed by RFLP and AFLP markers. Theor Appl Genet 96:219–227

Martinez OJ, Goodman MM, Timothy DH (1983) Measuring racial differentiation in maize using multivariate distance measures standardized by variation in F_2 populations. Crop Sci 23: 775–781

Melchinger AE (1993) Use of RFLP markers for analyses of genetic relationships among breeding materials and prediction of hybrid performance. In: Proceedings of the First International Crop Science Congress

Melchinger AE (1999) Genetic diversity and heterosis. In: Coors JG, Pandey S (eds) The genetics and exploitation of heterosis. ASA, CSSA, and SSSA, Madison, pp 99–118

Melchinger AE, Gumber RK (1998) Overview of heterosis and heterotic groups in agronomic crops. In: Lamkey KR, Staub JE (eds) Concepts and breeding of heterosis in crop plants. CSSA, Madison

Melchinger AE, Lee M, Lamkey KR, Woodman WL (1990) Genetic diversity for restriction fragment length polymorphisms: Relation to estimated genetic effects in maize inbreds. Crop Sci 30:1033–1040

Melchinger AE, Gumber RK, Leipert RB, Vuylsteke M, Kuiper M (1998) Prediction of test cross means and variances among F_3 progenies of F_1 crosses from test cross means and genetic distances of their parents in maize. Theor Appl Genet 96:503–512

Meuwissen THE, Hayes BJ, Goddard ME (2001) Prediction of total genetic value using genome-wide dense marker maps. Genetics 157:1819–1829

Miedaner T, Schneider B, Oettler G (2006) Means and variances for Fusarium head blight resistance of F_2-derived bulks from winter triticale and winter wheat crosses. Euphytica 152:405–411

Moser H, Lee M (1994) RFLP variation and genealogical distance, multivariate distance, heterosis, and genetic variation in oats. TheorAppl Genet 87:947–956

Munhoz REF, Prioli AJ, Amaral Júnior AT, Scapim CA, Simon GA (2009) Genetic distances between popcorn populations based on molecular markers and correlations with heterosis estimates made by diallel analysis of hybrids. Genet Mol Res 8:951–962

Nei M, Li WH (1979) Mathematical model for studying genetic variation in terms of restriction endonucleases. Proc Natl Acad Sci U S A 76:5269–5273

Ni XL, Zhang T, Jiang KF, Yang L, Yang QH, Cao CY, Wen CY, Zheng JK (2009) Correlations between specific combining ability, heterosis and genetic distance in hybrid rice. Yi Chuan 31:849–854

Qi X, Kimatu JN, Li Z, Jiang L, Cui Y, Liu B (2010a) Heterotic analysis using AFLP markers reveals moderate correlations between specific combining ability and genetic distance in maize inbred lines. Afr J Biot 9:1568–1572

Qi X, Li ZH, Jiang LL, Yu XM, Ngezahayo F, Liu B (2010b) Grain-yield heterosis in *Zea mays* L. shows positive correlation with parental difference in CHG methylation. Crop Sci 50: 2338–2346

Quinby JR (1963) Manifestation of hybrid vigor in sorghum. Crop Sci 3:288–291

Reif JC, Melchinger AE, Xia XC, Warburton ML, Hoisington DA, Vasal SK, Beck D, Bohn M, Frisch M (2003a) Use of SSRs for establishing heterotic groups in subtropical maize. Theor Appl Genet 83:628–634

Reif JC, Melchinger AE, Xia XC, Warburton ML, Hoisington DA, Vasal SK, Srinivasan G, Bohn M, Frisch M (2003b) Genetic distance based on simple sequence repeats and heterosis in tropical maize populations. Crop Sci 43:1275–1282

Reif JC, Melchinger AE, Xia XC, Warburton ML (2003c) Use of SSRs for establishing heterotic groups in subtropical maize. Theor Appl Genet 107:947–957

Reif JC, Melchinger AE, Frisch M (2005) Genetical and mathematical properties of similarity and dissimilarity coefficients applied in plant breeding and seed bank management. Crop Sci 41:1–7

Riday H, Brummer EC, Campbell TA, Luth D, Cazcarro PM (2003) Comparisons of genetic and morphological distance with heterosis between *Medicago sativa* subsp. *sativa* and subsp. *falcata*. Euphytica 131:37–45

Robinson HF, Harvey PH (1955) Genetic variances in open-pollinated crops varieties of corn. Genetics 40:45–60

Rojas BA, Sprague GF (1952) A comparison of variance components in corn yield trials: III. General and specific combining ability and their interaction with locations and years. Agron J 44:462–466

Saghai-Maroof MA, Yang GP, Zhang Q, Gravois KA (1997) Correlation between molecular marker distance and hybrid performance in U.S. Southern long grain rice. Crop Sci 37:145–150

Sarawat P, Stoddard FL, Marshall DR, Ali SM (1994) Heterosis for yield and related characters in pea. Euphytica 80:39–48

Schnell FW (1982) A synoptic study of the methods and categories of plant breeding. Z Pflanzenzücht 89:1–18

Schnell FW, Utz HF (1975) F_1-Leistung und Elternwahl in der Züchtung von Selbstbefruchtern. Berichtüber die Arbeitstagung der Vereinigungösterreichischer. Z Pflanzenzüchter 243–248

Schrag TA, Maurer HP, Melchinger AE, Piepho H-P, Peleman J, Frisch M (2007) Prediction of single-cross hybrid performance in maize using haplotype blocks associated with QTL for grain yield. Theor Appl Genet 114:1345–1355

Schrag TA, Möhring J, Kusterer B, Melchinger AE, Dhillon BS, Piepho H, Frisch M (2010) Prediction of hybrid performance in maize using molecular markers and joint analyses of hybrids and parental inbreds. Theor Appl Genet 120:451–461

Selvaraj I, Nagarajan P, Thiyagarajan K, Bharathi M (2010) Predicting the relationship between molecular marker heterozygosity and hybrid performance using RAPD markers in rice (*Oryza sativa* L.). Afr J Biot 9:7641–7653

Sneath PHA, Sokal RR (1973) Numerical taxonomy. W H Freeman & Co, San Francisco, 573 pp

Song S, Qu H, Chen C, Hu S, Yu J (2007) Differential gene expression in an elite hybrid rice cultivar (*Oryzasativa*, L) and its parental lines based on SAGE data. Plant Biol 7:1–15

Souza E, Sorrells ME (1989) Pedigree analysis of North American oat cultivars released from 1951 to 1985. Crop Sci 29:595–601

Souza E, Sorrells ME (1991) Relationships among 70 North American oat germplasms. I. Cluster analysis using quantitative characters. Crop Sci 31:599–605

Springer NM, Stupar RM (2007) Allelic variation and heterosis in maize: how do two halves make more than a whole? Genome Res 17:264–275

Springer NM, Stupar RM (2011) Allelic variation and heterosis in maize: how do two halves make more than a whole? Genome Res 17:264–275

Sun GL, William M, Liu J, Kasha KJ, Pauls KP (2004) Microsatellite and RAPD polymorphisms in Ontario corn hybrids are related to the commercial sources and maturity ratings. Mol Breed 7:13–24

Tams SH, Bauer E, Oettler G, Melchinger AE, Schön CC (2006) Prospects for hybrid breeding in winter triticale: II. Relationship between parental genetic distance and specific combining ability. Plant Breed 125:331–336

Tao Z, Xian-lin N, Kai-Feng J, Qianhua Y, Li Y, Xian-Qi W, Yingjiang C, Jiakui Z (2010) Correlation between heterosis and genetic distance based on molecular markers of functional genes in rice. Rice Sci 17:288–295

Thiemann A, Fu J, Schrag TA, Melchinger AE, Frisch M, Scholten S (2010) Correlation between parental transcriptome and field data for the characterization of heterosis in *Zea mays* L. Theor Appl Genet 120:401–413

Utz HF, Bohn M, Melchinger AE (2001) Predicting progeny means and variances of winter wheat crosses from phenotypic values of their parents. Crop Sci 41:1470–1478

Wong CK, Bernardo R (2008) Genomewide selection in oil palm: increasing selection gain per unit time and cost with small populations. Theor Appl Genet 116:815–824

Wright AJ (1974) A genetic theory of general varietal ability for diploid crops. Theor Appl Genet 45:163–169

Xiao J, Li J, Yuan L, McCouch SR, Tanksley SD (1996) Genetic diversity and its relationships to hybrid performance and heterosis in rice as revealed by PCR-based markers. Theor Appl Genet 92:637–643

Xin Q, Kimatu JN, Li Z, Jiang L, Cui Y, Liu B (2010) Heterotic analysis using AFLP markers reveals moderate correlations between specific combining ability and genetic distance in maize inbred lines. Afr J Biotechnol 9:1568–1572

Yang GX, Jan A, Shen SH, Yazaki J, Ishikawa M, Shimatani Z, Kishimoto N, Kikuchi S, Matsumoto H, Komatsu S (2004) Microarray analysis of brassinosteroids- and gibberellin regulated gene expression in rice seedlings. Mol Genet Genomics 271:468–478

Zhang PJ, Cai HW, Li HC, Yang LS, Bai YS, Hu XM, Xu CW (2000) RAPD molecular markers of rice genetic distance and its relationship with heterosis. J Anhui Agric Univ 28:697–700 (in Chinese with English abstract)

Zheng D, Van K, Wang L, Lee S (2008) Molecular genetic distance and hybrid performance between Chinese and American maize (*Zea mays* L.) inbreds. Aust J Agri Res 59:1010–1020

Zhong S, Jannink J (2007) Using quantitative trait loci results to discriminate among crosses on the basis of their progeny mean and variance. Genetics 177:567–576

Zhong S, Dekkers JCM, Fernando RL, Jannink JL (2009) Factors affecting accuracy from genomic selection in populations derived from multiple inbred lines: a barley case study. Genetics 182:355–364

Chapter 18
Variety Protection and Plant Breeders' Rights in the 'DNA Era'

Huw Jones, Carol Norris, James Cockram, and David Lee

The development of new crop varieties offers potential benefits, in terms of yield to growers, and in quality improvements to end users. A new variety represents a considerable investment by plant breeders and this can be sustained by commercial returns. A robust system to protect a new variety, and thus the plant breeders' intellectual property, is part of the infrastructure needed to promote the flow of new varieties. Here we describe current plant variety protection systems and discuss how DNA based markers may be used within those legal and administrative provisions.

New Plant Varieties

A variety is a taxonomic subdivision of a species consisting of naturally occurring or selectively bred populations or individuals that differ from the remainder of the species in certain minor characteristics. A fuller definition of a variety is given by The International Union for the Protection of New Varieties of Plants (UPOV) as

> "variety" means a plant grouping within a single botanical taxon of the lowest known rank, which grouping, irrespective of whether the conditions for the grant of a breeder's right are fully met, can be defined by the expression of the characteristics resulting from a given genotype or combination of genotypes, distinguished from any other plant grouping by the

---

H. Jones (✉) • J. Cockram • D. Lee
John Bingham Laboratory, National Institute of Agricultural Botany (NIAB), Cambridge, UK
e-mail: huw.jones@niab.com; James.cockram@niab.com; David.lee@niab.com

C. Norris
Bayer CropScience Limited, 230 Cambridge Science Park, Milton Road, Cambridge, CB4 0WB, United Kingdom

Agricultural Crop Characterisation, National Institute of Agricultural Botany (NIAB), Cambridge, UK
e-mail: spanglesownhat@gmail.com

expression of at least one of the said characteristics and considered as a unit with regard to its suitability for being propagated unchanged; International Convention for the Protection of New Varieties of Plants Article 1 (vi) (UPOV 1991)

Early attempts to produce 'improved varieties' were undertaken by farmers who selected and propagated superior plants from a varied crop. A farmer would select plants that showed resistance to pathogens or plants that yielded more grain per head than others. Planting stock could be improved by this process with each generation and the seed trade was initially established as a result of individual farmers passing improved stock to their neighbours and later selling the seed on a wider scale. More systematic plant breeding by selection was established by the end of the nineteenth century. By the twentieth century, following the discovery of Mendel's laws of heredity, some breeders started to employ deliberate cross-pollination techniques in a much more scientific way to aid selection. This directed breeding produced better varieties, not only by individual farmers, but by private industry and government sponsored institutes.

New varieties of plants, with improved yield, disease resistance and quality traits, improve agricultural productivity for a growing global population. The conventional breeding processes take many years, so it is important that plant breeders have a mechanism in place that ensures they receive a return on the investments of time, resources, intellectual property and money that are required to produce a new variety. There are various ways to do this including plant patents and plant variety protection (PVP). The International Union for the Protection of New Varieties of Plants (UPOV) state, in their mission statement:

> It is, therefore, important to provide an effective system of plant variety protection, with the aim of encouraging the development of new varieties of plants, for the benefit of society. (UPOV 2011a)

UPOV is an intergovernmental organization with headquarters in Geneva, whose system of PVP is intended to encourage innovation in the field of plant breeding. The origin of Intellectual Property Protection (IPP) for agricultural crop varieties goes back to the late nineteenth century with the growth of the European seed trade and development of breeders' associations and the enactment of the Plant Patent Act in the USA in 1930. The Plant Patent Act only covered asexually propagated plants and not the major grain species. The pressure to provide PVP on a global scale finally led to the formation of The International Union for the Protection of New Varieties of Plants, known as UPOV, in 1961.

The UPOV Convention

The World Trade Organization's Agreement on Trade-Related Aspects of Intellectual Property Rights (TRIPs) requires member states to provide protection for plant varieties either by patents or by an effective stand-alone system, or a combination of the two (World Trade Organization: Agreement on Trade-Related

Aspects of Intellectual Property Rights 1994). Most countries meet this requirement through UPOV Convention-compliant legislation. The Convention requires member countries to provide an intellectual property right specifically for plant varieties, and membership is voluntary. Plant Breeders' Rights (PBR) schemes are stand-alone systems for intellectual property protection, applicable to newly developed varieties of all agricultural and horticultural species.

> The UPOV Convention provides a *sui generis* form of intellectual property protection which has been specifically adapted for the process of plant breeding and has been developed with the aim of encouraging breeders to develop new varieties of plants. (UPOV 2011b).

The UPOV Convention ensures that member states acknowledge the achievements of breeders of new plant varieties, whilst at the same time allowing access to plant varieties for breeding purposes.

UPOV has been established by the International Convention for the Protection of New Varieties of Plants (the UPOV Convention), which was signed in Paris in 1961. The Convention entered into force in 1968, being initially ratified by the United Kingdom, Netherlands and Germany, and was revised in Geneva in 1972, 1978 and 1991. The UPOV Convention currently (2012) has 70 signatories. The 1991 Act entered into force on April 24, 1998 and states and certain intergovernmental organizations wanting to accede to the UPOV Convention have laws on plant variety protection in line with the 1991 Act of the Convention.

The purpose of the UPOV Convention is to ensure that the members of the Union acknowledge the achievements of breeders of new varieties of plants, by granting to them an intellectual property right, on the basis of a set of clearly defined principles. To be eligible for protection, varieties have to be (i) new in the sense that they must not have been commercialized prior to certain dates established by reference to the date of the application for protection (ii) distinct (D), distinct from existing, commonly known varieties, (iii) uniform (U), it is sufficiently uniform in its relevant characteristics and (iv) stable (S), if its relevant characteristics remain unchanged after repeated propagation (UPOV Convention Articles 5–9 (UPOV 1991)). The system for establishing distinctness, uniformity and stability is known as DUS testing.

The UPOV Convention defines acts concerning propagating material in relation to which the holder's authorization is required. Exceptionally, but only where the holder has had no reasonable opportunity to exercise his right in relation to the propagating material, his authorization may be required in relation to any of the specified acts done with harvested material of the variety.

Like all intellectual property rights, PVP grants rights for a limited period of time (not less than 20 years from the date of grant and not less than 25 years for trees and vines), at the end of which varieties protected by them pass into the public domain and the varieties are free to be used without permission from the breeder. The rights are also subject to controls, in the public interest, against any possible abuse. It is also important to note that the authorization of the holder of a plant breeder's right is not required for the use of their variety for private and non-commercial purposes, for research purposes, or for use in the breeding of further new varieties.

At the time of writing there are 70 member states represented within UPOV from all over the world. By becoming a member of UPOV, a state or an intergovernmental organization signals its intention to protect plant breeders on the basis of principles that have gained world-wide recognition and support. It offers its own plant breeders the possibility of obtaining protection in the territories of other members and provides an incentive to foreign breeders to invest in plant breeding and the release of new varieties on its own territory. Plant Variety Protection based on the UPOV Convention also allows breeders to use material of protected varieties for further breeding (breeder's exemption), whereas for plant material covered by a patent, such an exemption does not exist.

The examination of plant varieties for protection purposes involves close co-operation between members. It is based on arrangements whereby one member can conduct tests on behalf of others or whereby one member accepts the test results produced by others as the basis for its decision on the grant of a breeder's right. Through such arrangements, members are able to reduce the cost of operating their protection systems and breeders are able to obtain protection in several territories at relatively low cost.

Community Plant Variety Office (CPVO)

Within the European Community a scheme operates whereby intellectual property rights, valid throughout the Community, may be granted for plant varieties. The system is based on the 1991 Act of the UPOV Convention and is managed by the Community Plant Variety Office (CPVO). The CPVO has been operating since 1995 and is a self-financing Community body. Applications for plant variety rights are decided on the basis of a formal DUS examination and rights are valid in all 27 member states of the EU, for 25 or 30 years, depending on the species. CPVO is itself a member of the UPOV Convention and plays a prominent role in promoting awareness of PVP and encouraging the development of enforcement tools. In 2004, ten new states became members of the European Union followed by another two in 2007, which meant that additional DUS testing capacity had to be integrated into the existing system, provoking a strategic discussion about how to maintain quality within the scheme. The outcome of this was that providers of DUS testing (examination offices) are now regulated by CPVO for the quality and competence of their service by regular audits. Examination offices meeting the quality requirements become 'entrusted' by CPVO to carry out variety tests on their behalf. The advantage of this system for breeders is that they are able to substantially reduce the number of applications required for wider protection within the European Community. The centralised administrative procedure also means applications for breeders outside the European Community who wish to apply for protection in more than one country have been simplified.

Patents and Farmers' Rights

Although most countries meet the World Trade Organisation Trade-Related Aspects of Intellectual Property Rights (TRIPs) requirement to provide protection of plant varieties through UPOV Convention-compliant legislation, patents and Farmers' Rights are also used as a means of protection in some countries.

In some jurisdictions, including the United States, Australia and Europe Union, plants can be covered by patent claims provided that they meet all the necessary standards existing in that country for patentability. In the United States, any organism that is the subject of human intervention (such as plant breeding) can be patented. The United States patent laws extend to plants produced by sexual or asexual reproduction and to plant parts such as seeds and tissue culture. In Australia, the Patent Act (1990) allows all technologies to be patented (except "human beings and the biological processes for their production") provided that there is an 'invention', defined as "an innovative idea which provides a practical solution to a technological problem". This Act applies to biotechnology and Genetically Modified Organisms (GMOs) where the genetic modification rather than the plant itself is the subject of the patent. It is possible to apply for patent grants in nearly all European countries at the European Patent Office and the claims are registered in the national patent offices. The European Patent Convention governs the granting of patents within Europe and the introduction of Directive 98/44/EC (effective in all European Union Member States from 30th July 2000) aims to harmonize protection for biotechnological inventions (including plant protection) amongst the European Union members (Biological Innovation for Open Society Tutorial; "Can IP Rights Protect Plants?": http://www.patentlens.net/daisy/bios/1234.html).

In an attempt to protect the Intellectual Property Rights (IPRs) of non-commercial breeders and maintainers, the International Treaty on Plant Genetic Resources for Food and Agriculture (ITPGRFA) came into being in November 2001, giving recognition to Farmers' Rights. Farmers' Rights enable farmers and growers to maintain and develop crop genetic resources and their rights to use, exchange and sell farm-saved seed. There are two general perceptions of Farmers' Rights which are "the ownership approach" and the "stewardship approach" (Andersen 2006). In the ownership approach

> the right of farmers is rewarded for genetic material obtained from their fields and used in commercial varieties and/or protected through intellectual property rights. Such a reward system is necessary to enable equitable sharing of the benefits arising from the use of agro-biodiversity and to establish an incentive structure for continued maintenance of this diversity. Access and benefit-sharing legislation and farmers' intellectual property rights are suggested as central instruments. (Andersen 2006).

The stewardship approach refers to

> the rights that farmers must be granted in order to enable them to continue as stewards and as innovators of agro-biodiversity. The idea is that the 'legal space' required for farmers to continue this role must be upheld and that farmers involved in maintaining agro-biodiversity – on behalf of our generation, for the benefit of all mankind – should be recognized and rewarded for their contributions. (Andersen 2006).

Impact of Plant Variety Protection (PVP)

The introduction of a PVP system within a country provides many benefits, both on a national and international scale. Benefits differ from country to country and species to species but the overall impact of a PVP system is an advance in plant breeding technology and increasing scope and incentive for plant variety improvement. PVP has resulted in an increase in the overall number of varieties produced by plant breeders, and the varieties have improved yields, agronomic quality, nutritional quality, disease resistance and stress tolerance. In some countries, varieties must be demonstrably better than existing varieties in order to be accepted on their National List. All these factors encourage economic development and competition, especially in developing countries, where PVP can help to develop the economy in such a way that enables farmers to break out of subsistence farming, particularly when the PVP protection system recognises the contribution of 'farmer breeding' (Tripp et al. 2007; Salazar et al. 2006; Dutfield 2011). The breeders' exemption allows both public and private breeders access to breeding material which encourages innovation and increases the diversity of breeding programs. With access to international varieties and breeding lines, breeders have more incentive to improve varieties, increasing production and removing barriers to trade. New types of breeders, not just the large corporate breeding companies, are encouraged to compete, enabling innovation in all plant genera and species.

DUS Testing

According to the 1991 Act of the UPOV Convention, Articles 5–9, protection can only be granted in respect of a new plant variety after the examination of the variety has shown "that the variety is distinct (D) from any other variety whose existence is a matter of common knowledge at the time of the filing of the application....and that it is sufficiently uniform (U) and stable (S), or "DUS" in short" (UPOV 2002).

This examination, or the DUS test, is carried out based on the guidelines produced by UPOV where they are available. The growing tests are carried out by the

> authority competent for granting plant breeders' rights or by separate institutions.... acting on behalf of that authority (UPOV 2002).

In some cases the growing tests can be carried out by the breeder.

Current Systems: Test Guidelines

DUS testing is largely based on morphological description and measurement of plant characteristics. To ensure harmonization of testing procedures and internationally recognised descriptions of protected varieties, the general principles for DUS examination are laid out in the UPOV document TG/1/3: "General Introduction to the Examination of Distinctness, Uniformity and Stability and the Development of Harmonized Descriptions of New Varieties of Plants" (UPOV 2002). The international harmonization of testing procedures is an important aspect of the UPOV Convention and is emphasized in the supporting guidance documents (UPOV Test Guidelines are freely available at http://www.upov.int/en/publications/tg_rom/).

An important activity of UPOV is to produce guidelines for conducting DUS testing in a wide range of crop species. The individual Test Guidelines for the different crop species are prepared by the appropriate UPOV technical working party. The working parties (for agricultural, ornamental, vegetable crops and fruit trees) are composed of government appointed experts and invited experts from other interested states and observer organizations. The invited experts and non-governmental organizations are given the opportunity to comment on the Test Guidelines before they are adopted, making the preparation of Test Guidelines very much an interactive process. Most Test Guidelines are prepared for individual species. However, it may sometimes be appropriate to produce Test Guidelines covering a wider or narrower grouping of varieties. If no Test Guideline exists for a species, offices are encouraged to develop their own testing procedures, aligning them with the principles outlined in the General Introduction to DUS (TGP/7/2) (UPOV 2010a).

The design of growing trials or other tests, the layout of trials, number of plants to be examined, the number of growing cycles required, and methods of observation, are largely determined by the nature of the variety to be examined. Guidance on design is given in the Test Guidelines for individual species (UPOV Test Guidelines available at http://www.upov.int/en/publications/tg_rom/) (UPOV 2011c). At the time of writing, 264 Test Guidelines have been adopted by UPOV, whilst there are more than 2,750 genera and species with varieties examined for Plant Breeders' Rights (PBR). DUS tests are carried out according to the assessment of morphological characteristics detailed in the Test Guidelines and a formal description of the variety using its relevant characteristics is produced at the end of the examination. This formal description defines the variety.

For any variety to be protected it must first be clearly defined. According to the UPOV, a variety is

> defined by its characteristics and that those characteristics are therefore the basis on which a variety can be examined for DUS. (UPOV 2002)

Characteristics must fulfill certain criteria to be selected for use in the DUS examination. Characteristics should be:

(a) a result of a given genotype or combination of genotypes;
(b) sufficiently consistent and repeatable in a particular environment;
(c) exhibit sufficient variation between varieties to be able to establish distinctness;
(d) capable of precise definition and recognition (this requirement is specified in Article 6 of the 1961/1972 and 1978 Acts of the UPOV Convention, but is a basic requirement in all cases);

(e) allow uniformity requirements to be fulfilled;
(f) allow stability requirements to be fulfilled, meaning that it produces consistent and repeatable results after repeated propagation or, where appropriate, at the end of each cycle of propagation. (UPOV 2002)

The characteristics selected for use in the Test Guidelines are generally phenotypic characteristics that are unrelated to quality or commercial value or merit. The range of expression of each characteristic in the Test Guideline is divided into a number of states, sometimes termed notes, for the purpose of producing a detailed description. The number of states depends upon the type of expression of the characteristics. Characteristics can be quantitative (expression covers the full range of variation from one extreme to another), qualitative (expressed in discontinuous states), or pseudo-qualitative, where the range of expression is at least partly continuous, but varies in more than one dimension (e.g., shape: ovate, elliptic, circular or obovate; and cannot be adequately described by a simple linear range). Chemical tests may be used to measure characteristics such as herbicide tolerance as long as they meet the criteria specified above. Combined characteristics, such as ratios derived from image analysis data, may also be accepted, where they are found to be biologically meaningful.

Under the UPOV Convention, all new varieties should be compared to all other varieties of 'common knowledge' to determine whether they are a new variety or whether they exist already. In order to establish which varieties can be classified as in 'common knowledge', the General Introduction to the Examination of Distinctness, Uniformity and Stability and the Development of Harmonized Descriptions of New Varieties of Plants (UPOV 2002) gives guidance as to which specific aspects should be considered. A variety of common knowledge should be one that is in commercial production, has a published description, has had an application filed or is listed on an official register of varieties or in a publicly accessible plant register.

Distinctness should be assessed by comparing applicants (or candidates) with all varieties of common knowledge. These varieties of common knowledge comprise the 'variety collection' against which new candidates will be compared for distinctness. Because varieties of common knowledge need to be considered on a worldwide basis, variety collections are likely to be very large and it is, therefore, necessary for examination offices to take a pragmatic approach when selecting varieties for the collection. Guidance on acceptable ways of managing the variety collection is given in TGP/4 Constitution and Maintenance of Variety Collections (UPOV 2008a). It may not always be necessary to make a direct comparison of the candidate with all varieties of common knowledge as long as the candidate variety is sufficiently different in its characteristics. Where variety descriptions can be compared, and the candidate can be distinguished in a reliable way from varieties of common knowledge, it is not necessary to compare the two directly in a growing trial. Where variety descriptions are insufficient to distinguish candidate varieties from those of common knowledge, distinctness would be tested by side-by-side comparison in a growing trial or by statistical analysis of growing trial data using the criteria given the TGP/8: Trial design and techniques used in the examination of Distinctness, Uniformity and Stability (UPOV 2010b) and TGP/9: Examining Distinctness (UPOV 2008b).

The practical application of this system has its problems. In some species (particularly cross-pollinating species), database and variety description information cannot be used reliably to exclude reference varieties from the growing trial. The result of this is that as more and more varieties are added to National Lists, Common Catalogue and awarded Plant Variety Rights (PVR), they need to be included in the growing trial for comparison with new candidates. This can lead to variety collections increasing by hundreds of varieties every year as more varieties are included as varieties of common knowledge. This situation may not be sustainable in the long-term and solutions to management of the variety collections are being sought.

Essentially Derived Varieties (EDVs)

Although National Listing affords some protection to plant varieties, breeders are allowed to utilize the genetic resources of the 'common catalogue' to improve varieties. It has been recognized that the interest of breeders must be protected from variety 'plagiarism'. To this end, the concept of 'essentially derived varieties' (EDVs) was introduced in the UPOV 1991 Act of the UPOV Convention Article 14 (5) (UPOV 1991). In addition to the protected variety itself, the scope of the breeders' right also covers:

1. varieties which are essentially derived from the protected variety, where the protected variety is not itself an essentially derived variety;
2. varieties which are not clearly distinguishable from the protected variety; and
3. varieties whose production requires the repeated use of the protected variety.

The purpose of the provision on EDVs is to ensure that the Convention encourages sustainable plant breeding development by providing effective protection for plant breeders and by encouraging cooperation among breeders. The Convention contains some further clarification of what an EDV is, and defines mutants (natural and induced), including somaclonal variants produced by tissue culture, and genetically modified crops as EDVs, but fails to define what is meant by the term 'repeated use'.

The introduction of new genes, for example disease resistance from 'exotic' germplasm (landrace), often involves crossing it with an elite variety. Through multiple backcrosses with the elite variety and selection of the desired trait, weedy characteristics from the landrace can be eliminated leading to a new variety that has the agronomic advantages of the elite variety but with the extra disease resistance of the landrace. One presumes that the repeated use of the protected variety in backcrossing is what the concept of EDV was introduced to prevent.

While it is possible to use morphological characteristics to determine the similarity between two varieties, it is possible for the same DUS trait to result from the expression of genetic information that is not identical. For example, the 6-row head characteristic in barley can result from any one of three mutations from the 2-row wild-type (Komatsuda et al. 2007). In this case, two 6-row cultivars could be described as closely related based on morphology alone, when they are, in fact, more distantly related. In contrast, DNA markers can be used to measure genetic relatedness or distance in an objective manner and there have been many studies in this area (e.g. Borchert et al. 2008; Heckenberger et al. 2003, 2005). When quantifying similarities among EDVs, somaclonal variants or GM crops, there may only be a single genetic difference between derived and original variety. This difference should be easy to determine using DNA marker systems. Backcrossing presents a different challenge. The process of backcrossing reduces the genetic distance between the progeny and the recurrent parent at each generation. When the F_1 is backcrossed to the recurrent parent, then the progeny (BC1) has, on average, 75% recurrent parent genome. After a second backcross, BC2 progeny have average recurrent parental contribution of 87.5% and so on for further BC generations. The effect of the breeding system on derivation (F_2 versus BC1 or BC2) has been modelled to predict the genetic similarities among derived varieties (Heckenberger et al. 2005). These calculations may form the basis for defining a threshold for defining an EDV based on genetic similarities, though ultimately those thresholds are subjective and their implementation would require agreement among plant breeders.

The practice developed by the International Seed Federation around EDVs entails genotyping groups of varieties and breeders lines of known genealogy (Bruins 2009). The aim is to establish a working knowledge of inter- and intra-variety variability. A threshold is set by determining genetic distances within a collection of varieties that includes both unrelated and closely related varieties. The distribution of genetic distances is divided such that the region below the threshold includes those pairs of varieties or lines with close kinship and unrelated varieties fall in the region above the threshold. These thresholds would be set on a crop by crop basis. When variety pairs fall beneath that threshold, there would be a presumption of essential derivation. This method makes no allowance for loci that harbour identical alleles but the alleles are not identical by descent. There has not, as yet, been an acceptance of this methodology by UPOV though it should be noted that this methodology could be used as the basis of a new system under UPOV Working Group on Biochemical and Molecular Techniques and DNA-Profiling in Particular (UPOV BMT) Model 3 (see below).

The failure to set genetic distances for EDV may have been due to the inability to measure relatedness accurately when the 1991 Act of the UPOV Convention was drafted. The advent of high throughput genotyping and 'next generation sequencing' have made it possible to measure genetic distances accurately and the prospects for use of molecular methods in determination of EDV have been reported to the UPOV BMT (Table 18.1). It should be noted that essential derivation issues are currently decided by courts, not PVP offices.

Table 18.1 Survey on progress towards the implementation of molecular methods for DUS and EDV reported to UPOV BMT meetings (BMT/9: Washington, D.C., 2005, BMT/10: Seoul, 2006, BMT/11: Madrid, 2008, BMT/12: Ottawa, Canada, 2010, BMT/13: Brasilia, 2011)

Crop	Marker system	Report for UPOV
DUS		
Barley	SNP	BMT/9/9 BMT/12/5 BMT/12/7 BMT/13/5 MT/13/6
	SSR	BMT/11/21, BMT/12/19
Carnations	SSR	BMT/10/17
Grapevine	SSR	BMT/9/11, BMT/11/8
Japanese Barberry	AFLP	BMT/12/12
Lettuce	SSR	BMT/13/12
Maize	SSR	BMT/9/5 BMT/10/14 BMT/12/9
	SNP	BMT/12/14 BMT/12/15
Oilseed rape	SSR	BMT/9/8 BMT/10/11 BMT/11/7 BMT/13/7
Peach	SSR	BMT/13/11
Peas	SSR	BMT/12/11
Potato	SNP	BMT/9/13
	SSR	BMT/10/5 BMT/10/9 BMT/11/9 BMT/12/10 BMT/13/10
Quince	SSR	BMT/9/6
Rice	SSR	BMT/13/8
Rose varieties	SSR	BMT/9/12 BMT/10/16 BMT/11/14 BMT/13/21
Soybean	SSR	BMT/10/15 BMT/11/19 BMT/13/9 BMT/13/13 BMT/13/15 BMT/13/26
	SSR, AFLP, RAPD	BMT/12/18
Tomato	SNP, SCAR, CAPS	BMT/11/6
Wheat	SSR	BMT/11/7 BMT/13/14
EDV		
Various crops	AFLP, SSR, SNP	BMT/11//22
	AFLP	BMT/12/22
		BMT/12/23
Grapevine	SSR	BMT/11/6
Oilseed rape	SSR	BMT/9/7
Maize	SSR	BMT/13/19
Wheat	SSR	BMT/9/10

DNA-Based Marker Systems

Since each variety has a unique DNA sequence, any marker system that is capable of highlighting differences between varieties can be used (Doveri et al. 2008). The advent of DNA amplification technologies utilising polymerase chain reaction (PCR) (Saiki et al. 1988) has reduced the time required to genotype a DNA sample. Since the amount of DNA required for PCR is small, it is possible to genotype directly from DNA extracted from individual seeds without the need to grow plants, or from a small fragment of the plant, if such testing was required.

There are a number of requirements for any marker system: it must be able to identify polymorphisms, and data must be reliable and reproducible. Since there is the potential that these data may be used to resolve disputes, they have to be legally robust (UPOV 2010c). It is therefore unlikely that methods that target anonymous DNA polymorphisms such as randomly amplified polymorphic DNAs (RAPDs) (Williams et al. 1992) can be used for statutory testing. This method relies on the binding of short primers to genomic DNA to prime DNA synthesis. The binding of primers is sensitive to salt concentration and temperature, and even when amplification from specific primers is reliable (Lee et al. 1996), inter-laboratory studies suggest there is a high chance of variable results when reactions are performed in different labs (Jones et al. 1997).

Amplified fragment length polymorphisms (AFLP) (Vos et al. 1995) also amplifies anonymous DNA fragments. It does this by ligating short DNA sequences in the form of adapters from which to prime DNA synthesis. Extra base(s) to that of the adapter sequence at the 3' end of the primers serve to select only a subset of all the possible fragments. The major advantage of AFLP is that no sequence data are required for their use. Though AFLP have been shown to work for varietal discrimination (Law et al. 1998), one major hurdle to their adoption has been the requirement to pay royalties. Even though patent protection will cease in 2015, AFLP are of less interest in this context, as marker technology has evolved rapidly. For the same reason, retrotransposon-based markers such as REMAP, IRAP (Kalendar et al. 1999) and RBIP (Flavell et al. 1998) are unlikely to fulfil requirements for a molecular DUS testing system.

Genotyping using anonymous markers can be automated, further reducing costs. For example, DaRT markers (Jaccoud et al. 2001) apply a microarray technology platform to the analysis of DNA polymorphisms. This system is capable of providing high density marker coverage in species where the genome has not been sequenced. Informative polymorphisms are discovered in pooled DNA extracts derived from a diverse set of accessions, described as the diversity panel. This pooled sample is digested with combinations of frequent and rare cutters and fragments ligated to adapters. Polymorphic fragments are identified and these are added to a detection system known as a 'hybridisation array'. This array detects the presence or absence of specific DNA sequences in any sample. The system is limited in that it will only detect polymorphisms previously discovered. Thus discrimination power may be reduced in novel germplasm that was not represented in the diversity panel, which may lead to ascertainment bias.

Simple sequence repeats (SSRs), also known as microsatellites, are tandem reiterations of 1–5 bases, considered as candidate marker system for DUS testing. Although they require effort to isolate, characterise, and develop primers to assay, the virtue of SSRs is that they are highly polymorphic and thereby require fewer markers than bi-allelic marker systems, to obtain good separation of varieties (Leigh et al. 2003). The main reason for the polymorphic nature of SSRs is DNA replication slippage which alters repeat lengths and results in multiple alleles per SSR locus. This instability may also make them unsuitable for statutory testing.

Single nucleotide polymorphisms (SNPs) may be considered as the front runners for any molecular DUS testing system. Single base changes occur randomly and are found throughout the genome. These polymorphisms are found in all regions of the plant genome, including within those genes that are responsible for the morphological characters measured for DUS. Rapid, automated methods are available for SNP genotyping. SNP detection arrays and assays are offered by a number of companies as a cost effective screening tool, while 'next generation sequencing' is developing at a pace such that whole genome re-sequencing is on the cusp of becoming a routine, cost effective, high throughput technique. SNP arrays and assays depend on identification of SNPs within a diversity panel and, using sequence information flanking the polymorphism, detection arrays can be synthesised (Close et al. 2009), or SNP specific assays may be devised (http://www.kbioscience.co.uk/). In either case, the quality of data will depend on the diversity of the panel used to discover the SNPs and may be subject to ascertainment bias.

'Next generation sequencing' is a catch-all phrase applied to a number of technologies that have emerged to replace 'Sanger sequencing'. Mass sequencing, which is delivering the complete genome sequences of many organisms with shortening regularity, offers the opportunity to describe varieties in terms of DNA sequence (Rogers and Venter 2005). The '$1000 dollar genome', i.e., the cost to sequence a genome, was the goal for much of the early part of the twenty-first century, is now close to reality. The dramatic reduction in the cost of sequencing means that it is currently cheaper to sequence a plant genome than to put a variety through statutory field assessment. However, the ability to describe a variety based on its DNA sequence may pose as many problems as it addresses, in particular the problems of reproducible SNP calling (Neilson et al. 2011) and setting a statistically and biologically acceptable definition of intra- and inter-varietal variability within a whole genome.

The utility and discrimination power of molecular markers can be compared using Polymorphism Information Content (PIC) values, a measure of allelic diversity at each locus. The PIC value may be calculated as:

$$PIC = 1 - \sum_{i=1}^{n} p_i^2,$$

where p is the frequency of the ith allele.

DNA Markers for Varietal Identification

DNA is the genetic material that determines the phenotype / morphology of a plant. Differences (polymorphisms) in coding sequences within DNA are responsible for those morphological differences that make each variety distinct and it is possible to exploit those polymorphisms to distinguish between varieties. The requirement for field comparisons between candidates and variety collections imposes a relatively high cost, due to the labour costs of planting and managing trials, the cost of space to grow plants and the labour required to record the characteristics. The use of DNA markers could remove the necessity to grow plants beyond the need for a 'basic description' of the variety. In 'novel' systems, marker data can be collected for each variety and stored in a database allowing comparisons to be made between candidates and current registered varieties. DNA fingerprints can be collected using any part of the plant, at any stage of the plant's life cycle and the results are not affected by the growing environment.

The way plants are propagated is important in terms of variety maintenance. Crops that are produced and propagated through somatic means, by cuttings, runners or tissue culture, e.g., potatoes and strawberries, should present few problems with regards to uniformity since clonal propagation produces genetically identical samples.

Varieties of plants that reproduce sexually are intrinsically more variable than asexually produced plants and fertilisation may result from self-pollination (autogamy) or from out-crossing (allogamy). The method of reproduction has an intrinsic effect on the genetics of the crop: in autogamous crops, such as wheat and barley, the repeated generations of selfing produces varieties that are both homozygous and phenotypically uniform. Allogamy leads to heterozygosity and heterogeneity as a consequence of out-crossing in case of population or synthetic cultivars.

However, DUS testing makes no distinction between allogamous or autogamous species; distinctness is determined by comparing physical attributes in 'populations of plants', while uniformity is determined by measuring these characteristics over several generations.

Sampling strategies in population varieties are designed to represent a variety by a bulk of plants, where bulked samples from a pool of individuals are used for microsatellite genotyping in an effort to reduce costs. Pooled samples offer a cost effective strategy for maximising the information derived from microsatellite (SSR) genotyping. The data generated can be interpreted in three ways. In the first analysis, the microsatellite amplicon with the most intense instrumental response is recorded as the predominant allele and would represent the genotype. This approach has the advantage of being simple, easy to implement and would usually represent the most frequent allele found in a variety. The alternative of assaying individuals separately would require many more assays to ensure that the majority allele is represented. The second approach is described as 'thresholding' (Jones et al. 2008a). This approach establishes rules that allow 'major alleles' to be recorded, resulting in a multi-allele genotype. The thresholding rules are established to ensure consistent scoring between batches or between laboratories. The majority genotype will always be recorded, as will common 'minority alleles'. This approach is more complex and may change the Polymorphism Information Content (PIC) values at each locus when differing thresholding levels are selected. The significance of minor alleles may be overemphasised. When the threshold is set at a low level, a rare allele (e.g., $p = 0.1$) will be attributed the same significance as an allele with a higher frequency (e.g., $p = 0.4$). A third approach, described as 'calibration', is to 'calibrate' allele frequencies in pooled samples (LeDuc et al. 1995; Dubreuil et al. 2006). This system is intended to minimise within laboratory errors when generating allele frequency profiles from microsatellite assays in pooled DNA, and offers the possibility of representing heterogeneous populations with data suitable for population studies.

The Use of DNA-Profiling in PVP

The use of DNA profiling in PVP has been extensively considered by the Biochemical and Molecular Techniques (BMT) Working Group of UPOV, which was established in 1993. The role of the BMT is to maintain an awareness of relevant applications of biochemical and molecular techniques in plant breeding, consider their possible application in DUS testing and to establish guidelines for biochemical and molecular methodologies and their harmonization. Progress in 16 crop species has been reported by the BMT within the past 5 years (Table 18.1). The BMT is open to DUS experts, biochemical and molecular biology specialists, and plant breeders. The BMT guidelines provide guidance for developing harmonized methodologies with the aim of generating high quality molecular data for a range of applications. Models for possible application are:

1. Molecular characteristics as a predictor of traditional characteristics: Use of molecular characteristics which are directly linked to traditional characteristics (gene specific markers),
2. Calibrated molecular distances in the management of variety collections. Calibration of threshold levels for molecular characteristics against the minimum distance in traditional characteristics, and
3. Use of molecular marker characteristics. Development of a new system. (UPOV 2011d)

When a new model is proposed, the BMT Review Group assesses the potential application within the examination of DUS on the basis of conformity with the UPOV convention and potential impact on the strength of protection compared to that provided by current examination methods, and advises, if this could undermine the effectiveness of protection offered under the UPOV system. The BMT also provides a forum for discussion on the use of biochemical and molecular techniques in the consideration of essential derivation and variety identification, where there is potential for use in enforcement issues and legal disputes. We will consider the implementation of DNA based methods in case studies categorised by the three UPOV BMT models outline above.

UPOV BMT Model 1

Here we discuss molecular characteristics as a predictor of traditional phenotypic characteristics on a trait by trait basis. These functional markers are derived from true causative polymorphisms in the genome and the link between morphology and the marker cannot be broken by recombination. Advances in molecular genetic approaches have meant that a number of genes controlling morphological traits evaluated during the awarding of Plant Breeders' Rights (PBR) have been isolated in various crop species. UPOV BMT Model 1 provides an opportunity for PVP test centres to deploy molecular markers to aid DUS assessments. A case study of the temperate cereal crop barley (*Hordeum vulgare* ssp. *vulgare* L.) will illustrate the progress made towards the deployment of diagnostic molecular markers in the context of both crop breeding and PBR.

The key to implementation of UPOV BMT Model 1 is the identification of genomic regions associated with DUS traits. Improvement in grain harvesting by seed retention is a key domestication gene, one that has been selected for during adaptation of wild to crop plants. Domestication genes will be associated with regions of extended linkage disequilibrium within the genome, a zone of non-random association among loci, described as 'selective sweep'. The process of domestication reduces the genetic diversity of crops when compared to their wild ancestor. Selection on key domestication genes reduces the genetic diversity beyond that of domestication alone and the method has been validated by measurements of genetic diversity around key domestication quantitative trait loci (QTL) (Clark et al. 2004; Tian et al. 2009).

A more general approach to identifying QTL for traits used in statutory testing comes from using genome-wide association studies (GWAS). In conventional genetic mapping two parents are crossed and the traits and genetic markers are followed in subsequent generations. Linkage is defined as the non-random segregation of two characters and the degree of non-randomness is proportion to extent of linkage between them. GWAS utilises this principle but instead of having to create a cross and progenies, it can use historical recombination that has occurred during the production of varieties. This method requires a large set of varieties with data available for phenotypic traits (DUS morphological traits in this case) and DNA based markers that randomly sample the genomes of these varieties (Mackay and Powell 2007). GWAS uses the assumption that closely linked markers, whether they are morphological or DNA-based, will be co-inherited. This non-random inheritance is described as linkage disequilibrium (LD). As such genetic variation within a region of the genome usually exhibits restricted genetic variation, where markers are inherited in blocks and each variant within a block is called a 'haplotype'. Haplotypes can change with new mutations or by being broken up and shuffled by recombination.

GWAS has emerged as a very effective approach for identifying loci underlying complex human diseases (The Wellcome Trust Case Control Consortium 2007). More recently, it has been applied to identifying QTL for agronomic traits not only in sequenced but also un-sequenced crop plants (Huang et al. 2012; Cockram et al. 2010). Here we discuss recent molecular genetic investigations and GWAS of DUS characters in barley, as a case-study for the implementation of a UPOV BMT Model 1 approach.

DUS assessment of barley in the UK involves phenotypic evaluation of a set 28 morphological characters (Table 18.2). Fifteen of these characters have been shown

Table 18.2 Progress towards determining the genetic loci controlling currently used barley DUS traits

UPOV No.	Characteristic	Known genetic loci (Chr)	GWAS loci (No.)	Progress towards cloning underlying genes
1	**Plant growth habit**		Y (1)	
2	**Lower leaves: hairiness of leaf sheaths**	*HSH1* (4H)	Y (1)	Fine-mapping
3	**Flag leaf: anthocyanin colouration of auricles**	*ANT2* (2H)	Y (2)	Candidate gene, *HvbHLH1*
4	**Flag leaf: intensity of anthocyanin colouration of auricles**	*ANT2* (2H)	Y (2)	Candidate gene, *HvbHLH1*
5	Plant: frequency of plants with re-curved leaves		N	
6	Flag leaf: glaucosity of sheath	*Ecf* loci	N	
7	**Time of ear emergence (1st spike vis on 50% ears)**	*PPD-H1* (2H)	N	Cloned gene, *HvPRR7*
		PPD-H2 (1H)	N	Candidate gene, *HvFT3*
8	**Awbs: anthocyanin colouration of tips**	*ANT2* (2H)	Y (2)	Candidate gene, *HvbHLH1*
9	**Awns: intensity of anthocyanin colouration of awn tips**	*ANT2* (2H)	Y (2)	Candidate gene, *HvbHLH1*
10	Ear: glaucosity	*Ecf* loci	N	N/A
11	**Ear: attitude (at least 21 days after ear emergence)**	*VRN-H1* (5H)	Y	Candidate gene, *HvBM5a*
12	Plant: length (stem, ears and awns)	*HvBRI*, *Sdw1*	N	Candidate gene
13	**Ear: number of rows**	*VRS1* (2H)	Y (2)	Cloned gene, *HvHOX1*
		Int-c (4H)		Cloned gene, *HvTB1*
14	Ear: shape	N/A	N	N/A
15	Ear: density	N/A	N	N/A
16	ear length (excluding awns)	N/A	N	N/A
17	Awn length (compared to ear)	N/A	N	N/A
18	Rachis: length of first segment	N/A	N	N/A
19	Rachis: curvature of first segment	N/A	N	N/A
–	**Ear: development of sterile spikelets**	*VRS3* (1H)	Y (2)	Candidate locus
		VRS1 (2H)		Cloned gene, *HvHOX1*
20	**Sterile spikelet: attitude (mid 1/3 of ear)**	*VRS3* (1H)	Y (2)	Candidate locus
		VRS1 (2H)		Cloned gene, *HvHOX1*

(continued)

Table 18.2 (continued)

UPOV No.	Characteristic	Known genetic loci (Chr)	GWAS loci (No.)	Progress towards cloning underlying genes
21	Median spikelet: length of glume + awn cf grain	N/A	N	N/A
22	**Grain: rachilla hair type**	*SRH*	Y (1)	Fine-mapping
23	Grain: husk	N/A	N	N/A
24	**Grain: anthocyanin colouration of lemma nerves**	*ANT2* (2H)	Y (2)	Candidate gene, *HvbHLH1*
25	**Grain: spiculation of inner lateral nerves**	N/A	Y (1)	N/A
26	**Grain: ventral furrow – presence of hairs**	N/A	Y (1)	Fine-mapping
27	**Grain: disposition of lodicules**	*CLY1*	Y (1)	Cloned gene, *HvAP2*
28	**Kernel: colour of aleurone layer**	*BLX1*	Y (1)	Fine-mapping
29	**Seasonal growth habit**	*VRN-H1* (5H)	Y (2)	Cloned gene, *HvBM5a*
		VRN-H2 (4H)		Candidate genes, *ZCCT-H*

GWAS genome-wide association studies, Cockram et al. (2010)
Traits indicated in bold have had successful conversion of genetic markers to the KASPaR genotyping platform (Cockram et al. 2012)
Character 'Ear: development of sterile spikelets' is not included in UPOV, but is included under CPVO test guidelines (character 18*D)

to have a significant association with SNP markers mapped to particular regions of the barley genome (Fig. 18.1) (Cockram et al. 2010). This information, together with an earlier association study (Cockram et al. 2008) has been used to identify molecular markers useful as predictors of traditional characteristics, and convert these to a common genotyping platform (Cockram et al. 2012).

Phenotypic diversity for seasonal growth habit (UPOV character 29, also known as 'vernalization requirement') has been associated with specific molecular variants. In UK barley, this trait is controlled by two genetic loci, *VRN-H1* and *VRN-H2* (Laurie et al. 1995; Cockram et al. 2007a). *VRN-H1* is thought to encode a MADS-box transcription factor, with a range of deletions spanning a 'vernalization critical' region within intron 1 conferring the range of observed spring alleles (von Zitzewitz et al. 2005; Szűcs et al. 2007; Cockram et al. 2007b). The molecular characterisation of *VRN-H1* alleles within a large collection of European germplasm (Cockram et al. 2007b, c), allowed subsequent design of a molecular assay diagnostic for winter/spring alleles (Cockram et al. 2009). However, *VRN-H1* and *VRN-H2* interact epistatically with spring alleles being dominant over winter alleles. Thus, the allelic state at *VRN-H2* must be characterised in order to correctly predict the phenotype. Using a combination of fine-mapping and comparative analyses with the diploid wheat species *Triticum monococcum*, *VRN-H2* was shown to be tightly linked to a

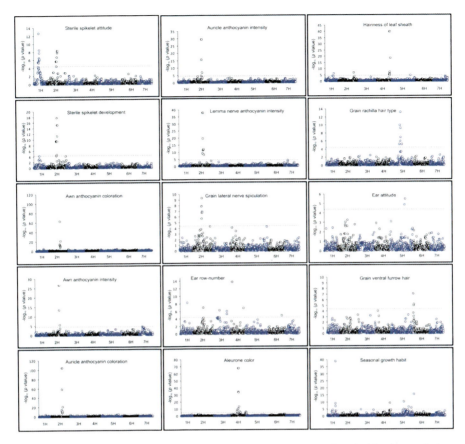

Fig. 18.1 Genome-wide association study scans of the 15 traits with significant associations ($P < 0.05$, Bonferroni corrected; indicated by a *dashed line*). Barley chromosomes 1H to 7H are shown. Taken from Cockram et al. (2010)

cluster of *ZCCT* genes (Yan et al. 2004; Karsai et al. 2005). A deletion spanning all three *ZCCT* genes is thought to result in the spring *vrn-H2* allele. An assay has been designed for presence/absence of the *ZCCT* genes (Karsai et al. 2005), and is diagnostic for allelic state at *VRN-H2* in all germplasm screened to date. Thus, simultaneous analysis of genetic markers for *VRN-H1* and *VRN-H2* allows prediction of winter or spring seasonal growth habit (Fig. 18.2a). It should be noted that a third phenotypic class ('alternative' seasonal growth habit) is also occasionally observed in UK barley. Varieties with the alternative seasonal growth habit display flowering time that is intermediate to that of spring and winter varieties; and the genetic determinants of the alternative phenotype are currently under investigation. Seasonal growth habit is not the only barley DUS characteristic, whose underlying genetic variation has been described at the gene level. The wild-type barley ear consists of a series of spikelets arranged along a central rachis. In the two-row barley form, each spikelet consists of single fertile floret, flanked on

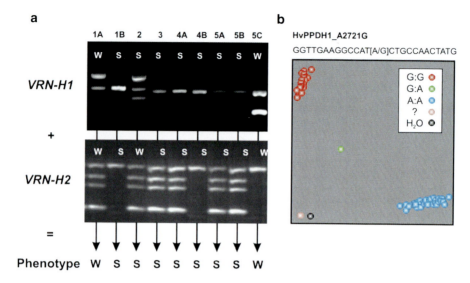

Fig. 18.2 Molecular markers for DUS characteristics in barley. (**a**) Character 'seasonal growth habit'. Allelic state predicted by the *VRN-H1* (Cockram et al. 2009) and *VRN-H2* (Karsai et al. 2005) assays are considered together to determine winter (*W*) or spring (*S*) phenotype, with spring alleles epistatic to winter alleles. *VRN-H1* haplotype is indicated, as described by Cockram et al. (2007b, 2009). (**b**) Character 'time of ear emergence'. Allelic state at the PPD-H1 long-day photoperiod response locus as predicted by the SNP within the CCT domain: G (responsive *Ppd-H1* allele), A (non-responsive *ppd-H1* allele), assayed using the KASPaR genotyping platform

each side by two sterile spikelets. However, in six-row barley all three florets are fertile. Ear row number (UPOV character 13) has long been known to be controlled by a genetic locus on chromosome 2H, termed *VRS1* (six-rowed spike 1) (Ubisch 1916; Lundqvist et al. 1997). Recently, *VRS1* was isolated by map-based cloning, and found to encode a homeodomain-leucine zipper class-I homeobox protein (Komatsuda et al. 2007). Three independent single-base mutations within *VRS1* are thought to have led to the suppression of lateral spikelet fertility. Two of these mutations result in a frameshift, leading to a truncated predicted protein, while the third results in a phenylalanine to leucine amino acid substitution within the HD-Zip domain at a highly conserved residue in plants, animals and yeasts. Accordingly, it is now possible to unambiguously determine ear row number by molecular analysis of the causative genetic polymorphisms.

The DUS character 'disposition of lodicules' (UPOV character 27) is controlled by the genetic locus *CLYSTOGAMOUS 1* (*CLY1*) on the long arm of chromosome 2H (Turuspekov et al. 2004). From the white anther stage onwards, varieties with the dominant wild type *CLY1* allele possess larger and fully formed lodicules which result in open florets, due to separation of the floret pallea and lemma as a result of lodicules swelling. Natural recessive variants at *CLY1* result in atrophied/small lodicules, resulting in closed flowering (also called cleistogamy). Positional cloning of *CLY1* identified the AP2 domain gene, *HvAP2*, as the underlying gene (Nair et al. 2010). Two synonymous SNPs within a putative micro-RNA binding domain within *HvAP2* were found to be associated with low *HvAP2* expression, small lodicules and closed flowering.

More recently, the genetic determinant likely to control the absence/presence of anthocyanin in the auricles, awn tips and lemma nerves (UPOV characters 3, 8 and 24, respectively) has been determined (Cockram et al. 2010). When the three related phenotypes were considered collectively, the ability to produce anthocyanin was found to be controlled by the *ANTHOCYANINLESS 2* (*ANT2*) locus on chromosome 2H. Using association mapping, *ANT2* was located to a 140 kb genomic region containing three genes. One of these, a basic helix-loop-helix domain gene, termed *HvbHLH1*, was found to be an orthologue of the genes underlying the *R/B* anthocyanin regulatory loci that control anthocyanin pigmentation in maize. A 16 bp deletion was identified within exon 2 that resulted in a frame shift and truncation of the predicted protein upstream of the bHLH domain. Analysis of ~500 UK barley varieties showed that the deletion was diagnostic for the lack of anthocyanin pigment in all tissues studied. However, the genetics underlying expression of anthrocyanins in specific tissues and the intensity of expression have not been elucidated. Thus, more work is needed before the anthocyanin related DUS characters (which score the intensity of the pigment on a 1-to-9 scale) can be replaced by molecular markers.

Finally, the *NUDUM* (*NUD*) locus, known to control the hulled/naked caryopsis phenotype has been positionally cloned, and encodes an ethylene-response factor (ERF) transcription factor (Taketa et al. 2008). All naked barleys investigated to date possess a 17 kb deletion that harbours the *ERF* gene. Deletion of the gene is associated with the loss of a lipid layer on the pericarp epidermis, preventing adhesion of the hull to the caryopsis, resulting in the naked phenotype. An assay has been developed by Takeda et al. (2008), that is diagnostic for naked/hulled barley. However, this trait is not used in all EU countries, as the occurrence of naked barley varietal submissions varies regionally.

Thus, a total of five DUS characters can currently be predicted from genotype, and diagnostic markers have been developed for many of these on a single genotyping platform (KASPaR, http//www.kbiosciences.co.uk/; Cockram et al. 2012) (Table 18.2, Fig. 18.2b). Furthermore, a recent association analysis of DUS traits in barley identified seven additional DUS traits for which the major locus controlling the phenotype will likely be identified in the near future (Fig. 18.1) (Cockram et al. 2010). These include sterile spikelet attitude (UPOV character 20), hairiness of leaf sheath (2), sterile spikelet development (CPVO character 18), grain rachilla hair type (22), grain inner lateral nerve spiculation (25), grain ventral furrow hair (26) and aleurone colour (28) (Table 18.1). Indeed, marker development, further association and linkage mapping have resulted in improved molecular markers for grain ventral furrow hair and hairiness of leaf sheath (Cockram and Jones, unpublished). Both of these characters have been reported by breeders and DUS assessment officers to be difficult or time-consuming to score, and would, therefore, benefit from the availability of appropriate diagnostic molecular markers.

The advances made in the identification of diagnostic markers for DUS-relevant traits inherited in a Mendelian manner are in sharp contrast to quantitative traits such as plant height. Perhaps the most promising quantitative DUS trait for which predictive markers may soon be developed is flowering time (UPOV character 7), as it is a highly heritable and easily scored trait. Flowering time is controlled to a large extent by the major photoperiod response loci *PHOTOPERIOD 1* (*PPD-H1*) and *PPD-H2* (Laurie et al. 1995). While *PPD-H1* has been positionally cloned (Turner et al. 2005), and its functional diversity dissected (Jones et al. 2008b), the gene underlying *PPD-H2* remains to be formally identified, although there is strong evidence for *HvFT3* being this gene (Faure et al. 2007). Thus, molecular markers predictive of allelic state at both of the major flowering time loci currently exist. The recent fine mapping of the barley flowering time QTL *FLT-2L* (Chen et al. 2009) supports the assumption that at least some of the additional genetic factors controlling flowering time will be characterised at the molecular level in the near future.

Increased understanding of the genes and causative genetic polymorphisms underlying barley DUS traits allow diagnostic genetic markers to be designed and implemented. Although such markers could potentially be deployed to determine distinctness for a number of DUS traits, limited work has been undertaken towards the genetic determination of uniformity (i.e., the detection of off-type individuals within a submitted variety). Furthermore, quantitatively controlled traits remain relatively uncharacterised at the molecular level. Until molecular genetic approaches have been shown to reliably address both of these areas, the replacement of DUS field assessment with genetic approaches is unlikely to be imminent.

For the purposes of statutory testing there are many obstacles to the adoption of molecular markers. For instance, although the molecular characterisation of some of the characters assessed in DUS examinations have been achieved, there are many DUS traits, where QTL have not been identified. Some traits are controlled by many genes (polygenic), each having an impact: the dissection of the effect of each would be too complex to define by our current understanding of plant genetics. Even for seemingly simple traits many different genes can have the same effect on the observed phenotype. For example, in rice four seed shattering genes have been identified, named as *sh1-4* (Nagao and Takahashi 1963; Oba et al. 1990; Eiguchi and Sano 1990; Fukuta and Yagi 1998), that improve seed retention on the plant and, therefore, help with harvesting the grain. In addition, QTL for shattering have been reported on chromosomes 1, 3, 4, 7, 8, and 11 (Xiong et al. 1999; Cai and Morishima 2000; Bres-Patry et al. 2001; Thomson et al. 2003).

Furthermore, even when the underlying mutation that gives rise to a specific phenotype is known, further mutations may change or even reverse the original mutation. For example in rice, there has been selection for white rice early during domestication. Two mutations in the *Rc* gene that cause truncation of the basic helix-loop-helix regulatory protein it encodes which leads to failure to express proanthocyanidin in seed, resulting in rice without pigmentation (Sweeney et al. 2006). Molecular diagnostic tests were developed to identify both mutations (Sweeney et al. 2007). However, for the predominant white rice allele, two mutations, leading to reversion to red rice, have been identified (Brooks et al. 2008;

Lee et al. 2009). Such mutations would confound a testing system based on UPOV BMT Model 1, and might be introduced by introgression of exotic germplasm, mutagenesis or genetic modification.

UPOV BMT Model 2

UPOV BMT Model 2 requires "Calibration of threshold levels for molecular characteristics against the minimum distance in traditional characteristics". This requirement is intended to ensure that decisions made under a new molecular testing system would be the same as those made under the existing morphological testing system. UPOV BMT Model 2 may be implemented in one of two categories: "Calibrated molecular distances in the management of variety collections" or "Combining phenotypic and molecular distances in the management of variety collections" (UPOV 2011d).

The costs of genotyping plant material have fallen dramatically with the advent of capillary based DNA analysis equipment, SNP arrays and 'next generation' sequencing. By comparison the costs of phenotyping to determine plant morphologies for DUS testing have remained relatively high. As the costs of genotyping decline in relative terms, the attraction of UPOV BMT Model 2 will increase, provided the quality of variety protection remains similar or improves. Ideally there would be a perfect relationship between morphological and molecular distances such that the decisions made using a molecular system would exactly mirror those made under the current system (Fig. 18.3, upper graph). Should the relationship between the two testing methods be anything less than perfect, there would be a zone of 'uncertainty' where ambiguous decisions might be made (Fig. 18.3, lower graph). The resolution of this ambiguity is fraught with difficulty. Should a candidate variety within the zone of uncertainty be distinct by molecular methods but non distinct by morphology, the candidate's breeder would be unfairly advantaged by the new method. One suggested solution would be to raise the distinctness threshold for molecular methods, the so-called 'Molecular threshold distinct plus' (Button 2008) effectively disadvantaging candidate varieties. It is useful to note that a calibration between molecular and morphological methods only needs to be true close to the distinctness thresholds for both methods. Where varieties are 'super distinct' by both methods there is no need for a precise relationship as the quality of decision making would be unaffected by the choice of method, allowing molecular methods to be adopted.

The utility of UPOV BMT Model 2 has been investigated in vegetatively propagated (grapevine), allogamous (maize and oilseed rape) and in predominantly selfing (durum wheat and barley) crops. The outcomes of these investigations have been mixed.

In a study of grapevine (Ibáñez et al. 2009), 991 cultivars were assayed at nine microsatellite loci. These markers offered unique identification for 352 accessions by pair-wise comparisons. The remaining 639 accessions were assayed at 16 further

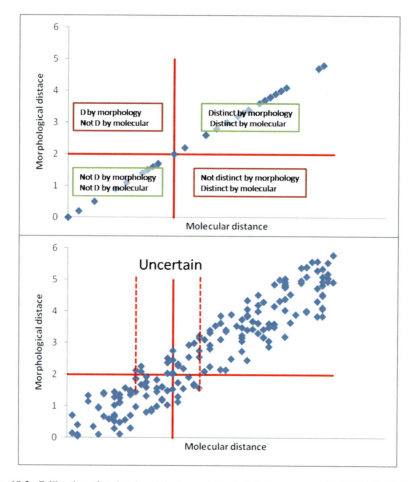

Fig. 18.3 Calibration of molecular against morphological distances under UPOV BMT Model 2. The *upper* graph illustrates decision marking under a perfect correlation between molecular and morphological distances. The *lower* graph illustrates possible uncertainty where the correlation between molecular and morphological distances is sub optimal

loci. The authors conclude that it is possible to calibrate a minimum distance using microsatellites between varieties produced by sexual reproduction in an accession set that included closely related varieties (parents, progeny, full sibs, half sibs, grandparents) with a difference greater than four alleles in all but 10 out of 119,316 pair-wise comparisons. However, varieties arising as 'sports' among clonally propagated varieties could not be differentiated in the same way, where a minimum inter-variety distance of two alleles was observed. The robustness of decisions made using an inter-variety distance of two alleles needs to be tempered by the observation of intra variety differences of one allele. The authors concluded that variety pairs that exceed a minimum threshold using molecular methods may be declared distinct, but in case of no or few differences in molecular profiles, further testing is required either by the use of additional markers or by comparing morphologies. This equates to a 'Super D' approach which would allow an initial screen that would increase efficiency of DUS testing by eliminating the number of comparisons that would need to be made in the field but would not allow full replacement of the current test system. The authors recommend that minimum distances can only be established experimentally, on a crop by crop basis, with full account taken of the inter- and intra-variety variability of the test system used.

An alternative experimental approach was used to study durum wheat lines (Noli et al. 2008) where a collection of 69 advanced lines from seven crosses were assessed for distinctness using 17 morphological markers from CPVO protocols selected as variable among the parental lines, a suite of 99 SSR markers and AFLP assays using combinations of two and three selective bases in seven primer combinations. The correlation between the molecular markers (SSRs and AFLP) was good ($r = 0.89$), while the correlation between morphological traits and molecular markers was moderate (SSRs, $r = 0.66$; AFLP, $r = 0.62$) (Fig. 18.2). Notwithstanding these correlations, the authors recognised difficulties in assessing distinctness using a 'Model 2' approach because of the wide range of variation for molecular marker differences among accessions around or beneath the distinctness threshold using morphological markers. Once more, the authors concluded that the calibration of molecular and morphological methods would allow a declaration of distinctness, where molecular profiles differ greatly in the style of a 'Super D' approach, but field testing could not be eliminated.

Gunjaca et al. (2008) examined correlations between morphology and molecular based distances in maize in a collection of 41 inbred lines comprising 13 publicly available varieties and 28 breeders' lines. Morphological descriptions were calculated using 34 characters from the UPOV guidelines and molecular distances calculated using data for 28 SSR loci. In this study, the correlation between morphological traits and molecular markers was poor ($r = 0.21$). The authors concluded that molecular markers are a possible addition to the DUS testing procedures but their implementation depends upon deciding on the type and number of markers to be used as well as setting the threshold values for distinctness.

A large, international set of varieties was examined in a study of oilseed rape (CPV5766 Final Report 2008) using 335 records from DUS testing authorities in Denmark, France, Germany and the UK. The collection was genotyped using a suite of 29 SSR markers. The outcome of this study was far more disappointing, with the correlation between morphological and molecular marker based distances between 0.03 and 0.08, depending on the methods used to calculate the distances. Clearly these results offer little prospect for successfully implementing a UPOV BMT Model 2 approach for oilseed rape.

Increasing marker density should improve the correlations between morphological and molecular marker based distances. An SSR study conducted in a set of 40 winter wheat varieties showed that pair-wise discrimination increased as more SSR loci were considered (unpublished data). The initial increase in discrimination was rapid but the rate of increase in discrimination tailed off as marker numbers

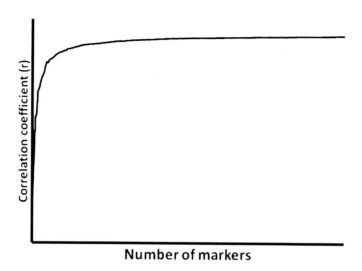

Fig. 18.4 Modelling the minimum number of markers needed to obtain the optimum correlation between morphological and molecular marker based distances. The plot shows the expected correlation between morphological and molecular distances vs. the number of markers used to calculate the molecular distance

increased. This observation can be explained by linkage between markers and population structure within the accession set under consideration. There is an expectation that the correlations between morphological and molecular marker based distances would improve in a similar way (Fig. 18.4), reaching a plateau when an optimum number of markers have been used to calculate molecular distances (Jones et al. 2012). While the minimum number of markers required should be determined empirically for each species, it is possible that the marker numbers used in the SSR studies described may be sub-optimal.

While larger data sets may offer the prospect of greater correlations, improvements may result from use of genomic prediction. Genomic prediction aims to quantify the contribution of each marker within the genotype data to each characteristic within the phenotypic data. Regression analysis in a 'training set' allows quantification of the contribution of each and every marker to expression of a characteristic. The results of this regression can be used to predict the expression of that characteristic in a 'test set' of varieties where genotypic data are available but phenotypic data are not.

Although morphological characteristics are the foundation for DUS testing, one possible way of reducing the number of varieties from the reference collection that are included in the growing trials is to incorporate the use of molecular markers into the DUS test. The use of molecular markers in plant breeding and plant identification has increased considerably in recent years and the advances in technology could increase the efficiency and speed of the current DUS test while maintaining the strength and scope of protection provided by the PVP system. There is much interest in approaches that could reduce the workload and costs of testing, by eliminating unnecessary comparisons between existing and candidate varieties prior to more formal testing, and biochemical markers in the form of seed proteins and isozymes are already included in the UPOV guidelines of certain crops such as wheat, barley and maize. One possible way in which this problem might be approached is to use DNA-profiling of varieties as a management tool. By comparing the profiles of candidate varieties with those of existing varieties maintained in a central database, it might be possible both to eliminate those varieties from further testing, which do not require comparison in a growing trial (according to an agreed set of criteria) and to select the varieties most similar to the candidate for close comparison in field tests (Jones et al. 2003; Tommasini et al. 2003). In order for such a scheme to work, it is necessary to have an agreed set of molecular markers to generate the DNA profiles, and an agreed means of utilising the profiling data. The data shown in the studies of grapevine, maize and durum wheat discussed above suggest that use of DNA profiles as a grouping tool would succeed.

A variation on this approach has already been implemented by the testing authorities in France. Candidate varieties are grown and a description prepared. The descriptions are entered into GAÏA software. Each measure is assigned an appropriate weight and inter-variety distances are calculated (TGP/8 PART II: 1: The Gaia Methodology; UPOV 2010b). Candidate varieties deemed to have achieved the 'Distinctness Plus' threshold are said to be distinct and no further distinctness testing is required. Conceptually 'Distinctness Plus' is similar to 'Super D' (Button 2008) and is used as a threshold used to eliminate candidates from further field testing and is set by local experts based on their experience gained with the crop. Remaining candidates may be examined using biochemical markers and / or quantitative measures; the data are entered into GAÏA software, each measure is assigned an appropriate weight and inter-variety distances re-calculated. Once more, candidates achieving 'Distinctness Plus' may be eliminated from costly growing trials. The remaining candidates will then be compared with a set of most similar varieties selected from a database for direct comparison in subsequent years. The philosophy underpinning GAÏA makes it suitable as a platform for novel testing methods based on molecular markers.

Distinctness testing of hybrids requires verification of the hybrid formula, a test to determine whether the submitted hybrid is the product of a cross between the declared parent lines (TGP/8/1: Part II: 2: Parent Formula of Hybrid Varieties (UPOV 2010b)). Hybrid purity can be compromised if there is a high proportion of self pollination by the maternal parent or the pollen originates from something other than the stated paternal parent. Confirmation of the hybrid formula could be readily achieved by use of a suite of co-dominant markers such as SSRs or SNPs.

Roses are a high value vegetatively propagated crop, widely traded across the world (Vosman et al. 2006; Smulders et al. 2009). It is estimated that there are in excess of 13,000 cultivars worldwide, making realistic assessment against 'varieties in common knowledge' a near impossibility. A systematic approach is needed to select varieties for side by side comparison in the growing test. Currently, selection requires comparison of the breeder's submission with reference collections, comparison with the collections in public rosariums, literature, database searches, and by reference to experts (so called 'walking reference collections'). Vosman et al. (2006) described a system that relies on a database that includes a simplified variety description taken from the breeder's submission, high quality images and molecular data collected at 12 loci. They proposed a database that had five possible uses:

- Characterization and cataloguing of the reference collection;
- Pre-screening and selection of appropriate reference varieties;
- Exchange of data on current candidate varieties between testing stations;
- Strong reduction or replacement of permanent living reference collections at testing stations;
- Quality assurance: verification of identity and authenticity of reference varieties.

A similar database is proposed as a tool for managing potato reference collections (Reid et al. 2011).

In conclusion, studies to test UPOV BMT Model 2 approaches requiring the use of "Calibrated molecular distances in the management of variety collections" have, as yet, not been rewarded with success. The large datasets and novel statistical approaches that are becoming available offer an opportunity to effectively test the model and until those studies are complete, it is difficult to reach any judgement on the prospects for UPOV BMT Model 2 in this way. On the other hand, studies evaluating UPOV BMT Model 2 approaches "Combining phenotypic and molecular distances in the management of variety collections" have been rewarded with greater success. Taken together, these studies strongly suggest that DNA profiling has a role to play in the management of reference collections, a use that would make distinctness testing more cost effective and this would allow the rapid and rational selection of reference varieties for comparison in growing trials, adding to the effectiveness of PVP.

UPOV BMT Model 3

UPOV BMT Model 3 suggests development of new systems complete replacement of the current system by molecular markers. Complete replacement of the current system by the use of molecular markers has its attractions. Variety registration could be completed in a matter of weeks or months with field inspections becoming a matter of historical interest. It is in this area that high-throughput DNA sequencing could have the greatest impact on future DUS statutory testing using molecular markers. There can be no fuller description of a variety than its entire DNA sequence. However, the ability to describe a variety based on its DNA sequence may pose as many problems as it addresses: should the use of data be based on UPOV BMT Model 1 or 2 as discussed above or should a new system be created following UPOV BMT Model 3? How should uniformity (U) be treated given that polymorphisms may exist between monozygotic (identical) twins? How should stability (S) be addressed when the probability of mutation at any base is of the order of 10^{-8} per base per generation? Even if a satisfactory outcome can be agreed, where would the boundaries for minimum distance and essential derivation be set for distinctness (D) testing? Given that no agreement has yet been reached with the currently available markers, it is unlikely that these problems will be resolved easily when whole genome sequencing becomes cost effective for DUS testing.

Conclusions

The prospects for molecular methods in variety registration are generally promising. Progress has been made in the use of functional markers to replace or augment traditional characteristics in barley. To progress from a whole genome scan identifying QTL for DUS traits to implementing tests based on perfect functional markers requires painstaking work over a number of years. The speed of research in plant genomics will lead to novel testing methods implementing UPOV BMT Model 1 in a range of crops in the near future. High throughput, high density genotyping platforms are increasingly used in plant sciences and breeding. The datasets produced using these methods will be investigated for correlations with the current morphological testing systems and may allow implementation of UPOV BMT Model 2. It is certain that novel systems are being implemented. Use of molecular databases to group varieties is an obvious next step, with an immediate cost benefit for enhanced protection of varieties, and this change can be made without major re-thinking of 'what is a variety?'

The impact of novel testing systems would extend beyond the field of DUS testing and could have consequences in other areas of statutory testing such as seed certification. Currently seed lots are certified by reference to their variety description, thus the use of variety descriptions based wholly, or in part, on molecular data in DUS testing would impose molecular testing on seed certification authorities. There would be clear advantages to this change: using the reference sequence as the varietal description, seed lots could be certified as pure and true to type by assaying samples of seeds without the need for repeated field inspections and the purity of hybrid seed lots with respect to the hybrid formula could be put beyond doubt. This revised system would require a review of the sampling techniques used in seed certification. A revised system may place small scale seed producers at a disadvantage and it could discourage seed production in nations without an infrastructure of sophisticated laboratory facilities.

A radical revision of PVP to utilise the data production potential of 'next generation sequencing' is almost inevitable. There should be urgency in the discussions to redefine 'varieties' with reference to the available data types and a managed transition to a new system that can be implemented in all nations, regardless of their economic status.

Acknowledgements We would like to thank Dr Robert Cooke and Dr John Law for guidance over many years, and Dr Lydia Smith for useful comments. Thanks to FERA (formerly MAFF, DEFRA) and the CPVO for financial support of NIAB research into innovation in DUS testing.

References

Andersen R (2006) Realising Farmers' Rights under the international treaty on plant genetic resources for food and agriculture: summary of findings from the Farmers' Rights project, phase 1. Fridtjof Nansen Institute, Lysaker. ISBN 82-7613-496-3

Biological Innovation for Open Society (BIOS) Tutorial. Can IP rights protect plants? A. Utility Patents. (online) Available at: http://www.patentlens.net/daisy/bios/1234.html. Accessed 18 July 2011

Borchert T, Krueger J, Hohe A (2008) Implementation of a model for identifying essentially derived varieties in vegetatively propagated Calluna vulgaris varieties. BMC Genet 9:56. doi:10.1186/1471-2156-9-56

Bres-Patry C, Lorieux M, Clement G, Bangratz M, Ghesquiere A (2001) Heredity and genetic mapping of domestication-related traits in a temperate japonica weedy rice. Theor Appl Genet 102:118–126

Brooks SA, Yan W, Jackson AK, Deren CW (2008) A natural mutation in rc reverts white-rice-pericarp to red and results in a new, dominant, wild-type allele: Rc-g. Theor Appl Genet 117:575–580. doi:10.1007/s00122-008-0801-8

Bruins M (2009) Essentially derived varieties. FleuroSelect: the International Organisation for the Ornamental Plants Industry Leiden, March. (online) Available at: http://www.fleuroselect.com/uploads/2009.03_02.EDV_ISF_Marcel%20Bruins.ppt.pdf. Accessed 18 July 2011

Button P (2008) Situation in UPOV concerning the use of molecular techniques in plant variety protection. Presented at symposium on the application of molecular techniques for plant breeding and in plant variety protection, Seoul, Korea

Cai HW, Morishima H (2000) Genomic regions affecting seed shattering and seed dormancy in rice. Theor Appl Genet 100:840–846

Chen A, Baumann U, Fincher GB, Collins NC (2009) FLT-2 L, a locus in barley controlling flowering time, spike density and plant height. Funct Integr Genomics 9:243–254

Clark RM, Linton E, Messing J, Doebley JF (2004) Pattern of diversity in the genomic region near the maize domestication gene $tb1$. Proc Natl Acad Sci U S A 101:700–707

Close TJ et al (2009) Development and implementation of high-throughput SNP genotyping in barley. BMC Genomics. doi:10.1186/1471-2164-10-582

Cockram J, Jones H, Leigh FJ, O'Sullivan D, Powell W, Laurie DA, Greenland AJ (2007a) Control of flowering time in temperate cereals: genes, domestication and sustainable productivity. J Exp Bot 58:1231–1244

Cockram J, Chiapparino E, Taylor SA, Stamati K, Donini P, Laurie DA, O'Sullivan DM (2007b) Haplotype analysis of vernalization loci in European barley germplasm reveals novel VRN-$H1$ alleles and a predominant winter VRN-$H1$/VRN-$H2$ multi-locus haplotype. Theor Appl Genet 115:993–1001

Cockram J, Mackay IJ, O'Sullivan DM (2007c) The role of double-stranded break repair in the creation of phenotypic diversity at cereal $VRN1$ loci. Genetics 177:1–5

Cockram J, White J, Leigh FJ, Lea VJ, Chiapparino E, Laurie DA, Mackay IJ, Powell W, O'Sullivan DM (2008) Association mapping of partitioning loci in barley. BMC Genet 9:16. doi:10.1186/1471-2156-9-16

Cockram J, Norris C, O'Sullivan DM (2009) PCR-based markers diagnostic for spring and winter seasonal growth habit in barley. Crop Sci 49:403–410

Cockram J, White J, Zuluaga DL, Smith D, Comadran J, Macaulay M, Luo Z, Kearsey MJ, Werner P, Harrap D et al (2010) Genome-wide association mapping to candidate polymorphism resolution in the unsequenced barley genome. Proc Natl Acad Sci U S A 107:21611–21616

Cockram J, Jones H, Norris C, O'Sullivan DM (2012) Evaluation of diagnostic molecular markers for DUS phenotypic assessment in the cereal crop, barley (*Hordeum vulgare ssp. vulgare* L.). Theor Appl Genet 125:1735–1749

CPV5766 Final Report (2008) Management of winter oilseed rape reference collections. NIAB, Cambridge, CB3 0LE on behalf of Community Plant Variety Office (CPVO), Anger, France

Doveri S, Lee D, Maheswaran M, Powell W (2008) Molecular markers – history, features and applications. In: Kole C, Abbott AG (eds) Principles and practices of plant genomics, vol I, Genome mapping, Chapter 2. Science Publishers Inc., Enfield

Dubreuil P, Warburton M, Chastanet M, Hoisington D, Charcosset A (2006) More on the introduction of temperate maize into Europe: large-scale bulk SSR genotyping and new historical elements. Maydica 51:281–291

Dutfield G (2011) Food, biological diversity and intellectual property: the role of the International Union for the Protection of New Varieties of Plants (UPOV). Global Economic Issue Publications: Intellectual Property Issue Paper No. 9, Quaker United Nations Office. Available at www.quno.org/economicissues/food-sustainability/foodLinks.htm#QUNOPUB

Eiguchi M, Sano Y (1990) A gene complex responsible for seed shattering and panicle spreading found in common wild rices. Rice Genet Newsl 7:105–107

Farmers' Rights: two approaches to Farmers' Rights. Resource pages for decision-makers and practitioners. (online) Available at: http://www.farmersrights.org/about/fr_contents_1.html. Accessed 18 Jul 2011

Faure S, Higgins J, Turner A, Laurie DA (2007) The FLOWERING LOCUS-T-like family in barley (*Hordeum vulgare*). Genetics 176:599–609

Flavell AJ, Knox MR, Pearce SR, Ellis THN (1998) Retrotransposon-based insertion polymorphisms (RBIP) for high throughput marker analysis. Plant J 16:643–650

Fukuta Y, Yagi T (1998) Mapping of a shattering resistance gene in a mutant line SR-5 induced from an indica rice variety, Nan-jing11. Breed Sci 48:345–348

Gunjaca J, Buhinicek I, Jukic M, Sarcevic H, Vragolovic A, Kozic Z, Jambrovic A, Pejic I (2008) Discriminating maize inbred lines using molecular and DUS data. Euphytica 161:165–172

Heckenberger M, van der Voort JR, Peleman J, Bohn M (2003) Variation of DNA fingerprints among accessions within maize inbred lines and implications for identification of essentially derived varieties: II. Genetic and technical sources of variation in AFLP data and comparison with SSR data. Mol Breed 12:97–106. doi:10.1023/A:1026040007166

Heckenberger M, Bohn M, Frisch M, Maurer HP, Melchinger AE (2005) Identification of essentially derived varieties with molecular markers: an approach based on statistical test theory and computer simulations. Theor Appl Genet 111:598–608. doi:10.1007/s00122-005-2052-2

Huang X, Zhao Y, Wei X, Li C, Wang A, Zhao Q, Li W, Guo Y, Deng L, Zhu C, Fan D, Lu Y, Weng Q, Liu K, Zhou T, Jing Y, Si L, Dong G, Huang T, Lu T, Feng Q, Qian Q, Li J, Han B (2012) Genome-wide association study of flowering time and grain yield traits in a worldwide collection of rice germplasm. Nat Genet 44:32–39

Ibáñez J, Vélez MD, de Andrés MT, Borrego J (2009) Molecular markers for establishing distinctness in vegetatively propagated crops: a case study in grapevine. Theor Appl Genet 119:1213–1222

Jaccoud D, Peng K, Feinstein D, Kilian A (2001) Diversity Arrays: a solid state technology for sequence information independent genotyping. Nucleic Acids Res 29:e25

Jones CJ, Edwards KJ, Castaglione S, Winfield MO, Sala F, van de Wiel C, Bredemeijer G, Vosman B, Matthes M, Daly A et al (1997) Reproducibility testing of RAPD, AFLP and SSR markers in plants by a network of European laboratories. Mol Breed 3:381–390

Jones H, Jarman RJ, Austin L, White J, Cooke RJ (2003) The management of variety reference collections in distinctness, uniformity and stability testing of wheat. Euphytica 132:175–184

Jones H, Bernole A, Jensen LB, Horsnell RA, Law JR, Cooke RJ, Norris CE (2008a) Minimising inter-laboratory variation when constructing a unified molecular database of plant varieties in an allogamous crop. Theor Appl Genet 117:1335–1344

Jones H, Leigh FJ, Mackay I, Bower MA, Smith LMJ, Charles MP, Jones G, Jones MK, Brown TA, Powell W (2008b) Population-based resequencing reveals that the flowering time adaptation of cultivated barley originated east of the fertile crescent. Mol Biol Evol 25:2211–2219

Jones H, Norris C, Smith D, Cockram J, Lee D, O'Sullivan DM, Mackay I (2012) Evaluation of the use of high-density SNP genotyping to implement UPOV Model 2 for DUS testing in barley. Theor Appl Genet. doi:10.1007/s00122-012-2024-2

Kalendar R, Grob T, Regina M, Suoniemi A, Schulman A (1999) IRAP and REMAP: two new retrotransposon-based DNA fingerprinting techniques. Theor Appl Genet 98:704–711

Karsai I, Szűcs P, Mészáros K, Filichkina T, Hayes PM, Skinner JS, Láng L, Bedö Z (2005) The *Vrn-H2* locus is a major determinant of flowering time in a facultative X winter growth habit barley (*Hordeum vulgare* L.) mapping population. Theor Appl Genet 110:1458–1466

Komatsuda T, Pourkheirandish M, He C, Azhaguvel P, Kanamori H et al (2007) Six-rowed barley originated from a mutation in a homeodomain-leucine zipper I-class homeobox gene. Proc Natl Acad Sci U S A 104:1424–1429

Laurie DA, Pratchett N, Bezant JH, Snape JW (1995) RFLP mapping of five major genes and eight quantitative trait loci controlling flowering time in a winter X spring barley (*Hordeum vulgare* L.) cross. Genome 38:575–585

Law JR, Donini P, Koebner RMD, Reeves JC, Cooke RJ (1998) DNA profiling and plant variety registration. III: The statistical assessment of distinctness in wheat using amplified fragment length polymorphisms. Euphytica 102:335–342

LeDuc C, Miller P, Lichter J, Parry P (1995) Batched analysis of genotypes. PCR Methods Appl 4:331–336

Lee D, Reeves JC, Cooke RJ (1996) DNA profiling and plant variety registration: 1. The use of random amplified polymorphisms to discriminate between varieties of oilseed rape. Electrophoresis 17:261–265

Lee D, Lupotto E, Powell W (2009) G-string slippage turns white rice red. Genome 52:490–493

Leigh F, Lea V, Law J, Wolters P, Powell W, Donini P (2003) Assessment of EST- and genomic microsatellite markers for variety discrimination and genetic diversity studies in wheat. Euphytica 133:359–366. doi:10.1023/A:1025778227751

Lundqvist U, Franckowiak JD, Konishi T (1997) New and revised descriptions of barley genes. Barley Genet Newslett 26:22–516

Mackay I, Powell W (2007) Methods for linkage disequilibrium mapping in crops. Trends Plant Sci 12:57–63

Nagao S, Takahashi M (1963) Trial construction of twelve linkage groups in Japanese rice. J Fac Agric Hokkaido Univ 53:72–130

Nair SK, Wang N, Turuspekov Y, Pourkheirandish M, Sinsuwongwat S et al (2010) Cleistogamous flowering in barley arises from the suppression of microRNA-guided *HvAP2* mRNA cleavage. Proc Natl Acad Sci U S A 107:490–495

Nielsen R, Paul JS, Albrechtsen A, Song YS (2011) Genotype and SNP calling from next-generation sequencing data. Nat Rev Genet 12:443–451

Noli E, Teriaca MS, Sanguineti MC, Conti S (2008) Utilization of SSR and AFLP markers for the assessment of distinctness in durum wheat. Mol Breed 22:301–313

Oba S, Kikuchi F, Maruyama K (1990) Genetic analysis of semidwarfness and grain shattering of Chinese rice (*Oryza sativa*) variety "Ai-Jio-Nan-Te". Jpn J Breed 40:13–20

Reid A, Hof L, Felix G, Rücker B, Tams S, Milczynska E, Esselink D, Uenk G, Vosman B, Weitz A (2011) Construction of an integrated microsatellite and key morphological characteristic database of potato varieties on the EU common catalogue. Euphytica. doi:10.1007/s10681-011-0462-6

Rogers YH, Craig Venter JC (2005) Massively parallel sequencing. Nature 437:326–327

Saiki RK, Gelfand DH, Stoffel S, Scharf SJ, Higuchi R, Horn GT, Mullis KB, Erlich HA (1988) Primer-directed enzymatic amplification of DNA with a thermostable DNA polymerase. Science 239:487–491

Salazar R, Louwaars NP, Visser B (2006) On protecting farmers' new varieties: new approaches to rights on collective innovations in plant genetic resources. CGIAR system wide program on collective action and property rights working paper 45. International Food Policy Research Institute, Washington, DC

Smulders MJM, Esselink D, Voorrips RE, Vosman B (2009) Analysis of a database of DNA profiles of 734 hybrid tea rose varieties. Acta Hort (ISHS) 836:169–175, http://www.actahort.org/books/836/836_24.htm

Sweeney MT, Thomson MJ, Pfeil BE, McCouch S (2006) Caught red-handed: *Rc* encodes a basic helix–loop–helix protein conditioning red pericarp in rice. Plant Cell 18:283–294. doi:10.1105/tpc.105.038430

Sweeney MT, Thomson MJ, Cho YG, Park YJ, Williamson SH, Bustamante CD, McCouch SR (2007) Global dissemination of a single mutation conferring white pericarp in rice. PLoS Genet 3:e133. doi:10.1371/journal.pgen.0030133

Szűcs P, Skinner JS, Karsai I, Cuesta-Marcos A, Haggard KG, Corey AE, Chen THH, Hayes PM (2007) Validation of the *VRN-H2/VRN-H1* epistatic model in barley reveals that intron length variation in *VRN-H1* may account for a continuum of vernalization sensitivity. Mol Genet Genomics 277:249–261

Taketa S, Amano S, Tsujino Y, Sato T, Saisho D, Kakeda K, Nomura M, Suzuki T, Matsumoto T, Sato K, Kanamori H, Kawasaki S, Takeda K (2008) Barley grain with adhering hulls is controlled by an ERF family transcription factor gene regulating a lipid biosynthesis pathway. Proc Natl Acad Sci U S A 105:4062–4067

Thomson MJ, Tai TH, McClung AM, Lai XH, Hinga ME et al (2003) Mapping quantitative trait loci for yield, yield components and morphological traits in an advanced backcross population between *Oryza rufipogon* and the *Oryza sativa* cultivar Jefferson. Theor Appl Genet 107:479–493

Tian F, Stevens NM, Buckler ES IV (2009) Tracking footprints of maize domestication and evidence for a massive selective sweep on chromosome 10. Proc Natl Acad Sci U S A 106:9979–9986

Tommasini L, Batley J, Arnold GM, Cooke RJ, Donini P, Lee D, Law JR, Lowe C, Moule C, Trick M, Edwards KJ (2003) The development of multiplex simple sequence repeats (SSR) markers to complement distinctness, uniformity and stability testing of rape (*Brassica napus* L.) varieties. Theor Appl Genet 106:1091–1101

Tripp R, Louwaars N, Eaton D (2007) Plant variety protection in developing countries. A report from the field. Food Policy 32:354–371

Turner A, Beales J, Faure S, Dunford RP, Laurie DA (2005) The pseudo-response regulator *Ppd-H1* provides adaptation to photoperiod in barley. Science 310:1031–1034

Turuspekov Y, Mano Y, Honda I, Kawada N, Watanabe Y, Komatsuda T (2004) Identification and mapping of cleistogamy genes in barley. Theor Appl Genet 109:480–487

Ubisch G (1916) Beitrag zu einer Faktorenanalyse von Gerste. Z Indukt Abs Ver 17:120–152

UPOV (1991) International convention for the protection of new varieties of plants. International Union for the Protection of New Varieties of Plants, Geneva. (online) Available at: http://www.upov.int/upovlex/en/conventions/1991/act1991.html. Accessed 8 Aug 2012

UPOV (2002) TG/1/3 General introduction to the examination of distinctness, uniformity and stability and the development of harmonized descriptions of new varieties of plants. International Union for the Protection of New Varieties of Plants, Geneva. (online) Available at: http://www.upov.int/en/publications/tg-rom/tg001/tg_1_3.pdf, http://www.upov.int/upovlex/en/conventions/1991/act1991.html. Accessed 8 Aug 2012

UPOV (2008a) TGP/4: constitution and maintenance of variety collections. International Union for the Protection of New Varieties of Plants, Geneva. (online) Available at: http://www.upov.int/export/sites/upov/en/publications/tgp/documents/tgp_4_1.pdf. Accessed 19 July 2011

UPOV (2008b) TGP/9: examining distinctness. International Union for the Protection of New Varieties of Plants, Geneva. (online) Available at: http://www.upov.int/export/sites/upov/en/publications/tgp/documents/tgp_4_1.pdf. Accessed 8 Aug 2012

UPOV (2010 a) TGP/7/2. Development of test guidelines. International Union for the Protection of New Varieties of Plants, Geneva. (online) Available at: http://www.upov.int/export/sites/upov/en/publications/tgp/documents/tgp_7_2.pdf. Accessed 19 July 2012

UPOV (2010 b) TGP/8. Trial design and techniques used in the examination of distinctness, uniformity and stability. International Union for the Protection of New Varieties of Plants, Geneva. (online) Available at: http://www.upov.int/export/sites/upov/en/publications/tgp/documents/tgp_8_1.pdf. Accessed 28 July 2011

UPOV (2010c) INF/17/1: Guidelines for DNA-profiling: molecular marker selection and database construction ("BMT Guidelines"). International Union for the Protection of New Varieties of Plants, Geneva. (online) Available at: http://www.upov.int/export/sites/upov/en/publications/pdf/upov_inf_17_1.pdf. Accessed 19 July 2011

UPOV (2011a) Mission statement. International Union for the Protection of New Varieties of Plants, Geneva. (online) Available at: http://upov.int/about/en/index.html. Accessed 19 July 2012

UPOV (2011b) About UPOV. International Union for the Protection of New Varieties of Plants, Geneva. (online) Available at: http://www.upov.int/about/en/overview.html. Accessed 8 Aug 2012

UPOV (2011c) Test guidelines. International Union for the Protection of New Varieties of Plants, Geneva. (online) Available at: http://www.upov.int/en/publications/tg_rom/. Accessed 8 Aug 2012

UPOV (2011d) INF/18/1 Possible use of molecular markers in the examination of distinctness, uniformity and stability (DUS). International Union for the Protection of New Varieties of Plants, Geneva. (online) Available at: http://www.upov.int/edocs/infdocs/en/upov_inf_18_1.pdf. Accessed 8 Aug 2012

von Zitzewitz J, Szűcs P, Dubcovsky J, Yan L, Francia E, Pecchioni N, Casas A, Chen THH, Hayes P, Skinner J (2005) Molecular and structural characterization of barley vernalization genes. Plant Mol Biol 59:449–467

Vos P, Hogers R, Bleeker M, Reijans M, van de Lee T, Hornes M, Friters A, Pot J, Paleman J, Kuiper M, Zabeau M (1995) AFLP: a new technique for DNA fingerprinting. Nucleic Acids Res 23:4407–4414

Vosman B, Barendrecht J, Esselink D, Jones H, Scott E, Spellerberg B, Tams S (2006) A European reference collection of rose varieties. Plant Research International B.V., Wageningen, The Netherlands. On behalf of Community Plant Variety Office (CPVO), Anger, France. Available at: http://www.cpvo.europa.eu/documents/techreports/RD_rose_project_final_report.pdf. Accessed 19 July 2011

Wellcome Trust Case Control Consortium (2007) Genome-wide association study of 14,000 cases of seven common diseases and 3,000 shared controls. Nature 447;7145:661–678

Williams JGK, Kubelik AR, Livak KJ, Rafalski JA, Tingey SV (1992) DNA polymorphisms amplified by arbitrary primers are useful as genetic markers. Nucleic Acids Res 18:6531–6535

World Trade Organization (1994) Agreement on trade-related aspects of intellectual property rights. (online) Available at: http://www.wto.org/english/tratop_e/trips_e/t_agm0_e.htm. Accessed 18 July 2011

Xiong LZ, Liu KD, Dai XK, Xu CG, Zhang Q (1999) Identification of genetic factors controlling domestication-related traits of rice using an F2 population of a cross between *Oryza sativa* and *O. Rufipogon*. Theor Appl Genet 98:243–251

Yan LL, Loukoianov A, Blechl A, Tranquilli G, Ramakrishna W, San-Miguel P, Bennetzen JL, Echenique V, Dubcovsky J (2004) The wheat *VRN2* gene is a flowering repressor down-regulated by vernalization. Science 303:1640–1644

Part IX
Examples

Chapter 19
Qualitative and Quantitative Trait Polymorphisms in Maize

Qin Yang and Mingliang Xu

Introduction

Maize (*Zea mays* L.) is an important cereal worldwide, serving as staple food, livestock feed, or as raw materials for industrial purposes. Over the last century, breeders have increased grain yields eightfold (Troyer 2006), partly by harnessing heterosis that can increase yields of hybrids by 15–60% relative to inbred parents (Duvick 1999). It is estimated that the demand for corn is expected to grow over the next decade by about 15%, or roughly 200 million metric tons per year (Service 2009), due to population growth and usage of maize grain for biofuel production. To meet this growing demand, maize yield has to be increased, more arable land has to be employed for maize cultivation, or both. With rapidly increasing maize genomic information, implementation of molecular tools will accelerate maize breeding efforts that exploit the high levels of morphological and nucleotide diversity in maize (Flint-Garcia et al. 2005). Indeed, unrelated maize inbred lines have on average a greater divergence than hominids (Springer et al. 2009). Over 47,000 accessions of maize exist in gene banks around the world, about 27,000 of which are stored at the International Maize and Wheat Improvement Center (CIMMYT) (http://www.croptrust.org/documents/web/Maize-Strategy-FINAL-18Sept07.pdf), including inbred lines, improved populations, traditional farmer's populations (landraces), and wild relatives. It is estimated that less than 5% of the available maize germplasm is used in commercial breeding programs world-wide, and less than 1% in the U.S. (Hoisington et al. 1999). Lack of characterization of germplasm stored in gene banks is one major impediment of increased usage (Yan et al. 2009).

Q. Yang • M. Xu (✉)
National Maize Improvement Center of China, China Agricultural University,
2 West Yuanmingyuan Rd., Beijing 100193, P.R. China
e-mail: yq_smile_745@163.com; mxu@cau.edu.cn

Molecular characterization of the functional variation in both known and unknown genes associated with important agronomic traits provides a powerful tool for investigating the contribution of specific genes to the overall genetic architecture underlying complex traits. Furthermore, identification of causative polymorphisms underlying these traits and development of respective functional markers will accelerate deployment of marker-assisted selection (MAS) in maize breeding, such as for disease resistance, nutritive value, and drought tolerance. To date, identifying the genetic components that contribute to quantitative traits in maize has mainly been carried out using linkage analysis in biparental mapping populations or association mapping panels. Linkage analysis usually relies on crosses between two inbred lines, thus capturing only a fraction of genetic diversity within maize. By contrast, association mapping widely samples genetic diversity, but has less power to detect quantitative trait loci (QTL) when alleles are not common (Mackay 2009). Recently, a nested association mapping (NAM) approach was developed to combine the benefits of linkage and association mapping and take advantage of both historic and recent recombination events in a single population. A NAM population consisting of 25 families with 200 RILs for each family has been developed and released as a genetic resource for identification of genome regions affecting various traits in maize (Yu et al. 2008; McMullen et al. 2009). It maximizes the genetic diversity in maize and provides genome-wide coverage with high resolution and is robust to genetic heterogeneity with representation of several alleles per locus.

This chapter provides a comprehensive overview on qualitative or quantitative trait polymorphisms (QTPs) for agronomic traits in maize. Information on the underlying genes and sequence motifs, genetic effects, as well as derived markers assay (if established) will be presented and thus made readily available to the maize community.

Genes Controlling Developmental and Morphological Characters of Maize

The genetic variation in plant growth rate and morphological traits have long been of interest for plant breeders. Maize was domesticated from teosinte by strong artificial selection on particular alleles of genes controlling traits of agronomic importance to become the world's largest production grain crop. Cultivated maize exhibits a profound morphological difference from its progenitor, which includes highly modified inflorescences, plant architecture, and a striking increase in yield. The majority of these genes controlling growth and development in maize encode transcription factors, or molecules that regulate transcription factors.

Several candidate regulatory genes have been tested in association analyses. Some of these genes exhibit pleiotropic effects (Table 19.1). The *zea floricaula leafy1* (*zfl1*) gene and its paralogue *zfl2* were associated with internodes in the lateral branch and tiller number, respectively (Weber et al. 2008). In another study, *zfl2* was also identified to be associated with plant height (Weber et al. 2007).

19 Qualitative and Quantitative Trait Polymorphisms in Maize

Table 19.1 Summary of association studies on candidate genes controlling developmental and morphological characters in maize

Gene	Affected traits	Plant material	Sequence motif	Genetic effect (%)	Reference
Dwarf8	Plant height	71 European inbred lines	1 Indel		Andersen et al. (2005)
	Flow erring time	92 maize inbred lines	9 polymorphisms		Thornsberry et al. (2001)
	Flow erring time	375 inbred lines and 275 landraces	1 polymorphisms		Camus-Kulandaivelu et al. (2006)
bif2	Tassel length	377 diverse maize inbreds	7 polymorphisms		Pressoir et al. (2009)
	Plant height	277 diverse maize inbreds	4 polymorphisms		Pressoir et al. (2009)
	Node number	277 diverse maize inbreds	7 polymorphisms		Pressoir et al. (2009)
	Leaf length	277 diverse maize inbreds	8 polymorphisms		Pressoir et al. (2009)
	Flowering time	277 diverse maize inbreds	9 polymorphisms		Pressoir et al. (2009)
4CL2	Plant height	39 European elite maize lines	2 SNPS	14.30	Chen et al. (2010)
PAL	Days to silking	39 European elite maize lines	17 polymorphisms	7.00	Chen et al. (2010)
CCoAOMT2	Plant height	39 European elite maize lines	1 Indel	25.80	Chen et al. (2010)
	Days ti silking	39 European elite maize lines	1 Indel	18.50	Chen et al. (2010)
4CL1	Days to silking	39 European elite maize lines	Indel454	20.20	Chen et al. (2010)
	Days to silking	39 European elite maize lines	Indel810	6.00	Chen et al. (2010)
COMT	Days silking	39 European elite maize lines	1 Indel	10.30	Chen et al. (2010)
F5H	Days to silking	39 European elite maize lines	1 SNP	22.40	Chen et al. (2010)
Vgt1	Flowerring time	375 maize inbred lines	4 polymorphisms		Ducrocq et al. (2008)
	Days to pollen shed	95 inbred lines	3 polymorphisms	32.00	Salvi et al. (2007)
	Leaf number	95 inbred lines	3 polymorphisms	29.00	Salvi et al. (2007)
tb1	Branch length	584 Balsas teosinte individuals			Weber et al. (2007, 2008)
	Female ear length	1612 Balsas teosinte plants			Weber et al. (2009)

(continued)

Table 19.1 (continued)

Gene	Affected traits	Plant material	Sequence motif	Genetic effect (%)	Reference
	Lateral inflorescence branch number	1612 Balsas teosinte plants			Weber et al. (2009)
	Tassel branch number	1612 Balsas teosinte plants			Weber et al. (2009)
te1	Female ear length	584 Balsas teosinte individuals			Weber et al. (2007)
	Female ear length	817 Balsas teosinte plants			Weber et al. (2008)
tga1	Naked grains	16 maize landraces and 12 teosinte plants	1 polymorphism		Wang et al. (2005)
ZmGS3	Kernel length	121 maize inbred lines	1 polymorphism	5.21–5.77	Li et al. (2010c)
	Hundred kernel weight	121 maize inbred lines	P319	6.29–7.73	Li et al. (2010c)
ZmGW2-CHR4	Kernel width	121 maize inbred lines	S40	8.00	Li et al. (2010b)
	Kernel width	121 maize inbred lines	S27, S304, and S1730		Li et al. (2010b)
	Kernel thickness	121 maize inbred lines	S628		Li et al. (2010b)
	Hundred Kernel weight	121 maize inbred lines	S40	11.40	Li et al. (2010b)
ZmGW2-CHR5	Kernel width	121 maize inbred lines	S1789		Li et al. (2010b)
	Kernel thickness	121 maize inbred lines	S1632		Li et al. (2010b)
zagl1	Ear shattering	817 Balsas teosinte plants			Weber et al. (2008)
	Female ear length	817 Balsas teosinte plants			Weber et al. (2008)
	Leaf number	817 Balsas teosinte plants			Weber et al. (2008)
	tassel branch number	817 Balsas teosinte plants			Weber et al. (2008)
ebm1	Leaf number	817 Balsas teosinte plants			Weber et al. (2008)
ZmCIR1	Leaf number	817 Balsas teosinte plants			Weber et al. (2008)
ZmG1	Leaf number	817 Balsas teosinte plants			Weber et al. (2008)

Gene	Trait	Sample	Reference
zfl1	Internodes in the lateral branch	817 Balsas teosinte plants	Weber et al. (2008)
zfl2	Tiller number	817 Balsas teosinte plants	Weber et al. (2008)
	Plant height	584 Balsas teosinte individuals	Weber et al. (2007)
zen1	Length of the lateral branches	584 Balsas teosinte individuals	Weber et al. (2007)
zap1	Inflorescence branching	584 Balsas teosinte individuals	Weber et al. (2007)
te1	Female ear length	584 Balsas teosinte individuals	Weber et al. (2007)
	Female ear length	817 Balsas teosinte plants	Weber et al. (2008)
ra1	Female ear length	817 Balsas teosinte plants	Weber et al. (2008)
	Short branch	30 maize inbred lines	Vollbrecht et al. (2005)
	Branching architecture	22 maize landraces and 21 teosintes	Sigmon and Vollbrecht (2010)
ra2	Ear structure	817 Balsas teosinte plants	Weber et al. (2008)
ids1	Inflorescence branching	817 Balsas teosinte plants	Weber et al. (2008)
lg1	Upper leaf angle	NAM population and 282 diverse maize inbred lines	Tian et al. (2011)
lg2	Upper leaf angle	NAM population and 282 diverse maize inbred lines	Tian et al. (2011)
idh	IDH enzyme activity	100 maize inbred lines	Zhang et al. 2010

The *teosinte branched1* (*tb1*) gene, a transcription factor whose regulatory changes have been proposed to underlie QTL of large effect for morphological differences (Clark et al. 2006), was found to be significantly associated with variation in both plant and inflorescence architecture (Weber et al. 2007, 2008, 2009). Two *ramosa* genes (*ra1* and *ra2*) that control branching architecture in tassel and ear of maize were validated in different populations (Vollbrecht et al. 2005; Bortiri et al. 2006; Sigmon and Vollbrecht 2010; Weber et al. 2008). The *zea apetala homolog1* (*zap1*) gene, encoding a transcription factor homologous to *APETALA1* of Arabidopsis, was associated with inflorescence branching (Weber et al. 2007). Using 584 Balsas teosinte individuals, Weber et al. (2007) observed that genes *zen1* (*zea centroradialis1*) and *te1* (*terminal ear1*) correlated with length of the lateral branches and female ear length, respectively. The MADS-box gene, *zagl1* (*zea agamous-like1*), was identified to be associated with ear shattering, female ear length, leaf number, and tassel branch number (Weber et al. 2008). Moreover, *elm1* (*elongated mesocotyl1*), *ZmCIR1* (*zea mays circadian1*), and *ZmGI* (*zea mays gigantea*) were all associated with leaf number across a panel of 817 Balsas teosinte plants (Weber et al. 2008). All observed effects in terms of the percentage of phenotypic variation explained were less than 10% (Weber et al. 2008). Pressoir et al. (2009) identified allelic variation at *barren inflorescence2* (*bif2*) in maize, which affected tassel architecture through association, linkage, and mutagenesis analysis. Association mapping using 277 diverse maize inbreds identified numerous SNPs, spanning the entire *bif2* region, which were associated with plant height, node number, leaf length, and flowering time (Pressoir et al. 2009).

One of the utmost important gene in maize domestication, *tga1* (*teosinte glume architecture 1*), responsible for the 'naked' grain phenotype of maize, has been positionally cloned (Wang et al. 2005), and encodes a SBP (SQUAMOSA promoter binding protein) transcriptional regulator. An amino acid mutation in *tga1* between maize and teosinte was shown to be the causative site in a set of 16 diverse maize landraces and 12 teosinte individuals. Grain size, a key component of grain yield, is another important domestication trait in maize. Recently, Li et al. (2010c) isolated the maize orthologue of *GS3,* which controls grain size in rice, named *ZmGS3*. A panel of 121 maize inbred lines was investigated for associations between *ZmGS3* and corn grain size. One polymorphism in the fifth exon and one polymorphism in the promoter of *ZmGS3* were found to be significantly associated with kernel length (KL) and 100-kernel weight (HKW), respectively (Li et al. 2010c). Two homologs of *GW2*, *ZmGW2-CHR4* and *ZmGW2-CHR5*, were both related with corn grain size or weight. One SNP in the promoter region of *ZmGW2-CHR4* was found to be significantly associated with kernel width (KW) and HKW. There are other polymorphisms within *ZmGW2-CHR4* and *ZmGW2-CHR5*, which showed significant associations with at least one of the four yield-related traits, including KL, KW, HKW, and kernel thickness (KT) (Li et al. 2010b).

Leaf architecture, including leaf angle, leaf length, and leaf width, is one of the critical factors affecting maize yield by influencing photosynthetic efficiency. Using the maize NAM population, Tian et al. (2011) observed significant associations around the *lg1* (*liguleless1*) and *lg2* (*liguleless2*) genes with upper leaf angle, while no correlations around candidate genes for leaf length and width were identified. Functional markers derived from *lg1* and *lg2* could be used to improve maize production by obtaining optimum leaf architecture for high plant density (Tian et al. 2011).

Flowering time is a complex trait that is fundamental for regional climatic adaptation of elite germplasm in maize production. Although a large number of QTL for flowering time were mapped, only one major QTL, *Vgt1* (*Vegetative to generative transition1*) has been cloned using a map-based approach. *Vgt1* functions as a cis-regulatory element of the floral repressor gene *ZmRap2.7*, which is located 70 kb downstream of *Vgt1*. Three polymorphisms within *Vgt1*, *G/A/indel324*, *Mite*, and *ATindel434*, were shown to be strongly associated with flowering time using a panel of 95 inbred lines (Salvi et al. 2007). The correlation between flowering time variation and polymorphisms in the *Vgt1* region was confirmed in a larger population consisting of 375 maize inbred lines. Another polymorphism within *Vgt1*, *CGindel587*, showed significant association with flowering time variation (Ducrocq et al. 2008). Allele frequencies at *Vgt1* were highly correlated with geographical origin in a 256-landrace collection, which suggests that functional markers in *Vgt1* could be used by breeders for targeted genetic modification of flowering time (Ducrocq et al. 2008; Jung and Muller 2009).

Nine sequence motifs across the *Dwarf8* gene, involved in the gibberellin pathway (Peng et al. 1999), were shown to be associated with variation for flowering time but not for plant height in a 92 maize inbred line panel representing North American temperate and subtropical modern origins (Thornsberry et al. 2001). Furthermore, one of the nine polymorphisms adjacent to the SH2-like domain was evaluated to be flowering time related in 375 inbred lines and 275 landraces (Camus-Kulandaivelu et al. 2006). Additionally, Andersen et al. (2005) tested the general applicability of the polymorphisms as functional markers (FMs) in 71 elite European inbred lines and identified a 2-bp insertion/deletion (indel) polymorphism in the *Dwarf8* promoter, which was associated with plant height but not with flowering time. It was proposed to use isogenic backgrounds to estimate differential allelic effects (Andersen et al. 2005). Recently, Cassani et al. (2009) characterized an amino acid insertion in the VHYNP domain of the maize *Dwarf8* gene that also leads to a dwarf phenotype but less severe than other mutants with mutations in DELLA domains. The *Dwarf8* gene is one of the limited successful studies to resolve allelic function at the sequence level. It is suggested that *Dwarf8* affects not only flowering time but also plant height. Allelic differences in different regions of the *Dwarf8* gene contribute to different traits. Application of these polymorphisms as functional markers will enable selection of optimal alleles for different traits, as QTPs were generally not pleiotropic (Chen and Lubberstedt 2010).

Recently, it was reported that several genes involved in monolignol biosynthesis were associated with morphological traits such as plant height and flowering time in a population of 39 European elite maize lines (Chen et al. 2010). A gene coding for PAL (phenylalanine ammonia lyase) showed 17 polymorphisms correlated with days to silking. Two SNPs in the *4CL2* (4-coumarate: CoA ligase2) locus were found to be associated with plant height that could explain 14.3% of the phenotypic variation. One indel within the *CCoAOMT2* (caffeoyl-CoA O-methyltransferase2) locus explained 25.8 and 18.5% of variation for plant height and days to silking, respectively. Two indels in the *4CL1* locus were both associated with days to silking. Two genes encoding for respective COMT (caffeic acid O-methyltransferase) and F5H (ferulate 5-hydroxylase) were both related to days to silking. Chen et al. (2010) found that independent polymorphisms are responsible for forage quality and morphological trait except for *4CL1* and *F5H*. It suggests that obtaining optimal alleles for both forage quality and yield traits using functional markers in these polymorphisms should be possible.

Genes Controlling Forage Quality of Maize

Maize is one of the most important forage crops worldwide. Cell wall digestibility is a limiting factor for improving feeding value of maize. Both lignin content and lignin structure influence cell wall digestibility (Barriere et al. 2003). The phenylpropanoid pathway is responsible for lignin biosynthesis (Boerjan et al. 2003). Comprehensive knowledge of lignin biogenesis is crucial towards breeding of highly digestible maize (Table 19.2).

The first step in lignin biosynthesis is removing ammonia from L-phenylalanine to produce p-coumaric acid catalyzed by phenylalanine ammonia lyase (PAL) (Winkel 2004). The following steps involve a set of enzymes including cinnamate 4-hydroxylase (C4H), 4-coumarate:CoA ligase (4CL), hydroxycinnamoyl-CoA transferase (HCT), p-coumarate 3-hydroxylase (C3H), caffeoyl-CoA *O*-methyltransferase (CCoAOMT), cinnamoyl-CoA reductase (CCR), ferulate 5-hydroxylase (F5H), caffeic acid *O*-methyltransferase (COMT), and cinnamyl alcohol dehydrogenase (CAD) catalyzing the biosynthesis of monolignols (Brenner et al. 2010) (Fig. 19.1). Genes encoding these enzymes have been isolated based on sequence homology or expression profiling. Several studies identified causative polymorphisms for forage quality traits in these genes using association analyses. Thus, functional markers can be derived from these polymorphic sites and applied in breeding programs to improve forage quality (Andersen and Lubberstedt 2003).

Two *brown midrib* genes (*bm1* and *bm3*) related to forage quality have been isolated, and shown to be involved in monolignol biosynthesis. The *bm1* mutation leads to decrease of CAD activity (Halpin et al. 1998), while the *bm3* mutation reduces the COMT activity (Vignols et al. 1995). Plants with the *bm1* and *bm3* alleles show lower lignin content and higher forage quality than normal genotypes, but negative effects have been observed for agronomic performance, such as reduced yield and stalk lodging (Barriere and Argillier 1993). COMT controls the biosynthesis of syringyl lignin units (Boerjan et al. 2003). To obtain optimal *bm3* alleles for improving forage quality, an association study was performed between the *COMT* gene and digestible neutral detergent fiber (DNDF) in a panel of 42 European maize elite inbred lines (Lubberstedt et al. 2005). One Indel polymorphism located in the intron was shown to be significantly associated with

Table 19.2 Summary of association studies on candidate genes controlling forage quality in maize

Gene	Affected traits	Plant material	Sequence motif	Genetic effect (%)	Reference
COMT	Digestible Neutral Detergent Fiber (DNDF)	42 European maize elite inbred lines	1 Indel		Lubberstedt et al. (2005)
COMT	Digestible Neutral Detergent Fiber	40 European maize elite inbred lines	16 SNPs and 9 Indels		Brenner et al. (2010)
	OMD, Organic Matter Digestibility	40 European maize elite inbred lines	12 polymorphic sites		Brenner et al. (2010)
	NDF, Neutral Detergent Fiber	40 European maize elite inbred lines	1 polymorphic site		Brenner et al. (2010)
	WSC, Water Soluble Carbohydrate	40 European maize elite inbred lines	6 polymorphic sites		Brenner et al. (2010)
	Cell wall digestibility	34 maize inbred lines	1 Indel		Guillet-Claude et al. (2004a)
CCoAOMT1	NDF, Neutral Detergent Fiber	40 European maize elite inbred lines	2 SNPs		Brenner et al. 2010
	OMD, Organic Matter Digestibility	40 European maize elite inbred lines	1 SNP		Brenner et al. (2010)
	WSC, Water Soluble Carbohydrate	40 European maize elite inbred lines	1 SNP and 2 Indels		Brenner et al. (2010)
CCoAOMT2	WSC, Water Soluble Carbohydrate	40 European maize elite inbred lines	2 Indels		Brenner et al. (2010)
	Cell wall digestibility	34 maize inbred lines	an 18-bp indel		Guillet-Claude et al. (2004a)
F5H	Dry matter content, DMC	39 European elite maize lines	2 SNPs and 1 Indel	23.50	Chen et al. (2010)
	Dry matter content, DMC	39 European elite maize lines	1 Indel	10.50	Chen et al. (2010)
	Dry matter yield, DMY	39 European elite maize inbred lines	2 SNPs in complete LD		Chen et al. (2010)
	NDF, Neutral Detergent Fiber	40 European forage maize inbred lines	1 intron SNP		Andersen et al. (2008)
4CL1	In vitro digestibility of organic matter, IVDOM	40 European forage maize inbred lines	1 Indel		Andersen et al. (2008)
C3H	In vitro digestibility of organic matter, IVDOM	40 European forage maize inbred lines	1 SNP		Andersen et al. (2008)
PAL	In vitro digestibility of organic matter, IVDOM	32 maize inbred lines	1 polymorphic site		Andersen et al. (2007)
ZmPox3	Cell wall digestibility	31 maize inbred lines	1 MITE insertion		Guillet-Claude et al. (2004b)
ZmPox3	Cell wall digestibility	25 flint maize inbred lines	1 MITE insertion		Guillet-Claude et al. (2004b)

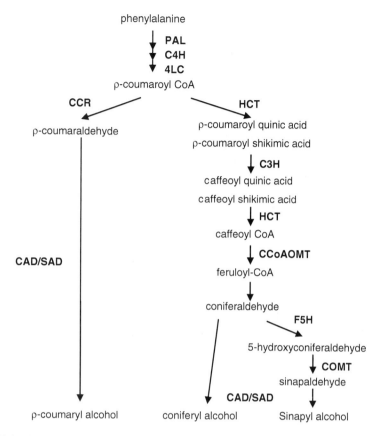

Fig. 19.1 Simplified monolignols biosynthesis pathway in grasses. Enzymes are shown in *bold*

DNDF which could be converted into a PCR-based functional marker for cell wall digestibility in maize. In another study, conducted by Guillet-Claude et al. (2004a), a single-base pair deletion located in the intron of the *COMT* gene was found to show significant association with cell wall digestibility. Brenner et al. (2010) investigated a collection of 40 European forage maize inbred lines to identify useful alleles of *COMT* for the improvement of maize digestibility. A total of 16 SNPs and 9 Indels showed significant associations with DNDF, while 12, 1 and 6 polymorphic sites were significantly associated with organic matter digestibility (OMD), neutral detergent fiber (NDF), and water soluble carbohydrate (WSC), respectively. Eight significantly associated Indels in the *COMT* intron impacted five motifs, which represent binding sites for transcription factors RAV1, GAmyb, and DOFs 1, 2, and 3, that have been identified as putative regulators of lignin biosynthetic genes (Rogers et al. 2005; Brenner et al. 2010). Polymorphisms located in the *COMT* intron might play an important role in gene regulation contributing to lignin content.

CCoAOMT plays an essential role in the synthesis of guaiacyl units as substrates for the synthesis of syringyl (Boerjan et al. 2003). Two *CCoAOMT* genes, named

CCoAOMT1 and *CCoAOMT2*, located on chromosomes 6 and 9, respectively, were identified in maize (Collazo et al. 1992; Roussel et al. 2002). Thereafter, another three CCoAOMT genes have been identified in maize (Guillaumie et al. 2007). An 18-bp indel located in the first exon of *CCoAOMT2* was found to be significantly associated with cell wall digestibility in a panel of 34 diverse lines from Europe and the U.S. (Guillet-Claude et al. 2004a). Two indels within the second intron of the *CCoAOMT2* gene were associated with WSC (Brenner et al. 2010). Chen et al. (2010) identified two indels starting at position 75 and 663, respectively, and two SNPs at positions 144 and 406 to be associated with dry matter content (DMC). One SNP located in the third intron of the *CCoAOMT1* gene was associated with NDF, while another SNP located in the fourth intron was associated with both NDF and OMD. In the fifth exon of the *CCoAOMT1* gene, a SNP was associated with WSC (Brenner et al. 2010).

PAL catalyzes the first step of lignin biosynthesis. To identify causative polymorphisms affecting forage quality at the *PAL* locus, a set of 32 maize inbred lines consisting of 19 Flints and 13 Dents were investigated in an association study. A 1-bp deletion in the second exon of *PAL*, introducing a premature stop codon, was associated with in vitro digestibility of organic matter (IVDOM) when considering population structure (Andersen et al. 2007).

Six putative phenylpropanoid pathway genes, including *C4H*, *4CL1*, *4CL2*, *C3H*, *F5H*, and *CAD*, were investigated for associations with forage quality traits in a panel of 40 European forage maize inbred lines (Andersen et al. 2008). A 1-bp indel at position 810 in the *4CL1* gene was associated with IVDOM. An intron SNP in the *F5H* gene was associated with NDF, while a non-synonymous SNP in the *C3H* gene was associated with IVDOM. However, when considering multiple testing, the associations between *F5H*, *C3H* and forage quality traits were no longer significant. No significant associations were found between polymorphisms in genes *4CL2*, *C4H*, and *CAD* with forage quality traits (Andersen et al. 2008). Two SNPs in the *F5H* gene, leading to a substitution from Proline to Arginine, were associated with dry matter yield (DMY) (Chen et al. 2010). Finally, a miniature inverted-repeat transposable elements (MITE) insertion in the second exon of *ZmPox3* was associated with cell wall digestibility in 56 inbreds (Guillet-Claude et al. 2004b).

In summary, genes encoding enzymes in lignin biosynthesis have been tested for associations with forage quality traits. Several polymorphic sites within these genes showed significant association with cell wall digestibility, and are thus promising targets for development of functional markers to improve forage digestibility by marker-assisted selection.

Genes Controlling Kernel Quality of Maize

Maize is a good source of carbohydrates, proteins, and lipids, as well as an excellent source of vitamins and minerals (Prasanna et al. 2001). An average maize kernel composition is 73% starch, 9% protein, 4% oil, and 14% other constituents on a dry matter basis (Laurie et al. 2004). The compositional quality of maize kernels, important to human health, has long been a major focus in breeding programs and scientific studies (Table 19.3).

Starch production in maize is of great importance to both grain yield and quality. Starch in "normal" maize is composed of 21% amylase, consisting of linear chains of glucose, and 79% amylopectin, consisting of branched glucose chains. In waxy maize kernels, starch is composed of almost 100% amylopectin (Wilson et al. 2004). Amylase contributes to an increase in pasting temperatures and shear stress stability, while amylopectin results in granule swelling (Tester and Morrison 1990). Pasting properties and amylose levels are important starch composition quality traits that affect eating and cooking quality.

Several key genes involved in the starch synthesis have been isolated through well-known endosperm mutants (James et al. 2003). The *shrunken1* (*sh1*) gene, encoding maize sucrose synthase, is responsible for the conversion of UDP-glucose and fructose from sucrose transported into the maize kernel (Chourey and Nelson 1976). ADP-glucose pyrophosphorylase (AGPase) is a key enzyme, consisting of two large and two small subunits, which catalyze the first reaction in starch synthesis, converting ADP-glucose into glucose-1-phosphate (James et al. 2003). The *sh2* gene encodes the large subunits of ADP-glucose pyrophosphorylase, while *brittle endosperm2* (*bt2*) encodes the small subunits (Bhave et al. 1990; Bae et al. 1990). An endosperm mutant, named *amylase extender1* (*ae1*), produces maize kernels with high amount of amylose proportion of the starch to 50% and higher. The *ae1* gene encodes starch branching enzyme IIb isoform (SBEIIb) in maize (Kim et al. 1998). The *sugary1* (*su1*) gene in maize encodes an isoamylase-type starch debranching enzyme (DBE), which increases water-soluble phytoglycogen content in maize kernels and has been adopted to produce sweet corn (James et al. 1995). *Waxy* (*wx*), coding for granule-bound starch synthase, is responsible for eliminated amylase as well as high amylopectin content, which has been widely used in modern breeding (Shure et al. 1983). The genes *sh1*, *sh2*, and *bt2*, located in the upstream of the pathway, take part in the formation of glucose. Furthermore, *ae1*, *su1*, and *wx* modify the amylase-to-amylopectin ratio in maize starch (Fig. 19.2).

Six key genes, *sh1*, *sh2*, *bt2*, *ae1*, *su1*, and *wx*, which are involved in kernel starch biosynthesis, have been studied at the level of allelic diversity or association analysis. Whitt et al. (2002) investigated nucleotide diversity for these six genes by sequencing 30 diverse maize inbred lines, and found low genetic diversity, which suggests that these loci have been a target of artificial selection. A polymorphism in the *su1* gene, which leads to an amino acid change from tryptophan to arginine at the conserved residue 578, was significantly associated with the sweetness (Whitt et al. 2002). An association study was carried out, where the genes *sh1*, *sh2*, *bt2*, *ae1*, *su1*, and *wx* were studied in a panel of 102 maize genotypes. Several causative polymorphisms within these genes were associated with starch concentration and kernel composition (Wilson et al. 2004). Two SNPs, Ae1-1509 and Ae1-1689, in the *ae1* gene were associated with pasting temperature and amylase content, respectively. A SNP, Bt2-925, in the *bt2* gene converting Pro to Leu at amino acid 22 was significantly associated with oil content. The *sh1* gene related to amylase

Table 19.3 Summary of association studies on candidate genes controlling kernel quality in maize

Gene	Affected traits	Plant material	Sequence motif	Genetic effect	Derived markers assay	Reference
Psy1	Yellow or white endosperm	75 maize inbred lines	78 polymorphic sites			Palaisa et al. (2003)
lcyE	The ratio of α- and β-carotene branches	288 maize lines	1 indel in promoter and 1 SNP in exon 1	36%; 5.2-fold	5′indels/TE	Harjes et al. (2008)
	the ratio of α- and β-carotene branches	288 maize lines	1 indel in 3′-UTR	3.3-fold	3′indel	Harjes et al. (2008)
	The ratio of α- and β-carotene branches	288 maize lines	1 SNP in intron4	2.5-fold	SNP216	Harjes et al. (2008)
	The ratio of α- and β-carotene branches	245 diverse maize inbred lines	5′TE	48%		Yan et al. (2010)
	The ratio of α- and β-carotene branches	245 diverse maize inbred lines	SNP216	37%		Yan et al. (2010)
	The ratio of α- and β-carotene branches	155 diverse maize inbred lines	5′TE	36–40%		Yan et al. (2010)
	The ratio of α- and β-carotene branches	155 diverse maize inbred lines	SNP216	36–38%		Yan et al. (2010)
	The ratio of α- and β-carotene branches	155 diverse maize inbred lines	3′indel	3.6%		Yan et al. (2010)
crtRB1 (*HYD3*)	Ratios of β-carotene to β-cryptoxanthin	281 maize inbred lines	5′TE	1–27%		Yan et al. (2010)
	ratios of β-carotene to β-cryptoxanthin	281 maize inbred lines	InDel4	3.38%		Yan et al. (2010)
	Ratios of β-carotene to β-cryptoxanthin	281 maize inbred lines	3′TE	2.29%		Yan et al. (2010)
	Ratios of β-carotene to β-cryptoxanthin	245 diverse maize inbred lines	3′TE	40%		Yan et al. (2010)
	ratios of β-carotene to β-cryptoxanthin	155 diverse maize inbred lines	InDel4	27–45%		Yan et al. (2010)

(continued)

Table 19.3 (continued)

Gene	Affected traits	Plant material	Sequence motif	Genetic effect	Derived markers assay	Reference
	Ratios of β-carotene to β-cryptoxanthin	155 diverse maize inbred lines	3′TE	14.32%		Yan et al. (2010)
HYD3	Ratios of β-carotene to β-cryptoxanthin	51 maize lines	1 indel	78%	Yes	Vallabhaneni et al. (2009)
Opaque2	Lysine content	375 maize inbred lines	SNP O3988	1.48%		Manicacci et al. (2009)
	Protein-starch ration	375 maize inbred lines	SNP O3988	1.23%		Manicacci et al. (2009)
CyPPDK1	Protein content	375 maize inbred lines	SNP C817	1.94%		Manicacci et al. (2009)
	Protein-starch ratio	375 maize inbred lines	SNP C817	1.75%		Manicacci et al. (2009)
	Protein content	375 maize inbred lines	SNP O3988 and CP125 /CP161/CP509/CP5	15.9–7.1%		Manicacci et al. (2009)
	Protein content	375 maize inbred lines	SNP O3988 and CP125 /CP161/CP209/CP5	16.7–7.9%		Manicacci et al. (2009)
	Lysine content	375 maize inbred lines	SNP O3988 and C2252			Manicacci et al. (2009)
	Lysine content	375 maize inbred lines	SNP O1866 and CP125			Manicacci et al. (2009)
fad2	Oil content	553 maize inbred lines	L71T	25–40%		Belo et al. (2008)
DGAT1-2	Oil content	71 maize lines	F469			Zheng et al. (2008)
	Oleic-acid content	71 maize lines	F469			Zheng et al. (2008)

Gene	Trait	Sample	Polymorphism	Effect	Reference
ae1	Pasting characteristics	102 maize inbred lines	Ae1 – 1509(G/A)		Wilson et al. (2004)
	Amylose content	102 maize inbred lines	Ae1 – 1689(T/C)	7–14%	Wilson et al. (2004)
bt2	Oil and protein content	102 maize inbred lines	Bt2-925(T/C)		Wilson et al. (2004)
sh1	Amylose content	102 maize inbred lines	Sh1 775(T/C)	12%	Wilson et al. (2004)
	Kernel composition factor	102 maize inbred lines	Sh1 – 1210(A/G)		Wilson et al. (2004)
sh2	Pasting characteristics	102 maize inbred lines	Sh2-3674(INDEK1)		Wilson et al. (2004)
	Kernel composition factor	102 maize inbred lines	Sh2-3674(INDEK1)		Wilson et al. (2004)
	Amylose content	102 maize inbred lines			Wilson et al. (2004)
sh2	Starch content	50 accessions of maize and teosinte	2 SNPs		Manicacci et al. (2007)
	Kernel weight	50 accessions of maize and teosinte	2 SNPs		Manicacci et al. (2007)
su1	Oil content	817 Balsas teosinte plants			Weber et al. (2008)
su1	Sweet phenotype	30 diverse lines	1 polymorphism		Whitt et al. (2002)
su1	Sweet phenotype	57 sugary 1 accessions	4 polymorphisms		Tracy et al. 2006
wx	Glutinous phenotype	55 Chinese waxy accessions	D7 and D10		Fan et al. (2009)

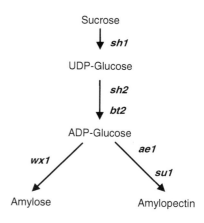

Fig. 19.2 A simplified starch synthesis pathway in maize. The six important genes in the pathway are shown in *bold*

content and kernel composition was associated with two polymorphisms Sh1-775 and Sh1-1210, respectively (Wilson et al. 2004). The *sh2* gene was significantly correlated with both amylase content and kernel composition, as well as starch content and kernel weight (Wilson et al. 2004; Manicacci et al. 2007). The genetic diversity of *sh1, sh2, bt2, ae1, su1*, and *wx1* has recently been surveyed in extensive maize panels, including a population of 40 waxy, 26 dent, and 15 sweet corns from Korea (Shin et al. 2006), 55 Chinese waxy accessions (Fan et al. 2009), and 67 Chinese elite maize inbred lines (Cao et al. 2009), respectively. A 30-bp Indel (D7) and a 15-bp Indel (D10) in the *wx1* gene were shown to be associated with a glutinous phenotype (Fan et al. 2009).

After starch, the next largest component of maize kernel is protein, 80% of which is stored in the endosperm (Flint-Garcia et al. 2009). The nutritional quality of maize kernels as food depends on the essential amino acid content of storage proteins. Zeins are a family of alcohol-soluble prolamin storage proteins, consisting of 50–70% the endosperm protein, abundant in glutamine, proline, alanine, and leucine (Prasanna et al. 2001). Unfortunately, the essential amino acids lysine and tryptophan are completely lacking in zeins. Therefore, maize proteins are nutritionally poor because of insufficient amounts of essential amino acids. Suppression of poor quality zeins has been seen as a feasible approach to improve the quality of maize protein.

There are four types of zeins, classified as α-, β-, γ-, and δ-zeins (Prasanna et al. 2001). Maize mutants with reduced levels of zeins have shown improved nutritional quality. The best-characterized zein mutants, *opaque2* (*o2*) and *floury2* (*fl2*), have substantially higher lysine and tryptophan content than wild type maize (Mertz et al. 1964; Nelson et al. 1965). The *o2* mutants usually have more lysine than *fl2*, and are found to be associated with a major reduction in the lysine poor α- and β-zein storage proteins (Prasanna et al. 2001). The *o2* maize kernels were found to be almost as effective as milk protein when fed to Guatemalan children. Genetically dissection of the *o2* mutant has been conducted to develop functional markers for development of Quality Protein Maize (QPM) varieties. The

O2 gene encodes a basic leucine zipper transcriptional activator (Schmidt et al. 1990), and was shown to activate the transcription of various genes during kernel development, including 22-kDa α-zein genes (Schmidt et al. 1992), 14-kDa β-zein genes (Neto et al. 1995), *b-32* gene (Lohmer et al. 1991), and *CyPPDK1* genes (Maddaloni et al. 1996). Allele sequencing of the *O2* gene from 17 inbred lines, 6 landraces, and 9 accessions of annual teosinte demonstrated that the molecular diversity was quite high (Henry et al. 2005). An association study was carried out to investigate the roles of *O2* and *CyPPDK1* on kernel quality traits across a panel of 375 maize inbred lines (Manicacci et al. 2009). The SNP O3988 located in the *O2* coding sequence individually explains 1.48% of lysine content as well as 1.23% of protein-starch ratio variation. In *CyPPDK1*, the SNP C817 is associated with several kernel traits such as protein content and protein-starch ratio but not lysine content. The combination of the SNP O3988 and one of the SNPs in the *CyPPDK1* promoter (CP125, CP161, CP509, or CP515) explains up to 7.9% of protein-starch ratio variation as well as 7.1% of protein content. Lysine content is associated with two combinations of alleles including O1866 and CP125, O3988 and C2252, respectively (Manicacci et al. 2009). There are many other zein mutants accumulating lysine in the seed in relation to wild type corn, including *o1*, *o5*, *o7*, *o10*, *o11*, *o13*, *o16*, and *fl1*, which exhibit a lower lysine content than observed in the *o2* endosperm (Yang et al. 2005; Azevedo et al. 2004; Landry et al. 2005). Great effort has been spent to develop QPM varieties using the *o2* allele singly or in combination with other high-lysine mutants (Babu et al. 2005; Danson et al. 2006). Unfortunately, the high-lysine trait was correlated with some undesirable agronomic characters, such as reduced yield and greater susceptibility to diseases, which slowed down the application in breeding (Prasanna et al. 2001).

Maize oil is valued for human nutrition as well as important feedstock for biodiesel. Moreover, high-oil maize usually has more protein, lysine, and carotenoids than regular maize varieties. Therefore, it is also good source for livestock feed (Han et al. 1987). A high-oil corn selection originated from an open-pollinated variety in 1896. And the oil concentration of the Illinois high oil (IHO) populations changed from 4.7 to ~20% after 100 generations of selection (Moose et al. 2004). However, the high oil content is often associated with poor yield, which affects application of IHO in commercial production (Laurie et al. 2004). It might be possible to dissect genetic factors underlying oil composition at the sequence level and improve IHO's utility by eliminating its negative effects on agronomic traits in breeding program.

Five fatty acids constitute 99% of maize oil, including palmitic (16:0), stearic (18:0), oleic (18:1), linoleic (18:2), and linolenic (18:3) acids (Browse and Somerville 1991). Oils with higher content of monounsaturated fatty acids (oleic-acid) are healthy for human nutrition and have many cooking benefits. Biosynthesis of storage oil in seed is genetically complex, while only a limited number of genes related to lipid metabolism were identified in maize so far (Lee and Huang 1994; Berberich et al. 1998; Zheng et al. 2008; Shen et al. 2010). A number of studies focused on QTL mapping of oil concentration and composition in maize kernel, with many chromosomal regions being identified using linkage mapping or association mapping, which suggests the genetic architecture of oil concentration in maize kernel is considerably complex (Alrefai et al. 1995; Mangolin et al. 2004; Laurie et al. 2004; Yang et al. 2010b). Recently, Li et al. (2010a) identified 147 lipid-related genes co-locating with 59 mapped QTL clusters for maize fatty acid composition and oil concentration, which provide good targets for QTL cloning and association mapping to identify favorable alleles in natural populations.

Until now, three important genes for maize kernel oil content have been isolated and validated. A high-oil QTL (*qHO6*) which encodes an acyl-CoA:diacylglycerol acyltransferase (DGAT1-2) that catalyzes the final step of oil synthesis was isolated by map-based cloning. An insertion of phenylalanine in *DGAT1-2* at position 467 (F469) was responsible for increased oil and oleic-acid contents. The genetic effect of F469 was validated in a set of 71 maize lines differing in oil and oleic-acid concentrations, which also showed that the high-oil allele is present in maize wild relatives and was subsequently lost. Transgenic evaluation showed that the high oil *DGAT1-2* allele increased seed-oil content and oleic-acid content by up to 42 and 107%, respectively (Zheng et al. 2008). Using a whole genome-wide association mapping across a panel of 553 maize inbreds, a fatty acid desaturase, *fad2*, was identified to be associated with oleic-acid content. A non-conservative amino acid substitution of a leucine by a threonine in *fad2* at position 71 (L71T) was proposed to account for the effect on oleic acid content (Belo et al. 2008). *ZmWri1* (*maize Wrinkled1*), encoding an AP2 transcription factor, controls the expression of glycolysis and fatty acid pathway genes. Overexpression of *ZmWri1* increases maize seed oil by an average of 30.6%, while the best line showed a 46% increase, without affecting germination, seedling growth, or grain yield (Shen et al. 2010). The causative polymorphisms of these three genes are promising targets for improving maize oil content and modify oil composition in important crops.

Vitamin A deficiency is one of the most prevalent nutritional deficiencies in developing countries, often resulting in night blindness or even complete blindness in severe conditions. The prevalence of night blindness from a deficiency of Vitamin A is also high among pregnant women in the developing world. Biofortification of micronutrient through plant breeding in staple crops such as maize is perhaps an economically and socially feasible approach to alleviate the problems caused by micronutrient deficiency. As the dominant subsistence crop for sub-Saharan African consumers, maize is an attractive vehicle for biofortification.

Provitamin A carotenoids, such as α-carotene and β-carotene, are major sources of Vitamin A for the majority of the world population. All steps of carotenoid biosynthesis occur in plastids (DellaPenna and Pogson 2006; Fig. 19.3). The first regulatory step of this pathway is mediated by phytoene synthase (PSY) and involves the formation of phytoene from geranylgeranyl diphophate (GGPP) (Hirschberg 2001; Li et al. 2007). Phytoene is converted to lycopene by a series of enzymes including phytoene desaturase (PDS), zetacarotene desaturase (ZDS), and carotene isomerase (CRTISO) (Fraser and Bramley 2004; Aluru et al. 2008). There are two major branches that occur at cyclization of lycopene by lycopene beta cyclase (LCYB) producing a molecule with two β-rings, as found in β-carotene.

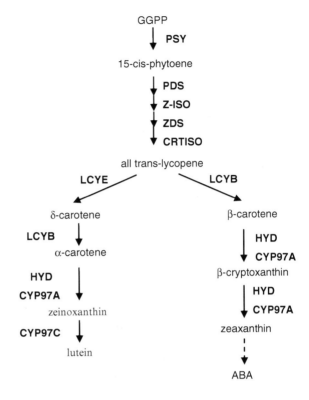

Fig. 19.3 Simplified carotenoid biosynthetic pathway in corn endosperm

Alternatively, lycopene may be cyclized by the coaction of LCYB and lycopene epsilon cyclase (LCYE) generating a molecule with one β-ring and one ε-ring, as found in α-carotene and its derivatives, zeinoxanthin and lutein (Pogson et al. 1996; Harjes et al. 2008). Relative activities of LCYB and LCYE are hypothesized to regulate the proportion of carotenes, directed to each branch of this pathway (Cunningham et al. 1996; Harjes et al. 2008). Several genes encoding enzymes involved in carotenoid biosynthesis pathway have recently been isolated based on sequence homology or expression profiling (Buckner et al. 1996; Li et al. 2007; Harjes et al. 2008; Vallabhaneni et al. 2009; Yan et al. 2010; Li et al. 2010d). Yellow maize varieties have been the target of breeding strategies for a long time which enhance accumulation of carotenoids in endosperm compare to white maize. It is reported that the *PSY1* gene underlies the shift of maize kernel from white to yellow (Buckner et al. 1996). Association analysis between *PSY1* and endosperm color in a collection of 75 maize inbred lines identified 78 significant polymorphisms, among which two SNPs showed complete associations as well as two Indels showed strong but incomplete associations with the phenotype (Palaisa et al. 2003). Recent studies found that sequence diversity within yellow maize lines at the *PSY1* locus is dramatically decreased as compared to white corn, which suggested that the *PSY1* locus has been the target of a broad selective sweep (Palaisa et al. 2004).

Considerable natural variation for kernel carotenoid concentrations in yellow maize has been observed, with some lines having high level of carotenoids as much as 66 mg/g (Menkir et al. 2008; Harjes et al. 2008). Recently, an association study identified the *lcyE* gene in maize, which is responsible for directing metabolic flux to the α-carotene versus the β-carotene branch of the carotenoid pathway. Joint linkage and association mapping, integrated with mutant analysis and expression profiling, showed that elevated expression of *lcyE* leads to high provitamin A compound levels. Four putative causative polymorphisms in the *lcyE* gene were associated with the ratio of α- and β-carotene branches in a set of 288 maize lines, which explained 58% of the total variation and showed a threefold difference in provitamin A compounds (Harjes et al. 2008). Significant associations between the *lcyE* gene and α/β carotene ratio were observed in another two populations consisting of 245 and 155 diverse maize inbred lines, respectively (Yan et al. 2010). The most favorable *lcyE* alleles, including the large promoter insertion and 3′ 8-bp insertion, were found based on performance of diverse germplasm. Simple PCR-based functional markers were derived that now enable breeders to increase levels of provitamin A compounds more efficiently in breeding program (Harjes et al. 2008).

Another gene important in the carotenoid biosynthesis pathway, *HYD3*, encoding β-carotene hydroxylase1, has recently been identified by transcript profiling. A polymorphic site located 40 bp adjacent to the transcript start site explained 78% of the variation in the β-carotene-to-β-cryptoxanthin ratio (Vallabhaneni et al. 2009). Yan et al. (2010) identified three polymorphisms (5′-TE, InDel4, and 3′TE) in the *HYD3* gene (or named *crtRB1*), showing significant associations with carotenoid variation across three different panels of maize inbred lines. The 5′-TE has the largest effect, resulting in an average increase of 6.5 μg/g β-carotene above the average effect of the unfavorable allelic class (1.5 μg/g). The effect of *crtRB1* on β-carotene concentration was further validated in five populations consisting of one recombinant inbred line (RIL) and four $F_{2:3}$ progenies. Because allelic variation at both *lcyE* and *crtRB1* affect β-carotene concentration, the combined effects of one major functional polymorphism of *lcyε* (5′TE) and two of *crtRB1* (5′TE and/or 3′TE) were investigated in three association panels (Yan et al. 2010). Combination of both optimal *crtRB1* and *lcyE* alleles lead to higher β-carotene concentration in maize kernels than that having allele of either one gene alone.

Ongoing studies attempt to identify further favorable alleles for other genes in carotenoid biosynthesis pathway (Harjes et al. 2008). Combination of optimal alleles by breeding to improve provitamin A content in maize would have a significant impact on world's poorest people.

Genes Conferring Tolerance to Abiotic Stress in Maize

Crop yield is greatly affected by abiotic stress factors, such as drought, submergence, salinity, and other soil toxicities. Climate change will increase drought in some areas and cause problems for soil fertility in other areas, which threatens crop growth, while the global water shortage is getting worse (Takeda and Matsuoka 2008; Boyer 2010). In response to these problems, crops with tolerance to respective environmental stress factors will be required. Maize is very sensitive to water stress at flowering as well as soil toxicity, with great losses being observed when stresses occur at key growth stages (Claassen and Shaw 1970; Pandey et al. 1994). Here, recent progress on an identification of genes involved in tolerance to abiotic stresses, especially to drought stress and soil stress in maize is reviewed (Table 19.4).

Extensive studies on the tolerance of maize to drought stress have been conducted. Maize plants often show an increase in the anthesis-silking interval (ASI) and leaf senescence, accompanied by reductions in leaf expansion, when treated with water stress (Bolanos and Edmeades 1996; Nelson et al. 2007). Selection for reduced ASI was effective for improvement of maize yield under drought (Bolanos et al. 1993; Bolanos and Edmeades 1996; Chapman and Edmeades 1999). Additionally, ASI and leaf expansion have usually been used as important physiological traits related to drought tolerance (Reymond et al. 2003; Welcker et al. 2007; Sadok et al. 2007).

Recently, several causative polymorphisms underlying genes for drought tolerance were identified by association mapping. Lu et al. (2010) adopted joint linkage disequilibrium and integrated mapping to investigate associations between drought-response candidate genes and ASI across 305 diverse inbred lines and 217 RILs from three populations. One haplotype, HP71, which includes 10 SNPs from two genes, was identified to be significantly associated with ASI under well-watered condition (ASI-WW). Phenotypic variation explained (PVE) by HP71 increased up to 34.7% when using haplotype-based integrated mapping. *SDG140* is one of the underlying genes of HP71, which encodes a SET domain protein involved in the control of flowering time that correlates with methyltransferase activity. Hp322 was found to be correlated with ASI under water-stressed conditions (ASI-WS), which includes two closely linked SNPs from a gene encoding aldo/keto reductase (AKR) and explains 28.81% of the phenotypic variation using integrated mapping (Lu et al. 2010). These two QTPs are expected to be effectively used in breeding programs because of their large effects. It is reported that ABA and sugar levels are correlated with water stress at flowering. Setter et al. (2011) performed a candidate gene based association mapping study to identify loci involved in accumulation of carbohydrates and ABA metabolites during water stress in a panel of 350 tropical and subtropical maize inbred lines. SNP465, within the gene *ZmMADS16*, a maize homologue of Arabidopsis regulatory gene *PISTILLATA*, was found to be significantly associated with ear phaseic acid. SNP 947, which is located in the exon 6 of the *ZmPDK2* gene encoding a regulatory protein kinase for mitochondrial pyruvate dehydrogenase, was identified to be correlated with total sugar concentration in silks. SNP186 located in the gene *ZmAO* encoding an aldehyde oxidase showed significant association with ABA levels in silks of WS plants. SNP1145 and SNP1198 were associated with total sugar level in silks and ear phaseic acid of WW plants, respectively (Setter et al. 2011). The causative polymorphisms need to be further validated, before they can be applied for development of drought-tolerant maize cultivars.

Table 19.4 Summary of association studies on candidate genes conferring tolerance to abiotic stresses in maize

Gene	Affected traits	Plant material	Sequence motif	Genetic effect (%)	Reference
SDG140	Anthesis-silking interval under well-watered condition (ASI-WW)	522 maize lines	HP71 (10 SNPs)	22.70–34.69	Lu et al. (2010)
aldp.keto reductcas (AKR)	Anthesis-silking interval under water-stressed condition (ASI-WS)	522 maize lines	HP322 (2 SNPs)	21.50–28.81	Lu et al. (2010)
	Anthesis-silking interval under water-stressed condition (ASI-WS)	305 maize inbred lines	HP153	3.66	Lu et al. (2010)
	Anthesis-silking interval under water-stressed condition (ASI-WS)	305 maize inbred lines	HP312	4.21	Lu et al. (2010)
aldehyde oxidase (AO)	ABA leves in silks in WS plants	305 maize inbred lines	SNP186 (A/G)		Setter et al. (2011)
pyruvate dehydrogenase kinase isoform 2 (PDK2)	Total sugar level in silks of WS plants	305 maize inbred lines	SNP947 (C/T)		Setter et al. (2011)
ZmMADS16	The level of PA in ears of WW plants	305 maize inbred lines	SNP465 (A/C)		Setter et al. (2011)
GRMZM2G173784	The level of sucrose in ears of WS plants	305 maize inbred lines	SNP251 (A/T)		Setter et al. (2011)
GRMZM2G021044	The level of sucrose in ears of WW plants	305 maize inbred lines	SNP255 (C/T)		Setter et al. (2011)
GRMZM2G092497	Total sugar level in silks of WW plants	305 maize inbred lines	SNP1145 (A/G)		Setter et al. (2011)
GRMZM2G125023	The level of PA in ears of WW plants	305 maize inbred lines	SNP1198 (T/G)		Setter et al. (2011)
ZmASL	Net root growth under Al stress	282 maize inbred lines and three F2 populations	7 SNPs and 4 Indels		Krill et al. (2010)
ZmSAHH	Net root growth under Al stress	282 maize inbred lines and three F2 populations	3 SNPs		Krill et al. (2010)
ZmME	Net root growth under Al stress	282 maize inbred lines and three F2 populations	3 SNPs		Krill et al. (2010)
ZmALMT2	Net root growth under Al stress	282 maize inbred lines and three F2 populations	2 SNPs		Krill et al. (2010)

Aluminium (Al) toxicity causes significant yield reduction on highly acidic soils, which constitute approximately 50% of the world's potentially arable lands (Kochian et al. 2004). Al is solubilized at low pH (<5.5), which inhibits root growth and thus leads to yield losses. Al tolerance in maize has been widely studied at the molecular and physiological levels, with considerable genetic variation being observed for both external and internal tolerance mechanisms (Kochian et al. 2004). The cell wall of the root apex is very sensitive to Al, with polymer callose formation when exposed to excess Al, thus leading to significant rigidification of the cell wall (Jones et al. 2006). Exudation of organic acid anions from roots, which bind with the Al^{3+} cations in the apoplast and detoxify them, was considered as a major mechanism of resistance to Al, which suggested that genes encoding the Al-chelating organic acid transporter might play important roles in Al tolerance (Kochian et al. 2004; Maron et al. 2010). Differences in cell wall pectin content and its degree of methylation in root apices have also been suggested to result in Al tolerance in maize (Eticha et al. 2005).

Maize Al tolerance is a complex quantitative trait. Several Al-regulated genes were identified in a survey of global transcriptional regulation under Al stress using microarrays with Al-tolerant and Al-sensitive genotypes, including cell wall related genes, oxidative stress responsive genes, low phosphate responsive genes, organic acid release related genes and transporters (Maron et al. 2008). Several QTL underlying Al tolerance in maize have been identified in biparental populations (Sibov et al. 1999; Ninamango-Cardenas et al. 2003). Recently, two Multidrug and Toxic Compound Extrusion (MATE) family members, *ZmMATE1* and *ZmMATE2*, were characterized from maize (Maron et al. 2010). *ZmMATE1* co-localized to a major Al tolerance QTL within bin 6.00, and explained 16.2% of the phenotypic variation for Al tolerance; while *ZmMATE2* co-localized to another Al tolerance QTL within bin 5.02/03, and explained 16% of the phenotypic variation (Maron et al. 2010). These two genes are good candidates for future laboratory and field-based studies on maize Al tolerance.

Krill et al. (2010) surveyed 21 candidate genes for associations with Al tolerance in a panel of 282 diverse maize inbred lines, from which six genes were found to be significantly associated with net root growth (NRG) under Al stress. Four of the six genes were confirmed to be associated with Al tolerance in three F_2 linkage populations. Eleven polymorphic sites, located in the introns of the *ZmASL* (*Zea mays Alt$_{SB}$like*) gene, which is homologous to the Al-activated citrate transporter *Alt$_{SB}$* from sorghum and is a member of the MATE family transporters, were significantly associated with NRG under Al stress. Two nonsynonymous SNPs and one triallelic SNP located in the first exon of the *SAHH* (S-adenosyl-L-homocysteine hydrolase) gene were associated with Al tolerance. Two SNPs found in the second intron as well as one SNP in the first exon of the *ME* gene were significantly associated with NRG under Al stress. Two SNPs in the *ZmALMT2* gene showed significant association with Al tolerance. Although a single polymorphism showed small genetic effect, with the most significant ones explaining less than 3% of genetic variation, combination of multiple causative polymorphisms enhanced the Al tolerance greatly (Krill et al. 2010).

Genes Conferring Resistance to Biotic Stresses in Maize

Crops are attacked by multiple pathogens, including fungi, viruses, bacteria, nematodes, and insects. It is estimated that more than 10% of the global food production are lost owing to plant diseases. Improved control of maize diseases by breeding resistant varieties is an environmentally friendly and cost-effective strategy. Plant disease resistance is generally divided into two categories: qualitative resistance based on a single resistance (R) gene with complete resistance, and quantitative disease resistance (QDR) conditioned by a single large effect gene or multiple small effect genes with partial resistance, respectively (Kou and Wang 2010). It is considered that QDR is more useful and effective in resistance breeding to protect yield due to its higher durable effect and broader specificity (Lindhout 2002; Parlevliet 2002). In maize, the majority of disease resistance belongs to QDR, with little knowledge about the underlying resistance mechanisms. A synthesis of 50 publications on the mapping of disease resistance genes in maize reported 437 QTL and 17 major genes, and QTL for resistance to different diseases were often clustered (Wisser et al. 2006).

Until now, only five major resistance genes in maize have been cloned. The *Hm1* gene, conferring resistance to maize leaf spot and ear mold caused by *Cochliobolus carbonum* race 1 (CCR1), was the first cloned plant resistance gene via transposon tagging (Ullstrup and Brunson 1947; Johal and Briggs 1992). It encodes an NADPH-dependent reductase, which inactivates HC-toxin (Johal and Briggs 1992). Chintamanani et al. (2008) isolated the second dominant resistance gene *Hm2*, a paralogue of *Hm1*, which protects maize against CCR1 over time, from having little or no impact in seedling tissues to providing complete immunity at anthesis. Two copies of *Hm2* confer a higher level of resistance than a single copy, suggesting that *Hm2* has a gene-dosage effect (Chintamanani et al. 2008). Characterization of both *Hm1* and *Hm2* in susceptible maize inbred lines, with a transposon insertion in *Hm1* and a deletion in *Hm2*, demonstrated that the disruption of both of the disease resistance genes resulted in susceptibility of maize to CCR1 (Multani et al. 1998).

Common rust, caused by *Puccinia sorghi* Schwein, is found in most subtropical, temperate, and highland environments with high humidity (deLeon and Jeffers 2004). The *Rp1* locus for resistance to maize common rust consists of nine homologous NBS-LRR (nucleotide binding site-leucine rich repeat) genes, which locate in the distal end of the short arm of chromosome 10 (Rhoades 1935; Collins et al. 1999). The *Rp1-D* gene was cloned from the *HRp1-D* haplotype using transposon tagging (Collins et al. 1999; Sun et al. 2001), and further validated via a complementation test (Ayliffe et al. 2004). The *Rp1* cluster varies widely in copy number among different maize haplotypes, in which a range of 1–52 copies has been estimated (Smith et al. 2004). The variation of the *Rp1* locus was likely due to mispairing and frequent unequal crossing over between a tandem array of NBS-LRR genes in meiosis (Collins et al. 1999; Sun et al. 2001; Smith and Hulbert 2005). Nearly 40% of commercial sweet corn hybrids carry the *Rp1-D* gene to protect corn against common rust from 1985 to 1999 (Pataky and Campana 2007), as it can be easily backcrossed into inbred lines and was effective against almost all biotypes of *P. sorghi* (Pataky et al. 2001). However, since 1999, many *Rp1D*-virulent isolates of *P. sorghi* have been identified throughout North America (Pataky and Tracy 1999; Pataky et al. 2000; Pate et al. 2000). Hence, it is essential to apply both qualitative and quantitative genes for more durable resistance to maize common rust.

The *Rcg1* gene, conferring resistance to anthracnose stalk rot (ASR) in maize, caused by the fungus *Colletotrichum graminicola* (Ces.), has recently been isolated by map-based cloning approach (Jung et al. 1994; Wolters et al. 2006). The *Rcg1* gene is the only cloned major QTL underlying disease resistance in maize and encodes a protein with nucleotide binding site and leucine rich repeats. Four independent *Mu* insertions into the candidate *Rcg1* gene contributed to phenotypic changes of these plants from resistant to susceptible (Wolters et al. 2006). Functional markers within the *Rcg1* gene are particularly useful for selecting elite resistance inbred lines. It is reported that *Rcg1* has been successfully applied in corn breeding programs to enhance resistance to *C. graminicola* (Broglie et al. 2011).

The *Rxo1* gene, located in the short arm of maize chromosome 6, was isolated which confers resistance to bacterial stripe of maize and sorghum caused by *Burkholderia andropogonis*, as well as to the non-host pathogen *Xanthomonas oryzae* pv. *Oryzicola* in rice (Zhao et al. 2004a, b, 2005). The *Rxo1* gene was cloned by using the Pic19 probe (Collins et al. 1998) which cosegregates perfectly with *Rxo1* (Zhao et al. 2005). It encodes a typical NBS-LRR structure protein, shared by many previously identified R genes. Transgenic rice expressing *Rxo1* showed resistance to *X. oryzae* pv. *oryzicola*, which suggests the feasibility of using non-host resistance genes to control diseases in other host plant species (Zhao et al. 2005).

Northern leaf blight is a serious foliar disease of maize in the Northeastern United States, in sub-Saharan Africa, Latin America, China, and India (Adipala et al. 1995; Dingerdissen et al. 1996). A number of dominant or partially dominant major genes have been discovered conferring race-specific resistance to NLB, including *Ht1* (Hooker et al. 1963), *Ht2* (Hooker 1977), *Ht3* (Hooker 1981), *Ht4* (Carson 1995), *Htn1* (Gevers 1975), *HtP* (Ogliari et al. 2005), and *Bx* (Couture et al. 1971). The *Ht* genes render delayed lesion development or chlorotic lesions rather than complete resistance, which suggests that the *Ht* genes should be considered as large-effect, race-specific QTL (Balint-Kurti and Johal 2008). A major QTL for NLB resistance located in bin 8.06 (designated *qNLB8.06*) was delimited to a region of 0.46 Mb, with three candidate genes, including two protein kinase-like genes (GRMZM2G135202 and GRMZM2G164612) and one serine-threonine specific protein phosphatase-like gene (GRMZM2G119720) (Chung et al. 2010a). The *Bx* locus on short arm of chromosome 4 also conditions resistance to NLB in maize seedlings, by releasing DIMBOA (2-4-dihydroxy-7-methoxy-1,4-benzoxazin-3-one) to result in a decrease in both NLB lesion number and lesion size (Couture et al. 1971; Park et al. 2004). Two polymorphisms in the *bx1* gene were identified to be significantly associated with DIMBOA content in a population of 282 diverse lines, explaining 12% of the phenotypic variation (Butron et al. 2010).

Two NLB QTL, *qNLB1.02* and *qNLB1.06*, were successfully validated in a set of 82 TBBC3 introgression lines and several derived NILs by crossing selected TBBC3 lines to B73, which revealed that the two QTL were equally effective in both juvenile and adult plants and could potentially be utilized to protect maize from NLB (Chung et al. 2010b). Poland et al. (2011) performed a genome-wide nested association mapping using NAM population for resistance to NLB. A set of 208 SNP loci were found to be associated with resistance to NLB, with multiple potential candidate genes being identified that might be involved in plant defense (Poland et al. 2011).

Southern leaf blight, caused by the ascomycete *Cochliobolus heterostrophus*, is prevalent in hot, humid maize-growing areas throughout the world, comprises a severe threat to corn production. One recessive major gene *rhm1*, for almost complete resistance at the seedling stage and partial resistance in the adult plant to race O of *C. heterostrophus*, is currently the only fine-mapped SLB resistance gene (Chang and Peterson 1995; Zaitlin et al. 1993). Genetic resistance of SLB is mostly polygenic, with quantitative resistance being used in commercially-grown maize. Two chromosomal locations in bins 3.04 and 6.01 were noted as 'hot-spots' for SLB resistance QTL, with their effects having been evaluated (Balint-Kurti and Carson 2006; Balint-Kurti et al. 2007; Carson et al. 2004; Zwonitzer et al. 2009). A genetic complementation test proved that a resistance QTL on bin 6.01 and *rhm1* represent the same resistance gene (Belcher 2009). A genome-wide association study was conducted to search for quantitative resistance loci for SLB using the maize NAM population (Kump et al. 2011). Thirty-two QTL for SLB resistance were identified in a joint linkage analysis across all families, each with relatively small effects. The largest-effect QTL mapped to bin 3.04, a hot-spot for SLB resistance. Fifty-one SNPs were found to be significantly associated with variation for SLB resistance, with several causative polymorphisms within or immediately adjacent to genes previously shown to be involved in disease resistance pathways (Kump et al. 2011).

Sugarcane mosaic virus (SCMV) is a major virus pathogen in maize worldwide, causing severe yield losses in susceptible cultivars. Two major resistance loci, *Scmv1* on bin 6.00/01 and *Scmv2* on bin 3.04/05, were identified to confer resistance to SCMV in different populations (Melchinger et al. 1998; Xia et al. 1999; Xu et al. 1999; Wu et al. 2007). The *Scmv2* was delimited to a region of 1.34 Mb covering four candidate genes by using a large isogenic population segregating for *Scmv2*, but not *Scmv1* (Ingvardsen et al. 2010). The Pic19 NBS-LRR gene family members were identified to be closely linked to *Scmv1*, but not involved in resistance to SCMV (Jiang et al. 2008).

Head smut, caused by the host-specific fungus *Sporisorium reiliana*, is a serious systemic disease worldwide, leading to significant yield losses in maize each year (Frederiksen et al. 1976; Stromberg 1981). Several resistance QTLs against head smut have been detected on almost all maize chromosomes (Lubberstedt et al. 1999; Li et al. 2008; Chen et al. 2008). A major QTL on bin 2.09 has been detected in two independent mapping populations, explaining 43.7 and 30% of the total phenotypic variation, respectively (Chen et al. 2008; Li et al. 2008). The major QTL (*qHSR1*) was fine mapped into a region of ~2 Mb. The genetic effect of *qHSR1* has been validated in large backcross populations, indicating that *qHSR1* could reduce the disease incidence by 25% (Chen et al. 2008). Molecular markers within the *qHSR1* region have been used to improve maize resistance to head smut (Zhao et al. 2012).

Gibberella stalk rot in maize, caused by the fungus *Fusarium graminearum* (teleomorph *Gibberella zeae*), is one of the most devastating diseases worldwide. QTLs conferring resistance to *F. graminearum* have been mapped on chromosomes 1, 3, 4, 5, 6, and 10 (Pe et al. 1993; Yang et al. 2004, 2010a). A major QTL *qRfg1*, located in bin 10.04, was delimited into an interval of ~500 kb (Yang et al. 2010a); while, a minor QTL, *qRfg2*, has been fine-mapped into a 300 kb interval with a putative candidate gene coding for an auxin-regulated protein (Zhang et al. 2012). The *qRfg1* and *qRfg2* loci could enhance the frequency of resistant plants by 32–43 and ~12% in the susceptible maize inbred 'Y331' background (Yang et al. 2010a; Zhang et al. 2012).

Recently, several studies reported that some loci in maize genome conferred multiple disease resistance (MDR) to protect plants from various biotic stresses. Two QTL in bins 3.04 and 6.01 from multiple disease-resistant inbred NC250P have been evaluated for resistance to different diseases using NILs in B73 background, which showed large genetic effects on Southern leaf blight (SLB) caused by *Cochliobolus heterostrophus*, as well as significant effects for Northern leaf blight (NLB) caused by *Setosphaeria turcica*, and grey leaf spot (GLS) caused by *Cercospora zeae-maydis* (Belcher 2009). In addition, these two genome regions were also found to confer resistance to multiple potyviruses, including SCMV, MDMV, ZeMV, and WSMV (Lubberstedt et al. 2006). To investigate genomic regions associated with MDR in maize, Chung et al. (2011) adopted heterogeneous inbred families (HIFs) and RILs to survey their response to eight diseases, discovering two MDR QTL, one QTL in bin 1.06/1.07 conferred resistance to NLB and Stewart's wilt, and another in bin 6.05 conferring resistance to NLB and ASR. Two reliable QTL conferring resistance to NLB were also found to be involved in MDR of maize, with $qNLB1.02_{B73}$ associated with resistance to Stewart's wilt and common rust, and $qNLB1.06_{Tx303}$ conferring resistance to Stewart's wilt as well (Chung et al. 2010b). A glutathione S-transferase (*GST*) gene was identified to be a pleiotropic gene correlated with modest levels of resistance to SLB, NLB, and GLS (Wisser et al. 2011). An amino acid substitution from histidine to aspartic acid in the encoded protein domain was found to be responsible for maize MDR, confirmed by re-sequencing the full-length *GST* gene across the panel of 253 maize inbred lines (Wisser et al. 2011). MDR phenotype in these studies may be due to cluster of genes with disease-specific effects, or to the presence of gene(s) with pleiotropic effects, which suggests resolving the genetic basis of MDR at the level of sequence polymorphisms is essential.

To date, only a limited number of disease resistance genes/QTLs have been isolated in maize. Herein, FMs available for a particular resistance are rare in practical breeding. As a multitude of disease resistance QTLs have been fine-mapped and validated, with many candidate genes underlying these QTLs being identified, functional markers will become available for most disease resistance genes/QTLs in the near future.

Perspectives

Maize is a highly polymorphic species, and its genome is the product of a segmental allotetraploid duplication event, which likely occurred ~11.4 million years ago (Gaut and Doebley 1997). Targeted re-sequencing studies revealed tremendous sequence diversity of maize inbred lines and non-collinearity between allelic regions (Wang and Dooner 2006). The B73 genome is characterized with complexity, diversity, and dynamics (Schnable et al. 2009). Apart from SNPs and small indels, copy number variation (CNV) and presence/absence variation (PAV) frequently underlie QTPs. Thus, in some cases, a complete new gene rather than an optimal allele needs to be integrated into genotypes lacking this gene. In addition, commercial maize varieties are hybrids, thus a new gene or an elite allele can also be introgressed into hybrid by combining inbreds that have the right combination of genes.

Advances by the Maize Genome Sequencing Project (Schnable et al. 2009), next-generation sequencing technology, high-throughput genotyping, and improved bioinformatics tools are yielding useful, large-scale genome data sets. About 1.6 million SNPs have been identified recently by the maize HapMap project (Gore et al. 2009). Several high throughput genotyping platforms have been developed for commercial use (Gupta et al. 2008). Availability of new technologies and information speeds up research on large scale diversity analysis, high-density linkage map construction, high-resolution QTL mapping, large-scale linkage disequilibrium (LD) analysis, and genome-wide association analysis, thus accelerating dissection of a QTL into QTPs in future.

References

Adipala E, Takan JP, Ogengalatigo MW (1995) Effect of planting density of maize on the progress and spread of northern leaf blight from Exserohilum-Turcicum infested residue source. Eur J Plant Pathol 101(1):25–33

Alrefai R, Berke TG, Rocheford TR (1995) Quantitative trait locus analysis of fatty acid concentrations in maize. Genome 38(5):894–901

Aluru M, Xu Y, Guo R, Wang ZG, Li SS, White W, Wang K, Rodermel S (2008) Generation of transgenic maize with enhanced provitamin A content. J Exp Bot 59(13):3551–3562

Andersen JR, Lubberstedt T (2003) Functional markers in plants. Trends Plant Sci 8(11):554–560

Andersen JR, Schrag T, Melchinger AE, Zein I, Lubberstedt T (2005) Validation of dwarf8 polymorphisms associated with flowering time in elite European inbred lines of maize (*Zea mays* L.). Theor Appl Genet 111(2):206–217

Andersen JR, Zein I, Wenzel G, Krutzfeldt B, Eder J, Ouzunova M, Lubberstedt T (2007) High levels of linkage disequilibrium and associations with forage quality at a Phenylalanine Ammonia-Lyase locus in European maize (*Zea mays* L.) inbreds. Theor Appl Genet 114(2):307–319

Andersen JR, Zein I, Wenzel G, Darnhofer B, Eder J, Ouzunova M, Lubberstedt T (2008) Characterization of phenylpropanoid pathway genes within European maize (*Zea mays* L.) inbreds. BMC Plant Biol. doi:10.1186/1471-2229-8-2

Ayliffe MA, Steinau M, Park RF, Rooke L, Pacheco MG, Hulbert SH, Trick HN, Pryor AJ (2004) Aberrant mRNA processing of the maize Rp1-D rust resistance gene in wheat and barley. Mol Plant Microbe Interact 17(8):853–864

Azevedo RA, Lea PJ, Damerval C, Landry J, Bellato CM, Meinhardt LW, Le Guilloux M, Delhaye S, Varisi VA, Gaziola SA, Gratao PL, Toro AA (2004) Regulation of lysine metabolism and endosperm protein synthesis by the opaque-5 and opaque-7 maize mutations. J Agric Food Chem 52(15):4865–4871

Babu R, Nair SK, Kumar A, Venkatesh S, Sekhar JC, Singh NN, Srinivasan G, Gupta HS (2005) Two-generation marker-aided backcrossing for rapid conversion of normal maize lines to quality protein maize (QPM). Theor Appl Genet 111(5):888–897

Bae JM, Giroux M, Hannah L (1990) Cloning and characterization of the brittle-2 gene of maize. Maydica 35(4):317–322

Balint-Kurti PJ, Carson ML (2006) Analysis of quantitative trait Loci for resistance to southern leaf blight in juvenile maize. Phytopathology 96(3):221–225

Balint-Kurti PJ, Johal GS (2008) Maize disease resistance. In: Bennetzen JL, Hake SC (eds) Handbook of maize: its biology. Springer, New York, pp 229–250

Balint-Kurti PJ, Zwonitzer JC, Wisser RJ, Carson ML, Oropeza-Rosas MA, Holland JB, Szalma SJ (2007) Precise mapping of quantitative trait loci for resistance to southern leaf blight, caused by Cochliobolus heterostrophus race O, and flowering time using advanced intercross maize lines. Genetics 176(1):645–657

Barriere Y, Argillier O (1993) Brown-midrib genes of maize – a review. Agronomie 13(10):865–876

Barriere Y, Guillet C, Goffner D, Pichon M (2003) Genetic variation and breeding strategies for improved cell wall digestibility in annual forage crops. Anim Res 52(3):193–228

Belcher AR (2009) The physiology and host genetics of quantitative resistance in maize to the fungal pathogen Cochliobolus heterostrophus. Dissertation, North Carolina State University, Raleigh

Belo A, Zheng PZ, Luck S, Shen B, Meyer DJ, Li BL, Tingey S, Rafalski A (2008) Whole genome scan detects an allelic variant of fad2 associated with increased oleic acid levels in maize. Mol Gen Genet 279(1):1–10

Berberich T, Harada M, Sugawara K, Kodama H, Iba K, Kusano T (1998) Two maize genes encoding omega-3 fatty acid desaturase and their differential expression to temperature. Plant Mol Biol 36(2):297–306

Bhave MR, Lawrence S, Barton C, Hannah LC (1990) Identification and molecular characterization of shrunken-2 cDNA clones of maize. Plant Cell 2(6):581–588

Boerjan W, Ralph J, Baucher M (2003) Lignin biosynthesis. Annu Rev Plant Biol 54:519–546

Bolanos J, Edmeades GO (1996) The importance of the anthesis-silking interval in breeding for drought tolerance in tropical maize. Field Crop Res 48(1):65–80

Bolanos J, Edmeades GO, Martinez L (1993) Eight cycles of selection for drought tolerance in lowland tropical maize. III. Responses in drought-adaptive physiological and morphological traits. Field Crop Res 31(3–4):269–286

Bortiri E, Chuck G, Vollbrecht E, Rocheford T, Martienssen R, Hake S (2006) ramosa2 encodes a LATERAL ORGAN BOUNDARY domain protein that determines the fate of stem cells in branch meristems of maize. Plant Cell 18(3):574–585

Boyer JS (2010) Drought decision-making. J Exp Bot 61(13):3493–3497

Brenner EA, Zein I, Chen YS, Andersen JR, Wenzel G, Ouzunova M, Eder J, Darnhofer B, Frei U, Barriere Y, Lubberstedt T (2010) Polymorphisms in O-methyltransferase genes are associated with stover cell wall digestibility in European maize (*Zea mays* L.). BMC Plant Biol. doi:10.1186/1471-2229-10-27

Broglie KE, Butler KH, Butruille MG et al. (2011) Method for identifying maize plants with Rcg1 gene conferring resistance to Colletotrichum infection. International Patent Publication No.: US 8,062,847B2

Browse J, Somerville C (1991) Glycerolipid synthesis – biochemistry and regulation. Annu Rev Plant Physiol Plant Mol Biol 42:467–506

Buckner B, Miguel PS, JanickBuckner D, Bennetzen JL (1996) The Y1 gene of maize codes for phytoene synthase. Genetics 143(1):479–488

Butron A, Chen YC, Rottinghaus GE, McMullen MD (2010) Genetic variation at bx1 controls DIMBOA content in maize. Theor Appl Genet 120(4):721–734

Camus-Kulandaivelu L, Veyrieras JB, Madur D, Combes V, Fourmann M, Barraud S, Dubreuil P, Gouesnard B, Manicacci D, Charcosset A (2006) Maize adaptation to temperate climate: relationship between population structure and polymorphism in the Dwarf8 gene. Genetics 172(4):2449–2463

Cao WB, Zheng LL, Zhang ZF, Li XB (2009) Genetic diversity of starch synthesis genes of Chinese maize (Zea mays L.) with SNAPs. Mol Biol 43(6):937–945

Carson ML (1995) A new gene in maize conferring the chlorotic halo reaction to infection by Exserohilum-Turcicum. Plant Dis 79(7):717–720

Carson ML, Stuber CW, Senior ML (2004) Identification and mapping of quantitative trait loci conditioning resistance to southern leaf blight of maize caused by Cochliobolus heterostrophus race O. Phytopathology 94(8):862–867

Cassani E, Bertolini E, Cerino Badone F, Landoni M, Gavina D, Sirizzotti A, Pilu R (2009) Characterization of the first dominant dwarf maize mutant carrying a single amino acid insertion in the VHYNP domain of the dwarf8 gene. Mol Breed 24:375–385

Chang RY, Peterson PA (1995) Genetic control of resistance to bipolaris maydis: one gene or two genes. J Hered 86(2):94–97

Chapman SC, Edmeades GO (1999) Selection improves drought tolerance in tropical maize populations: II. Direct and correlated responses among secondary traits. Crop Sci 39(5):1315–1324

Chen Y, Lubberstedt T (2010) Molecular basis of trait correlations. Trends Plant Sci 15(8):454–461

Chen YS, Chao Q, Tan GQ, Zhao J, Zhang MJ, Ji Q, Xu ML (2008) Identification and fine-mapping of a major QTL conferring resistance against head smut in maize. Theor Appl Genet 117(8):1241–1252

Chen YS, Zein I, Brenner EA, Andersen JR, Landbeck M, Ouzunova M, Lubberstedt T (2010) Polymorphisms in monolignol biosynthetic genes are associated with biomass yield and agronomic traits in European maize (Zea mays L.). BMC Plant Biol. doi:10.1186/1471-2229-10-12

Chintamanani S, Multani DS, Ruess H, Johal GS (2008) Distinct mechanisms govern the dosage-dependent and developmentally regulated resistance conferred by the maize Hm2 gene. Mol Plant Microbe Interact 21(1):79–86

Chourey PS, Nelson OE (1976) The enzymatic deficiency conditioned by the shrunken-1 mutations in maize. Biochem Genet 14(11–12):1041–1055

Chung CL, Jamann T, Longfellow J, Nelson R (2010a) Characterization and fine-mapping of a resistance locus for northern leaf blight in maize bin 8.06. Theor Appl Genet 121(2):205–227

Chung CL, Longfellow JM, Walsh EK, Kerdieh Z, Van Esbroeck G, Balint-Kurti P, Nelson RJ (2010b) Resistance loci affecting distinct stages of fungal pathogenesis: use of introgression lines for QTL mapping and characterization in the maize – Setosphaeria turcica pathosystem. BMC Plant Biol. doi:10.1186/1471-2229-10-103

Chung CL, Poland J, Kump K, Benson J, Longfellow J, Walsh E, Balint-Kurti P, Nelson R (2011) Targeted discovery of quantitative trait loci for resistance to northern leaf blight and other diseases of maize. Theor Appl Genet 123:307–326

Claassen MM, Shaw RH (1970) Water deficit effects on corn. II. Grain components. Agron J 62(5):652–655

Clark RM, Wagler TN, Quijada P, Doebley J (2006) A distant upstream enhancer at the maize domestication gene tb1 has pleiotropic effects on plant and inflorescent architecture. Nat Genet 38(5):594–597

Collazo P, Montoliu L, Puigdomenech P, Rigau J (1992) Structure and expression of the lignin O-Methyltransferase gene from Zea-Mays L. Plant Mol Biol 20(5):857–867

Collins NC, Webb CA, Seah S, Ellis JG, Hulbert SH, Pryor A (1998) The isolation and mapping of disease resistance gene analogs in maize. Mol Plant Microbe Interact 11(10):968–978

Collins N, Drake J, Ayliffe M, Sun Q, Ellis J, Hulbert S, Pryor T (1999) Molecular characterization of the maize Rp1-D rust resistance haplotype and its mutants. Plant Cell 11(7):1365–1376

Couture RM, Routley DG, Dunn GM (1971) Role of cyclic hydroxamic acids in monogenic resistance of maize to Helminthosporium-Turcicum. Physiol Plant Pathol 1(4):515–521

Cunningham FX, Pogson B, Sun ZR, McDonald KA, DellaPenna D, Gantt E (1996) Functional analysis of the beta and epsilon lycopene cyclase enzymes of Arabidopsis reveals a mechanism for control of cyclic carotenoid formation. Plant Cell 8(9):1613–1626

Danson JW, Mbogori M, Kimani M, Lagat M, Kuria A, Diallo A (2006) Marker assisted introgression of opaque2 gene into herbicide resistant elite maize inbred lines. Afr J Biotechnol 5(24):2417–2422

deLeon C, Jeffers D (2004) Maize diseases: a guide for field identification, 4th edn. CIMMYT Publications, Mexico City

DellaPenna D, Pogson BJ (2006) Vitamin synthesis in plants: tocopherols and carotenoids. Annu Rev Plant Biol 57:711–738

Dingerdissen AL, Geiger HH, Lee M, Schechert A, Welz HG (1996) Interval mapping of genes for quantitative resistance of maize to Setosphaeria turcica, cause of northern leaf blight, in a tropical environment. Mol Breed 2(2):143–156

Ducrocq S, Madur D, Veyrieras JB, Camus-Kulandaivelu L, Kloiber-Maitz M, Presterl T, Ouzunova M, Manicacci D, Charcosset A (2008) Key impact of Vgt1 on flowering time adaptation in maize: evidence from association mapping and ecogeographical information. Genetics 178(4):2433–2437

Duvick DN (1999) Hazard identification of agricultural biotechnology. Science 286(5439): 418–419

Eticha D, Stass A, Horst WJ (2005) Cell-wall pectin and its degree of methylation in the maize root-apex: significance for genotypic differences in aluminium resistance. Plant Cell Environ 28(11):1410–1420

Fan LJ, Bao JD, Wang Y, Yao JQ, Gui YJ, Hu WM, Zhu JQ, Zeng MQ, Li Y, Xu YB (2009) Post-domestication selection in the maize starch pathway. PLoS One. doi:10.1371/Journal.Pone.0007612

Flint-Garcia SA, Thuillet AC, Yu J, Pressoir G, Romero SM, Mitchell SE, Doebley J, Kresovich S, Goodman MM, Buckler ES (2005) Maize association population: a high-resolution platform for quantitative trait locus dissection. Plant J 44(6):1054–1064

Flint-Garcia SA, Bodnar AL, Scott MP (2009) Wide variability in kernel composition, seed characteristics, and zein profiles among diverse maize inbreds, landraces, and teosinte. Theor Appl Genet 119(6):1129–1142

Fraser PD, Bramley PM (2004) The biosynthesis and nutritional uses of carotenoids. Prog Lipid Res 43(3):228–265

Frederiksen RA, Berry RW, Foster JH (1976) Head smut of maize in Texas. Plant Dis Rep 60(7):610–611

Gaut BS, Doebley JF (1997) DNA sequence evidence for the segmental allotetraploid origin of maize. Proc Natl Acad Sci USA 94(13):6809–6814

Gevers HO (1975) New major gene for resistance to helminthosporium-turcicum leaf blight of maize. Plant Dis Rep 59(4):296–299

Gore MA, Chia JM, Elshire RJ, Sun Q, Ersoz ES, Hurwitz BL, Peiffer JA, McMullen MD, Grills GS, Ross-Ibarra J, Ware DH, Buckler ES (2009) A first-generation haplotype map of maize. Science 326(5956):1115–1117

Guillaumie S, San-Clemente H, Deswarte C, Martinez Y, Lapierre C, Murigneux A, Barriere Y, Pichon M, Goffner D (2007) MAIZEWALL. Database and developmental gene expression profiling of cell wall biosynthesis and assembly in maize. Plant Physiol 143(1):339–363

Guillet-Claude C, Birolleau-Touchard C, Manicacci D, Fourmann M, Barraud S, Carret V, Martinant JP, Barriere Y (2004a) Genetic diversity associated with variation in silage corn digestibility for three O-methyltransferase genes involved in lignin biosynthesis. Theor Appl Genet 110(1):126–135

Guillet-Claude C, Birolleau-Touchard C, Manicacci D, Rogowsky PM, Rigau J, Murigneux A, Martinant JP, Barriere Y (2004b) Nucleotide diversity of the ZmPox3 maize peroxidase gene: relationships between a MITE insertion in exon 2 and variation in forage maize digestibility. BMC Genet. doi:10.1186/1471-2156-5-19

Gupta PK, Rustgi S, Mir RR (2008) Array-based high-throughput DNA markers for crop improvement. Heredity 101(1):5–18

Halpin C, Holt K, Chojecki J, Oliver D, Chabbert B, Monties B, Edwards K, Barakate A, Foxon GA (1998) Brown-midrib maize (bm1) – a mutation affecting the cinnamyl alcohol dehydrogenase gene. Plant J 14(5):545–553

Han Y, Parsons CM, Alexander DE (1987) Nutritive-value of high oil corn for poultry. Poult Sci 66(1):103–111

Harjes CE, Rocheford TR, Bai L, Brutnell TP, Kandianis CB, Sowinski SG, Stapleton AE, Vallabhaneni R, Williams M, Wurtzel ET, Yan JB, Buckler ES (2008) Natural genetic variation in lycopene epsilon cyclase tapped for maize biofortification. Science 319(5861):330–333

Henry AM, Manicacci D, Falque M, Damerval C (2005) Molecular evolution of the Opaque-2 gene in *Zea mays* L. J Mol Evol 61(4):551–558

Hirschberg J (2001) Carotenoid biosynthesis in flowering plants. Curr Opin Plant Biol 4(3):210–218

Hoisington D, Khairallah M, Reeves T, Ribaut JV, Skovmand B, Taba S, Warburton M (1999) Plant genetic resources: what can they contribute toward increased crop productivity? Proc Natl Acad Sci USA 96(11):5937–5943

Hooker AL (1977) 2nd major gene locus in corn for chlorotic-lesion resistance to Helminithosporium-Turcicum. Crop Sci 17:132–135

Hooker AL (1981) Citation classic – reaction of corn seedlings with male-sterile cytoplasm to Helminthosporium-Maydis. Curr Content/Agric Biol Environ Sci 52:18–18

Hooker AL, Johnson PE, Shurtleff MC (1963) Soil fertility and Northern corn leaf blight infection. Agron J 55:411–412

Ingvardsen CR, Xing YZ, Frei UK, Lubberstedt T (2010) Genetic and physical fine mapping of Scmv2, a potyvirus resistance gene in maize. Theor Appl Genet 120(8):1621–1634

James MG, Robertson DS, Myers AM (1995) Characterization of the maize gene sugary1, a determinant of starch composition in kernels. Plant Cell 7(4):417–429

James MG, Denyer K, Myers AM (2003) Starch synthesis in the cereal endosperm. Curr Opin Plant Biol 6(3):215–222

Jiang L, Ingvardsen CR, Lubberstedt T, Xu ML (2008) The Pic19 NBS-LRR gene family members are closely linked to Scmv1, but not involved in maize resistance to sugarcane mosaic virus. Genome 51(9):673–684

Johal GS, Briggs SP (1992) Reductase-activity encoded by the Hm1 disease resistance gene in maize. Science 258(5084):985–987

Jones DL, Blancaflor EB, Kochian LV, Gilroy S (2006) Spatial coordination of aluminium uptake, production of reactive oxygen species, callose production and wall rigidification in maize roots. Plant Cell Environ 29(7):1309–1318

Jung C, Muller AE (2009) Flowering time control and applications in plant breeding. Trends Plant Sci 14(10):563–573

Jung M, Weldekidan T, Schaff D, Paterson A, Tingey S, Hawk J (1994) Generation means analysis and quantitative trait locus mapping of anthracnose stalk rot genes in maize. Theor Appl Genet 89:413–418

Kim KN, Fisher DK, Gao M, Guiltinan MJ (1998) Molecular cloning and characterization of the amylose-extender gene encoding starch branching enzyme IIB in maize. Plant Mol Biol 38(6):945–956

Kochian LV, Hoekenga OA, Pineros MA (2004) How do crop plants tolerate acid soils? – Mechanisms of aluminum tolerance and phosphorous efficiency. Annu Rev Plant Biol 55:459–493

Kou YJ, Wang SP (2010) Broad-spectrum and durability: understanding of quantitative disease resistance. Curr Opin Plant Biol 13(2):181–185

Krill AM, Kirst M, Kochian LV, Buckler ES, Hoekenga OA (2010) Association and linkage analysis of aluminum tolerance genes in maize. PLoS One. doi:10.1371/Journal.Pone.0009958

Kump KL, Bradbury PJ, Wisser RJ, Buckler ES, Belcher AR, Oropeza-Rosas MA, Zwonitzer JC, Kresovich S, McMullen MD, Ware D, Balint-Kurti PJ, Holland JB (2011) Genome-wide association study of quantitative resistance to southern leaf blight in the maize nested association mapping population. Nat Genet 43(2):163–168

Landry J, Damerval C, Azevedo RA, Delhaye S (2005) Effect of the opaque and floury mutations on the accumulation of dry matter and protein fractions in maize endosperm. Plant Physiol Biochem 43(6):549–556

Laurie CC, Chasalow SD, LeDeaux JR, McCarroll R, Bush D, Hauge B, Lai CQ, Clark D, Rocheford TR, Dudley JW (2004) The genetic architecture of response to long-term artificial selection for oil concentration in the maize kernel. Genetics 168(4):2141–2155

Lee KY, Huang AHC (1994) Genes encoding oleosins in maize kernel of inbreds Mo17 and B73. Plant Mol Biol 26(6):1981–1987

Li FQ, Murillo C, Wurtzel ET (2007) Maize Y9 encodes a product essential for 15-cis-zeta-carotene isomerization. Plant Physiol 144(2):1181–1189

Li XH, Wang ZH, Gao SR, Shi HL, Zhang SH, George MLC, Li MS, Xie CX (2008) Analysis of QTL for resistance to head smut (Sporisorium rediana). Field Crop Res 106(2):148–155

Li L, Li H, Li JY, Xu ST, Yang XH, Li JS, Yan JB (2010a) A genome-wide survey of maize lipid-related genes: candidate genes mining, digital gene expression profiling and co-location with QTL for maize kernel oil. Sci China Life Sci 53:690–700

Li Q, Li L, Yang X, Warburton ML, Bai G, Dai J, Li J, Yan J (2010b) Relationship, evolutionary fate and function of two maize co-orthologs of rice GW2 associated with kernel size and weight. BMC Plant Biol. doi:10.1186/1471-2229-10-143

Li Q, Yang X, Bai G, Warburton ML, Mahuku G, Gore M, Dai J, Li J, Yan J (2010c) Cloning and characterization of a putative GS3 ortholog involved in maize kernel development. Theor Appl Genet 120(4):753–763

Li QR, Farre G, Naqvi S, Breitenbach J, Sanahuja G, Bai C, Sandmann G, Capell T, Christou P, Zhu CF (2010d) Cloning and functional characterization of the maize carotenoid isomerase and beta-carotene hydroxylase genes and their regulation during endosperm maturation. Transgenic Res 19(6):1053–1068

Lindhout P (2002) The perspectives of polygenic resistance in breeding for durable disease resistance. Euphytica 124(2):217–226

Lohmer S, Maddaloni M, Motto M, Difonzo N, Hartings H, Salamini F, Thompson RD (1991) The maize regulatory locus Opaque-2 encodes a DNA-binding protein which activates the transcription of the B-32 gene. EMBO J 10(3):617–624

Lu Y, Zhang S, Shah T, Xie C, Hao Z, Li X, Farkhari M, Ribaut JM, Cao M, Rong T, Xu Y (2010) Joint linkage-linkage disequilibrium mapping is a powerful approach to detecting quantitative trait loci underlying drought tolerance in maize. Proc Natl Acad Sci USA 107(45):19585–19590

Lubberstedt T, Xia XC, Tan G, Liu X, Melchinger AE (1999) QTL mapping of resistance to Sporisorium reiliana in maize. Theor Appl Genet 99(3–4):593–598

Lubberstedt T, Zein I, Andersen J, Wenzel G, Krutzfeldt B, Eder J, Ouzunova M, Chun S (2005) Development and application of functional markers in maize. Euphytica 146(1–2):101–108

Lubberstedt T, Ingvardsen C, Melchinger AE, Xing Y, Salomon R, Redinbaugh MG (2006) Two chromosome segments confer multiple potyvirus resistance in maize. Plant Breed 125(4):352–356

Mackay TFC (2009) A-maize-ing diversity. Science 325(5941):688–689

Maddaloni M, Donini G, Balconi C, Rizzi E, Gallusci P, Forlani F, Lohmer S, Thompson R, Salamini F, Motto M (1996) The transcriptional activator Opaque-2 controls the expression of a cytosolic form of pyruvate orthophosphate dikinase-1 in maize endosperms. Mol Gen Genet 250(5):647–654

Mangolin CA, de Souza CL, Garcia AAF, Garcia AF, Sibov ST, de Souza AP (2004) Mapping QTLs for kernel oil content in a tropical maize population. Euphytica 137(2):251–259

Manicacci D, Falque M, Le Guillou S, Piegu B, Henry AM, Le Guilloux M, Damerval C, De Vienne D (2007) Maize Sh2 gene is constrained by natural selection but escaped domestication. J Evol Biol 20(2):503–516

Manicacci D, Camus-Kulandaivelu L, Fourmann M, Arar C, Barrault S, Rousselet A, Feminias N, Consoli L, Frances L, Mechin V, Murigneux A, Prioul JL, Charcosset A, Damerval C (2009) Epistatic interactions between opaque2 transcriptional activator and its target gene CyPPDK1 control kernel trait variation in maize. Plant Physiol 150(1):506–520

Maron LG, Kirst M, Mao C, Milner MJ, Menossi M, Kochian LV (2008) Transcriptional profiling of aluminum toxicity and tolerance responses in maize roots. New Phytol 179(1):116–128

Maron LG, Pineros MA, Guimaraes CT, Magalhaes JV, Pleiman JK, Mao CZ, Shaff J, Belicuas SNJ, Kochian LV (2010) Two functionally distinct members of the MATE (multi-drug and toxic compound extrusion) family of transporters potentially underlie two major aluminum tolerance QTLs in maize. Plant J 61(5):728–740

McMullen MD, Kresovich S, Villeda HS, Bradbury P, Li HH, Sun Q, Flint-Garcia S, Thornsberry J, Acharya C, Bottoms C, Brown P, Browne C, Eller M, Guill K, Harjes C, Kroon D, Lepak N, Mitchell SE, Peterson B, Pressoir G, Romero S, Rosas MO, Salvo S, Yates H, Hanson M, Jones E, Smith S, Glaubitz JC, Goodman M, Ware D, Holland JB, Buckler ES (2009) Genetic properties of the maize nested association mapping population. Science 325(5941):737–740

Melchinger AE, Kuntze L, Gumber RK, Lubberstedt T, Fuchs E (1998) Genetic basis of resistance to sugarcane mosaic virus in European maize germplasm. Theor Appl Genet 96(8):1151–1161

Menkir A, Liu WP, White WS, Mazlya-Dixon B, Rocheford T (2008) Carotenoid diversity in tropical-adapted yellow maize inbred lines. Food Chem 109(3):521–529

Mertz ET, Nelson OE, Bates LS (1964) Mutant gene that changes protein composition and increases lysine content of maize endosperm. Science 145(362):279–280

Moose SP, Dudley JW, Rocheford TR (2004) Maize selection passes the century mark: a unique resource for 21st century genomics. Trends Plant Sci 9(7):358–364

Multani DS, Meeley RB, Paterson AH, Gray J, Briggs SP, Johal GS (1998) Plant-pathogen microevolution: molecular basis for the origin of a fungal disease in maize. Proc Natl Acad Sci USA 95(4):1686–1691

Nelson OE, Mertz ET, Bates LS (1965) Second mutant gene affecting the amino acid pattern of maize endosperm proteins. Science 150(3702):1469–1470

Nelson DE, Repetti PP, Adams TR, Creelman RA, Wu J, Warner DC, Anstrom DC, Bensen RJ, Castiglioni PP, Donnarummo MG, Hinchey BS, Kumimoto RW, Maszle DR, Canales RD, Krolikowski KA, Dotson SB, Gutterson N, Ratcliffe OJ, Heard JE (2007) Plant nuclear factor Y (NF-Y) B subunits confer drought tolerance and lead to improved corn yields on water-limited acres. Proc Natl Acad Sci USA 104(42):16450–16455

Neto GC, Yunes JA, Dasilva MJ, Vettore AL, Arruda P, Leite A (1995) The involvement of Opaque-2 on beta-prolamin gene-regulation in maize and coix suggests a more general role for this transcriptional activator. Plant Mol Biol 27:1015–1029

Ninamango-Cardenas FE, Guimaraes CT, Martins PR, Parentoni SN, Carneiro NP, Lopes MA, Moro JR, Paiva E (2003) Mapping QTLs for aluminum tolerance in maize. Euphytica 130(2):223–232

Ogliari JB, Guimaraes MA, Geraldi IO, Camargo LEA (2005) New resistance genes in the *Zea mays* – exserohilum turcicum pathosystem. Genet Mol Biol 28(3):435–439

Palaisa KA, Morgante M, Williams M, Rafalski A (2003) Contrasting effects of selection on sequence diversity and linkage disequilibrium at two phytoene synthase loci. Plant Cell 15(8):1795–1806

Palaisa K, Morgante M, Tingey S, Rafalski A (2004) Long-range patterns of diversity and linkage disequilibrium surrounding the maize Y1 gene are indicative of an asymmetric selective sweep. Proc Natl Acad Sci USA 101(26):9885–9890

Pandey S, Ceballos H, Magnavaca R, Bahia AFC, Duquevargas J, Vinasco LE (1994) Genetics of tolerance to soil acidity in tropical maize. Crop Sci 34(6):1511–1514

Park WJ, Hochholdinger F, Gierl M (2004) Release of the benzoxazinoids defense molecules during lateral-and crown root emergence in *Zea mays*. J Plant Physiol 161(8):981–985

Parlevliet JE (2002) Durability of resistance against fungal, bacterial and viral pathogens; present situation. Euphytica 124(2):147–156

Pataky JK, Campana MA (2007) Reduction in common rust severity conferred by the Rp1D gene in sweet corn hybrids infected by mixtures of Rp1D-virulent and avirulent Puccinia sorghi. Plant Dis 91(11):1484–1488

Pataky JK, Tracy WF (1999) Widespread occurrence of common rust, caused by Puccinia sorghi, on Rp-resistant sweet corn in the Midwestern United States. Plant Dis 83(12):1177

Pataky JK, Natti TA, Snyder EB, Kurowski CJ (2000) Puccinia sorghi in Sinaloa, Mexico virulent on corn with the Rp1-D gene. Plant Dis 84(7):810

Pataky JK, Pate MC, Hulbert SH (2001) Resistance genes in the rp1 region of maize effective against Puccinia sorghi virulent on the Rp1-D gene in North America. Plant Dis 85(2):165–168

Pate MC, Pataky JK, Houghton WC, Teyker RH (2000) First report of Puccinia sorghi virulent on sweet corn with the Rp1-D gene in Florida and Texas. Plant Dis 84(10):1154

Pe ME, Gianfranceschi L, Taramino G, Tarchini R, Angelini P, Dani M, Binelli G (1993) Mapping quantitative trait loci (QTLs) for resistance to Gibberella Zeae infection in maize. Mol Gen Genet 241(1–2):11–16

Peng JR, Richards DE, Hartley NM, Murphy GP, Devos KM, Flintham JE, Beales J, Fish LJ, Worland AJ, Pelica F, Sudhakar D, Christou P, Snape JW, Gale MD, Harberd NP (1999) 'Green revolution' genes encode mutant gibberellin response modulators. Nature 400(6741):256–261

Pogson B, McDonald KA, Truong M, Britton G, DellaPenna D (1996) Arabidopsis carotenoid mutants demonstrate that lutein is not essential for photosynthesis in higher plants. Plant Cell 8(9):1627–1639

Poland JA, Bradbury PJ, Buckler ES, Nelson RJ (2011) Genome-wide nested association mapping of quantitative resistance to northern leaf blight in maize. Proc Natl Acad Sci USA 108(17):6893–6898

Prasanna BM, Vasal SK, Kassahun B, Singh NN (2001) Quality protein maize. Curr Sci India 81(10):1308–1319

Pressoir G, Brown PJ, Zhu WY, Upadyayula N, Rocheford T, Buckler ES, Kresovich S (2009) Natural variation in maize architecture is mediated by allelic differences at the PINOID co-ortholog barren inflorescence2. Plant J 58(4):618–628

Reymond M, Muller B, Leonardi A, Charcosset A, Tardieu F (2003) Combining quantitative trait loci analysis and an ecophysiological model to analyze the genetic variability of the responses of maize leaf growth to temperature and water deficit. Plant Physiol 131(2):664–675

Rhoades VH (1935) The location of a gene for disease resistance in maize. Proc Natl Acad Sci USA 21:243–246

Rogers LA, Dubos C, Surman C, Willment J, Cullis IF, Mansfield SD, Campbell MM (2005) Comparison of lignin deposition in three ectopic lignification mutants. New Phytol 168(1):123–140

Roussel V, Gibelin C, Fontaine AS, Barriere Y (2002) Genetic analysis in recombinant inbred lines of early dent forage maize. II – QTL mapping for cell wall constituents and cell wall digestibility from per se value and top cross experiments. Maydica 47(1):9–20

Sadok W, Naudin P, Boussuge B, Muller B, Welcker C, Tardieu F (2007) Leaf growth rate per unit thermal time follows QTL-dependent daily patterns in hundreds of maize lines under naturally fluctuating conditions. Plant Cell Environ 30(2):135–146

Salvi S, Sponza G, Morgante M, Tomes D, Niu X, Fengler KA, Meeley R, Ananiev EV, Svitashev S, Bruggemann E, Li B, Hainey CF, Radovic S, Zaina G, Rafalski JA, Tingey SV, Miao GH, Phillips RL, Tuberosa R (2007) Conserved noncoding genomic sequences associated with a flowering-time quantitative trait locus in maize. Proc Natl Acad Sci USA 104(27):11376–11381

Schmidt RJ, Burr FA, Aukerman MJ, Burr B (1990) Maize regulatory gene opaque-2 encodes a protein with a leucine-zipper motif that binds to zein DNA. Proc Natl Acad Sci USA 87(1):46–50

Schmidt RJ, Ketudat M, Aukerman MJ, Hoschek G (1992) Opaque-2 is a transcriptional activator that recognizes a specific target site in 22-kD zein genes. Plant Cell 4(6):689–700

Schnable PS, Ware D, Fulton RS et al (2009) The B73 maize genome: complexity, diversity, and dynamics. Science 326(5956):1112–1115

Service RF (2009) The promise of drought-tolerant corn. Science 326(5952):517

Setter TL, Yan JB, Warburton M, Ribaut JM, Xu YB, Sawkins M, Buckler ES, Zhang ZW, Gore MA (2011) Genetic association mapping identifies single nucleotide polymorphisms in genes that affect abscisic acid levels in maize floral tissues during drought. J Exp Bot 62(2):701–716

Shen B, Allen WB, Zheng PZ, Li CJ, Glassman K, Ranch J, Nubel D, Tarczynski MC (2010) Expression of ZmLEC1 and ZmWRI1 increases seed oil production in maize. Plant Physiol 153(3):980–987

Shin JH, Kwon SJ, Lee JK, Min HK, Kim NS (2006) Genetic diversity of maize kernel starch-synthesis genes with SNAPs. Genome 49(10):1287–1296

Shure M, Wessler S, Fedoroff N (1983) Molecular identification and isolation of the waxy locus in maize. Cell 35(1):225–233

Sibov ST, Gaspar M, Silva MJ, Ottoboni LMM, Arruda P, Souza AP (1999) Two genes control aluminum tolerance in maize: genetic and molecular mapping analyses. Genome 42(3):475–482

Sigmon B, Vollbrecht E (2010) Evidence of selection at the ramosa1 locus during maize domestication. Mol Ecol 19(7):1296–1311

Smith SM, Hulbert SH (2005) Recombination events generating a novel Rp1 race specificity. Mol Plant Microbe Interact 18(3):220–228

Smith SM, Pryor AJ, Hulbert SH (2004) Allelic and haplotypic diversity at the Rp1 rust resistance locus of maize. Genetics 167(4):1939–1947

Springer NM, Ying K, Fu Y, Ji TM, Yeh CT, Jia Y, Wu W, Richmond T, Kitzman J, Rosenbaum H, Iniguez AL, Barbazuk WB, Jeddeloh JA, Nettleton D, Schnable PS (2009) Maize inbreds exhibit high levels of copy number variation (CNV) and presence/absence variation (PAV) in genome content. PLoS Genet. doi:10.1371/journal.pgen.1000734

Stromberg EL (1981) Head smut of maize, a new disease in Minnesota. Phytopathology 71(8):906

Sun Q, Collins NC, Ayliffe M, Smith SM, Drake J, Pryor T, Hulbert SH (2001) Recombination between paralogues at the rp1 rust resistance locus in maize. Genetics 158(1):423–438

Takeda S, Matsuoka M (2008) Genetic approaches to crop improvement: responding to environmental and population changes. Nat Rev Genet 9(6):444–457

Tester RF, Morrison WR (1990) Swelling and gelatinization of cereal starches. I. Effects of amylopectin, amylose, and lipids. Cereal Chem 67(6):551–557

Thornsberry JM, Goodman MM, Doebley J, Kresovich S, Nielsen D, Buckler ES (2001) Dwarf8 polymorphisms associate with variation in flowering time. Nat Genet 28(3):286–289

Tian F, Bradbury PJ, Brown PJ, Hung H, Sun Q, Flint-Garcia S, Rocheford TR, McMullen MD, Holland JB, Buckler ES (2011) Genome-wide association study of leaf architecture in the maize nested association mapping population. Nat Genet 43(2):159–162

Tracy WF, Whitt SR, Buckler ES (2006) Recurrent mutation and genome evolution: example of sugary 1 and the origin of sweet maize. Crop Sci 46:S49–S54

Troyer AF (2006) Adaptedness and heterosis in corn and mule hybrids. Crop Sci 46(2):528–543

Ullstrup AJ, Brunson AM (1947) Linkage relationships of a gene in corn determining susceptibility to a Helminthosporium leaf spot. J Am Soc Agron 39(7):606–609

Vallabhaneni R, Gallagher CE, Licciardello N, Cuttriss AJ, Quinlan RF, Wurtzel ET (2009) Metabolite sorting of a germplasm collection reveals the hydroxylase3 locus as a new target for maize provitamin A biofortification. Plant Physiol 151(3):1635–1645

Vignols F, Rigau J, Torres MA, Capellades M, Puigdomenech P (1995) The brown midrib3 (bm3) mutation in maize occurs in the gene encoding caffeic acid O-methyltransferase. Plant Cell 7(4):407–416

Vollbrecht E, Springer PS, Goh L, Buckler ES, Martienssen R (2005) Architecture of floral branch systems in maize and related grasses. Nature 436(7054):1119–1126

Wang Q, Dooner HK (2006) Remarkable variation in maize genome structure inferred from haplotype diversity at the bz locus. Proc Natl Acad Sci USA 103(47):17644–17649

Wang H, Nussbaum-Wagler T, Li BL, Zhao Q, Vigouroux Y, Faller M, Bomblies K, Lukens L, Doebley JF (2005) The origin of the naked grains of maize. Nature 436(7051):714–719

Weber A, Clark RM, Vaughn L, Sanchez-Gonzalez JD, Yu JM, Yandell BS, Bradbury P, Doebley J (2007) Major regulatory genes in maize contribute to standing variation in teosinte (*Zea mays* ssp parviglumis). Genetics 177(4):2349–2359

Weber AL, Briggs WH, Rucker J, Baltazar BM, Sanchez-Gonzalez JD, Feng P, Buckler ES, Doebley J (2008) The genetic architecture of complex traits in Teosinte (*Zea mays* ssp parviglumis): new evidence from association mapping. Genetics 180(2):1221–1232

Weber AL, Zhao Q, McMullen MD, Doebley JF (2009) Using association mapping in Teosinte to investigate the function of maize selection-candidate genes. PLoS One. doi:10.1371/Journal.Pone.0008227

Welcker C, Boussuge B, Bencivenni C, Ribaut JM, Tardieu F (2007) Are source and sink strengths genetically linked in maize plants subjected to water deficit? A QTL study of the responses of leaf growth and of Anthesis-Silking Interval to water deficit. J Exp Bot 58(2):339–349

Whitt SR, Wilson LM, Tenaillon MI, Gaut BS, Buckler ES (2002) Genetic diversity and selection in the maize starch pathway. Proc Natl Acad Sci USA 99(20):12959–12962

Wilson LM, Whitt SR, Ibanez AM, Rocheford TR, Goodman MM, Buckler ES (2004) Dissection of maize kernel composition and starch production by candidate gene association. Plant Cell 16(10):2719–2733

Winkel BSJ (2004) Metabolic channeling in plants. Annu Rev Plant Biol 55:85–107

Wisser RJ, Balint-Kurti PJ, Nelson RJ (2006) The genetic architecture of disease resistance in maize: a synthesis of published studies. Phytopathology 96(2):120–129

Wisser RJ, Kolkman JM, Patzoldt ME, Holland JB, Yu JM, Krakowsky M, Nelson RJ, Balint-Kurti PJ (2011) Multivariate analysis of maize disease resistances suggests a pleiotropic genetic basis and implicates a GST gene. Proc Natl Acad Sci USA 108(18):7339–7344

Wolters P, Frey T, Conceicao A, Multani D, Broglie K, Davis S, Fengler K, Johnson E, Bacot K, Simcox K, Weldekidan T, Hawk J (2006) Map based cloning of a major QTL for anthracnose stalk rot resistance in maize. Plant and Animal Genomes XIV conference W 412, San Diego

Wu JY, Ding JQ, Du YX, Xu YB, Zhang XC (2007) Genetic analysis and molecular mapping of two dominant complementary genes determining resistance to sugarcane mosaic virus in maize. Euphytica 156(3):355–364

Xia XC, Melchinger AE, Kuntze L, Lubberstedt T (1999) Quantitative trait loci mapping of resistance to sugarcane mosaic virus in maize. Phytopathology 89(8):660–667

Xu ML, Melchinger AE, Xia XC, Lubberstedt T (1999) High-resolution mapping of loci conferring resistance to sugarcane mosaic virus in maize using RFLP, SSR, and AFLP markers. Mol Gen Genet 261(3):574–581

Yan JB, Shah T, Warburton ML, Buckler ES, McMullen MD, Crouch J (2009) Genetic characterization and linkage disequilibrium estimation of a global maize collection using SNP markers. PLoS One. doi:10.1371/Journal.Pone.0008451

Yan JB, Kandianis CB, Harjes CE, Bai L, Kim EH, Yang XH, Skinner DJ, Fu ZY, Mitchell S, Li Q, Fernandez MGS, Zaharieva M, Babu R, Fu Y, Palacios N, Li JS, DellaPenna D, Brutnell T, Buckler ES, Warburton ML, Rocheford T (2010) Rare genetic variation at *Zea mays* crtRB1 increases beta-carotene in maize grain. Nat Genet 42(4):322–327

Yang DE, Zhang CL, Zhang DS, Jin DM, Weng ML, Chen SJ, Nguyen H, Wang B (2004) Genetic analysis and molecular mapping of maize (*Zea mays* L.) stalk rot resistant gene Rfg1. Theor Appl Genet 108(4):706–711

Yang WP, Zheng YL, Zheng WT, Feng R (2005) Molecular genetic mapping of a high-lysine mutant gene (opaque-16) and the double recessive effect with opaque-2 in maize. Mol Breeding 15(3):257–269

Yang Q, Yin GM, Guo YL, Zhang DF, Chen SJ, Xu ML (2010a) A major QTL for resistance to Gibberella stalk rot in maize. Theor Appl Genet 121(4):673–687

Yang XH, Guo YQ, Yan JB, Zhang J, Song TM, Rocheford T, Li JS (2010b) Major and minor QTL and epistasis contribute to fatty acid compositions and oil concentration in high-oil maize. Theor Appl Genet 120(3):665–678

Yu JM, Holland JB, McMullen MD, Buckler ES (2008) Genetic design and statistical power of nested association mapping in maize. Genetics 178(1):539–551

Zaitlin D, Demars S, Ma Y (1993) Linkage of rhm, a recessive gene for resistance to Southern corn leaf blight, to RFLP marker loci in maize (*Zea mays*) seedlings. Genome 36(3):555–564

Zhang NY, Gur A, Gibon Y, Sulpice R, Flint-Garcia S, McMullen MD, Stitt M, Buckler ES (2010) Genetic analysis of central carbon metabolism unveils an amino acid substitution that alters maize NAD-dependent isocitrate dehydrogenase activity. PLoS ONE. doi:10.1371/Journal.Pone.0009991

Zhang D, Liu Y, Guo Y, Yang Q, Ye J, Chen S, Xu M (2012) Fine-mapping of qRfg2, a QTL for resistance to Gibberella stalk rot in maize. Theor Appl Genet 124(3):585–596

Zhao BY, Ardales E, Brasset E, Claflin LE, Leach JE, Hulbert SH (2004a) The Rxo1/Rba1 locus of maize controls resistance reactions to pathogenic and non-host bacteria. Theor Appl Genet 109(1):71–79

Zhao BY, Ardales EY, Raymundo A, Bai JF, Trick HN, Leach JE, Hulbert SH (2004b) The avrRxo1 gene from the rice pathogen Xanthomonas oryzae pv. oryzicola confers a nonhost Defense reaction on maize with resistance gene Rxo1. Mol Plant Microbe Interact 17(7):771–779

Zhao BY, Lin XH, Poland J, Trick H, Leach J, Hulbert S (2005) A maize resistance gene functions against bacterial streak disease in rice. Proc Natl Acad Sci USA 102(43):15383–15388

Zhao XR, Tan GQ, Xing YX, Wei L, Chao Q, Zuo WL, Lubberstedt T, Xu ML (2012) Marker-assisted introgression of qHSR1 to improve maize resistance to head smut. Mol Breed. doi:10.1007/s11032-011-9694-3

Zheng P, Allen WB, Roesler K, Williams ME, Zhang S, Li J, Glassman K, Ranch J, Nubel D, Solawetz W, Bhattramakki D, Llaca V, Deschamps S, Zhong GY, Tarczynski MC, Shen B (2008) A phenylalanine in DGAT is a key determinant of oil content and composition in maize. Nat Genet 40(3):367–372

Zwonitzer JC, Bubeck DM, Bhattramakki D, Goodman MM, Arellano C, Balint-Kurti PJ (2009) Use of selection with recurrent backcrossing and QTL mapping to identify loci contributing to southern leaf blight resistance in a highly resistant maize line. Theor Appl Genet 118(5): 911–925

Chapter 20
Molecular Diagnostics in Rice (*Oryza sativa*)

Wenhao Yan, Zhongmin Han, and Yongzhong Xing

Introduction

The global human population continues to grow and it is estimated that rice production will need to increase 40% by 2030 to meet the food demand. To confront the challenge of producing more rice from limited area of arable land, we need rice varieties with higher yield potential and greater yield stability (Khush 2005). Rice yield is a complex inherited trait, which consists of three components: number of panicles, number of grains per panicle, and grain weight. Grain yield is strongly affected by environmental factors such as light, temperature, nutrient availability, and biotic stresses.

Rice has a small genome size and high co-linearity with other members of the grass family (Gale and Devos 1998; Han et al. 2007). Rice was regarded as the model species for cereal crops, and studied with high priority in the last two decades. The rice whole genome sequence was released, and various rice mutant libraries were constructed in the past decade (Yu et al. 2002; Goff et al. 2002; Ito et al. 2002; Wu et al. 2003). These resources greatly promoted progress in rice genetic and functional genomic research (Han et al. 2007; Zhang 2007; Wu et al. 2008; Xue et al. 2008; Yan et al. 2011). For dissection of the genetic basis of yield and its three components, hundreds of mapping populations were investigated, derived from crosses between subspecies and within subspecies, as well as between cultivated and wild rice. Numerous QTL were identified for these traits, which provided resources for developing new cultivars. Several agronomically important genes were isolated by map-based cloning (Takahashi 2001; Ashikari 2005; Fan et al. 2006; Song et al. 2007; Jin et al. 2008; Tan et al. 2008; Xue et al. 2008; Jiao et al. 2010; Yan et al. 2011).

W. Yan • Z. Han • Y. Xing (✉)
National Key Lab of Crop Genetic Improvement, Huazhong University, 430070 Wuhan, China
e-mail: yzxing@mail.hzau.edu.cn

For rice, the most serious yield losses are caused by two major diseases (bacterial blight caused by *Xanthomonas oryzae pv. oryzae* and blast caused by *Pyricularia grisea*), and two major pests (a migratory pest, plant hopper (predominantly brown plant hopper), and stem borer). From conventional landraces to the current *Bt* modified crops, we could witness progress in the improvement of methods for protecting crops from pests and diseases. Among these methods, pesticides, including inorganic chemicals, organic chemicals and synthetic insecticides, played very important roles. However, the intensity of crop protection has increased considerably as exemplified by a 15–20 fold increase in the amount of pesticides used worldwide during the past 40 years (Oerke 2004, 2006). Pesticide abuse has caused environmental problems, and the effect of those chemicals decreased after a long-term usage. The most economical and environment-friendly method is application of host resistance, conferred by resistant plants themselves.

In this chapter, we will focus on the genetic basis of yield traits and biotic stress tolerance in rice. We emphasize implications for molecular diagnostics in using linked markers to major QTL, and functional markers derived from functionally characterized genes/QTL for these traits in rice. The fine mapped and cloned genes/QTL in rice for heading, yield components, quality, disease resistance, insect resistance, and plant architecture are reviewed in this chapter. We also compare the genome regions harboring QTL for yield related traits and biotic stress tolerance, and discuss how to develop varieties with higher quality, higher yield potential, and greater yield stability by breaking or avoiding genetic drag between yield traits and biotic stress tolerance with the help of rice diagnostic molecular markers.

Diagnostics for Yield Traits

Rice production is influenced by its genetic constitution as well as the environmental factors. Rice yield as a typical quantitative trait, is controlled by multiple minor QTL (Xing and Zhang 2010). However, recent progress in QTL analysis uncovered that rice grain yield is, in addition, controlled by few major QTL, which make molecular marker-aided selection for yield improvement feasible.

We compared the QTL mapping results from four populations (F_2, $F_{2:3}$, RIL, IF_2) derived from the same cross between Zhenshan 97 and Minghui 63, the best hybrid in the 1990s in China. About 20 distinct QTL were detected for each trait, including rice yield itself. Only few QTL, except for grain weight, could be mapped in more than one population or in more than 1 year (Xing et al. 2002; Xing and Zhang 2010), which indicated that the performance of yield traits was strongly dependent on environmental conditions. However, major QTL were repeatedly identified at least in two conditions among the four populations. Concurrently, a large number of QTL were mapped for various agronomic and biological traits (about 6,000 QTL for more than 200 traits) by scientists worldwide (http://www.gramene.org/). About one third of these QTL were directly related to agronomic traits (Table 20.1). For plant height alone, 475 QTL were detected. 262 QTL were

Table 20.1 Number of QTL detected for biotic stress resistance, grain yield, and rice quality

Traits[a]	Chromosomes												Total
	1	2	3	4	5	6	7	8	9	10	11	12	
BBR		2	4	2	1		1	2	1	1	2	1	17
BR	17	8	6	11	3	14	9	5	7	1	8	12	101
BPHR	7	6	4	6	4	9	1	5	2	4	5	5	58
GLR			1								1		2
SBR		4	4	2	2		2	2	5		2	1	24
TGW	4	1	4		2	1		3	1	1	3		20
GYP	22	18	11	13	14	13	11	10	6	6	11	5	140
TN	23	14	18	17	12	7	10	15	5	2	9	6	138
SNP	45	26	23	32	15	38	18	16	13	7	18	10	261
PL	21	24	22	25	13	23	12	16	15	17	7	10	205
PB	2	4	3	9	3	6	5	12		1	2	1	48
SB	6	2	6	3		6	3		1		1	3	31
PH	68	40	67	57	30	38	33	41	26	25	29	21	475
HD	26	21	40	17	14	28	29	27	13	17	14	16	262
GC	3	3				9	4						19
AC	4	3	8	1	2	11	4	3	2	2	1	2	43
GL	2	6	5	1	1	3	2			1	1	1	23
GW	3	2	1		5	4	1	1		1	3	1	22
LA	2	1	1	1	1	2	1		3		2		14
LAR	4	2	3	1	2	5		2	1	4	2	2	28

Information collected from http://www.gramene.org/
[a]*BBR* bacterial blight disease resistance, *BR* blast disease resistance, *BPHR* brown planthopper resistance, *GLR* green leafhopper resistance, *SBR* sheath blight disease resistance, *PH* plant height, *HD* heading date, *GC* gel consistency, *AC* amylose content, *GL* grain length, *GW* grain width, *SNP* spikelets number per panicle, *TGW* 1,000-grain weight, *GYP* grain yield per panicle, *PL* panicle length, *PB* primary branch, *SB* secondary branch, *LA* leaf angle, *LAR* leaf area, *TNP* tiller number per plant

detected for the adaptation trait heading date, indicating diverse genetic resources in the world. Among yield components, the highest number of QTL (262) was identified for spikelets per panicle. Most major QTL were repeatedly detected across environments or populations. Many major QTL were validated in near isogenic lines and then fine mapped or even cloned.

Panicle length and spikelets per panicle, which are controlled by the number of primary and secondary branches, determine panicle size. Spikelets per panicle displays substantial genetic variation, ranging from 50 to 500 (Xing and Zhang 2010). In addition, among the three yield components, grain number per panicle makes the highest contribution to rice yield. Thus, in the past decade, it received most attention, and significant progress on its genetic improvement has been made.

Gn1a was the first cloned rice QTL that affects grains per panicle (Ashikari 2005). A population of 96 backcross inbred lines (BILs) between Habataki and Koshihikari, two accessions with a large difference in grain number per panicle, was developed to detect QTL for grain number. Five QTL for grain number were obtained. Among them, *Gn1*, located on the short arm of chromosome 1, explains 44% of the trait variation. Further research indicated that this locus was the combination of two distinct QTL, *Gn1a* and *Gn1b*. *Gn1a* was narrowed to a 6.3-kb region with 13,000 near-isogenic line (NIL)-F_2 individuals. One gene encoding cytokinin oxidase/dehydrogenase (CKX) protein in this region was chosen as candidate gene. Sequencing of the candidate gene among rice varieties revealed that an 11-bp deletion, which caused a premature stop of the protein, was a key element for functional variation of *Gn1a*. Transformation experiments confirmed the function of this candidate gene. *Gn1a* (*OsCKX2*) encodes an active enzyme, which inhibits cell division by degrading cytokinin (CK). The inhibition of cell division in inflorescence meristems results in small panicles. In addition, another unidentified QTL, *Gn1b*, tightly linked to *Gn1a,* increases grain number in an additive manner. The NIL carrying the functional Habataki *Gn1a* and *Gn1b* alleles could produce 45% more grain than the control, showing great potential for increasing grain productivity (Ashikari 2005). *DEP1* was identified as QTL, affecting both grain number per panicle and panicle density (Huang et al. 2009). This locus also reduces the length of the inflorescence internode, resulting in erect panicle architecture. *DEP1* was mapped on chromosome 9 between SSR markers RM3700 and RM7424. Primary analysis of the QTL revealed that *dep1* acted as dominant negative regulator in regulating rice panicle architecture and grain number. Yet, it controlled panicle architecture and caused 40.9% more grain yield per plant without affecting other traits such as heading date and tiller number. Molecular characterization of the *DEP1* locus showed that *DEP1* encoded an unknown PEBP (phosphatidylethanolamine-binding protein) like domain containing protein. The functional allele was the mutant allele with a premature stop codon, which was caused by replacement of 637-bp of the middle of exon 5 by a 12-bp sequence. This variation can be converted into a PCR-based functional marker for breeding erect panicle rice varieties. Additionally, *dep1* also functions during vascular system establishment and sclerenchyma cell wall development, suggesting an important role in both water transport capacity and the mechanical strength of the stem, both of which are important factors for the breeding of high-yielding, lodging-resistant varieties (Huang et al. 2009).

Grain weight has the highest heritability among the three yield components, indicating that it is less affected by environment. Grain size (characterized by grain length, width, and thickness) is closely associated with grain weight. Thus, grain weight is easily manipulated for QTL analysis in the field. Major QTL affecting grain size were detected across multiple environments and even across different populations. *GS3*, a major QTL contributed to grain shape (mainly grain length), located in the centromeric region of chromosome 3, explained approximately 30% of grain length variation in F_2 and recombinant inbred line (RIL) populations from the cross between Zhenshan 97 and Minghui 63 (Fan et al. 2006). The *GS3* allele for short grain is dominant over the allele for long grain, which is present in indica rice. A NIL-F_2 population of 5,000 individuals was developed using Minghui 63 (with large grain size) as the recurrent and Chuan7 (with small grain size) as the donor parent to fine map the gene. *GS3* was narrowed to one single gene. The candidate gene contained four conserved domains, including A PEBP-like domain, a transmembrane domain, a putative TNFR (tumor necrosis factor receptor)/NGFR C-terminal and a cysteine-rich region similar to the von Willebrand factor type C (VWFC) domain (Fan et al. 2006). Later studies concluded that the N-terminal motif of 66 amino acids, which was previously predicted to encode a PEBP domain, could be defined as organ size regulation (OSR) domain, necessary and sufficient for functioning as a negative regulator (Mao et al. 2010). Comparative sequencing revealed that there was one common single nucleotide mutation (C–A) in the second exon of the *GS3* gene. This SNP changed a cysteine codon (TGC) in the small-grain group to a termination codon (TGA) in the large-grain group (Fan et al. 2006, 2009). The C–A mutation was used to develop a CAPS marker (SF28) to distinguish different grain lengths. There was a perfect correlation between the C–A genotype and grain length within 180 varieties. Genotype A has long grain length ranging from 8.8 to 10.7 mm, while the grain length in genotype C varies from 6.3 to 8.8 mm. Additionally, the varieties whose grain length were around 8.8 mm could not be distinguished by the marker SF28 because of the existence of other minor grain length QTL. However, SF28 still offers a very cheap, rapid, and easily operated tool for grain length selection. Specifically, longer grain as a quality trait was preferred by most of rice consumers and the *CAPS* marker provides the prediction with 100% accuracy regarding to the long/short grain genotype. In addition, another two genic SSR markers RGS1 and RGS2 in the last intron and the final exon of GS3, respectively, were found to be associated with grain length. Especially, RGS1 and RGS2 have strong power to predict medium and short grain (Wang et al. 2011b).

GW2, a major grain width QTL on chromosome 2, encodes a previously unknown RING-type protein with E3 ubiquitin ligase activity. A 1-bp deletion in exon 4 of the gene resulted in a premature stop codon. A loss-of-function mutation of *GW2* increased cell numbers and led to enhanced grain width, grain weight, as well as yield (Song et al. 2007). *GW5/qSW5* on chromosome 5 was the major QTL controlling rice grain width, which explained 38.5% of trait variation in a F_2 population (Shomura et al. 2008; Weng et al. 2008). A 1121-bp deletion of *GW5/qSW5* resulted in wider grain and this deletion was closely associated with the increase of grain width among more than 100 rice landraces. Two Indel markers were developed and implemented in rice breeding (Weng et al. 2008).

Diagnostics for Ideal Plant Architecture

Plant architecture forms the basis for rice yield. In the 1960s, rice yield was largely increased by introducing the recessive gene *sd1* which resulted in a decreased plant height and an enhancing rice harvest index (Spielmeyer 2002). The most famous semi-dwarf variety, IR8, represents popular rice plant architecture: short stature, high tillering ability, sturdy stems, dark green and erect leaves (Virk et al. 2004). However, the unproductive tillers and small panicles limits these semi-dwarf varieties in further improvement for yield potential. Rice breeders suggested a new ideal plant architecture with low tillering capacity (3–4 tillers when directly sown, 8–10 tillers when transplanted); large panicle size (200–250 grains per panicle), a suitable plant height (90–100 cm), and thick and sturdy stems (Virk et al. 2004). For this architecture, genes controlling plant height, stem strength and tiller growth are critical.

IPA1 (*ideal plant architecture1*), also called *WFP* (*wealthy farmer's panicle*), was independently identified by two groups recently (Jiao et al. 2010; Miura et al. 2010). It was regarded as the main QTL contributing to rice ideal architecture. *IPA1/WFP* encodes the transcription factor SQUAMOSA PROMOTER BINDING PROTEIN-LIKE (*OsSPL14*). It acts in a semi-dominant manner and is linked with RM149, RM1345, RM223, and RM264. Functional analysis revealed that *IPA1/WFP* is the target of OsmiR156. Mutated *OsSPL14* produced fewer tillers (4.1 tillers) than wild type plants (Jiao et al. 2010). The single nucleotide change from C to A at the OsmiR156-targeted site (292th amino acid) in *OsSPL14* might be used as the functional SNP marker for *IPA1* in rice breeding (Miura et al. 2010). Moreover, the expression level of *OsSPL14* is highly correlated with its function and could thus be used as expression marker (Jiao et al. 2010; Miura et al. 2010). *MOC1* is another gene affecting rice tillering. The spontaneous *moc1* mutant plants have only one main culm compared with multiple tillers in wild type rice. It encodes a putative GRAS family protein and is mainly expressed in axillary buds. Comparative sequencing between mutant and wild type revealed that a 1.9 kb retrotransposon inserted at position 948 causes a premature translation stop (Li et al. 2003). Moreover, *moc1* plants also produce much fewer rachis-branches and spikelets as compared to wild-type plants. In addition to its potential function in regulating tillering, *MOC1* seems to be a negative regulator of plant height (Li et al. 2003).

Compared with wild rice, cultivated rice usually shows erect growth (a smaller tiller angle) as adaptation mechanism to dense planting. Recently, two research groups independently cloned the QTL, *PROSTATE GROWTH 1* (*PROG1*) (Jin et al. 2008; Tan et al. 2008), which is involved in domestication of rice tillering. *PROG1* was mapped on the short arm of chromosome 7 between SSR markers RM298 and RM481. Transformation results confirmed that *PROG1* encodes a C2H2-type zinc-finger protein. Among the many SNPs in this gene between Teqing (*O. sativa* L. ssp. indica variety) and wild rice (*O. rufipogon*), a 1-bp substitution in Teqing transformed threonine into serine and the threonine genotype was responsible for wild-rice plant architecture. Cultivated rice possesses an erect growth habit instead of a prostrate growth pattern. *PROG1* could be a key tillering growth regulator according to the phenotype change caused by this important SNP (Jin et al. 2008; Tan et al. 2008). Another major QTL controlling rice tiller angle is *TAC1* (*Tiller Angle Control 1*), which was mapped on chromosome 9 using a large F_2 population from the cross between two rice cultivars, IR24 and Asominori. *TAC1* harbors four introns, the fourth of which is located in the 3'-untranslated region. Sequence analysis with more than 150 rice accessions, including 21 wild rice varieties, reveals that a mutation from 'AGGA' to 'GGGA' in the fourth intron of *TAC1* leads to a change in expression level, which affects gene function in a significant manner. Eighty-eight compact japonica rice accessions held a 'GGGA' type *TAC1* gene, while 'AGGA' type *TAC1* was present in 21 wild rice and 43 indica rice accessions, which showed a spread growth habit. This sequence variant might serve as a marker for *TAC1* gene selection (Yu et al. 2007). The cloning and functional analysis of the *LA1* gene, located near the centromere of chromosome 11 between AC35795 and AC6330, indicates that polar auxin transportation affects tiller angle variation. Furthermore, an 8-bp deletion in the fourth exon of the *LA1* (*LAZY1*) gene, which leads to premature termination, was the functional mutation for this gene (Li et al. 2007). Along with *PROG1* and *TAC1*, selection for the *LA1* gene could facilitate breeding for high stand density.

Both biomass and harvest index are the two factors contributing to further improvement of higher yielding varieties. Current high yielding rice varieties frequently are tall plants with large panicles. As more fertilizer is used, rice plants are susceptible to lodging, which causes a great loss of grain yield. Thus, genes controlling rice culm mechanical strength need to be detected and integrated into high yield rice. *STRONG CULM2* (*SCM2*) was isolated by positional cloning (Ookawa et al. 2010; Terao et al. 2010). *SCM2* was mapped to the long-arm of chromosome 6 and results from transformation demonstrate that *SCM2* is identical to *ABERRANT PANICLE ORGANIZATION1* (*APO1*), a gene reported to control panicle structure. Plants carrying *SCM2* show enhanced culm strength and increased spikelet number. This is consistent with previous studies showing that *APO1* enhances cell proliferation in inflorescence meristem and leads to an increase of spikelet number (Ikeda et al. 2005, 2009). *SCM2/APO1* could be a very important candidate for improving grain yield and culm strength.

Diagnostics for Rice Adaptation

Rice is a short day plant. Rice heading date is controlled by various QTL, whose functions are regulated by the photoperiod of its growing area and cropping season (Hayama et al. 2003). Genetic dissection in different populations revealed that at least 14 QTL contribute to rice heading (Yano et al. 2001). Among these QTL, some are related to photoperiod sensitivity (*Hd1*, *Hd2*, *Hd3*, *Hd5*, and *Hd6*) (Yano et al. 2001), while interacting with each other. For example, the epistatic interactions between *Hd1* and *Hd5*, as well as *Hd2* and *Hd6* were clearly proven in advanced populations (Yamamoto et al. 2000; Hong et al. 2003). Most of those important QTL or genes for rice heading date have been identified by map-based cloning (Yano et al. 2000; Takahashi 2001; Kojima et al. 2002; Doi et al. 2004; Matsubara et al. 2008; Wu et al. 2008; Xue et al. 2008; Yan et al. 2011). *Hd1* explained 67% of phenotypic variation in a F_2 population (Yano et al. 1997) and it encodes a homologue of an Arabidopsis CO protein containing a zinc finger domain and a CCT domain. Unlike the situation that CO promotes flowering under long day conditions in Arabidopsis, *Hd1* inhibits rice heading under long day and promotes flowering under short day conditions. *Hd1* has been a major player in the process of rice adaptation. Compared with the functional Nipponbare allele, the nonfunctional Kasalath allele contains various sequence variations such as a 36-bp insertion and a 33-bp deletion in the first exon, as well as two single-base substitutions and a two-base deletion in the second exon (Yano et al. 2000). Sequence analysis of 64 rice cultivars, which represent the genetic diversity of 332 accessions from around the world, showed a high diversity of *Hd1*. All the sequenced *Hd1* alleles were grouped into 15 distinct protein types. Compared with Kasalath, many deletions and SNPs were detected. Some of them even cause a defective CCT domain and, therefore, result in nonfunctional *Hd1* alleles. This sequence diversity, in addition to the expression of *Hd1*, might be one of the main reasons for the diverse heading date in rice (Takahashi et al. 2009).

Another two important rice flowering genes are *Ghd7* and *Ghd8* (Xue et al. 2008; Yan et al. 2011). They inhibit rice flowering under long day conditions, but have no effect under short day conditions. *Ghd7* and *Ghd8* are located on chromosomes 7 and 8, respectively. Rice flowering was delayed under long day condition for 20 days by *Ghd7* and 10 days by *Ghd8* in near isogenic lines. *Ghd7* encodes a CCT domain-containing protein without an obvious B-box zinc finger structure. Moreover, *Ghd7* does not have homologues in Arabidopsis, and thus, is a rice-specific flowering regulator. Being transformed into rice plants with different genetic background, *Ghd7* showed various genetic effects, and it delayed flowering from 30 days to more than 60 days. Five protein-encoding alleles were found among 19 genotypes representing cultivars from a wide geographic range of Asia. Two out of five alleles are supposed to be nonfunctional and mainly exist in early rice or varieties being grown in high latitude regions with a short rice growing season such as the Heilongjiang Province in Northeast China. Two *Ghd7* alleles having strong effects on the three traits were found in rice varieties from tropics and subtropics with a hot and long summer, where varieties taking full advantage of light resources for producing more grain. The fifth allele with weak effect was found in varieties growing in temperate areas with short and cool summer, which ensures completion of the life cycle of respective cultivars. These findings indicate the specific potential benefit of each *Ghd7* allele in rice production. The suitable *Ghd7* alleles should be considered according to the local temperature and light condition (Xue et al. 2008). *Ghd8* encodes a HEME ACTIVATOR PROTEIN3 (HAP3), a subunit of the HAP2-HAP3-HAP5 trimetric complex (also known as DTH8 in another study (Wei et al. 2010)). *Ghd8* was isolated in two different segregating populations. Within these two distinct genetic backgrounds, the same *Ghd8* allele showed different genetic effects. Specifically, *Ghd8* delayed heading under long day conditions in both populations and promoted flowering under short day conditions in only one population. In addition to genetic background, the genetic effect of *Ghd8* also varied with different of *Ghd8* alleles. For example, the strong-effect 93-11 *Ghd8* allele delayed rice heading for 20 days, while the Nipponbare *Ghd8* allele caused only 10 days delay of heading (Yan et al. 2011). The nucleotide diversity of *Ghd8* was investigated by comparing a 4 kb region that contained the promoter, the open reading frame (ORF), and the 3′ untranslated region (UTR) in a subset (94 accessions) of the Chinese mini-core collection of rice. Nine alleles of *Ghd8* were identified based on predicted proteins. Five main polymorphic sites were affecting phenotypic variation in association analysis. In addition, *Ghd8* is allelic with *Hd5*, which was reported to interact with *Hd1* (Lin et al. 2003). Understanding of *Ghd8* will provide better understanding of *Hd1* selection in rice breeding.

Ghd7 and *Ghd8* also contribute to plant height, grain production, as well as plant architecture. *Ghd7* increases rice plant height for about 30 cm and produces 60% more spikelets per panicle. *Ghd7* also has marked effects on stem growth and development. *Ghd7* leads to plants with more internodes and thicker stems (Xue et al. 2008). Similar phenotypes were observed in rice plants with functional *Ghd8* alleles. The dominant genotype of *Ghd8* NILs produces 50% more grain per plant than plants with a recessive allele. These major pleiotropic genes provide the opportunity to modify different aspects of rice plants simultaneously (Yan et al. 2011). It seems that yield related genes are frequently associated with heading date according to the characterization of these two pleiotropic QTL. The application of these genes are limited by climate and photoperiod of rice growing regions. Thus, specific alleles suitable to a given rice growing region needs to be identified in order to ensure positive contributions to rice production.

However, some rice yield related genes do not affect heading date. *Gn1a*, mentioned previously, affects grain number, but not heading date (Ashikari 2005). Application of these QTL is more flexible regardless of photoperiod in the rice growing season. These two kinds of yield related QTL agree with results from QTL mapping, where QTL for spikelets per panicle are either dependent or independent of heading date (Liu et al. 2011).

Diagnostics for Rice Quality

Rice grain production has been greatly increased since the first green revolution. Thus, preventing starvation is no longer the main crucial issue for rice production in the most places of the world. Grain quality has received increasing attention in rice breeding programs. Grain quality of rice consists of two components: appearance, as well as cooking and eating quality. Grain appearance is determined by grain shape measured by the ratio of grain length to width. Most of the rice consumers in China, USA, and most Asian countries prefer long and slender grain rice (Unnevehr et al. 1992). This can be achieved by introducing *GS3* (Fan et al. 2005, 2006), a major QTL regulating rice grain length and *qSW5* (Shomura et al. 2008; Weng et al. 2008) and *GW2* (Song et al. 2007), both controlling grain width. Cooking and eating quality is mainly affected by three components: amylose content (AC), gel consistency (GC), and gelatinization temperature (GT) (Cagampang et al. 1973). AC and GT were greatly affected by two tightly linked QTL, *Wx* and *ALK*, respectively, which are located on chromosome 6 (Isshiki et al. 1998; Gao 2003; Hirano and Sano 1991). *Wx* encodes a key enzyme in amylase synthesis, granule-bound starch synthase (GBSS). Two types of waxy alleles, *Wxa* and *Wxb*, exist in indica rice and japonica rice, respectively. A G-to-T mutation in the intron 1 splice donor differentiates between these two *waxy* alleles, and the expression of *Wxa* is much higher than that of *Wxb*. Additional evidence confirmed the major effect of *Wx* on AC and GC, and a minor effect on GT (Tian et al. 2009).

Fragrance is a key factor in determining rice market price (Kovach et al. 2009). Rice fragrance is mainly determined by the content of 2-acetyl-1-pyrroline (2AP) (Buttery et al. 1982). Its synthesis is inhibited by the *Badh2* gene, encoding (β) betaine aldehyde dehydrogenase (BADH2) (Chen et al. 2008b). An 8-bp deletion and a 7-bp deletion in exon 2 of this gene both result in recessive *badh2.1* alleles, which are responsible for increased rice fragrance. Both indels are valuable tools for fragrance allele selection (Shi et al. 2008).

Diagnostics for Disease and Pest Resistance

Host Resistance for Rice Diseases

Rice disease resistance can be classified into two categories: qualitative resistance and quantitative resistance (*QR*). Qualitative resistance is also known as complete resistance that is performed through a single resistance (*R*) gene, while quantitative resistance (*QR*) results in incomplete resistance, which is mediated by multiple genes of partial effect. R-gene-mediated qualitative resistance provides resistance to a specific pathogen race. The lifetime of R-gene-mediated resistance is generally short due to the rapid evolution of the pathogen. Compared with R-gene-mediated qualitative resistance, quantitative resistance contributes to broad-spectrum and durable disease resistance, although each QTL only has a minor effect (Poland et al. 2009; White and Yang 2009; Kou and Wang 2010; St.Clair 2010).

Bacterial blight is one of the most destructive diseases in rice, decreasing rice yield by 5–20% (Mew 1987). Up to now, more than 30 R genes have been mapped, of which six have been characterized (Hu et al. 2008; Yoshimura et al. 1998; Sun et al. 2004; Chu 2006; Xiang et al. 2006; Iyer and McCouch 2004; Gu et al. 2005; Song et al. 1995). Among these six genes, *Xa21* was the first identified R gene, tightly linked with marker RG103 on chromosome 11. *Xa21* confers major resistance to a broad spectrum of Xoo races, and especially bacterial pathogen strain PXO99. *Xa21* encodes a protein with 23 leucine-rich repeats (LRR) in the extracellular domain and a serine and threonine kinase in the intracellular domain (Song et al. 1995). A recent study showed that the Xa21 protein functions by auto-phosphorylating its residues, Ser686, Thr688, and Ser689, to prevent degradation (Xu et al. 2006). Similar to *Xa21*, *Xa26* also encodes a leucine-rich repeat (LRR) receptor kinase-like protein, and it has three more LRR domains than the Xa21 protein. The *Xa26* gene is located on the long arm of rice chromosome 11, where many other resistance genes are located, such as *Xa3*, *Xa4*, and *Xa22*(t), (a later study revealed that *Xa3* was *Xa26*). *Xa26* provides resistance at both seedling and adult stage, while the resistance effect of *Xa21* is developmentally controlled (Song et al. 1995). Therefore, *Xa26* has a broader application in disease resistance breeding than *Xa21*. Interestingly, the function of *Xa26* is influenced by genetic background. Compared with two indica rice varieties, Minghui 63 and IRBB3, *Xa26* shows enhanced resistance and a wider resistance spectrum in japonica background (Sun et al. 2004). In addition to *Xa21* and *Xa26*, other rice bacterial blight disease resistance genes include: *Xa1*, encoding a nucleotide binding-LRR protein (Yoshimura et al. 1998); the recessive allele of *Xa5*, encoding the gamma subunit of transcription factor IIA (Iyer and McCouch 2004); *Xa27,* encoding a novel protein (Gu et al. 2005); the fully recessive allele of *xa13*, encoding a novel plasma membrane protein (Chu et al. 2006). Comparative sequencing revealed that *Xa13* and *xa13* encode identical proteins, but have crucial sequence differences in their promoter regions. Resistance was both enhanced by suppressed the expression of the dominant and recessive allele. Thus, expression level of *xa13* rather than protein composition, is the key factor for *xa13*-mediated resistance (Yuan et al. 2009). *Xanthomonas oryzae* can induce the expression of the dominant but not the recessive *xa13* allele by binding to UPT (up-regulated by TALe) promoter boxes of *Xa13* by bacteria transcription activator-like effector (TALe) proteins (Römer et al. 2010; Yuan et al. 2011). XA13 interacted with two other proteins, COPT1 and COPT5, to transport copper (a very important component of pesticides) from xylem vessels, where Xoo multiplies and spreads to cause disease (Yuan et al. 2010). These studies on the molecular mechanism of XA13 offer new insights for breeding resistant varieties.

Rice blast (leaf and neck blast) caused by the fungus *Magnaporthe oryzae*, result in 10–30% of annual loss of rice grain yield. Genetic studies identified more than 85 R genes and approximately 300 QTL, which contribute to blast disease resistance (Ballini et al. 2008). To date, a total of 12 rice blast resistance genes have been cloned and characterized. Except for *Pi-d2* (Chen et al. 2006) and *Pi21* (Fukuoka et al. 2009), encoding a non-RD receptor-like kinase and a proline-rich protein, respectively, the remaining 10 genes encode NB-LRR type proteins (Ronald and Dardick 2006), including *Pib* (Wang et al. 1999), *Pita* (Bryan et al. 2000), *Pi9* (Qu et al. 2006), *Pi2* and *Piz-t* (Zhou et al. 2006), *Pi-d2* (Chen et al. 2006), *Pi36* (Liu et al. 2007), *Pi37* (Lin et al. 2007), *pi5*(Lee et al. 2009), pb1 (Hayashi et al. 2010), and *Pikm* (Ashikawa et al. 2008). Individual R genes lose their effectiveness within 1 or 2 years after they were deployed in the field, because of rapid evolution of pathogens (McDonald and Linde 2002). The first cloned QTL for blast resistance, *pi21*, provided durable blast disease resistance in rice. In a susceptible background, *pi21* showed consistent effects against all 10 widely distributed races of *M. oryzae*, although resistance was complete, compared to R gene mediated resistance. *Pi21* encodes a heavy metal–transport/detoxification-domain-containing protein. Sequence comparison revealed 21-bp and 48-bp deletions within the coding region, causing functional changes of the protein, resulting in disease resistance (Fukuoka et al. 2009).

Host Resistance for Rice Insect Pests

Climate change over years was always followed by outbreak of insect pest (Yamamura et al. 2006; Gregory et al. 2009). The brown planthopper (BPH) is one of the most destructive insect pests of rice. Planthopper can immigrate to fields from thousand miles away by wind. Thus the efficiency of chemical control to planthopper is low. Moreover, these insecticides are toxic to predators of rice planthoppers, such as spiders, the mirid bug and the dryinid wasp. Decrease in the amount of these predators rescinded the threat to planthoppers, and thus resulted in an increased amount of planthoppers (Tanaka 2000). Developing BPH resistant varieties is a promising way to control BPH. Host resistance of rice against BPH was first reported for the variety Mudgo in 1969 (Du et al. 2009). Until now, 19 BPH-resistance genes from cultivated and wild rice species have been mapped (Zhang 2007). Single loci or pyramiding of multiple QTL conveyed resistance to planthoppers. *Bph14* and *Bph15*, locating on the long arm of chromosome 3 and the short arm of chromosome 4, respectively, were two major loci that contributed to BPH resistance for 'B5', a highly resistant line, with resistance genes derived from wild rice (Huang et al. 2001). Recently, *Bph14* was cloned by map-based cloning. *Bph14* showed stable resistance in different genetic backgrounds and is, thus, valuable for development of resistant rice varieties. It encodes a coiled-coil, nucleotide-binding, and leucine-rich repeat (CC-NB-LRR) protein (Du et al. 2009). It is the first and only characterized BPH-resistance gene so far. More genes related to BPH-resistance need to be elucidated, and they will provide gene resources for BPH-resistance breeding.

Stem borers are serious pests of rice. They can cause two types of damage to rice plants. The first type of damage is called dead heart, caused by caterpillars entering into the leaf sheath at the base of young stems at the beginning of and during tillering. Dead heart results in heterogeneous maturity. The second type of damage is called white panicle, caused by caterpillars eating the stem below the panicle at flowering. White panicle decreases grain fill. Both types of damage cause annual losses of 11.5 billion yuan (US$1.69 billion) in China (Chen et al. 2011; Sheng et al. 2003). Similar to planthoppers, chemical control is not effective to control stem borers. Due to the lack of suitable germplasm with a good level of resistance against the widespread yellow stem borer, no major QTL/gene has been found in cultivated rice (Deka and Barthakur 2010). However, Bacillus thuringiensis, or Bt bacteria in soil produces protein crystals known to be toxic to several varieties of insect larvae, including moths, beetles, mosquitoes, black flies, nematodes and flatworms (Chen et al. 2005, 2008a; Tang et al. 2006; Tu et al. 2000).

The Role of the Conserved LRR Domain in Host Resistance

The products of pathogen avirulence (AVR) genes were first recognized by the products of R-genes, always possessing a R gene-pathogen specific recognition manner, and various defense responses such as the initiation of localized cell death were then triggered. This response is termed hypersensitive responses (HR). Numerous studies revealed a set of intracellular pathogen AVR proteins recognition genes, which were known as NBS-LRR (nucleotide binding site-leucine rich repeat) proteins. These plant NB-LRR proteins are characterized by a tripartite domain architecture consisting of an N-terminal coiled-coil (CC) or Toll/interleukin-1 receptor (TIR) domain, a central NBS domain, and a C-terminal LRR domain. Many known R genes encode a NBS-LRR protein (Table 20.1), except for a few exceptions, indicating their importance for rice biotic stress response. There are more than 400 NBS-LRR genes in rice. NBS-LRR genes are subdivided into two main groups according to the N-terminus, either a toll/interleukin-1 receptor (TIR) homologous region or a coiled-coil (CC) motif. All rice NBS-LRR genes belong to CC-type NBS-LRR proteins (Tamura and Tachida 2011). The LRR domain is thought to be responsible for recognizing pathogen-encoded ligands during HR reaction (Parniske et al. 1997). Genetic variation within 20 rice varieties and landraces showed that there is a much higher ratio of nonsynonymous-to-synonymous SNPs than for other genes, indicating a positive selection during evolution (McNally et al. 2009). A genome-wide variation survey of NBS-LRR genes within two sequenced varieties, 93-11 (indica) and Nipponbare (japonica), shows a high rate of diversity and strong human selection (Shang et al. 2009). These results indicate that during rice evolution, NBS-LRR genes, especially, the LRR domain, played a key role during pathogen recognition and co-evolved with avirulence proteins, thus providing permanent and wide resistance.

Molecular Diagnostics to Address Linkage Drag

Besides the above-mentioned cloned genes and QTL, hundreds of QTL for agronomic traits and biotic stress were mapped (http://www.gramene.org/). QTL and genes for biotic stress resistance, grain yield and quality are closely linked on 11 chromosomes (Fig. 20.1). The rice bacterial blight resistance gene *OsNPR1* is tightly linked with the grain number gene *Gn1a* on chromosome 1 (Yuan et al. 2007; Ashikari et al. 2005). On chromosome 3, three small regions each harbor two major QTL for distinct traits (Li et al. 1995, 1997; Wang et al. 2011a; Xu et al. 2002; Mei et al. 2005). Notably, *pi21*, a rice blast resistant gene is tightly linked to an eating quality gene, LOC-Os04_g32890 (Fukuoka et al. 2009) on chromosome 5. Heading date QTL *qHD7* is closely linked to sheath blast disease resistance QTL *qSB-7* on chromosome 7 (Zou et al. 2000). The ideal plant architecture gene, *IPA1*, is closely linked to the quality gene, *OsISA*, on chromosome 8 (Fujita et al. 2003; Jiao et al. 2010). These tight linkages of QTL need to be carefully considered during the process of molecular marker-assisted breeding (MAB). Once two favorable alleles have been recombined into coupling phase, it will facilitate MAB and joint transmission of favorable alleles to offspring. Negative correlations are frequently reported between yield and biotic stress resistance, as well as between

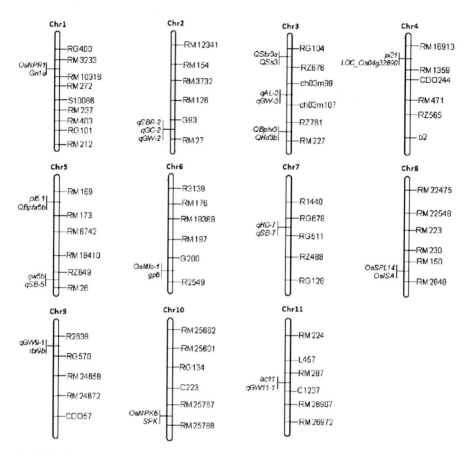

Fig. 20.1 Chromosome map showing regions of major QTL clusters for different traits. The major QTL positions were determined by the physical locations of their flanking molecular markers. Cloned genes were indicated on the basis of their physical positions

yield and quality traits (Fukuoka et al. 2009). When exotic germplasm is used as source of disease resistance, undesirable genes linked to the target gene might hitchhike and, thus, reduce the agronomic fitness of the improved cultivar. To avoid linkage of deleterious genes introduced along with the target gene, closely linked molecular markers are required to help identifying recombinants, in which linkage drag has been removed. One good example exists for two tightly linked genes on chromosome 4. The introduced *pi21* chromosome segment caused undesirable grain characteristics in addition to resistance to blast disease, indicating that likely another undesirable gene is tightly linked with *pi21* (Fukuoka et al. 2009). Further analysis showed that the loss-of-function protein Pi21 provides durable resistance to rice blast without affecting yield, and a gene causing defective seed development was tightly linked with *pi21*. In consequence, functional markers derived from known genes along with closely linked markers are very efficient to remove linkage drag (Table 20.2).

Table 20.2 Functional markers or potential functional markers derived from cloned major genes for rice diagnostics

Trait	Gene	Quantitative trait polymorphism (QTP)	Putative functional markers
Grain number	Gn1a	11-bp deletion caused premature termination	InDel
Erect panicle	DEP1	12-bp substitution caused premature termination	InDel
Tillering	PROG1	1-bp substitution at 152thAA caused Thr to Ser	CAPS or SNP
	TAC1	Mutation in the 3′-splicing site of the 4th intron	SNP
	Lazy	8-bp deletion caused premature termination	InDel
	MOC1	1.9-kb retrotransposon insertion	InDel
Plant architecture	IPA1	C-A mutation at the OsmiR156-targeted site	InDel
Stem texture	SCM2/APO1	Indels in both promoter and coding region	InDel or SNP
Grain length	GS3	1 C–A mutation caused premature termination	CAPS marker-SF28
Grain weight	GW2	1-bp deletion caused premature termination	Rare allele
	GW5	1212-bp deletion	Indel1, Indel2
Quality	Wx	G-T mutation in the 1st intron	CAPS or SNP
Fragrance	BADH2	8-bp deletion in exon 7 or 7-bp deletion in exon 2	FMbadh2-E2A, FMbadh2-E7
Heading date	Hd1	Substitution, insertion or deletion	InDel or SNP
	Ghd7	Substitution or absent, premature termination	InDel or SNP
	Ghd8	Substitution or absent, premature termination	InDel or SNP
Blast disease	Pib	A single amino acid difference at position 441	SNP
	Pi-d2	Amino acid change at S321L	dCAPS1
	Pi21	21 and 48-bp deletions caused disease resistance	InDel
	Pb1	Deletion in the gene	InDel
Blight disease	xa5	2 SNP causing V to E change at position 39	InDel
	xa13	Promoter mutation changing expression level	InDel
	Xa21	Origin from wild rice	Tightly linked SSR
Brown planthopper resistance	Bph14	Unique LRR domain and 6-bp deletion	CAPS or SNP

Future Perspectives

Currently, many major QTL and genes for yield components, disease and insect pest resistance have been cloned or fine mapped in rice. These major genes are the first targets for breeding by molecular design (Fig. 20.2). Multiple genes control panicle size, tiller number, and grain weight. Rational application of those genes will enhance grain yield. Genes controlling brown planthopper, bacterial blight, and rice blast can be introgressed into elite rice varieties to ensure stability of rice production. Genes for photoperiod control can be used to adapt varieties according to their growing season and geographic target regions. Rice quality genes can be selected to improve varieties according to consumer preference. Manipulation of such single genes by marker-aided selection is straightforward. Functional markers derived from these genes enable selection at the seedling stage (Table 20.2). Yield can be manipulated by selecting for yield related genes at the seedling stage using functional markers, which helps to avoid selection at the reproductive stage, when corresponding traits need to be differentiated and evaluated by eye. Functional marker selection for stress tolerance has particular advantages, because resistance can be predicted, even without disease or pest infection trials. Most causative polymorphisms, resulting from loss-of-function mutations or important functional modifications, were in many cases identified in biparental mapping populations. Nucleotide diversity analysis of functional genes will help to determine alleles with moderate functional effects. These alleles would produce a range of functional markers. One good example is molecular selection of *Ghd7* breeding

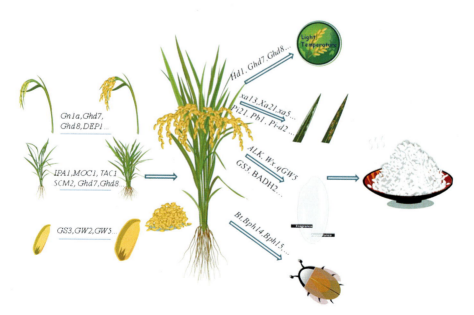

Fig. 20.2 Genes for rice molecular diagnostics to increase yield and yield stability

of high-yielding rice. As mentioned above, five main alleles of *Ghd7* exist in rice germplasm. Strong alleles, which could increase panicle size and delay rice heading, might be selected for varieties in the tropics and subtropics with extended period of sunlight, but not for varieties to be grown in areas with cool and short summers. A second example is selection of *GS3* for grain shape improvement. By combined selection of CAPS marker SF28 with two SSR markers RGS1 and RGS2, varieties with long, medium, and short grains would be easily obtained (Wang et al. 2011a). In addition, more genes for rice genetic improvement are needed for systematic molecular diagnostics. Especially genes conferring resistance to biotic and abiotic stresses would be valuable, as these traits are more difficult to be observed by breeders under field conditions. In order to support genome-wide selection, China National Seed Group Company (CNSGC) has developed a SNP array of 6,000 markers based on the causal mutations derived from hundreds of known functional genes and plus polymorphic sites of predicted genes between indica and japonica (unpublished results). These SNP sites are associated with pathways controlling yield components, plant architecture, adaptation, and resistance. Thus, hybridization with this array will evaluate any genotypes at all the targeted loci, and will assist in pyramiding favorable alleles by foreground selection. CNSGC is planning to develop a 60 K SNP array in order to cover all rice genes. With an increased number of functionally characterized alleles, more favorable information will be obtained from the SNP array hybridization. This SNP array will be suitable for both foreground and background selection, and greatly accelerate the process of rice breeding.

References

Ashikari M (2005) Cytokinin oxidase regulates rice grain production. Science 309(5735):741–745
Ashikari M, Sakakibara H, Lin SY, Yamamoto T, Takashi H, Nishimura A, Angeles EA, Qian Q, Kitana H, Matsuoka M (2005) Cytokinin oxidase regulates rice grain production. Science 309:741–745
Ashikawa I, Hayashi N, Yamane H, Kanamori H, Wu J, Matsumoto T, Ono K, Yano M (2008) Two adjacent nucleotide-binding site-leucine-rich repeat class genes are required to confer Pikm-specific rice blast resistance. Genetics 180(4):2267–2276
Ballini E, Morel JB, Droc G, Price A, Courtois B, Notteghem JL, Tharreau D (2008) A genome-wide meta-analysis of rice blast resistance genes and quantitative trait loci provides new insights into partial and complete resistance. Mol Plant Microbe Interact 21(7):859–868
Bryan GT, Wu KS, Farrall L, Jia Y, Hershey HP, McAdams SA, Faulk KN, Donaldson GK, Tarchini R, Valent B (2000) A single amino acid difference distinguishes resistant and susceptible alleles of the rice blast resistance gene *Pi-ta*. Plant Cell 12(11):2033–2046
Buttery RG, Ling LC, Juliano BO (1982) 2-Acety-1-pyrroline: an important aroma component of cooked rice. Chem Ind 12:958–959
Cagampang GB, Perez CM, Juliano BO (1973) A gel consistency test for eating quality of rice. J Sci Food Agric 24(12):1589–1594
Chen H, Xu C, Li X, Lin Y, Zhang Q (2005) Transgenic indica rice plants harboring a synthetic *cry2A** gene of *Bacillus thuringiensis* exhibit enhanced resistance against lepidopteran rice pests. Theor Appl Genet 111(7):1330–1337

Chen X, Shang J, Chen D, Lei C, Zou Y, Zhai W, Liu G, Xu J, Ling Z, Cao G, Ma B, Wang Y, Zhao X, Li S, Zhu L (2006) A B-lectin receptor kinase gene conferring rice blast resistance. Plant J 46(5):794–804

Chen H, Zhang G, Zhang Q, Lin Y (2008a) Effect of transgenic *Bacillus thuringiensis* rice lines on mortality and feeding behavior of rice stem borers (Lepidoptera: Crambidae). J Econ Entomol 101(1):182–189

Chen S, Yang Y, Shi W, Ji Q, He F, Zhang Z, Cheng Z, Liu X, Xu M (2008b) *Badh2*, encoding betaine aldehyde dehydrogenase, inhibits the biosynthesis of 2-acetyl-1-pyrroline, a major component in rice fragrance. Plant Cell 20(7):1850–1861

Chen M, Shelton A, Ye GY (2011) Insect-resistant genetically modified rice in China: from research to commercialization. Annu Rev Entomol 56:81–101

Chu Z (2006) Promoter mutations of an essential gene for pollen development result in disease resistance in rice. Genes Dev 20(10):1250–1255

Clair St DA (2010) Quantitative disease resistance and quantitative resistance loci in breeding. Annu Rev Phytopathol 48(1):247–268

Deka S, Barthakur S (2010) Overview on current status of biotechnological interventions on yellow stem borer *Scirpophaga incertulas* (Lepidoptera: Crambidae) resistance in rice. Biotechnol Adv 28(1):70–81

Doi K, Izawa T, Fuse T, Yamanouchi U, Kubo T, Shimatani Z, Yano M, Yoshimura A (2004) *Ehd1*, a B-type response regulator in rice, confers short-day promotion of flowering and controls FT-like gene expression independently of *Hd1*. Genes Dev 18(8):926–936

Du B, Zhang W, Liu B, Hu J, Wei Z, Shi Z, He R, Zhu L, Chen R, Han B, He G (2009) Identification and characterization of *Bph14*, a gene conferring resistance to brown planthopper in rice. Proc Natl Acad Sci U S A 106(52):22163–22168

Fan CC, Yu XQ, Xing YZ, Xu CG, Luo LJ, Zhang QF (2005) The main effects, epistatic effects and environmental interactions of QTLs on the cooking and eating quality of rice in a doubled-haploid line population. Theor Appl Genet 110:1445–1452

Fan C, Xing Y, Mao H, Lu T, Han B, Xu C, Li X, Zhang Q (2006) *GS3*, a major QTL for grain length and weight and minor QTL for grain width and thickness in rice, encodes a putative transmembrane protein. Theor Appl Genet 112(6):1164–1171

Fan C, Yu S, Wang C, Xing Y (2009) A causal C-A mutation in the second exon of *GS3* highly associated with rice grain length and validated as a functional marker. Theor Appl Genet 118(3):465–472

Fujita N, Kubo A, Suh DS, Wong KS, Jane JL, Ozawa K, Takaiwa F, Inaba Y, Nakamura Y (2003) Antisense inhibition of isoamylase alters the structure of amylopectin and the physicochemical properties of starch in rice endosperm. Plant Cell Physiol 44:607–618

Fukuoka S, Saka N, Koga H, Ono K, Shimizu T, Ebana K, Hayashi N, Takahashi A, Hirochika H, Okuno K, Yano M (2009) Loss of function of a proline-containing protein confers durable disease resistance in rice. Science 325(5943):998–1001

Gale MD, Devos KM (1998) Comparative genetics in the grasses. Proc Natl Acad Sci U S A 95(5):1971–1974

Gao Z (2003) Map-based cloning of the *ALK* gene, which controls the gelatinization temperature of rice. Sci China C Life Sci 46(6):661. doi:10.1360/03yc0099

Goff SA, Ricke D, Lan TH, Presting G, Wang R, Dunn M, Glazebrook J, Sessions A, Oeller P, Varma H, Hadley D, Hutchison D, Martin C, Katagiri F, Lange BM, Moughamer T, Xia Y, Budworth P, Zhong J, Miguel T, Paszkowski U, Zhang S, Colbert M, Sun WL, Chen L, Cooper B, Park S, Wood TC, Mao L, Quail P, Wing R, Dean R, Yu Y, Zharkikh A, Shen R, Sahasrabudhe S, Thomas A, Cannings R, Gutin A, Pruss D, Reid J, Tavtigian S, Mitchell J, Eldredge G, Scholl T, Miller RM, Bhatnagar S, Adey N, Rubano T, Tusneem N, Robinson R, Feldhaus J, Macalma T, Oliphant A, Briggs S (2002) A draft sequence of the rice genome (*Oryza sativa* L. ssp. japonica). Science 296(5565):92–100

Gregory P, Newton AC, Ingram JSI (2009) Integrating pests and pathogens into the climate change/food security debate. J Exp Bot 60(10):2827–2838

Gu K, Yang B, Tian D, Wu L, Wang D, Sreekala C, Yang F, Chu Z, Wang GL, White FF, Yin Z (2005) R gene expression induced by a type-III effector triggers disease resistance in rice. Nature 435(7045):1122–1125

Han B, Xue Y, Li J, Deng XW, Zhang Q (2007) Rice functional genomics research in China. Philos Trans R Soc Lond B Biol Sci 362(1482):1009–1021

Hayama R, Yokoi S, Tamaki S, Yano M, Shimamoto K (2003) Adaptation of photoperiodic control pathways produces short-day flowering in rice. Nature 422(6933):719–722

Hayashi N, Inoue H, Kato T, Funao T, Shirota M, Shimizu T, Kanamori H, Yamane H, Hayano-Saito Y, Matsumoto T, Yano M, Takatsuji H (2010) Durable panicle blast-resistance gene *Pb1* encodes an atypical CC-NBS-LRR protein and was generated by acquiring a promoter through local genome duplication. Plant J 64(3):498–510

Hirano H-Y, Sano Y (1991) Molecular characterization of the *waxy* locus of rice (*Oryza sativa*). Plant Cell Physiol 32(7):989–997

Hong X, Sasaki T, Yano M (2003) Fine mapping and characterization of quantitative trait loci *Hd4* and *Hd5* controlling heading date in rice. Breed Sci 53:51–59

Hu KM, Qiu DY, Shen XL, Li XH, Wang SP (2008) Isolation and manipulation of quantitative trait loci for disease resistance in rice using a candidate gene approach. Mol Plant 1(5):786–793

Huang Z, He G, Shu L, Li X, Zhang Q (2001) Identification and mapping of two brown planthopper resistance genes in rice. Theor Appl Genet 102:929–934

Huang X, Qian Q, Liu Z, Sun H, He S, Luo D, Xia G, Chu C, Li J, Fu X (2009) Natural variation at the *DEP1* locus enhances grain yield in rice. Nat Genet 41(4):494–497

Ikeda K, Nagasawa N, Nagato Y (2005) *ABERRANT PANICLE ORGANIZATION1* temporally regulates meristem identity in rice. Dev Biol 282(2):349–360

Ikeda K, Yasuno N, Oikawa T, Iida S, Nagato Y, Maekawa M, Kyozuka J (2009) Expression level of *ABERRANT PANICLE ORGANIZATION1* determines rice inflorescence form through control of cell proliferation in the meristem. Plant Physiol 150(2):736–747

Isshiki M, Morino K, Nakajima M, Okagaki RJ, Wessler SR, Izawa T, Shimamoto K (1998) A naturally occurring functional allele of the rice *waxy* locus has a GT to TT mutation at the 5′ splice site of the first intron. Plant J 15(1):133–138

Ito T, Motohashi R, Kuromori T, Mizukado S, Sakurai T, Kanahara H, Seki M, Shinozaki K (2002) A new resource of locally transposed Dissociation elements for screening gene-knockout lines in silico on the Arabidopsis genome. Plant Physiol 129(4):1695–1699

Iyer AS, McCouch SR (2004) The rice bacterial blight resistance gene *xa5* encodes a novel form of disease resistance. Mol Plant Microbe Interact 17(12):1348–1354

Jiao Y, Wang Y, Xue D, Wang J, Yan M, Liu G, Dong G, Zeng D, Lu Z, Zhu X, Qian Q, Li J (2010) Regulation of *OsSPL14* by *OsmiR156* defines ideal plant architecture in rice. Nat Genet 42(6):541–544

Jin J, Huang W, Gao J-P, Yang J, Shi M, Zhu M-Z, Luo D, Lin H-X (2008) Genetic control of rice plant architecture under domestication. Nat Genet 40(11):1365–1369

Khush GS (2005) What it will take to Feed 5.0 Billion Rice consumers in 2030. Plant Mol Biol 59(1):1–6

Kojima S, Takahashi Y, Kobayashi Y, Monna L, Sasaki T, Araki T, Yano M (2002) *Hd3a*, a rice ortholog of the Arabidopsis *FT* gene, promotes transition to flowering downstream of *Hd1* under short-day conditions. Plant Cell Physiol 43(10):1096–1105

Kou Y, Wang S (2010) Broad-spectrum and durability: understanding of quantitative disease resistance. Curr Opin Plant Biol 13(2):181–185

Kovach MJ, Calingacion MN, Fitzgerald MA, McCouch SR (2009) The origin and evolution of fragrance in rice (*Oryza sativa* L.). Proc Natl Acad Sci U S A 106(34):14444–14449

Lee SK, Song MY, Seo YS, Kim HK, Ko S, Cao PJ, Suh JP, Yi G, Roh JH, Lee S, An G, Hahn TR, Wang GL, Ronald P, Jeon JS (2009) Rice *Pi5*-mediated resistance to magnaporthe oryzae requires the presence of two coiled-coil-nucleotide-binding-leucine-rich repeat genes. Genetics 181(4):1627–1638

Li ZK, Pinson SRM, Marchetti MA, Stansel JW, Park WD (1995) Characterization of quantitative trait loci (QTLs) in cultivated rice contributing to field resistance to sheath blight (*Rhizoctonia solani*). Theor Appl Genet 91:382–388

Li ZK, Pinson SRM, Park WD, Paterson AH, Stansel JW (1997) Epistasis for three grain yield components in rice (*Oryza sativa* L.). Genetics 145:453–465

Li X, Qian Q, Fu Z, Wang Y, Xiong G, Zeng D, Wang X, Liu X, Teng S, Hiroshi F, Yuan M, Luo D, Han B, Li J (2003) Control of tillering in rice. Nature 422(6932):618–621

Li P, Wang Y, Qian Q, Fu Z, Wang M, Zeng D, Li B, Wang X, Li J (2007) *LAZY1* controls rice shoot gravitropism through regulating polar auxin transport. Cell Res 17(5):402–410

Lin F, Chen S, Que Z, Wang L, Liu X, Pan Q (2007) The blast resistance gene *Pi37* encodes a nucleotide binding site leucine-rich repeat protein and is a member of a resistance gene cluster on rice chromosome 1. Genetics 177(3):1871–1880

Lin H, Liang ZW, Sasaki T, Yano M (2003) Fine mapping and characterization of quantitative trait loci *Hd4* and *Hd5* controlling heading date in rice. Breeding science 53:51–59

Liu X, Lin F, Wang L, Pan Q (2007) The in silico map-based cloning of *Pi36*, a rice coiled-coil nucleotide-binding site leucine-rich repeat gene that confers race-specific resistance to the blast fungus. Genetics 176(4):2541–2549

Liu T, Zhang H, Xing Y (2011) Quantitative trait loci for the number of grains per panicle dependent on or independent of heading date in rice (*Oryza sativa* L.). Breed Sci 61(2): 142–150

Mao H, Sun S, Yao J, Wang C, Yu S, Xu C, Li X, Zhang Q (2010) Linking differential domain functions of the *GS3* protein to natural variation of grain size in rice. Proc Natl Acad Sci U S A A107(45):19579–19584

Matsubara K, Yamanouchi U, Wang ZX, Minobe Y, Izawa T, Yano M (2008) *Ehd2*, a rice ortholog of the maize *INDETERMINATE1* gene, promotes flowering by up-regulating *Ehd1*. Plant Physiol 148(3):1425–1435

McDonald BA, Linde C (2002) Pathogen population genetics, evolutionary potential, and durable resistance. Annu Rev Phytopathol 40:349–379

McNally KL, Childs KL, Bohnert R, Davidson RM, Zhao K, Ulat VJ, Zeller G, Clark RM, Hoen DR, Bureau TE, Stokowski R, Ballinger DG, Frazer KA, Cox DR, Padhukasahasram B, Bustamante CD, Weigel D, Mackill DJ, Bruskiewich RM, Ratsch G, Buell CR, Leung H, Leach JE (2009) Genomewide SNP variation reveals relationships among landraces and modern varieties of rice. Proc Natl Acad Sci U S A 106(30):12273–12278

Mei HW, Li ZK, Shu QY, Guo LB, Wang YP, Yu XQ, Ying CS, Luo LJ (2005) Gene actions of QTLs affecting several agronomic traits resolved in a recombinant inbred rice population and two backcross populations. Theor Appl Genet 110:649–659

Mew TW (1987) Current status and further prospects of research on bacterial blight of rice. Annu Rev Phytopathol 25:359–382

Miura K, Ikeda M, Matsubara A, Song XJ, Ito M, Asano K, Matsuoka M, Kitano H, Ashikari M (2010) *OsSPL14* promotes panicle branching and higher grain productivity in rice. Nat Genet 42(6):545–549

Oerke E (2004) Safeguarding production—losses in major crops and the role of crop protection. Crop Prot 23(4):275–285

Oerke E-C (2006) Crop losses to animal pests, plant pathogens, and weeds. Encycl Pest Manage. doi:10.1081/E-EPM-120009897

Ookawa T, Hobo T, Yano M, Murata K, Ando T, Miura H, Asano K, Ochiai Y, Ikeda M, Nishitani R, Ebitani T, Ozaki H, Angeles ER, Hirasawa T, Matsuoka M (2010) New approach for rice improvement using a pleiotropic QTL gene for lodging resistance and yield. Nat Commun 1(8):132

Parniske M, Hammond-Kosack KE, Golstein C, Thomas CM, Jones DA, Harrison K, Wulff BB, Jones JD (1997) Novel disease resistance specificities result from sequence exchange between tandemly repeated genes at the *Cf-4/9* locus of tomato. Cell 91(6):821–832

Poland JA, Balint-Kurti PJ, Wisser RJ, Pratt RC, Nelson RJ (2009) Shades of gray: the world of quantitative disease resistance. Trends Plant Sci 14(1):21–29

Qu S, Liu G, Zhou B, Bellizzi M, Zeng L, Dai L, Han B, Wang GL (2006) The broad-spectrum blast resistance gene *Pi9* encodes a nucleotide-binding site-leucine-rich repeat protein and is a member of a multigene family in rice. Genetics 172(3):1901–1914

Römer P, Recht S, Strauß T, Elsaesser J, Schornack S, Boch J, Wang S, Lahaye T (2010) Promoter elements of rice susceptibility genes are bound and activated by specific TAL effectors from the bacterial blight pathogen, *Xanthomonas oryzae* pv. oryzae. New Phytol 187(4):1048–1057

Ronald P, Dardick C (2006) Plant and animal pathogen recognition receptors signal through non-RD kinases. PLoS Pathog 2(1):e2

Shang J, Tao Y, Chen X, Zou Y, Lei C, Wang J, Li X, Zhao X, Zhang M, Lu Z, Xu J, Cheng Z, Wan J, Zhu L (2009) Identification of a new rice blast resistance gene, *Pid3*, by genomewide comparison of paired nucleotide-binding site-leucine-rich repeat genes and their pseudogene alleles between the two sequenced rice genomes. Genetics 182(4):1303–1311

Sheng CF, Sheng SY, Gao LD, Xuan JW (2003) Pest status and loss assessment of crop damage caused by the rice borers, Chilo suppressalis and Tryporyza incertulas in China. Entomol Knowl 40:289–294

Shi W, Chen S, Xu M (2008) Discovery of a new fragrance allele and the development of functional markers for the breeding of fragrant rice varieties. Mol Breed 22:185–192

Shomura A, Izawa T, Ebana K, Ebitani T, Kanegae H, Konishi S, Yano M (2008) Deletion in a gene associated with grain size increased yields during rice domestication. Nat Genet 40(8):1023–1028

Song WY, Wang GL, Chen LL, Kim HS, Pi LY, Holsten T, Gardner J, Wang B, Zhai WX, Zhu LH, Fauquet C, Ronald P (1995) A receptor kinase-like protein encoded by the rice disease resistance gene, *Xa21*. Science 270(5243):1804–1806

Song XJ, Huang W, Shi M, Zhu MZ, Lin HX (2007) A QTL for rice grain width and weight encodes a previously unknown RING-type E3 ubiquitin ligase. Nat Genet 39(5):623–630

Spielmeyer W (2002) Semidwarf (*sd-1*), "green revolution" rice, contains a defective gibberellin 20-oxidase gene. Proc Natl Acad Sci U S A 99(13):9043–9048

Sun X, Cao Y, Yang Z, Xu C, Li X, Wang S, Zhang Q (2004) *Xa26*, a gene conferring resistance to *Xanthomonas oryzae* pv. oryzae in rice, encodes an LRR receptor kinase-like protein. Plant J 37(4):517–527

Takahashi Y (2001) *Hd6*, a rice quantitative trait locus involved in photoperiod sensitivity, encodes the alpha subunit of protein kinase CK2. Proc Natl Acad Sci U S A 98(14):7922–7927

Takahashi Y, Teshima KM, Yokoi S, Innan H, Shimamoto K (2009) Variations in *Hd1* proteins, *Hd3a* promoters, and *Ehd1* expression levels contribute to diversity of flowering time in cultivated rice. Proc Natl Acad Sci U S A 106(11):4555–4560

Tamura M, Tachida H (2011) Evolution of the number of LRRs in plant disease resistance genes. Mol Genet Genomics 285(5):393–402

Tan L, Li X, Liu F, Sun X, Li C, Zhu Z, Fu Y, Cai H, Wang X, Xie D, Sun C (2008) Control of a key transition from prostrate to erect growth in rice domestication. Nat Genet 40(11):1360–1364

Tanaka K (2000) Toxicity of insecticides to predators of rice planthoppers: Spiders, the mirid bug and the dryinid wasp. Appl Entomol Zool 35(1):177–187

Tang W, Xu C, Li X, Lin Y, Zhang Q (2006) Development of insect-resistant transgenic indica rice with a synthetic *cry1C* * gene. Chin Sci Bull 18(1):1–10

Terao T, Nagata K, Morino K, Hirose T (2010) A gene controlling the number of primary rachis branches also controls the vascular bundle formation and hence is responsible to increase the harvest index and grain yield in rice. Theor Appl Genet 120(5):875–893

Tian Z, Qian Q, Liu Q, Yan M, Liu X, Yan C, Liu G, Gao Z, Tang S, Zeng D, Wang Y, Yu J, Gu M, Li J (2009) Allelic diversities in rice starch biosynthesis lead to a diverse array of rice eating and cooking qualities. Proc Natl Acad Sci U S A 106(51):21760–21765

Tu J, Zhang G, Datta K, Xu C, He Y, Zhang Q, Khush GS, Datta SK (2000) Field performance of transgenic elite commercial hybrid rice expressing bacillus thuringiensis delta-endotoxin. Nat Biotechnol 18(10):1101–1104

Unnevehr LJ, Duff B, Juliano BO (1992) Consumer demand for rice grain quality. International Rice Research Institute/International Development Research Center, Manila/Ottawa

Virk PS, Khush GS, Peng S (2004) Breeding to enhance yield potential of rice at IRRI: the ideotype approach. Int Rice Res Notes 29(1):5–9

Wang ZX, Yano M, Yamanouchi U, Iwamoto M, Monna L, Hayasaka H, Katayose Y, Sasaki T (1999) The *Pib* gene for rice blast resistance belongs to the nucleotide binding and leucine-rich repeat class of plant disease resistance genes. Plant J 19(1):55–64

Wang L, Wang A, Huang XH, Zhao Q, Dong GJ, Qian Q, Sang T, Han B (2011a) Mapping 49 quantitative trait loci at high resolution through sequencing-based genotyping of rice recombinant inbred lines. Theor Appl Genet 122:327–340

Wang C, Chen S, Yu S (2011b) Functional markers developed from multiple loci in GS3 for fine marker-assisted selection of grain length in rice. Theor Appl Genet 122:905–913

Wei X, Xu J, Guo H, Jiang L, Chen S, Yu C, Zhou Z, Hu P, Zhai H, Wan J (2010) *DTH8* suppresses flowering in rice, influencing plant height and yield potential simultaneously. Plant Physiol 153(4):1747–1758

Weng J, Gu S, Wan X, Gao H, Guo T, Su N, Lei C, Zhang X, Cheng Z, Guo X, Wang J, Jiang L, Zhai H, Wan J (2008) Isolation and initial characterization of *GW5*, a major QTL associated with rice grain width and weight. Cell Res 18(12):1199–1209

White FF, Yang B (2009) Host and pathogen factors controlling the rice-*Xanthomonas oryzae* interaction. Plant Physiol 150(4):1677–1686

Wu C, Li X, Yuan W, Chen G, Kilian A, Li J, Xu C, Zhou DX, Wang S, Zhang Q (2003) Development of enhancer trap lines for functional analysis of the rice genome. Plant J 35(3):418–427

Wu C, You C, Li C, Long T, Chen G, Byrne ME, Zhang Q (2008) *RID1*, encoding a Cys2/His2-type zinc finger transcription factor, acts as a master switch from vegetative to floral development in rice. Proc Natl Acad Sci U S A 105(35):12915–12920

Xiang Y, Cao Y, Xu C, Li X, Wang S (2006) *Xa3*, conferring resistance for rice bacterial blight and encoding a receptor kinase-like protein, is the same as *Xa26*. Theor Appl Genet 113(7): 1347–1355

Xing Y, Zhang Q (2010) Genetic and molecular bases of rice yield. Annu Rev Plant Biol 61: 421–442

Xing YZ, Tan YF, Hua JP, Sun XL, Xu CG, Zhang Q (2002) Characterization of the main effects, epistatic effects and their environmental interactions of QTLs in the genetic basis of yield traits in rice. Theor Appl Genet 105:248–257

Xu XF, Mei HW, Luo LJ, Cheng XN, Li ZK (2002) RFLP-facilitated investigation of the quantitative resistance of rice to brown planthopper (*Nilaparvata lugens*). Theor Appl Genet 104:248–253

Xu WH, Wang YS, Liu GZ, Chen X, Tinjuangjun P, Pi LY, Song WY (2006) The autophosphorylated Ser686, Thr688, and Ser689 residues in the intracellular juxtamembrane domain of XA21 are implicated in stability control of rice receptor-like kinase. Plant J 45(5):740–751

Xue W, Xing Y, Weng X, Zhao Y, Tang W, Wang L, Zhou H, Yu S, Xu C, Li X, Zhang Q (2008) Natural variation in *Ghd7* is an important regulator of heading date and yield potential in rice. Nat Genet 40(6):761–767

Yamamoto T, Lin H, Sasaki T, Yano M (2000) Identification of heading date quantitative trait locus *Hd6* and characterization of its epistatic interactions with *Hd2* in rice using advanced backcross progeny. Genetics 154(2):885–891

Yamamura K, Nishimori M, Ueda Y, Yokosuka T (2006) How to analyze long-term insect population dynamics under climate change: 50-year data of three insect pests in paddy fields. Popul Ecol 48:31–48

Yan WH, Wang P, Chen HX, Zhou HJ, Li QP, Wang CR, Ding ZH, Zhang YS, Yu SB, Xing YZ, Zhang QF (2011) A major QTL, *Ghd8*, plays pleiotropic roles in regulating grain productivity, plant height, and heading date in rice. Mol Plant 4(2):319–330

Yano M, Nagamura Y, Kurata N, Minobe Y, Sasaki T (1997) Identification of quantitative trait loci controlling heading date in rice using a high-density linkage map. Theor Appl Genet 95: 1025–1032

Yano M, Katayose Y, Ashikari M, Yamanouchi U, Monna L, Fuse T, Baba T, Yamamoto K, Umehara Y, Nagamura Y, Sasaki T (2000) *Hd1*, a major photoperiod sensitivity quantitative trait locus in rice, is closely related to the Arabidopsis flowering time gene *CONSTANS*. Plant Cell 12(12):2473–2484

Yano M, Kojima S, Takahashi Y, Lin H, Sasaki T (2001) Genetic control of flowering time in rice, a short-day plant. Plant Physiol 127(4):1425–1429

Yoshimura S, Katayose Y, Toki S, Wang ZX, Kono I, Yano M, Iwata N, Sasaki T (1998) Expression of Xa1, a bacterial blight resistance gene in rice, is induced by bacterial inoculation. Proc Natl Acad Sci U S A 95:1663–1668

Yu J, Hu S, Wang J, Wong GK, Li S, Liu B, Deng Y, Dai L, Zhou Y, Zhang X, Cao M, Liu J, Sun J, Tang J, Chen Y, Huang X, Lin W, Ye C, Tong W, Cong L, Geng J, Han Y, Li L, Li W, Hu G, Li J, Liu Z, Qi Q, Li T, Wang X, Lu H, Wu T, Zhu M, Ni P, Han H, Dong W, Ren X, Feng X, Cui P, Li X, Wang H, Xu X, Zhai W, Xu Z, Zhang J, He S, Xu J, Zhang K, Zheng X, Dong J, Zeng W, Tao L, Ye J, Tan J, Chen X, He J, Liu D, Tian W, Tian C, Xia H, Bao Q, Li G, Gao H, Cao T, Zhao W, Li P, Chen W, Zhang Y, Hu J, Liu S, Yang J, Zhang G, Xiong Y, Li Z, Mao L, Zhou C, Zhu Z, Chen R, Hao B, Zheng W, Chen S, Guo W, Tao M, Zhu L, Yuan L, Yang H (2002) A draft sequence of the rice genome (*Oryza sativa* L. ssp. indica). Science 296(5565):79–92. doi:10.1126/science

Yu B, Lin Z, Li H, Li X, Li J, Wang Y, Zhang X, Zhu Z, Zhai W, Wang X, Xie D, Sun C (2007) *TAC1*, a major quantitative trait locus controlling tiller angle in rice. Plant J 52(5):891–898

Yuan YX, Zhong SH, Li Q, Zhu ZR, Lou YG, Wang LY, Wang JJ, Wang MY, Li QL, Yang DL, He ZH (2007) Functional analysis of rice *NPR1-like* genes reveals that *OsNPR1/NH1* is the rice orthologue conferring disease resistance with enhanced herbivore susceptibility. Plant Biotechnol 5:313–324

Yuan M, Chu Z, Li X, Xu C, Wang S (2009) Pathogen-induced expressional loss of function is the key factor in race-specific bacterial resistance conferred by a recessive R gene *xa13* in rice. Plant Cell Physiol 50(5):947–955

Yuan M, Chu Z, Li X, Xu C, Wang S (2010) The bacterial pathogen Xanthomonas oryzae overcomes rice defenses by regulating host copper redistribution. Plant Cell 22(9):3164–3176

Yuan T, Li X, Xiao J, Wang S (2011) Characterization of Xanthomonas oryzae-responsive cis-acting element in the promoter of rice race-specific susceptibility gene *Xa13*. Mol Plant 4(2):300–309

Zhang Q (2007) Strategies for developing Green Super Rice. Proc Natl Acad Sci U S A 104(42):16402–16409

Zhou B, Liu G, Dolan M, Sakai H, Guodong L, Bellizzi M, Wang GL (2006) The eight amino-acid differences within three leucine-rich repeats between Pi2 and Piz-t resistance proteins determine the resistance specificity. Mol Plant Microbe Interact 19(11):1216–1228

Zou JH, Pan XB, Chen ZX, Xu JY, Lu JF, Zhai WX, Zhu LH (2000) Mapping quantitative trait loci controlling sheath blight resistance in two rice cultivars (*Oryza sativa* L.). Theor Appl Genet 101:569–573

Chapter 21
Functional Marker Development Across Species in Selected Traits

Hélia Guerra Cardoso and Birgit Arnholdt-Schmitt

Introduction

Breeding is a dynamic and adaptive process due to variable environments. Domestication of wild plants to modern cultivated crops involved a suite of changes in morphological, physiological, and biochemical traits (Doebley et al. 2006). Implementation of new mechanic methods of harvesting, human inclination towards valuing novelty and high quality exigency, the increase of the human population and global as well as local changes in climate and soil conditions are factors that require adaptation. One well known example for illustration is tomato, where wild forms bear small (1–2 g), round, seed dense berries, ideal for reproduction and dispersal, which are quite different from cultivated varieties, typically producing fruit that weigh 50–1,000 g, come in a wide variety of shapes (e.g., round, oblate, pear-shaped, torpedo-shaped), and are not well adapted for seed dispersal. The high production demand and relatively lower value of processing tomatoes led to the development of machine-harvestable varieties starting in the 1960s. Typical round-fruited tomatoes are too soft for harvest by machines, so selection for firmer fruited varieties resulted in a shape change from round to elongated or torpedo-shaped tomatoes. Additionally, some of the more extreme tomato fruit shapes, such as extremely long-fruited, pear-shaped, or bell pepper–shaped tomatoes, may reflect the human inclination toward valuing novelty.

While there is an increased demand for sustainably produced agricultural products of high quality, availability of agricultural land and natural resources such as water and fertilizers are limited. Climate change can alter frequency and

---

H.G. Cardoso (✉) • B. Arnholdt-Schmitt
ICAAM – Instituto de Ciências Agrárias e Ambientais Mediterrânicas, Universidade de Évora, Núcleo da Mitra, Ap. 94, 7002-554 Évora, Portugal
e-mail: hcardoso@uevora.pt

severity of abiotic and subsequently biotic constrains. In view to support the growing global population, breeders need to create crop plants that are able to grow under unfavorable natural conditions.

Plants as sessile organisms are able to acquire during individual and genetic evolution an unique capacity for developmental plasticity. Its manifestation depends on environmental conditions in interaction with the genotype and its developmental stage. Environmental factors affecting plant development include light, temperature, wind, humidity, soil structure, water and nutrient availability as well as biotic components, from pathogens to competitors (Tonsor et al. 2005). The capacity for phenotypic variation is genetically determined (Jungk 2001). Genetic variation is thought to arise spontaneously by genetic mutations (Alonso-Blanco et al. 2005). However, repetitive events in evolution were observed (Feldman et al. 2009; Rokas and Carroll 2008; Rundle et al. 2000; Wood et al. 2005; Zhang and Kumar 1997) and closer similarities between genotypes from distant regions than from within regions were recognized (e.g., Coelho et al. 2006). Thus, functional polymorphisms might be conserved for key genes of common traits across species. The identification of polymorphisms within gene and regulatory sequences are relevant with regard to developing FMs for plant breeding (Andersen and Lübberstedt 2003). FMs are expected to reliably predict the potential for a phenotype based on genotypic information (Brenner et al. in this edition).

Functional polymorphisms can be located in both protein-coding and non-coding regions of a gene. The higher frequency of polymorphisms, functional or non-functional, in non-coding parts reflects the strict functional requirements of the protein-coding regions (Wang et al. 2005). Polymorphisms occurring in coding sequences can affect protein sequences due to amino acid changes (non-synonymous polymorphism) which can interfere with the function of the protein. Interruption of the protein sequence by a nonsense mutation can create a truncated protein with consequent loss-of-function. In introns, polymorphisms can be functionally critical in view of its potential to influence binding of transcription factors (Xie et al. 2005), alternative splicing (Baek et al. 2008), the coding of intronic regulatory elements, such as micro- or small nucleolar- RNAs (Li et al. 2007), as well as nonsense-mediated mRNA decay (Jaillon et al. 2008). Genetic and epigenetic regulation of the organization of DNA into condensed structures and loops in eukaryotic chromosomes plays an important role for gene expression, DNA synthesis, recombination, and repair by modulating the accessibility of DNA (Arnholdt-Schmitt 2004; Fransz and De Jong 2011; Shaposhnikov et al. 2007).

Mutations through spontaneous insertion/deletion (InDel) and single nucleotide polymorphism (SNPs) are thought to be the major driving forces in genome evolution besides retroelements (Gregory 2004; Zhang and Gerstein 2003). They are highly abundant and distributed throughout the genomes in various plant species (Batley et al. 2003; Costa et al. 2009a, b; Drenkard et al. 2000; Frederico et al. 2009b; Nasu et al. 2002). Functional SNPs and InDels can contribute directly to a phenotype (Thornsberry et al. 2001), which makes both types of polymorphisms suitable for FMs development. SNPs are becoming important genetic markers for major crop species for genetic research and breeding (Fan et al. 2009; Jia et al.

2004; Lagudah et al. 2009; Lata et al. 2011; Ramkumar et al. 2010; Wang et al. 2011a; Yeam et al. 2005). Particularly, non-synonymous coding SNPs (nsSNPs) which together with SNPs in regulatory regions are believed to have the highest impact on phenotype determination (Ramensky et al. 2002). InDels have also been successfully exploited as FMs (Bradbury et al. 2005b; Chen et al. 2010; Juwattanasomran et al. 2012; Lagudah et al. 2009; Shi et al. 2008).

Here, we review functional polymorphisms identified in genes that are linked to selected traits across species. For some traits, research efforts were more focused on cereals and sometimes exclusively on rice, which is explained not only by the importance of this species for human food but also as it is used as a model for domesticated plants due to the small genome (Alonso-Blanco et al. 2005). Traits only explored in rice will not be considered in this chapter in order to avoid repetitions in the present edition (see Chapter 20 on rice).

Selected Traits for Yield and Quality Parameters

Yield Parameters in Grain Cereals

Grain Weight

Grain weight is determined by different mechanisms that regulate grain size through its length, width, and/or thickness (Sakamoto and Matsuoka 2008; Takano-Kai et al. 2009; Xing and Zhang 2010).

Grain Length

Gene sequences or QTL for grain length have been identified in common regions from rice (*Oryza sativa* L.), wheat (*Triticum aestivum* L.), and maize (*Zea mays* L.). In rice, the gene *GS3* (*GRAIN SIZE*3) overlaps with a major QTL for grain length and weight and also a minor QTL for grain width and thickness (Fan et al. 2006). Recent advances in comparative sequence analysis between wheat and rice genomes have confirmed extensive synteny between the two species (Quraishi et al. 2009). This enables to assess the positional correspondence between QTL identified in wheat and known QTL or loci that affect grain morphology in rice. The *O. sativa GS3* (*OsGS3*) corresponds to the strong *T. aestivum* QTL for grain size (*TaGS3*), which cosegregates consistently with grain width (Gegas et al. 2010). In contrast, *OsGS3* effects primarily grain length and rather than width (Fan et al. 2006).

OsGS3 encodes a transmembrane protein consisting of four putative domains: a plant-specific organ size regulator (OSR) domain of 66 aa at the N terminus (Mao et al. 2010), which substitutes the Phosphatidylethanolamine-Binding Protein (PEBP)-like domain previously proposed by Fan et al. (2006); a transmembrane domain at 97–117 aa, a TNFR (Tumor Necrosis Factor Receptor)/NGFR (Nerve

Growth Factor Receptor) family cysteine-rich domain at sites 116–155 aa, and a Von Willebrand Factor type C (VWFC) 60–80 aa in length in the C-terminal region (Mao et al. 2010). The VWFC domain of the OsGS3 functional protein is reported to be important for protein-protein interaction and signaling (Zhang et al. 2007). Mao et al. (2010) identified four alleles, three associated with differences in grain length (Table 21.1). The allele *OsGS3*-4 with a one bp deletion is characterized by the loss-of-function mutation of the C-terminal TNFR/NGFR and VWFC domains with a consequent inhibitory effect on the OSR function and production of very short grain (Mao et al. 2010). The allele *OsGS3-3* is characterized by a nsSNP, which leads to a nonsense mutation and consequently elimination of part of the OSR domain and all the other three conserved domains. According to Mao et al. (2010), the OSR domain is both necessary and sufficient for functioning as a negative regulator. The nonsense mutation is shared among all the large-grain varieties of *O. sativa* sequenced in comparison with small-to medium-grain varieties (Fan et al. 2006, 2009; Takano-Kai et al. 2009). These findings suggest that *OsGS3* acts as a negative regulator of grain length, in agreement with the recessive nature of the long-grain phenotype (Fan et al. 2006).

Different molecular markers which target the functional SNP at the *OsGS3-3* allele were already developed and can be used as a tool for routine and large-scale genotyping and selection of long/short grain length genotypes at the seedling stage which is vital for long grain breeding plant materials (Table 21.1).

Recently, Wang et al. (2011a) identified two new polymorphisms in other regions of *OsGS3* defining two new alleles (RGS1 and RGS2) which confer in combinations moderate/short grain (Table 21.1). However, the development of specific markers which allow the use of RGS1 and RGS2 motifs in breeding programs was not developed until now.

In *Z. mays*, to date only one cytosolic Glutamine Synthetase isoenzyme (GS1), product of *Gln1-4* has been shown by mutant analysis to have an impact on grain length (Martin et al. 2006), although an ortholog of *OsGS3*, named *ZmGS3*, was isolated. *ZmGS3* encodes a protein sharing common domains with OsGS3, including one transmembrane domain and two overlapping TNFR/ NGFR family, cysteine-rich domains (Li et al. 2010b). Expression analysis revealed that *ZmGS3* is primarily expressed in immature ears and kernel and transcript decreases rapidly after pollination (Li et al. 2010b), suggesting a role in kernel development, as in rice (Fan et al. 2006, 2009). However, different functional polymorphisms were identified suggesting different mechanisms from that of *OsGS3* (Table 21.1). A polymorphism in the promoter region of this gene was found to affect 'hundred grain weight' (HGW) (Li et al. 2010b).

GS3 is a major gene for grain length in rice, and it can explain up to 72% of the variation in grain length (Fan et al. 2006, 2009). However, *ZmGS3* was marginally significant and the phenotypic variation explained by the identified polymorphism is less than 8%, indicating that *ZmGS3* is only a minor gene for variation in maize grain traits (Li et al. 2010b). Nevertheless, it cannot be excluded that it may play an important role in maize grain development, and, thus, hold potential for yield improvement in maize.

Table 21.1 Genes, polymorphisms and FMs identified as related with phenotypic variation in physical and nutritional parameters of grains and fruits

sp	Gene	Allele	Mutation	Phenotype	Marker	Validation	References
Grain lenght							
O. sativa	OsGS3 Chr.3	OsGS3-1	Wild allele	Medium grain	ND	ND	Mao et al. (2010)
		OsGS3-2	No-functional InDel in exon5 and in-frame insertion of 3 bp	Medium grain	ND	ND	Fan et al. (2009)
		OsGS3-3	Nonsense mutation due to a nsSNP in exon2 (C/A)	Large grain	SF28	180 Chinese var and 10 var. from other countries	Wang et al. (2011a)
						2RIL and 34 NIL	
					DRR-GL	F$_2$ sgp (200 ind), 274 rice vars, 30 basmati vars	Ramkumar et al. (2010)
					GS3-PstI SR17	172 cultivars 2RIL and 34 NIL	Yan et al. (2009) Wang et al. (2011a)
		OsGS3-4	Frameshift of C-terminal domains (TNFR and VWFC) due to 1 bp InDel in exon5	Very short grain	ND	ND	Mao et al. (2010)
		OsGS3-RGS1	Seq. repeat (AT)n motif in intron4	Allelic combination of both alleles confers moderate/short grain	RGS1	2 RIL and 34 NIL	Wang et al. (2011a)
		OsGS3-RGS2	Seq. repeat (TCC)n motif in exon5		RGS2	2 RIL and 34 NIL	

(continued)

Table 21.1 (continued)

sp	Gene	Allele	Mutation	Phenotype	Marker	Validation	References
Zm	ZmGS3	ZmGS3-1	unknown	unknown	ND	ND	Li et al. (2010b)
		ZmGS3-2	Insertion of 2 bp exon 5(CCC/CCCCC)	Associated with grain lenght	ND	ND	
Grain width
O. sativa	OsGW2 Chr2	OsGW2	Wild allele	Low widht	ND	ND	Song et al. (2007)
		Osgw2	Nonsense mutation due to 1 bp deletion in exon4	Enhanced grain width, weight and yield	GW2-HapI	172 cultivars	Yan et al. (2009)
	OsGW5/ qSW5 Chr5	OsGW5	Wild allele	Low width	ND	ND	Shomura et al. (2008)
		Osgw5	1.2Kb deletion in ORF	Increase in sink size swing to an increase in cell number in the outer glume of rice flower	ND	ND	Weng et al. (2008)
Ta	TaGW2	Hap-6A-A	SNP$_{-593}$ (A/G) at the promoter	Higher grain width, weight and TGW, early heading and maturity	Hap-6A-P1 + Hap-6A-P2	265 Chinese wheat vars	Su et al. (2011)
		Hap-6A-G	SNP$_{-739}$ (G/A) at the promoter	Low width	ND	ND	
Zm	ZmGW2 Chr6A	ZmGW2	SNP$_{40}$ (C/T) at the promoter	Increase GW and HGW	ND	ND	Li et al. (2010a)
	ZmGW2 Chr4						
	ZmGW2 Chr5	ZmGW2	unknown	unknown	ND	ND	

Fragrance/Aroma						
O. sativa	OsBadh1 Chr. 4	Osbadh1-PH1	Wild allele	Without fragrance	ND	Singh et al. (2010)
		Osbadh1-PH2	nsSNP$_{1371}$ (A/T) and nsSNP$_{3493}$(A/C)	Loss of function of BADH2 (fragrance)	ND	
	OsBadh2 Chr. 8	OsBADH	Wild allele	Without fragrance	ND	Bradbury et al. (2005a) Shi et al. (2008)
		Osbadh2-E7/ Osbadh2.1	Nonsense mutation due by 8 bp deletion (5'-GATTATGG-3') and 3 SNPs exon7	Enhancement of 2AP (fragrance)	ESP/IFAP/ INSP/EAP	F_2 sgp (300 ind)	Bradbury et al. (2005b)
					FMbadh2-E7 nksbad2F/ nksbad2R	F_2 sgp (106 ind) 209 RILs	Shi et al. (2008) Amarawathi et al. (2008)
					BADEX7-5 Fgr1F + frg1R	F_2 sgp (196 ind) 416 lines	Sakthivel et al. (2009a) Jin et al. (2010)
		Osbadh2-E2/ Osbadh2.2	Frameshift due by 7 bp deletion (5'-CGGGCGC-3') exon2	Enhancement of 2AP (fragrance)	FMbadh2-E2A FMbadh2-E2B	F_2 sgp (106 ind)	Shi et al. (2008)
		Osbadh2.3	Nonsense mutation due by 2 bp deletion exon1	Loss of function of BADH2 (fragrance)	ND	Kovach et al. (2009)
		Osbadh2.4	Nonsense mutation due by 1 bp insertion exon10	Loss of function of BADH2 (fragrance)	ND	
		Osbadh2.5	Nonsense mutation due by 1 bp deletion exon10	Loss of function of BADH2 (fragrance)	ND	
		Osbadh2.6	Nonsense mutation due by SNP (G/T) exon10	Loss of function of BADH2 (fragrance)	ND	

(continued)

Table 21.1 (continued)

sp	Gene	Allele	Mutation	Phenotype	Marker	Validation	References
		Osbadh2.7	Nonsense mutation due by 1 bp insertion exon14	Loss of function of BADH2 (fragrance)	ND	ND	Kovach et al. (2009)
		Osbadh2.8	Additional in-frame aa due by 3 bp insertion exon13	Loss of function of BADH2 (fragrance)	ND	ND	
		Osbadh2.9	nsSNP (G/T) exon14	Loss of function of BADH2 (fragrance)	ND	ND	
		Osbadh2.10	nsSNP (C/T) exon13	Loss of function of BADH2 (fragrance)	ND	ND	
G. max	GmBadh1	GmBADH1	Wild allele	Without fragrance	ND	ND	Juwattanasomran et al. (2011)
		GmBadh1	Nonfunctional SNP exon1 (T/C)	Without fragrance	ND	ND	
	GmBadh2	Gmbadh2-1	Nonsense mutation due by nsSNP2214(G/A) exon10	Loss of function of BADH2 (fragrance)	GmBADH2-G1 GmBADH2-G2 GmBADH2-G3 GmBADH2-A1 GmBADH2-A2	F2 sgp (82 ind)	
		Gmbadh2-2	Nonsense mutation due by 2 bp InDel in exon10	Loss of function of BADH2 (fragrance)	GmBADH2_EX10	F1 hybrids F2 sgp (55 ind)	Juwattanasomran et al. (2012)
					Gm2AP	10 var + F2 sgp (60 ind)	Arikit et al. (2011a)

sp	gene	allele	polymorphism	phenotype			reference
Grain lipids content							
Br	Brfad2	Brfad2	Wild allele	Low-oleic-acid content	ND	ND	Tanhuanpää et al. (1998)
		brfad2	3 synonymous SNPs + nsSNP$_{484}$(T/C)	High-oleic-acid content	co-dominant	F$_2$ sgp (100 ind.)	
	BraA.GL2.a	unknown	unknown	High-oleic-acid content	ND	ND	Chai et al. (2010)
Bn	BnaA.GL2.a	unknown	unknown	High-oleic-acid content	ND	ND	Doganlar et al. (2002)
	BnaA.GL2.b	unknown	unknown	High-oleic-acid content	ND	ND	
Bo	BolC.GL2.a	unknown	unknown	High-oleic-acid content	ND	ND	
Fruit size and fruit shape							
Le	fw2.2	FW2.2	Wild allele	Small size fruit	ND	ND	Frary et al. (2000)
	Chr.2	fw2.2	35SNPs 5'UTR	Increase in fruit size	ND	ND	
	OVATE	OVATE	Wild allele	Round fruit	ND	ND	Liu et al. (2002)
		ovate	nsSNP493(G/T) exon2	Pear-shape fruit	ND	ND	
S	fw2.1	unknown	unknown	unknown	ND	ND	Doganlar et al. (2002)

sp species, *ND* not developed, *sgp* segregating population, *var* varieties, *ind* individuals, *RIL* recombinant inbred line, *NIL* near-isogenic lines, *Zm* Zea mays, *Br* Brassica rapa, *Bn* Brassica napus, *Bo* Brassica olearacea, *Le* Lycopersicon esculentum, *S* Solana melongena

Grain Width

GW2 (*GRAINWIDTH2*) was the first gene cloned in rice controlling grain width and weight (Song et al. 2007). *OsGW2* encodes a cytoplasm RING-type protein with intrinsic E3 ubiquitin ligase activity. Its function is related to the degradation step in the ubiquitin-proteosome pathway. Homologous genes to *OsGW2* with high aa sequence identities (86.5 and 81%, respectively) were identified in *T. aestivum* and *Z. mays* (Song et al. 2007).

In rice, the loss of *GW2* function, due to a polymorphism in the ORF (Table 21.1) increased cell numbers, resulting in a larger (wider) spikelet hull. Furthermore, the loss-of-function accelerated the grain milk filling rate, resulting in enhanced grain width, weight and yield (Song et al. 2007). These findings suggest that *OsGW2* functions as a negative regulator for grain width through the control of cell division in the spikelet hull by targeting unknown substrate(s) for the ubiquitin-dependent degradation by the 26S proteasome (Song et al. 2007). Pleiotropic effects were attributed to this gene, at least on the panicle number per plant, days to heading and main panicle length, in addition to on the grain numbers per main panicle (Song et al. 2007). The development of a FM based on the polymorphism identified was reported (Table 21.1) (Yan et al. 2009).

Li et al. (2010a) describe identification of two genes as chromosomal duplicates co-orthologous of rice *GW2* in *Z. mays*, *ZmGW2-CHR4* and *ZmGW2-CHR5*. Expression and candidate gene-based association analyses suggested that both genes play a role in kernel size and weight variation, as does rice *GW2*. From all the 70 fixed polymorphic sites identified, covering different regions of *ZmGW2* genes, the SNP_{40}(C/T) located in the promoter region of *ZmGW2-CHR4* was of great interest, because it was significantly associated with phenotypic differences in GW and HGW (Li et al. 2010a). However, no further studies were reported, confirming this hypothesis, or the possible application of this polymorphism in the development of molecular markers for plant breeding applications.

In *T. aestivum* an orthologous gene of *OsGW2* was identified (*TaGW2*). Nucleotide sequence analysis of *TaGW2* led to the identification of two *TaGW2* haplotypes, Hap-6A-A and Hap-6A-G (Su et al. 2011). Expression analysis revealed a negative correlation between grain width and the expression level of *TaGW2*. Moreover, the average expression level of *TaGW2* in varieties with Hap-6A-G was higher than in varieties with Hap-6A-A, indicating association of the latter haplotype with higher grain width and weight (Su et al. 2011). The effect of *TaGW2*-6A Hap-6A-A in wheat was similar to a loss-of-function mutation in *OsGW2* in rice, leading to increased grain width and weight and higher TGW, and it was associated with earlier heading and maturity. A CAPS marker has already been developed and validated (Table 21.1). Association analysis revealed that Hap-6A-A was significantly associated with wider grains and higher one-thousand grain weight (TGW) in two crop seasons (Su et al. 2011).

A second gene identified in rice controlling grain width is *GW5* (*GRAINWIDTH5*) (Weng et al. 2008) or *qSW5* (QTL for SEED WIDTH on chromosome 5) (Shomura et al. 2008). *GW5* encodes an unknown nuclear protein containing a predicted

NLS and an arginine-rich domain, which physically interacts with polyubiquitin, indicating that GW5 may be involved in the ubiquitin-proteosome pathway to regulate cell division during seed development (Weng et al. 2008). Recent studies have also pointed to a critical role of the ubiquitin pathway in seed development. As example, in *Arabidopsis thaliana* L. an induced mutation in the *DA1* gene (DA means "large" in Chinese) is related with loss-of-function of a predicted ubiquitin receptor and, consequently, with an increase in seed and organ size (Li et al. 2008).

A functional polymorphism (Table 21.1) was associated with differences for grain width and weight. Loss of function of GW5 resulted in a significant increase in sink size owing to an increase in cell number in the outer glume of the rice flower (Shomura et al. 2008). A negative regulation as reported in *GW2* (Song et al. 2007) was also described for *GW5* (Weng et al. 2008). However, no FMs were reported based in this functional polymorphic difference.

Quality Parameters in Cereals and Fruits

Lipid Content

Plants are a vital source of renewable oils for food (representing 25% of human caloric intake in developed countries) but also nonfood applications, which represents a third of plant oil harvested. Controlling the composition and maximizing the energy-efficient yield of oils within diverse crop species have been recognized as major goals for plant breeders and the biotechnology industry. Rapeseed (*Brassica napus* L.), soybean (*Glycine max* L.), oil palm (*Elaeis guinensis* Jacq.), and sunflower (*Helianthus annuus* L.) account for more than 65% of vegetable oil production world-wide (Gunstone 2001).

fad2 (*FATTYACIDDESATURATION2*) encodes the enzyme responsible for the desaturation of oleic acid to linolenic acid in *A. thaliana* (Okuley et al. 1994). High-oleic-acid content in seed of *Brassica* is a current breeding objective because it increases the thermostability and cooking quality of oil. Cloning of homologous *Atfad2* in *B. rapa* allowed to make a comparison between *Brfad2* sequences from the wild-type and the high-oleic-acid allele. From SNPs, which differentiate high-oleic-acid allele from the wild-type allele, only SNP_{484}(T/C) creates an aa change (L131P) (Tanhuanpää et al. 1998). Based on that nsSNP, FMs for selection of plants with high-oleic-acid were developed (Table 21.1).

Another gene related to seed oil accumulation is the class IV homeodomain-ZIP transcription factor GLABRA 2 (*GLABRA2, GL2*), characterized in *A. thaliana* and related to the regulation of seed oil accumulation (Chai et al. 2010). Chai et al. (2010) reported the cloning of four orthologues of *AtGL2*. From *B. napus*, the homologous genes were named *BnaA.GL2.a* and *BnaA.GL2.b*, from *Brassica rapa* L. *BraA.GL2.a*, and from *Brassica olearace* L. *BolC.GL2.a*. The existence of four orthologous *GL2* genes is explained by the origin of that species, since *B. napus* (genome AACC, 2n = 38) results from spontaneous hybridization between

B. rapa (AA, 2n = 20) and *B. olearacea* (CC, 2n = 18), comprising two sets of homologous chromosomes from the two species. Eleven non-synonymous point mutations were identified among the four gene sequences, and responsible for aa changes (Q238H, L269V, A404V, M412V, -418T, A419S, K477R, M660L, A697-, M709L, C745S). Higher levels of variation were reported among intron sequences, specifically for introns 5 and 7, which were used to develop three PCR based markers to distinguish *B. napus*, *B. rapa*, and *B. olearacea* (Table 21.1) (Chai et al. 2010). A CAPS marker was developed, based on SNP_{3486}(A/C), where A is present in A-genomes (*BnaA.GL2.a* and *BraA.GL2.a*) and C in C-genomes (*BnaC.GL2.b* and *BolC.GL2.a*) (Table 21.1). Both PCR and CAPS markers can be used as FMs to distinguish *Brassica* homoelogues and A- and C-genomes (Chai et al. 2010).

Fragrance/Aroma

Rice grains with a fragrance, like Basmati and Jasmine rice varieties, are appealing to consumers due to their superior grain qualities and pleasant aroma, which increase the retail price when compared to conventional rice (Shi et al. 2008). Also aromatic vegetable soybean (known as "Chamame" or green soybean) yields higher prices than normal varieties (Statistics Department, Ministry of Agriculture, Forestry and Fisheries 2009).

In soybean as in rice, aroma has been associated with increased levels of 2-Acetyl-1-Pyrroline (2AP) (Arikit et al. 2011b; Fushimi and Masuda 2001). In both species, a single gene was suggested to be responsible for fragrance (Bradbury et al. 2005a; AVRDC 2003). In rice, that gene is known as *Os2AP* (Vanavichit et al. 2008), but also *OsBad2* (Bradbury et al. 2005a, 2008) or *OsBadh2* (Niu et al. 2008), and in soybean as *GmBadh2*, since BADH2 was proposed to encode the BETAINE ALDEHYDE DEHYDROGENASE 2, which inhibits the biosynthesis of 2AP (Chen et al. 2008; Juwattanasomran et al. 2012, 2011; Shi et al. 2008).

The involvement of *Badh2* in fragrance of rice was initially reported by Bradbury et al. (2005a, b) with the identification of the recessive allele *Osbadh2-E7* carrying a 8 bp deletion in exon 7, which causes a premature stop codon, and, consequently, generates a non-functional OsBADH2 allele causing fragrance. Shi et al. (2008) identified a novel recessive allele, *Osbadh2-E2*, differing from the *Osbadh2-E7* by having an intact exon 7 but a 7-bp deletion in exon 2, which causes a frame shift leading to a non-functional OsBADH2. Chen et al. (2008) showed that only the intact 503 aa protein encoded from full-length transcript of *OsBadh2* inhibits 2AP synthesis. The functional OsBADHs protein contains two peptide sequences – VSLELGGKSP and EGCRLGSVVS – and a cysteine residue (28 aa away from VSLELGGKSP) in exons 8, 9, and 10, respectively, highly conserved in aldehyde dehydrogenases. This suggests that these peptide sequences are essential for functional activity of OsBADHs (Bradbury et al. 2005a). In *OsBadh2*, exons 8, 9, and 10 also contain coding regions for these elements, respectively (Bradbury et al. 2005a).

A transgenic approach also demonstrated the involvement of *badh2* in rice fragrance. Silencing the *OsBADH2* gene in non-fragrant rice varieties resulted in increased 2AP biosynthesis and, thus, fragrance in those varieties (Niu et al. 2008; Vanavichit et al. 2008). Similarly, transferring the functional *OsBadh2* gene into fragrant rice resulted in non-fragrant rice (Chen et al. 2008). Additionally, other polymorphisms were identified in *Osbadh2*, absence of MITE (miniature interspersed transposable element) in promoter region (Bourgis et al. 2008), two new SNPs in the central section of intron 8 (Sun et al. 2008), a TT deletion in intron 2 and a repeated (AT)n insert in intron 4 (Chen et al. 2008).

The development of FMs for rice fragrance to perform the discrimination between fragrant and non-fragrant varieties and the identification of homozygous fragrant, homozygous non-fragrant and heterozygous non-fragrant individuals were firstly reported by Bradbury et al. (2005b), based on the InDel of *Osbadh2-E7* allele (Table 21.1). Based on the same polymorphism new FMs have been developed (Amarawathi et al. 2008; Jin et al. 2010; Sakthivel et al. 2009a; Shi et al. 2008). Shi et al. (2008) also reported the development of efficient FMs based on the InDel polymorphism involving the *Osbadh2-E2* allele, which has recently been applied in a breeding program to improve the quality of line II-32B (Jin et al. 2010).

In rice, in addition to *OsBadh2* that strongly affects strength of fragrance, Lorieux et al. (1996) reported two minor QTL on chromosomes 4 and 12, while Amarawathi et al. (2008) identified two minor QTL on chromosomes 3 and 4, influencing the level of fragrance and suggesting *OsBadh1*, a homologous of *OsBadh2*, as a candidate gene for the QTL on chromosome 4. The biochemical function and substrate specificity of the BADH enzymes encoded by the two genes is similar (Bradbury et al. 2008).

Recent studies on *OsBadh1* suggest involvement of two nsSNPs in substrate binding capacity of OsBADH1 towards Gamma-Aminobutyr-Aldehyde (GABald), a precursor of 2AP (Chen et al. 2008; Singh et al. 2010). Based on nsSNPs, Chen et al. (2008) defined two haplotypes (*Badh1_PH1* and *Badh1-PH2*) (Table 21.1). By *in silico* analysis, Singh et al. (2010) discriminated only 8 out of the 18 binding sites as sites for GABald binding in *Badh1-PH2*, suggesting a drastic reduction in the affinity of that haplotype for GABald. Thus, *Badh1-PH2* could be a loss-of-function allele of the *Badh1* gene with implications in rice aroma similar to the loss-of-function alleles of the *Badh2* (Kovach et al. 2009; Singh et al. 2010). Singh et al. (2010) associated the BADH1 haplotype PH2 with aromatic rice varieties and observed a significant association between that haplotype and aroma score. Nevertheless, this association awaits validation in segregating populations for potential utilization in rice breeding programs.

In soybean, cloning of two *Badh* genes, *GmBadh1* and *GmBadh2* has been reported (Juwattanasomran et al. 2011). As in rice, silencing of *GmBadh2* resulted in 2AP biosynthesis in non-fragrant soybean varieties (Arikit et al. 2011b). Comparison of gene sequences obtained from fragrance and non-fragrance varieties allowed the identification of several polymorphisms (Table 21.1), two of which resulted in loss-of-function of OsBadh2 and, consequently, enhanced fragrance (Bradbury et al. 2005a; Juwattanasomran et al. 2011; Niu et al. 2008; Shi et al. 2008; Vanavichit et al. 2008). Both mutations (distinguishing both alleles *Gmbadh2.1* and *Gmbadh2.2*) occur in exon 10 of *GmBadh2*, which contains the conserved motif EGCRLGPIVS, similar to the motif for the essential functional activity of *OsBadhs*. The first SNP (G/A) in *GmBadh2* causes a change of the conserved motif, which may be associated with the loss of functional activity (Juwattanasomran et al. 2011). A 2 bp InDel causes a truncation of the protein and consequently lacks the peptide sequence EGCRLGPIVS, resulting in 2AP accumulation (Arikit et al. 2011b; Juwattanasomran et al. 2012).

Nevertheless, there are several studies showing that the fragrance is not totally explained by the identified polymorphisms at the single gene *Badh2*, but could be under control of other genetic loci, which explains quantitative inheritance of this trait (Amarawathi et al. 2008; Sakthivel et al. 2009a; Singh et al. 2010).

Fruit Size and Shape

Fruit Size

In tomato, 28 QTL were identified for fruit size by Grandillo et al. (1999). Seven QTL explained more than 20% of the phenotypic variance (Grandillo et al. 1999; Tanksley 2004). Nevertheless, up to now *fw2.2* (*SECOND FRUIT WEIGHT* QTL on chromosome 2), is the only locus for which the underlying gene has been identified (Frary et al. 2000). *fw2.2* encodes a protein with similarity to a human oncogene RAS protein (Frary et al. 2000), known to belong to a family which includes proteins with wide regulatory functions, including control of cell division (Sprang 1997).

Natural genetic variation at *fw2.2* locus (3 SNPs in the 5'-UTR, 35 SNPs within the two predicted introns, and 4 SNPs representing silent mutations) alone can change fruit size up to 30% (Frary et al. 2000). Control of tomato fruit size seems to be mediated by a cis-regulatory mechanism due to 5'-regulatory regions and gene expression patterns, rather than variation in protein sequences of different alleles (Frary et al. 2000; Nesbitt and Tanksley 2002). The described changes in the 5'-UTR are associated with lower total transcript levels during cell division phase of fruit development, as well as a shift in the timing of expression (Cong et al. 2002). Changes in gene regulation, rather than protein function, have long been hypothesized as a major mode of evolutionary change, especially concerning morphological differentiation. In this regard, *fw2.2* is one of a growing number of examples, in which natural variation associated with morphological changes can be traced to regulatory mutations (Cong et al. 2002). The most striking evidence in support of this notion came from the fact that the coding sequence of a small-fruit wild tomato species *Lycopersicon cheesmanii* Riley is identical to that of the large-fruit domestication species *Lycopersicon esculentum* P. Mill., indicating that the *fw2.2* coding sequence cannot be the reason for fruit size variation (Nesbitt and

Tanksley 2002). Thus, the hypothesis emerged that the 3 SNPs identified in the 5'-UTR of *fw2.2* may be a cause of the observed phenotypic difference (Frary et al. 2000). However, no reference was made to any of the polymorphisms at the intron level and/or the potential role in regulating gene expression. Frary et al. (2000) also proposed that the differences in fruit size imparted by the different *fw2.2* alleles may be modulated by combinations of the 35 SNPs identified within the two predicted introns.

Orthologous genes to *fw2.2* were identified in other *Solanaceae* species, (Ben Chaim et al. 2001; Doganlar et al. 2002). However, no research has been published, establishing a correlation between sequence variation and fruit size.

Fruit Shape

Like fruit size, fruit shape is a quantitative trait. In tomato, the major loci affecting shape are *OVATE, SUN, FRUITSHAPE CHR8.1* (*fs8.1*), *FASCIATED* (*f*) and *LOCULE NUMBER* (*lc*) (Tanksley 2004). The three major loci *OVATE, SUN* and *fs8.1* modulate fruit shape with a minimal effect on fruit size (Tanksley 2004). *OVATE* from tomato is the only locus that had been characterized at molecular level (Liu et al. 2002).

OVATE controls transition from round to pear-shape fruits in tomato (Liu et al. 2002) and also in eggplant (*fl2.1*) (Doganlar et al. 2002). The *OVATE* gene from tomato encodes a hydrophilic protein with a putative bipartite Nuclear Localization Signal (NLS; Robbins et al. 1991), two putative VWFC protein–protein interaction domains (Hunt and Barker 1987), and a 70 aa carboxyl-terminal domain conserved in tomato, *Arabidopsis* and rice (Liu et al. 2002).

The similarity in morphological change and DNA sequence deletion between rice grain and tomato fruit strongly suggests that the putative VWFC domain may have a role in regulating fruit/grain shape by negatively affecting growth. It is known that the VWFC domain, also referred to as Chordin-like cysteine-rich (CR) repeats, is present in a growing number of extracellular matrix proteins, and binds to members of the transforming growth factor-ß (TGF-ß) superfamily (Abreu et al. 2002). It has been proposed that the general function of VWFC is to regulate growth factor signaling by disrupting the receptor binding sites in the TGF-ß superfamily of the extracellular matrix, thus preventing activation of the TGF-ß receptor (O'Leary et al. 2004). Such inhibitory activity of VWFC on growth factor signaling is clearly consistent with the mechanism of negative regulation in the development of grain size and fruit shape hypothesized for the *GS3* and *OVATE* genes.

OVATE presents a nonsense mutation (Table 21.1) in the C terminus of the predicted protein which eliminates most of the conserved carboxyl-terminal domain of the protein. It may account for the loss-of-function (recessive) phenotypes. Sequence comparison revealed that all varieties of tomato with pear-shaped fruits had this mutation in the *OVATE* gene (Liu et al. 2002).

Stress Tolerance

Abiotic Stress Tolerance

Drought Tolerance

DREB (DEHYDRATION-RESPONSIVE ELEMENT-BINDING) proteins are important transcription factors that belong to the APETALA 2/ETHYLENE RESPONSIVE FACTOR (AP2/ERF) family. The AP2/ERF domain specifically binds to the CRT/DRE (C-REPEAT/DEHYDRATION-RESPONSIVE ELEMENT) of downstream genes, regulating their expression and consequently enhancing plant tolerance to abiotic stresses like low temperature, drought, and high-salinity (Agarwal et al. 2006; Liu et al. 1998; Sakuma et al. 2002; Yamaguchi-Shinozaki and Shinozaki 1994). DREBs were identified in a wide range of herbaceous and woody plant species (Agarwal et al. 2006; Benedict et al. 2006; Kitashiba et al. 2004; Yang et al. 2011).

Differences between *DREB* alleles in relation to drought were reported by Chen et al. (2005) with the identification of nine haplotypes whitin 20 accessions of *T. aestivum*. Two haplotypes (1 and 3) were identified as conferring drought tolerance. Combining the gene expression data, which show an induction under drought treatment, and the existence of gene sequence polymorphisms with drought tolerance in the two haplotypes, *TaDREB* was considered as a useful gene for improving drought-tolerance in wheat. FMs were developed by Wei et al. (2009) based on genome-specific primers for each of the orthologous DREB loci on chromosome 3A, 3B and 3D based on InDels and SNPs previously identified as locus-specific (Chen et al. 2005).

Salinity Tolerance

Salinity tolerance is a polygenic trait, controlled by interaction of genes involved in different pathways, such as ion compartmentation, ion extrusion, ion selectivity, compatible solute synthesis and Reactive-Oxygen-Species (ROS) scavenging (Munns and Tester 2008; Zhu 2001). Although many of these mechanisms are probably universal in most plants, their relative importance in salt tolerance may vary from species to species, depending on the metabolic background (Sun et al. 2010).

The impact of salinity on plant growth is due to effects of dehydration (osmotic toxicity) and interference with cellular metabolism caused by high levels of Na^+ in the cytoplasm (ion-specific toxicity) (Munns and Tester 2008). Na^+ can inhibit K^+ uptake (Rains and Epstein 1965), and in cytoplasm, Na^+ readily displaces K^+ in many enzymes that require K^+ as a co-factor for their activity (Tester and Davenport 2003). Using genetic approaches, many genes have been identified and associated with enhanced salinity tolerance in diverse plant species. These genes are generally divided into three groups, according to their function:

1. Genes that enhance osmotic protection and ROS scavenging such as OSMOREG-ULATORYTREHALOSESYNTHESIS (*OTS*) (Garg et al. 2002), MANNITOL-1-PHOSPHATEDEHYDROGENASE (*M1PD*) (Abebe et al. 2003) and the Δ^1-PYRROLINE-5-CARBOXYLATESYNTHETASE (*P5CS*) (Hong et al. 2000). The *P5CS* gene has been cloned from several higher plants (Armengaud et al. 2004; Chen et al. 2009; Hu et al. 1992) and encodes a rate-limiting enzyme (P5CS) involved in the biosynthesis of proline from glutamate (Yoshiba et al. 1995). Proline in turn is an important osmo-protectant present in higher plants that is thought to be critical for adaptation to several abiotic stresses such as drought and salt (Verslues et al. 2006). Expression studies demonstrated that the transcript level of *P5CS* increases significantly by salt and drought treatments (Chen et al. 2009; Dombrowski et al. 2008; Hu et al. 1992; Igarashi et al. 1997; Strizhov et al. 1997). Natural variation of *P5CS* was recently reported among 27 common bean accessions of *Phaseolus vulgaris* L. (Table 21.2), which was used for FM development (Chen et al. 2010).

2. Genes involved in Na^+ and K^+ transport, including the HIGH-AFFINITYK^+ TRANSPORTER family of genes (*HKT*), that are involved in K^+ transport (Horie et al. 2009) and the Na^+/H^+ EXCHANGERS (*NHX*) genes family (e.g., *NHX1*) or SALT-OVERLY-SENSITIVE genes (e.g., *SOS1*) involved in Na^+/H^+ antiport systems (Shi et al. 2003). From this group, natural variation in *HKT* genes is known to be related with salinity tolerance (Qiu et al. 2011). Two alleles were identified (*HvHKT1* and *HvHKT2*) by sequence comparison in 40 different Tibetan wild barley accessions (see Table 21.2).

3. Regulatory genes such as transcription factors (i.e., *DREB/CBF* (C-REPEAT BINDINGFACTOR) family) that function in signaling pathways, regulating the expression of downstream genes (Morran et al. 2011) involved in salinity tolerance in plants (Cong et al. 2008; Liu et al. 1998).

At least 20 *CBF* genes were identified in barley (*Hordeum vulgare* L.), classified as subgroup *HvCBF1*, *HvCBF3*, and *HvCBF4* (Skinner et al. 2005). *HvCBF4* encodes a protein closely homologous to *DREB1/CBF* in *A. thaliana* and *Vitis* sp. (Haake et al. 2002; Xiao et al. 2008). Transgenic overexpression of *HvCBF4* in rice has been demonstrated to enhance tolerance to drought, high-salinity, and low-temperature (Oh et al. 2007). Natural variation within that gene sequence was reported by Wu et al. (2011) and Rivandi et al. (2011), who developed a FM for application in breeding programs of barley.

In the species *Setaria italic* L. (foxtail millet) a *DREB* gene was characterized that belong to the A2 subgroup (*SiDREB2*) related with dehydration and salinity (NaCl) tolerance (Lata et al. 2011). Sequence variation in this gene correlated with differences in salinity tolerance, and its use for FM development was reported (Table 21.2).

Table 21.2 Genes, polymorphisms and FMs identified as related with phenotypic variation in abiotic stress resistance/tolerance

sp	Gene	Allele	Mutation	Phenotype	Marker	Validation	References
Drought tolerance							
Ta	TaDREB	Dreb-A1	InDels and SNPs	Drought tolerant	P25/PR P21F/P21R	RIL (115 ind)	Wei et al. (2009)
		Dreb-B1	InDels and SNPs	Drought tolerant	P18F/P18R	RIL (115 ind)	
		Dreb-D1	InDels and SNPs	Drought tolerant	P22F/PR P20F/P20R	RIL (115 ind)	
Salinity tolerance							
H. vulgare	HvHKT	HvHKT1	17 SNPs	Associated with root [Na$^+$] in salinity, explain 10–22% of variability	ND	ND	Qiu et al. (2011)
			9 SNPs	associated with shoot [K$^+$] under salinity	ND	ND	
			nsSNP$_{760}$(A/G)	associated with shoot [K$^+$] under salinity, explain 26% of variability	ND	ND	
		HvHKT2	SNP$_{1637}$ and SNP$_{1768}$ in intron2	associated with shoot and root [K$^+$] under salinity, explain 31–10% of variability	ND	ND	
			SNP$_{1267}$ more 8 polymorphic sites	associated with shoot and root [Na$^+$] under salinity, explain 13 and 10–16% of variability	ND	ND	

21 Functional Marker Development Across Species in Selected Traits

sp	Gene	Polymorphism	Trait	Marker		Accessions	Reference
	HvCBF	Several SNPs	Salinity tolerant	ND	ND	ND	Wu et al. (2011)
	HvCBF1	$nsSNP_{610}$(T/C)	Salinity tolerant	ND	ND	ND	Rivandi et al. (2011)
	HvCBF4	$nsSNP_{111}$(A/T)	Potential implications in structure and function of protein	ND	ND	ND	
Pv	PvP5CS	63 SNPs and 183 InDels including one of 97 bp InDel at intron1	Salinity tolerant	Pv-97	ND	221 common bean acc	Chen et al. (2010)
Si	SiDREB2	Wild allele	Dehydration sensitive	ND	ND	ND	Lata et al. (2011)
	Sidreb2	SNP558(G/A)	Dehydration tolerance	DREB.AP2-F SNP.Tol-R		47 foxtail millet acc	
Low temperature tolerance							
Hv	Hvcbf14	Hvcbf14-10	SNP at 3'-UTR	Frost resistance	ND	ND	Fricano et al. (2009)
		Hvcbf14-7	SNP at 3'-UTR	Frost resistance	ND	ND	

sp species, ND not developed, ind individuals, acc accessions, Ta T. aesticum, Hv Hordeum vulgare, Pv P. vulgaris, Si S. italic

Low Temperature Tolerance

Low temperature is one of the primary stresses limiting growth and productivity in winter. To cope with low-temperature stress, plants have evolved adaptive mechanisms that are temperature regulated. Low-temperature acclimation and vernalization response are the most important (Fowler et al. 1996).

Cold acclimation is coordinated by a complex process of up- or down- regulation of hundreds of *COLD-REGULATED* (*COR*) genes which, in turn, are controlled by a complex regulatory network (Fowler and Thomashow 2002). In many plant species, the *CBF* genes are key regulators of a signal cascade that leads to the expression of *COR* genes. There are several examples showing a positive correlation between freezing tolerance and *DREB/CBF* transcript accumulation (Chen et al. 2009; Fricano et al. 2009; Yang et al. 2011). Heterologous over-expression of *CBF* sequences in transgenic plants resulted in increased levels to frost tolerance (Takumi et al. 2008; Oh et al. 2007). Association of *HvCbf14* with frost tolerance has been reported (Table 21.2) (Fricano et al. 2009). However, it was not established, which polymorphic sites are responsible for variation in frost tolerance.

Beside *DREB* genes, other cold-responsive genes have been identified. Fowler and Thomashow (2002) revealed the existence of 306 cold-reponsive genes in *A. thaliana*. A gene family already reported to be involved in response to low temperature is the *ALTERNATIVE OXIDASE* (AOX) gene family. AOX is an inner mitochondrial membrane protein that functions as terminal oxidase in the alternative (cyanide-resistant) pathway of respiration, where it generates water from ubiquinol (Umbach et al. 2002). AOX serves to relieve oxidative stress originating from environmental stresses by limiting mitochondrial reactive oxygen species (ROS) formation and preventing specific components of the respiration chain from over-reduction (Popov et al. 1997) and canalizing ROS signals (Amirsadeghi et al. 2007). AOX activity can support the homeostasis of plant growth to changing environmental conditions (Hansen et al. 2002; Arnholdt-Schmitt et al. 2006; Vanlerberghe et al. 2009). A sharp increase in *AOX* transcript and/or protein content after transfer to or growth under low temperatures has been reported for several species (Umbach et al. 2009; Wang et al. 2011b; Watanabe et al. 2008). Abe et al. (2002) showed by site-direct mutagenesis that $nsSNP_{297}$(G/T) of *AOX1a* in *O. sativa* (*OsAOX1a*), leading to substitution of K71N, affects a quantitative trait locus (QTL) for thermo tolerance. *AOX* genes have been proposed for FM development related to multi-stress-tolerance and plant plasticity (Arnholdt-Schmitt 2009; Arnholdt-Schmitt et al. 2006; Polidoros et al. 2009, see also under section "Plant Plasticity – A New Trait Across Species and Plant Systems" in this chapter).

Biotic Stress Tolerance

Race/cultivar-specific resistance involves recognition of a pathogen avirulence (*avr*) gene product by the complementary host resistance (*R*) gene protein.

This recognition initiates a signal transduction cascade and defense response (Martin et al. 2003). The largest family of *R* genes is the *NBS-LRR* gene family, which encodes cytoplasmic proteins with nucleotide binding site (NBS) and leucine-rich repeat (LRR) domains. The *NBS-LRR* gene family can be divided into the TIR and non-TIR subfamilies, depending on the presence of a domain at the N-terminal end with similarity to the Toll/Interleukin-1Receptor (TIR) domain (Meyers et al. 1999). R proteins of the non-TIR class are often predicted to have a coiled-coil (CC) structure near their N-terminus and are referred to as the CC-NBS-LRR class.

R genes that confer resistance to different types of pathogens encode very similar proteins. However, resistance genes that control closely related or identical pathogens are only rarely located in corresponding positions in different genera. This is particularly true for dominant resistance factors that are involved in gene-for-gene interactions and are characterized by a NBS and/or a LRR domain (Ruffel et al. 2005). The LRR region plays a major role in pathogen recognition specificity (Yahiaoui et al. 2006). Interestingly, this feature appears not to be shared by recessive resistance genes that control viruses belonging to the genus *Potyvirus*. In comparison with resistance genes controlling other pathogens, recessive resistance to potyviruses is relatively common, comprising about half of all known resistances against this viral genus (Diaz-Pendon et al. 2004).

Additionally, identification of extensive intra- and inter-specific genetic variation within *NBS-LRR* genes makes it difficult to identify functional polymorphisms of *NBS-LRR* alleles based on sequence homology (Ingvardsen et al. 2008). In the following paragraphs *R* genes across different plant species and the nucleotide polymorphisms related with resistance to different pathogens will be described.

Fungus Resistance

Resistance Against *Blumeria graminis* f.sp. *tritici*, *Puccinia triticina* and *P. striiformis* sp. *tritici*

Blumeria graminis (DC.) Speer f. sp. *tritici* Marchal, *Puccinia triticina* Erikss. and *Puccinia striiformis* Westend. f. sp. *tritici* are three of the most devastating pathogens in wheat production causing powdery mildew, leaf or brown rust and yellow or stripe rust diseases, respectively. Substantial investments were made in order to develop FMs to assist plant breeding, and nowadays several *Pm* (*POWDERY MILDEW*) genes have been identified (Maxwell 2008). *Pm3* encodes a protein belonging to the CC-NBS-LRR family that confers race-specific resistance to *B. graminis* f.sp. *tritici* (Yahiaoui et al. 2004). *Pm3* carries a higher number of alleles than other *Pm* genes (Tommasini et al. 2006). In hexaploid wheat, seven resistance alleles (*Pm3a*, *Pm3b*, *Pm3c*, *Pm3d*, *Pm3e*, *Pm3f* and *Pm3g*) have been described, all derived from one susceptible allele, *Pm3CS*, which is widespread among hexaploid bread wheat lines (Tommasini et al. 2006; Yahiaoui et al. 2006) (see details in Table 21.3). Only three nsSNPs differentiate the resistant alleles *Pm3d* and *Pm3e* from *Pm3CS*, suggesting that the specific resistance they confer might be based on these few polymorphic

Table 21.3 Genes, polymorphisms and FMs identified as related with phenotypic variation in fungi and nematode stress resistance/tolerance

Fungi resistance

sp	Gene	Allele	Mutation	Phenotype	Marker	Validation	References
T. aestivum	Pm3	Pm3CS	Wild allele	Sensitive to B. graminis	ND	ND	Yahiaoui et al. (2006)
	Chr. 1A	Pm3a, Pm3b, Pm3c, Pm3f	Group A: short block of polymorphic nucleic acid sequences in NBS and LRR encoding regions	Resistant to B. graminis	UP3B + UP1A	93 genotypes including cultivars and breeding lines of wheat and T. spelta	Tommasini et al. (2006)
		Pm3d, Pm3e, and Pm3g	Group B: nsSNPs related with aa changes at LRR domain, except the R659W in Pm3d and Pm3e	Resistant to B. graminis			Yahiaoui et al. (2006)
		Pm3k	SNP$_{1332}$C	Resistant to B. graminis	ND	ND	Yahiaoui et al. (2009)
	Pm38/ LR34/ Lm/ Yr18	-Lr34	Hapl: A—C define the wild allele	Braod-spectrum sensitive	ND	ND	Krattinger et al. (2009)
			Hapl: T—C, SNP (A/T) intron 4	Braod-spectrum sensitive	ND	ND	Lagudah et al. (2009)
		+Lr34	Hapl: TTTCT, SNP(A/T) intron4	Braod-spectrum resistant	cssfr1, cssfr2, cssfr3, cssfr4, cssfr5, cssfr6	33 genotypes	
			3 bp insertion in exon 11 (TTC) and SNP (C/T) exon12 SNP exon 22 (cv. Jagger)	Braod-spectrum resistant	cssfr7		
O. sativa	Pi-ta Chr. 12	pi-ta	Wild allele	Susceptible to M. grisea	YL153/YL154	15 rice var, 141 germplasm acc and lines	Jia et al. (2002) Wang et al. (2007)

	Pi-ta	nsSNP2752(T/G) located at the C-terminal	Resistant to M. grisea	YL100/YL102	15 rice var 2 F_2 sgp (1722 ind)	Jia et al. (2002)
				YL155/YL87	141 germplasm acc	Jia et al. (2004)
				YL183/YL87	2 F_2 sgp (1722 ind)	Wang et al. (2007)
				YL155 (Pi-ta)/YL200 YL183 (pi-ta)/YL200	4 F_2 ind	Jia et al. (2004)
	Pit	nsSNP2338(G/A) dDart (DNA transposon) in the promoter Renovator (LTR-retrotransposon) in the promoter		st780A + st780B tk59 + tkpb tdDN + tdDK	68 germplasm acc	Hayashi et al. (2010)
	Pi54/Pik^h	144 bp deletion in ORF	Resistant to blast disease	Pi54MAS	F_2 sgp (200 ind)	Ramkumar et al. (2011)
C. melo	Fom-1	Not described	Resistant to F. oxysporum races 0 and 2	ND	ND	Wang et al. (2011c)
	Fom-2	nsSNP271(G/A) nsSNP1076(G/A)	Resistant to F. oxysporum races 0 and 1	2–3 F3 + 2–3R2	F_2 sgp (100 ind)	Wang et al. (2011c)
Gm	rgh1	nsSNP1216(C/T)	SCN resistant	CAPS-3 F + CAPS-3R	34 var	Li et al. (2009)
	rgh1-hp1	nsSNP689(C/T), SNP757 (T/C) exon1 SNP2564 (A/G) exon2 SNP3995 (C/A) 3'-UTR		ACAS-PCR	62 genotypes from China and USA	Ruben et al. (2006)

sp. species, ND not developed, sgp segregating population, var varieties, ind individuals, acc accessions, Gm Glycine max residues. Direct mutagenesis studies indicated that aa W659 is required for *Pm3d*-dependent resistance, and the replacement of E1334V in *Pm3CS* was sufficient to convert the susceptible to a resistant phenotype (Yahiaoui et al. 2006).

The identified polymorphisms (including single and multiple nucleotide polymorphisms and a small InDel) were mainly located in the terminal part of the *Pm3* coding region (encoding the LRR region of protein) and in the 3'-UTR (Tommasini et al. 2006). They were used for FM development in order to distinguish the allelic series of powdery mildew resistance (Table 21.3). The FMs were used to characterize the seven haplotypes present in 1,005 accessions (Bhullar et al. 2009).

The *Pm3* locus is conserved in tetraploid wheat, in which Yahiaoui et al. (2009) identified 61 allelic sequences that corresponded to 21 different haplotypes (H1-H21) and one additional resistance allele (*Pm3k*, H22) with only a single polymorphism, not found in a susceptibility allele ($SNP_{1332}C$), a predicted solvent-exposed residue of LRR27. As in hexaploid wheat, the highest sequence diversity was located in the LRR-encoding region, and a change in a solvent-exposed residue of LRR27 was sufficient to convert the susceptible *Pm3CS* into a functional allele (Yahiaoui et al. 2006).

A second gene related to fungus resistance in wheat is *Pm38* (*POWDERY MILDEW38*), which confers a high level of broad-spectrum resistance (Spielmeyer et al. 2005). This gene has been identified in several genetic backgrounds, and is inherited as a gene complex, which also confers resistance to leaf or brown rust (*P. triticina*), stem or black rust and leaf tip necrosis (*P. graminis*), stripe or yellow rust disease (*P. striiformis* sp. *tritici*), and moderate resistance to powdery mildew (*B. graminis*). This is also the reason why the same gene *Pm38* acquired different synonyms: *Lr34* (*LEAFRUST34*), *Ltn* (*LEAFTIPNECROSIS*) or *Yr18* (*YELLOWRUST18*) (Liang et al. 2006; Schnurbusch et al. 2004; Singh 1992; Spielmeyer et al. 2005, 2008). This gene is known to encode a pleiotropic drug resistance (PDR)-like ATP-binding cassette (ABC) transporter (Krattinger et al. 2009).

Lr34 sequence comparison of resistant and susceptible wheat cultivars revealed the existence of three haplotypes, two susceptible haplotypes differing in only one nucleotide (−Lr34) and one resistant (+Lr34) haplotype, due to the existence of three polymorphic sites among the gene sequences (Krattinger et al. 2009; Lagudah et al. 2009) (see details in Table 21.3). Lagudah et al. (2009) also reported a SNP (G/T) in exon 22 of the wheat cv. Jagger (susceptible), which results in a premature stop codon lacking 185 aa of the C-terminus. The same authors developed a co-dominant functional marker to detect this nsSNP.

Resistance Against *Magnaporthe oryzae*

Rice blast disease caused by the pathogenic fungus *Magnaporthe grisea* (T.T. Hebert) M.E. Barr recently renamed as *Magnaporthe oryzae* B. Couch (anamorph *Pyricularia grisea* Sacc. or *Pyricularia oryzae* Cavara, respectively) is one of the most devastating diseases in rice production (Zeigler et al. 1994). To date, more than 80 blast *R* genes have been identified (Ballini et al. 2008), but fewer than 20 are characterized at the molecular level.

Pi-ta

Pi-ta encodes a protein with unique features when compared with other proteins of the NBS-containing class *R* genes (see Bryan et al. 2000). It includes a C-terminal LRD (Leucine-Rich Domain) instead of the characteristic LRR motif found in other genes of this class (Bryan et al. 2000).

Bryan et al. (2000) characterized two *Pi-ta* alleles with a nsSNP located at the C-terminal region (T2752G), which is responsible for the aa change A918S and the subsequent change from the susceptible phenotype (allele *pi-ta* with A918) to a resistance phenotype (allele *Pi-ta* with S918). Functional analysis by transforming the susceptible rice variety Nipponbare with genomic and cDNA of the *Pi-ta* allele confirmed the identity of the gene as resistant to *M. grisea* (Bryan et al. 2000). The importance of A918 in determining *in vivo* specificity in the *Pi-ta* gene-for-gene system was also demonstrated by transient expression assays (Bryan et al. 2000).

Four additional nsSNPs were outlined when the variety Yashiro-mochi (resistant) was compared with the susceptible variety Tsuyuake: G17T: S6I, G444C: S148R, G474C: Q158H, T527A: V176D (Bryan et al. 2000). The five polymorphic sites were confirmed by Jia et al. (2003), who included eight new rice cultivars. The same authors reported additional SNPs at 5'-UTR (G2040A) and 3'-UTR (T6808A) and in intron sequences (A3536CC, G4234A, G4270A, C4391T, T4394A and GCC4426-4428CTAT). The identified polymorphisms were used to develop dominant and co-dominant functional markers for identification and incorporation of the *Pi-ta* gene by MAS (Table 21.3).

Pit

The *Pit* gene belongs to the CC-NBS-LRR family of resistance genes (Hayashi and Yoshida 2009). Sequence comparison of *Pit* alleles between a susceptible (Nipponbare) and a resistant cultivar (K59) revealed that the resistance-conferring allele contains four aa substitutions (G143R, I176M, T720A and V780M), a DNA transposon *dDart*, and a long terminal repeat (LTR)-retrotransposon (*Renovator*), both inserted in the promoter. The effect of *Renovator* was verified by gene expression analysis and in a transgenic approach. The level of *Pit* mRNA was up-regulated by its insertion, and the effect of *Renovator* on the *Pit* promoter activity was greater than that of the aa substitutions (Hayashi and Yoshida 2009).

Based on that knowledge Hayashi et al. (2010) studied the variability of the *Pit* coding sequence in ten rice cultivars, identifying the same $nsSNP_{2338}$ (G/A) located at the LRR region. PCR-based markers were developed to detect the nsSNP, the *Renovator* and the *dDart* (Hayashi et al. 2010).

Pi54 (Pik^h)

Pik^h, recently renamed *Pi54* (Sharma et al. 2005) is one of the major blast resistance genes identified as encoding a NBS-LRR protein. A recent study including 27 landraces collected in the north-eastern part of India, report several polymorphisms in Pi54 in which an InDel of 144 bp in the coding sequence is related with a resistant phenotype (Ramkumar et al. 2011). A functional co-dominant marker was developed to identify the resistance allele (Table 21.3).

Resistance Against *Fusarium oxysporum f. sp. melonis* Snyder and Hansen

Fusarium wilt caused by the fungus *Fusarium oxysporum f. sp. melonis* Snyder and Hansen has become one of the most destructive diseases of melon (*Cucumis melo* L.) crops throughout the world (Leach 1933). To date four races (0, 1, 2, and 1–2) of this fungus have been defined, and two resistance genes were identified to control the resistance of races: *Fom-1* (for races 0 and 2) and *Fom-2* (for races 0 and 1) (Risser et al. 1976). *Fom-2* is predicted to encode a protein belonging to the NBS-LRR type of *R* genes. Wang et al. (2011c) analyzed the LRR region in order to identify functional polymorphisms useful for FM development. Sequence comparison between resistant and one susceptible genotype revealed three candidate polymorphic sites (Table 21.3).

Nematodes Resistance

FM development for nematode resistance is most advanced in *G. max*. The cyst nematode (SCN; *Heterodera glycines* Ichinohe) is an important pathogen of soybean worldwide. Resistance is controlled by three recessive genes (*rhg1*, *rhg2*, *rhg3*) (Caldwell et al. 1960) and a dominant resistant gene (*Rhg4*) (Matson and Williams 1965). Nevertheless, the allele for partial resistance at the *rhg1* resistance locus has been demonstrated to control more than 50% of the variation for resistance and appears to effectively control a number of SCN races (Concibido et al. 1997). The *rhg1* gene family encodes a PROTEIN-RECEPTOR-LIKE KINASE (RLK) (Hauge et al. 2001), with an N-terminal signal peptide (1–61), an extracellular domain with ten extracellular Leucine-Rich Repeats (LRR, 141–471), two trans-membrane domains (TM, 40–60; 485–507), and a cytoplasmic Serine/Threonine/Tyrosine Kinase domain (STYKc, 569–840) (Ruben et al. 2006). LRR-containing RLKs, which form the largest group of RLKs in plants, were predicted to play a central role in signaling during pathogen recognition in plant defense mechanisms and in developmental regulation.

DNA sequencing from 112 SCN-resistant Plant Introductions (PIs) and 34 derived cultivars inferred nine *rhg1* haplotypes, four of which were SCN resistant (Hauge et al. 2001; Ruben et al. 2006). Relatively few nucleotide substitutions resulted in aa changes so that only five protein allotypes were predicted with two of them potentially useful for FM development: one alters A47V, and the second alters H297N. Both substitutions may alter protein transport or function or both since A47 was only associated with resistance in the presence of H297 (Ruben et al. 2006).

Very recently, Li et al. (2009) demonstrated that the gene *rhg1* is essential for the development of resistant soybean cultivars. Polymorphisms in that gene were responsible for sensitive phenotypes. Four SNPs discriminated a haplotype present in five resistant soybean genotypes and another haplotype in the susceptible genotypes Suinong 14 and Guxin. From the four SNPs, three are located in the coding region, two in exon 1 between the N-terminal signal peptide and the LRR domain, and one in exon 2 located in the LRR domain. The fourth SNP is located in the 3'-UTR (Table 21.3). The two SNPs in exon 1 forming one haplotype (689C–757C), were perfectly associated with SCN resistance and allowed development of functional co-dominant markers to separate resistant from susceptible genotypes (Table 21.3).

Virus Resistance in Dicots

Plant viruses are obligate parasites that multiply within their hosts by establishing specific interactions between viral factors and macromolecules, structures and processes of the plant, which determine plant susceptibility to viral infection (Maule et al. 2002). A deleted or defective host protein that is essential for viral infection, but is dispensable for the host may result in resistance to the virus. In this case, resistance is based on the 'negative model', where resistance is expected to be genetically recessive (Fraser 1992). Recessive resistance was found to be relevant for viruses belonging to the *Potyviridae* (Ruffel et al. 2005) and the *Tombusviridae* family (Nieto et al. 2006). Although the recessive resistance mechanism conserved in dicotyledonous plants (Kanyuka et al. 2005) and also described in some monocotyledonous (Albar et al. 2006; Stein et al. 2005), resistance against *Potyvirus* can also be dominant. In maize, *Wheat streak mosaic virus* (WSMV) and *Sugarcane mosaic virus* (SCMV; previously called MDMV-B) resistance was described as controlled by dominant genes (McMullen et al. 1994; Xu et al. 1999).

Characterization of recessive resistance genes in several dicot and monocot plant species such as pepper (*pvr1*, Ruffel et al. 2002), lettuce (*mo1*, Nicaise et al. 2003), pea (*sbm1*, Gao et al. 2004), tomato (Ruffel et al. 2005), barley (*rym4/5*, Stein et al. 2005), and rice (*Rymv1*, Albar et al. 2006), and the mutagenesis assays performed in *A. thaliana* (Duprat et al. 2002) implicate a component of the eukaryotic translational initiation complex, i.e., eIF4E, eIF(iso)4E, eIF4G, and eIF(iso)4G as being responsible for conferring resistance in plant systems to RNA viruses (for reviews see, Kang et al. 2005b; Maule et al. 2007). eIF4E is a component of the eIF4F complex and provides the 5' cap-binding function during formation of translation initiation complexes for most eukaryotic mRNAs (Strudwick and Borden 2002). In plant cells, this complex is composed of only two proteins, eIF4E and eIF4G (Browning 1996), and an additional cap-binding complex, eIF(iso)4F, in which a second cap-binding protein [eIF(iso)4E] binds with eIF(iso)4G (Bailey-Serres 1999).

Diversity in *eIF4E* conferring multiallelic recessive virus resistance to plant viruses seems to be a widespread mechanism. Often just a single or a limited number of aa changes in the eIF4E protein, that disrupts the direct interaction with the potyviral VPg protein, have been shown to result in virus resistance (Charron et al. 2008; German-Retana et al. 2008; Maule et al. 2007; Naderpour et al. 2010; Robaglia and Caranta 2006; Yeam et al. 2007). Thus screening for natural diversity in homologous *eIF4E* genes in different crop species could provide an option to identify underlying genes and causative polymorphisms which define new resistance alleles useful for breeding programs.

Pepper

In pepper (*Capsicum annum* L.), the homolog of *eIF4E* located at the locus *pvr1* on chromosome 3 (Murphy et al. 1998), was demonstrated to confer resistance against several *Potyvirus* species including *Tabacco etch virus* (TEV), *Potato Y virus* (PVY), and *Pepper mottle virus* (Pepmov) (Kang et al. 2005a; Ruffel et al. 2002). Kang et al. (2005a) reported the existence of four alleles, which encode the eIF4E protein: $Pvr1^+$ defined as the allele for susceptibility, and the three resistance alleles, *pvr1*, *pvr1¹*, and *pvr1²*. The resistance alleles, due to aa changes (see details of aa changes sites in Table 21.4), encode a protein that fails to interact with the viral protein VPg (Kang et al. 2005a; Ruffel et al. 2002). In order to understand the biochemical effect of each aa substitution, Yeam et al. (2007) generated alleles containing each aa substitution separately. The results indicate that the loss of VPg binding ability of eIF4E encoded by the *pvr1¹* and *pvr1²* alleles is the result of an additive effect of the V67E and L79R changes. In the case of *pvr1*, it is caused by the single change G107R. Amino acid 107 is adjacent to R171, an aa that interacts directly with the negative charge of the cap phosphate group and is known to be important for cap binding (Marcotrigiano et al. 1997). It is striking to note that the critical aa substitution in *pvr1*, G107R, also exists at the homologous sites in several other recessive resistance genes, including *sbm1* and *sub-1* (G107R) from pea (Smýkal et al. 2010), *mo1¹* (QGA108-110H) from lettuce (Nicaise et al. 2003), and *pot¹* (M109I) from tomato (Ruffel et al. 2005), which will be described below.

This capacity of the G107R aa change in eIF4E- *pvr1* alone to be sufficient to abolish the capacity of eIF4E to bind VPg was supported by a yeast two-hydrid assay (Yeam et al. 2007), and using recombinant *Capsicum*-eIF4E proteins produced in *Escherichia coli* (Kang et al. 2005a). A transgenic approach overexpressing *pvr1* in tomato also resulted in gain of viral resistance (Kang et al. 2007).

Yeam et al. (2005) developed allele-specific CAPS markers for the three recessive viral resistance alleles from 13 *Capsicum* genotypes known to be homozygous for each of the four *pvr1* alleles (Table 21.4). Three exceptions were observed in genotypes showing resistance to PepMoV with the absence of the *pvr1* allele.

21 Functional Marker Development Across Species in Selected Traits

Table 21.4 Amino acid polymorphisms in eukaryotic translation factor 4E (eIF4E) found in susceptible and resistant/tolerant genotypes of different dicot plant species

Genotype	Phenotype	Amino Acid Position at polymorphic sites																										Other polymorphisms			Marker	Validation	References
		48	51	53	62	65	66	67	69	70	71	73	74	76	77	79	81	93	107	108	109	110	111	169	186	228	3'UTR	Ind	Ind3 size			
Capsicum annuum (pepper) (687bp/228aa, Ruffel et al. 2004)																																	
Pvr1	TEV, PVY suscep	T				P	V								L		G		D											Pvr1-S	15 breeding lines provided from a breeding program	Yeam et al. 2005
pvr1	TEV, PVY resist	A				T	.										R													Pvr1-R1		
pvr1¹	TEV, PVY resist	.				.	E			R							.												Pvr1-R2		
pvr1²	TEV, PVY resist	.				.	E			R							.		N												
Lycopersicon hirsutum (tomato) (696bp/231aa, Ruffel et al. 2005)																																	
pot-1⁺	PVY, TEV suscep	L							N			A											M							eIF4E-SpeI	F₃ sgp (17 ind)	Ruffel et al. 2005
pot-1	PVY, TEV resist	F							K			D											I									
Lactuca sativa (lettuce) (690bp/230aa, Nicaise et al. 2003)																																	
Lse1F4E¹ (mo¹⁻)	LMV suscep									A								F	Q		G		A		A		C			eIF4E-PagI	2 F₃ sgp (25 ind)	Nicaise et al. 2003
Lse1F4E¹ (mo1^f)	LMV resist									.								F	H		A		S									
Lse1F4E¹ (mo1^g)	LMV tolerant									P											.		.				T					
Pisum sativum (pea) (687bp/228aa, Bruun-Rasmussen et al. 2007)																																	
sbm1⁺	PSbMV suscep				W							A	A						G				N							1201	W62-CTG/CTT-218R 218F+SNP-G107-CCGCCA N169-ATT/AAG-586gR W62-CTG/CTT-218R W62-CTG/CTT+218R W62-CTG/CTT+218R_SstI Intron3750F+586gR	60 acc	Smýkal et al. 2010
sbm1¹	PSbMV resist				L							D	D						R				K							1151			
sbm1²	PSbMV resist											P	D		Δ				.				.							1151			
Cucumis melo (melon) (708bp/235aa, Nieto et al. 2006)																																	
NSV	MNSV suscep																											H			ND	ND	Nieto et al 2006
nsv	MNSV resist																																
Citrullus lanatus (watermelon) (708bp/235aa, Ling et al. 2009)																																	
Cl-eIF4E¹	ZYMV suscep											D					T														CAPS1_MseI, CAP2_PasI CAPS3_BstEII	6 PIs, F₂sgp (70 ind), 114 BC₁R	Ling et al. 2009
Cl-eIF4E²	ZYMV resist											G					P																
Phaseolus vulgaris (common bean) (only partial sequences available, Naderpour et al. 2010)																																	
PveIF4E¹	BCMV, BCMNV sus		N			F								A										D							ENMFWe/RVe_RsaI	F₂ sgp (96 ind)	Naderpour et al. 2010
PveIF4E²	BCMV, BCMNV resist		K			Y								E										G									

Δ means deletion, *sus* and *suscep* susceptible, *resist* resistant, *sgp* segregating population, *ind* individuals, *acc* accessions, *ND* not developed, *PIs* plant introductions, *BC* reciprocal backcrossing F1

Tomato

Comparison of resistant *Lycopersicum hirsutum* and susceptible genotypes of *L. hirsutum* and *L. esculentum* revealed the existence of two alleles, *pot-1*$^+$ (characterizing the susceptible phenotype) and *pot-1* which confers resistance to the *Potato virus Y* (PVY) and *Tobacco etch virus* (TEV). Both alleles were distinguished by the existence of four nsSNPs, which were used for FM development for application in plant breeding (see Table 21.4) (Ruffel et al. 2005). Additional confirmation of the involvement of the *pot-1* allele in resistance and *pot-1*$^+$ in the susceptibility to PVY and TEV was achieved by a transgenic approach. Transient expression of the dominant allele restored susceptibility to both PVY and TEV, whereas expression of *pot-1* did not support potyvirus infection (Ruffel et al. 2005).

Lettuce

In lettuce (*Lactuca sativa* L.), *mo1*^{1} and *mo1*^{2} are known as recessive alleles of a single gene (Nicaise et al. 2003), associated with reduced accumulation and lack of symptoms (tolerance) or absence of accumulation (resistance) of common isolates of potyvirus *Lettuce mosaic virus* (LMV; Dinant and Lot 1992). Resistance or tolerance depends on virus isolate and genetic background (Revers et al. 1997). However, *mo1*^{1} is generally associated with resistance and *mo1*^{2} with tolerance (Revers et al. 1997).

Nicaise et al. (2003) characterized the *eIF4E* cDNA sequence in eight lettuce genotypes and classified three lettuce *eIF4E* alleles: *Ls-eIF4E*^{o} (susceptibility), *Ls-eIF4E*^{1} (resistance) and *Ls-eIF4E*^{2} (tolerance) (see Table 21.4). The aa that discriminates the three Ls-eIF4E alleles were all mapped near the cap recognition pocket (Nicaise et al. 2003). A strict correlation between *Ls-eIF4E*^{1} and presence of *mo1*^{1} and between *Ls-eIF4E*^{2} and presence of *mo1*^{2} was reported by Nicaise et al. (2003). A functional co-dominant marker was developed to identify the resistance allele *Ls-eIF4E*^{1} (Table 21.4).

Pea

Two homologous *eIF4E* and *eIF(iso)4E* genes were identified in the *Pisum sativum* (pea) genome to be responsible for *Pea seed-borne mosaic virus* (PSbMV) and white lupin strain of *Bean yellow mosaic virus* BYMV-W resistance respectively at the *sbm1* and *sbm2* locus (Bruun-Rasmussen et al. 2007; Gao et al. 2004).

Resistance to the common strains of PSbMV is conferred by a single recessive allele (*sbm1*) encoding a mutation that fails to interact with the PSbMV avirulence protein (VPg). This difference at the protein level is caused by five polymorphisms (see Table 21.4), which were described as highly conserved between different plant species (Smýkal et al. 2010). However, only W62 and N169 displayed resistance when analyzed by direct mutagenesis (Ashby et al. 2011).

In addition to the polymorphisms within exon sequence, resistant and susceptible genotypes can also be differentiated by polymorphisms in intron sequences. InDels of 50 and 56 bp cause shorter intron 3 sequences (1,151 bp) characteristic of all resistant (*sbm1*) accessions compared to susceptible genotypes with longer intron 3 sequences (1,201 bp) (Table 21.4) (Smýkal et al. 2010). Functional dominant and co-dominant markers were developed based on nsSNPs and intron length polymorphisms (Table 21.4).

Melon and Watermelon

In melon (*C. melo*) a gene coding for eIF4E was identified, in which Nieto et al. (2006) found SNP mutations involved in potyviruses resistance, mostly located in the N-terminal region (Table 21.4). Nevertheless, a nsSNP in the C-terminal region of eIF4E (H228L) causes resistance to *Melon necrotic spot virus* (MNSV), a virus belonging to the *Tombusviridae* (Nieto et al. 2006). Susceptible genotypes carry the H228 *allele* (*NSV*) and resistant genotypes carry the L228 allele (*nsv*). Genetic transformation demonstrated that the expression of the *nsv* allele caring H228 in resistant melon is sufficient to restore susceptibility to the NRB strain of MNSV.

In watermelon, (*Citrullus lanatus* [Thunb.] Matsum. & Nakai var. *lanatus*) the *Zucchini yellow mosaic virus* (ZYMV) is one of the most economically devastating potyviruses (Ma et al. 2005). According to Ling et al. (2009) ZYMV resistance is controlled by different SNP mutations in the same *Citrullus eIF4E* gene resulting in an allelic series. Two SNPs are located in intron 1, and one nsSNP in exon 1, which result in a T81P substitution, unique for the ZYMV-resistant PI 595203 genotype (Ling et al. 2009). T81P is predicted to be located in the critical area for cap recognition and binding. SNPs are close to SNPs in other plant species causing resistance, like L79R in *pvr1¹* and *pvr1²* in pepper and A77D in *pot-1* in tomato (see Table 21.4). An additional nsSNP$_{171}$(A/G) responsible for aa substitution D71G was identified in four ZYMV-resistant *C. lanatus* var. *citroides* accessions. Functional co-dominant markers were developed to differentiate between ZYMV-resistant and susceptible plants (Table 21.4).

Common Bean

In common bean (*P. vulgaris*), four recessive genes have been proposed to control resistance to the potyviruses *Bean common mosaic virus* (BCMV): *bc-1*, *bc-2*, *bc-3* and *bc-u* (Naderpour et al. 2010). *PveIF4E* gene cloning and sequence analysis revealed the existence of four nsSNPs responsible for aa changes at positions N53K, F65Y, A76E, and D111G, defining susceptibility allele *PveIF4E¹* and resistance allele *PveIF4E²*. Bean genotypes reported to carry *bc-3* resistance were found to have a set of mutations, known to determine potyvirus resistance in other species, which make *PveIF4²* a strong candidate gene for *bc-3* (Table 21.4). Existence of polymorphisms directly related to BCMV resistance allowed to development of a co-dominant FM. Its application in a segregating F_2 population revealed that only plants homozygous for the $PveIF4E^2$ allele resisted virus infection (Naderpour et al. 2010).

Plant Plasticity – A New Trait Across Species and Plant Systems

Plants as sessile organisms learned during evolution to respond to diverse environmental constraints and opportunities in terms of adaptive growth and development. The potential for adaptive plasticity influences the stability of plant biomass and yield production, and the capacity for efficient adventitious morphological responses upon stress, such as adventitious rooting (Macedo et al. 2009) or formation of somatic embryos (Zavattieri et al. 2010), there are important traits for cost- and time-efficient plant production. For example, in spite of extensive studies to improve olive propagation, success rates are still limited, with formation of adventitious roots being an important factor (Peixe et al. 2007). Development of FMs for adventitious rooting would be of great value. Somatic embryogenesis is also used as a propagation method for clonal testing and selection of superior genotypes as in the case of *Pinus pinea* L.. Development of FMs to select genotypes easily inducible for somatic embryos are obviously of interest.

Differences in the robustness of plant genotypes to grow under diverse environmental conditions and in recalcitrant behavior related to inducing conditions for adventitious organogenesis or somatic embryogenesis are well described across species. However, so far the capacity for plasticity, although known as a main driver in evolution for organisms to occupy ecological niches, has not been explored as a trait *per se* for molecular plant breeding.

The involvement of mitochondria as a physical platform for networks, signal perception and signal canalization play a central role in plant reacquiring homeostasis (Amirsadeghi et al. 2007; Fernie et al. 2004; Noctor et al. 2007; Raghavendra and Padmasree 2003; Rhoads and Subbaiah 2007; Sweetlove et al. 2007). Significance of mitochondria for cell fate decisions by dedifferentiation and *de novo* differentiation is recognized (Amirsadeghi et al. 2007; Sheahan et al. 2005). The alternative respiration pathway is localized in mitochondria and is increasingly getting into the focus of research on stress acclimation and adaptation. AOX plays a key role in regulating the process of cell-reprogramming by ameliorating metabolic transitions related with the cellular redox state and the flexible carbon balance (Arnholdt-Schmitt et al. 2006; Rasmusson et al. 2009). Clifton et al. (2005, 2006) pointed to the importance of this pathway as an early-sensoring system for cell programming. AOX may coordinate phenotypic changes related to adaptation to environmental changes. AOX is involved in biotic and abiotic stress responses (McDonald and Vanlerberghe 2006; Plaxton and Podestá 2006, see section "Stress Tolerance" in this chapter: involvement of *AOX* to low temperature tolerance), including morphogenic responses (Campos et al.

2009; Fiorani et al. 2005; Frederico et al. 2009a; Ho et al. 2007; Macedo et al. 2009, 2012).

The ability of plants to adapt growth to varying conditions is genetically determined (Jungk 2001) and genetic variation has been shown to affect alternative respiration related to growth behaviour (Hilal et al. 1997; Millenaar et al. 2001). Recently, AOX was proposed in a hypothesis-driven approach as target to develop FMs for efficient cell reprogramming. Respective FMs would be valuable for general stress tolerance across species and stresses, and include responses such as adventitious root hair development under nutrient stress (Arnholdt-Schmitt et al. 2006; Arnholdt-Schmitt 2009; Polidoros et al. 2009; see also www.aox2008.uevora.pt and Physiologia Plantarum 2009, special issue: alternative oxidase Vol. 137 (issue 4)). Involvement of a SNP in *OsAOX1a* in rice tolerance to low temperature (Abe et al. 2002) is an example of the putative involvement of *AOX* sequence polymorphisms and phenotypic variability. Cardoso et al. (2009, 2011) reported the existence of sequence variation in two *AOX* genes between genotypes of *Daucus carota* L. (*DcAOX2a* and *DcAOX2b*), putatively related to pre-miRNAs in intronic regions. Involvement of *AOX* genes in somatic embryogenesis was found in *Daucus carota* L. (Frederico et al. 2009a), which can be interpreted as a stress-induced morphogenic response (Fehér 2005; Kikuchi et al. 2006; Pasternak et al. 2002; Potters et al. 2007). Involvement of *AOX* genes in stress effects on cell reprogramming was also demonstrated with the inoculation of differentiated secondary root phloem explants in a cytokinin-containing nutrient solution that induces tissue redifferentiation and callus growth (Campos et al. 2009). Sequence polymorphisms in protein coding and non-coding regions of *DcAOX1a* were related to phenotypes with growth potential and temperature adaptation (Nogales et al. 2012).

Macedo et al. (2009) and Ferreira et al. (2009) published SNPs between genotypes in *Olea europaea* L. and *Hypericum perforatum* L., respectively. Involvement of the *OeAOX2* gene in the adventitious root formation in microshoots and semi-hardwood shoot cuttings of olive submitted to a treatment with auxins was demonstrated (Macedo et al. 2009, 2012). Growth responses to stress are plant acclimation strategies to diminish stress exposure (Potters et al. 2007). Polymorphisms across genotypes with different acclimation responses (e.g., ± easy rooting or development of adventitious roots) maybe good candidates for FM development (Holtzapffel et al. 2003; Macedo et al. 2009).

An alternative candidate for FM development for stress-tolerant behaviour was proposed by Arnholdt-Schmitt et al. (2006). The group of mitochondrial inner membrane uncoupling proteins (UCPs), first reported in mammals (1976) and afterwards in plants (1995), is involved in an energy-dissipating pathway, like AOX. UCPs along with AOX may have a role in controlling energy metabolism by serving as safety "valves" in case of overloads in the redox and/or phosphate potential (Vercesi et al. 2006). Whereas AOX dissipates the redox potential, UCPs dissipate the proton motive force. Thus, both gene families are involved in tuning the capacity of oxidative phosphorylation (Arnholdt-Schmitt et al. 2006). UCPs act in a complementary fashion with AOX during fruit ripening, seed and embryo development (Costa et al. 1999; Considine et al. 2001; Daley et al. 2003; Nogueira et al. 2011). As AOX, UCPs protect cells against reactive oxygen species (ROS) during biotic and abiotic stresses (Borecký et al. 2006; Brandalise et al. 2003; Nogueira et al. 2005, 2011; Van Aken et al. 2009; Watanabe et al. 1999). UCPs and AOX are differentially expressed in a time-dependent manner (Daley et al. 2003). Results from knockout studies using *aox1a* and *ucp1* (Giraud et al. 2008; Sweetlove et al. 2006) indicate coordination of endogenous levels of both energy-dissipating proteins.

Several reports suggest AOX as more promising candidate for FM development for cell reprogramming under stress conditions in plants. Tissue-enriched expression profiling in monocot and dicot model species showed that UCP genes were expressed more ubiquitously than AOX genes (Borecký et al. 2006). In Arabidospis, UCPs were not among the most stress-responsive mitochondrial proteins in contrast to AOX1a (Van Aken et al. 2009). In mammals, SNPs within *UCP1* (promoter, 5' flanking region and exon 2) were associated with obesity phenotypes, diabetes mellitus and lipid/lipoprotein-related disease, body fat accumulation and body weight gain or body mass index (Hamann et al. 1998; Heilbronn et al. 2000; Herrmann et al. 2003; Kiec-Wilk et al. 2002; Kim et al. 2005; Kotani et al. 2008).

Searching for FMs directly linked with the capacity to react with efficient phenotypic plasticity across species is a novel strategy in molecular plant breeding and functional domains in target genes need to be identified. Beside the mutations which create genetic variation underlying phenotypic traits, epigenetic events should also be considered as source of variation for selection (Tsaftaris et al. 2005) and several reports describe the involvement of epigenetic changes in plant stress tolerance (Alvarez et al. 2010; Boyko and Kovalchuk 2008; Chen et al. 2010; Chinnusamy and Zhu 2009; Yaish et al. 2011). Physiological and morphological plasticity can be reflected by plasticity at genome level due to the flexibility in linear sequence modulation, DNA and histone modifications and the structural organization of genomic DNA in the chromatin (Arnholdt-Schmitt 2004; Arnholdt-Schmitt 2005; Fransz and De Jong 2011). In eukaryotes, global genome regulation refers to the structural and compositional organization of chromatin in the nucleus that defines coordinated accessibility to the DNA. Accessibility of DNA sequences to the transcription machinery is crucially determined by the degree of packaging of the DNA into condensed and open chromatin domains. A high degree of condensation demonstrates highly compact structures that are generally inaccessible to DNA-binding factors, leading to the silencing of underlying sequences. A flux from decondensed to more condensed state and *vice versa* is crucial for all kinds of cellular differentiation. This flexibility in chromatin structure, named chromatin remodeling, can be achieved by many different mechanisms and might refer exclusively to nucleosomes or may involve inhibition-repression complexes (Gendall et al. 2001). Stress-mediated effects in chromatin remodeling were described in plants (Chua et al. 2003; Gendall et al. 2001; Grandbastien 1998; Steward et al. 2002; Tsaftaris et al. 2005). Especially important events in chromatin remodeling with an effect on gene transcription are (i) histone modifications (include methylation, acetylation, ADP-ribosylation, glycosylation, phosphorylation, ubiquitination or SUMOylation), (ii) DNA methylation, which in most of the cases is correlated to the inhibition of transcription, at transcriptional (if methylation occurs in promoters) or post-transcriptional level (if methylation occurs in protein-coding sequences) (Okamoto and Hirochika 2001), and (iii) ATPases, which alter conformation and positioning of the nucleosome (Jeddeloh et al. 1999).

Conclusions

In the last 20 years routine protocols have been developed to identify and characterize genetic loci that contribute to quantitative traits. However, the capacity to zoom into natural segregating loci with quantitative or qualitative effects to find the molecular base of phenotypic variability has been accelerated only recently, due to new DNA sequencing technologies. This explains the low number of FMs compared to QTL that have been identified so far. Nevertheless, the number of promising candidate genes for FM development is increasing. Genes of interest for FM development can be identified by high-throughput sequencing or differential gene analysis or by hypothesis-driven research approaches.

However, it might be challenging to find conserved FMs across species. In contrast to genes, causative polymorphic sites within genes seem to have a low degree of conservation. Loss-of-function polymorphisms seem to play an important role for trait variability. Phenotypic variation in the same trait across species may be linked to a diversity of sequence polymorphisms in the according orthologous genes. Our review confirms Risch (2000) who concluded that SNPs located in coding regions causing non-synonymous non-conservative amino acid changes are more likely to be functional, than non-synonymous conservative, and synonymous amino acid substitutions.

We would thus like to propose, that FM development for adaptive phenotypic variation for selected traits across species should better focus on polymorphic patterns in functional domains of genes involved in superimposed metabolic pathways and in the capacity for reorganizing genome structures through epigenetic mechanisms, rather than on conserved polymorphisms in individual areas of downstream genes.

Acknowledgements The authors would like to thank to Luz Muñoz for the critical revision and comments.

References

Abe F, Saito K, Miura K, Toriyama K (2002) A single nucleotide polymorphism in the alternative oxidase gene among rice varieties differing in low temperature tolerance. FEBS Lett 527:181–185

Abebe T, Guenzi AC, Martin B, Cushman JC (2003) Tolerance of mannitol accumulating transgenic wheat to water stress and salinity. Plant Physiol 131:1748–1755

Abreu JG, Coffinier C, Larraın J, Oelgeschlager M, Robertis EMD (2002) Chordin-like CR domains and the regulation of evolutionarily conserved extracellular signaling systems. Gene 287:39–47

Agarwal PK, Agarwal P, Reddy MK, Sopory SK (2006) Role of DREB transcription factors in abiotic and biotic stress tolerance in plants. Plant Cell Rep 25:1263–1274

Albar L, Bangratz-Reyser M, Hebrard E, Ndjiondjop M-N, Jones M, Ghesquiere A (2006) Mutations in the eIF(iso)4G translation initiation factor confer high resistance of rice to *Rice yellow mottle virus*. Plant J 47:417–426

Alonso-Blanco C, Mendez-Vigo B, Koornneef M (2005) From phenotypic to molecular polymorphisms involved in naturally occurring variation of plant development. Int J Dev Biol 49:717–732

Alvarez ME, Nota F, Cambiagno DA (2010) Epigenetic control of plant immunity. Mol Plant Pathol 11:563–576

Amarawathi Y, Singh R, Singh AK, Singh VP, Mohapatra T, Sharma TR, Singh NK (2008) Mapping of quantitative trait loci for basmati quality traits in rice (*Oryza sativa* L.). Mol Breed 21:49–65

Amirsadeghi S, Robson CA, Vanlerberghe GC (2007) The role of the mitochondrion in plant responses to biotic stress. Physiol Plant 129:253–266

Andersen JR, Lübberstedt T (2003) Functional markers in plants. Trends Plant Sci 8:554–560

Arikit S, Yoshihashi T, Wanchana S, Tanya P, Juwattanasomran R, Srinives P, Vanavichit A (2011a) A PCR-based marker for a locus conferring aroma in vegetable soybean (*Glycine max* L.). Theor Appl Genet 122:311–316

Arikit S, Yoshihashi T, Wanchana S, Uyen TT, Huong NTT, Wongpornchai S, Vanavichit A (2011b) Deficiency in the amino aldehyde dehydrogenase encoded by *GmAMADH2*, the homologue of rice *Os2AP*, enhances 2-acetyl-1-pyrroline biosynthesis in soybeans (*Glycine max* L.). Plant Biotechnol J 9:75–87

Armengaud P, Thiery L, Buhot N, March GG, Savoure A (2004) Transcriptional regulation of proline biosynthesis in *Medicago truncatula* reveals developmental and environmental specific features. Physiol Plant 120:442–450

Arnholdt-Schmitt B (2004) Stress-induced cell reprogramming. A role for global genome regulation? Plant Physiol 136:2579–2586

Arnholdt-Schmitt B (2005) Efficient cell reprogramming as a target for functional marker strategies? Towards new perspectives in applied plant nutrition research. J Plant Nutr Soil Sci 168:617–624

Arnholdt-Schmitt B (2009) Alternative oxidase (AOX) and stress tolerance–approaching a scientific hypothesis. Physiol Plant 137:314–315

Arnholdt-Schmitt B, Costa JH, Fernandes de Melo D (2006) AOX – a functional marker for efficient cell reprogramming under stress? Trends Plant Sci 11:281–287

Ashby J, Stevenson C, Jarvis G, Lawson D, Maule A (2011) Structure-based mutational analysis of *eIF4E* in relation to *sbm1* resistance to *Pea seed-borne mosaic virus* in pea. PLoS One 6(e15873):1–13

AVRDC (2003) AVRDC progress report 2002. AVRDC-The World Vegetable Center, Shanhua

Baek JM, Han P, Iandolino A, Cook DR (2008) Characterization and comparison of intron structure and alternative splicing between *Medicago truncatula*, *Populus trichocarpa*, *Arabidopsis* and rice. Plant Mol Biol 67:499–510

Bailey-Serres J (1999) Selective translation of cytoplasmic mRNAs in plants. Trends Plant Sci 4:142–148

Ballini E, Morel JB, Droc G, Price A, Courtois B et al (2008) A genome-wide meta-analysis of rice blast resistance genes and quantitative trait loci provides new insights into partial and complete resistance. Mol Plant Microbe Interact 21:859–868

Batley J, Barker G, O'Sullivan H, Edwards KJ, Edwards D (2003) Mining for single nucleotide polymorphisms and insertions/deletions in maize expressed sequence tag data. Plant Physiol 132:84–91

Ben Chaim A, Paran I, Grube R, Jahn M, van Wijk R, Peleman J (2001) QTL mapping of fruit related traits in pepper (*Capsicum annuum*). Theor Appl Genet 102:1016–1028

Benedict C, Skinner JS, Meng R, Chang Y, Bhalerao R, Huner NPA, Finn CE, Chen THH, Hurry V (2006) The CBF1-dependent low temperature signalling pathway, regulon, and increase in freeze tolerance are conserved in *Populus* spp. Plant Cell Environ 29:1259–1272

Bhullar NK, Street K, Mackay M, Yahiaoui N, Keller B (2009) Unlocking wheat genetic resources for the molecular identification of previously undescribed functional alleles at the *Pm3* resistance locus. Proc Natl Acad Sci U S A 106:9519–9524

Borecký J, Nogueira FTS, de Oliveira KAP, Maia IG, Vercesi AE, Arruda P (2006) The plant energy-dissipating mitochondrial systems: depicting the genomic structure and the expression profiles of the gene families of uncoupling protein and alternative oxidase in monocots and dicots. J Exp Bot 57:849–864

Bourgis F, Guyot R, Gherbi H, Tailliez E, Amabile I, Salse J et al (2008) Characterization of the major fragarnce gene from an aromatic *japonica* rice and analysis of its diversity in Asian cultivated rice. Theor Appl Genet 117:353–368

Boyko A, Kovalchuk I (2008) Epigenetic control of plant stress response. Environ Mol Mutagen 49:61–72

Bradbury LMT, Fitzgerald TL, Henry RJ, Jin Q, Waters DLE (2005a) The gene for fragrance in rice. Plant Biotechnol J 3:363–370

Bradbury LMT, Henry RJ, Jin Q, Reinke RF, Waters DLE (2005b) A perfect marker for fragrance genotyping in rice. Mol Breed 16:279–283

Bradbury LMT, Gillies SA, Brushett DJ, Waters DLE, Henry RJ (2008) Inactivation of an aminoaldehyde dehydrogenase is responsible for fragrance in rice. Plant Mol Biol 68:439–449

Brandalise M, Maia IG, Borecký J, Vercesi AE, Arruda P (2003) Overexpression of plant uncoupling mitochondrial protein in transgenic tobacco increases tolerance to oxidative stress. J Bioenerg Biomembr 35:203–209

Browning KS (1996) The plant translational apparatus. Plant Mol Biol 32:107–144

Bruun-Rasmussen M, Møller I, Tulinius G, Hansen J, Lund O, Johansen I (2007) The same allele of translation initiation factor 4E mediates resistance against two Potyvirus spp. in *Pisum sativum*. Mol Plant Microbe Interact 20:1075–1082

Bryan GT, Wu K-S, Farrall L, Jia Y, Hershey HP, McAdams SA, Donaldson GK, Tarchini R, Valent B (2000) A single amino acid difference distinguishes resistant and susceptible alleles of the rice blast resistance gene *Pi-ta*. Plant Cell 12:203–2046

Caldwell BE, Brim CA, Ross JP (1960) Inheritance of resistance of soybeans to the cyst nematode, *Heterodera glycines*. Agron J 52:635–636

Campos MD, Cardoso H, Linke B, Costa JH, Fernandes de Melo D, Justo L, Frederico AM, Arnholdt-Schmitt B (2009) Differential expression and co-regulation of carrot *AOX* genes (*Daucus carota*). Physiol Plant 137:578–591

Cardoso H, Campos MD, Costa AR, Campos MC, Nothnagel T, Arnholdt-Schmitt B (2009) Carrot alternative oxidase gene *AOX2a* demonstrates allelic and genotypic polymorphisms in intron 3. Physiol Plant 137:592–608

Cardoso H, Campos MD, Nothnagel T, Arnholdt-Schmitt B (2011) Polymorphisms in intron 1 of carrot *AOX2b* – a useful tool to develop a functional marker? Plant Genet Resour: Charact Util 9:177–180

Chai G, Bai Z, Wei F, King GJ, Wang C, Shi L, Dong C, Chen H, Liu S (2010) *Brassica GLABRA2* genes: analysis of function related to seed oil content and development of functional markers. Theor Appl Genet 120:1597–1610

Charron C, Nicolai M, Gallois JL, Robaglia C, Moury B, Palloix A, Caranta C (2008) Natural variation and functional analyses provide evidence for co-evolution between plant eIF4E and potyviral VPg. Plant J 54:56–68

Chen JB, Jing RL, Yuan HY, Wei B, Chang XP (2005) Single nucleotide polymorphism of *TaDREB1* gene in wheat germplasm. Sci Agric Sin 38:2387–2394

Chen S, Yang Y, Shi W, Ji Q, He F, Zhang Z, Cheng Z, Liu X, Xu M (2008) *Badh2*, encoding betaine aldehyde dehydrogenase, inhibits the biosynthesis of 2-Acetyl-1-pyrroline, a major component in rice fragrance. Plant Cell 20:1850–1861

Chen M, Xu Z, Xia L, Li L, Cheng X, Dong J, Wang Q, Ma Y (2009) Cold-induced modulation and functional analyses of the DRE-binding transcription factor gene, *GmDREB3*, in soybean (*Glycine max* L.). J Exp Bot 60:121–135

Chen J, Zhang X, Jing R, Blair MW, Mao X, Wang S (2010) Cloning and genetic diversity analysis of a new *P5CS* gene from common bean (*Phaseolus vulgaris* L). Theor Appl Genet 120:1393–1404

Chinnusamy V, Zhu JK (2009) Epigenetic regulation of stress responses in plants. Curr Opin Plant Biol 12:133–139

Chua YL, Watson LA, Gray JC (2003) The transcriptional enhancer of the pea plastocyanin gene associates with the nuclear matrix and regulates gene expression through histone acetylation. Plant Cell 15:1468–1479

Clifton R, Lister R, Parker KL, Sappl PG, Elhafez D, Millar AH, Day DA, Whelan J (2005) Stress-induced co-expression of alternative respiratory chain components in *Arabidopsis thaliana*. Plant Mol Biol 58:193–212

Clifton R, Millar AH, Whelan J (2006) Alternative oxidases in *Arabidopsis*: a comparative analysis of differential expression in the gene family provides new insights into function of nonphosphorylating bypasses. Biochim Biophys Acta 1757:730–741

Coelho AC, Lima MB, Neves D, Cravador A (2006) Genetic diversity of two evergreen oaks (*Quercus suber* L. and *Q.* (*ilex*) *rotundifolia* Lam.) in Portugal using AFLP markers. Silvae Genet 55:105–118

Concibido VC, Lange DA, Denny RL, Orf JH, Young ND (1997) Genome mapping of soybean cyst nematode resistance genes in 'Peking', PI 90763, and PI 88788 using DNA markers. Crop Sci 37:258–264

Cong B, Liu J, Tanksley SD (2002) Natural alleles at a tomato fruit size quantitative trait locus differ by heterochronic regulatory mutations. Proc Natl Acad Sci U S A 99:13606–13611

Cong L, Chai TY, Zhang YX (2008) Characterization of the novel gene *BjDREB1B* encoding a DRE-binding transcription factor from *Brassica juncea* L. Biochem Biophys Res Commun 371:702–706

Considine MJ, Daley DO, Whelan J (2001) The expression of alternative oxidase and uncoupling protein during fruit ripening in mango. Plant Physiol 126:1619–1629

Costa ADT, Nantes IL, Jezek P, Leite A, Arruda P, Vercesi AE (1999) Plant uncoupling mitochondrial protein activity in mitochondria isolated from tomatoes at different stages of ripening. J Bioenerg Biomembr 31:527–533

Costa JH, Cardoso HC, Campos MD, Zavattieri A, Frederico AM, Fernandes de Melo D, Arnholdt-Schmitt B (2009a) *D. carota* L. – an old model for cell reprogramming gains new importance through a novel expansion pattern of *AOX* genes. Plant Physiol Biochem 47:753–775

Costa JH, Fernandes de Melo D, Gouveia Z, Cardoso HG, Peixe A, Arnholdt-Schmitt B (2009b) The alternative oxidase family of *Vitis vinifera* reveals an attractive model for genomic design. Physiol Plant 137:553–556

Daley DO, Considine MJ, Howell KA, Millar AH, Day DA, Whelan J (2003) Respiratory gene expression in soybean cotyledons during post-germinative development. Plant Mol Biol 51:745–755

Diaz-Pendon JA, Truniger V, Nieto C, Garcia-Mas J, Bendahmane A, Aranda MA (2004) Advances in understanding recessive resistance to plant viruses. Mol Plant Pathol 5:223–233

Dinant S, Lot H (1992) *Lettuce mosaic virus*: a review. Plant Pathol 41:528–542

Doebley J, Gaut BS, Smith BD (2006) The molecular genetics of crop domestication. Cell 127:1309–1321

Doganlar S, Frary A, Daunay MC, Lester RN, Tanksley SD (2002) Conservation of gene function in the *Solanaceae* as revealed by comparative mapping of domestication traits in eggplant. Genetics 161:1713–1726

Dombrowski JE, Baldwin JC, Martin RC (2008) Cloning and characterization of a salt stress-inducible small *GTPase* gene from the model grass species *Lolium temulentum*. J Plant Physiol 165:651–661

Drenkard E, Richter BG, Rozen S et al (2000) A simple procedure for the analysis of single nucleotide polymorphism facilitates map-based cloning in *Arabidopsis*. Plant Physiol 124:1483–1492

Duprat A, Caranta C, Revers F, Menand B, Browning KS, Robaglia C (2002) The *Arabidopsis* eukaryotic initiation factor (iso)4E is dispensable for plant growth but required for susceptibility to potyviruses. Plant J 32:927–934

Fan C, Xing Y, Mao H, Lu T, Han B, Xu C, Zhang XLQ (2006) *GS3*, a major QTL for grain length and weight and minor QTL for grain width and thickness in rice, encodes a putative transmembrane protein. Theor Appl Genet 112:1164–1171

Fan C, Yu S, Wang C, Xing Y (2009) A causal C–A mutation in the second exon of *GS3* highly associated with rice grain length and validated as a functional marker. Theor Appl Genet 118:465–472

Fehér A (2005) Why somatic plant cells start to form embryos? In: Mujib A, Samaj J (eds) Somatic embryogenesis, vol 2, Plant cell monographs. Springer, Heidelberg, pp 85–101

Feldman CR, Brodie ED Jr, Pfrender ME (2009) The evolutionary origins of beneficial alleles during the repeated adaptation of garter snakes to deadly prey. Proc Natl Acad Sci U S A 106:13415–13420

Fernie AR, Carrari F, Sweetlove LJ (2004) Respiration: glycolysis, the TCA cycle and the electron transport chain. Curr Opin Plant Biol 7:254–261

Ferreira A, Cardoso H, Macedo ES, Breviario D, Arnholdt-Schmitt B (2009) Intron polymorphism pattern in *AOX1b* of wild St John's Wort (*Hypericum perforatum* L) allows discrimination between individual plants. Physiol Plant 137:520–531

Fiorani F, Umbach AL, Siedow J (2005) The alternative oxidase of plant mitochondrial is involved in the acclimation of shoot growth at low temperature. A study of Arabidopsis *AOX1a* transgenic plants. Plant Physiol 139:1795–1805

Fowler S, Thomashow MF (2002) Arabidopsis transcriptome profiling indicates that multiple regulatory pathways are activated during cold acclimation in addition to the CBF cold response pathway. Plant Cell 14:1675–1690

Fowler DB, Limin AE, Wang S, Ward RW (1996) Relationship between low-temperature tolerance and vernalization response in wheat and rye. Can J Plant Sci 76:37–42

Fransz P, De Jong H (2011) From nucleosome to chromosome: a dynamic organization of genetic information. Plant J 66:4–17

Frary A, Nesbitt TC, Frary A et al (2000) *fw2.2*: a quantitative trait locus key to the evolution of tomato fruit size. Science 289:85–88

Fraser RSS (1992) The genetics of plant-virus interactions: implications for plant breeding. Euphytica 63:175–185

Frederico AM, Campos MD, Cardoso HCG, Imani J, Arnholdt-Schmitt B (2009a) Alternative oxidase involvement in *Daucus carota* L. somatic embryogenesis. Physiol Plant 137:498–508

Frederico AM, Zavattieri MA, Campos MD, Cardoso H, McDonald AE, Arnholdt-Schmitt B (2009b) The gymnosperm *Pinus pinea* contains both AOX gene subfamilies, *AOX1* and *AOX2*. Physiol Plant 137:566–577

Fricano A, Rizza F, Faccioli P, Pagani D, Pavan P, Stella A, Rossini L, Piffanelli P, Cattivelli L (2009) Genetic variants of *HvCbf14* are statistically associated with frost tolerance in a European germplasm collection of *Hordeum vulgare*. Theor Appl Genet 119:1335–1348

Fushimi T, Masuda R (2001) 2-Acetyl-1-pyrroline concentration of the aromatic vegetable soybean "Dadacha-Mame". In: Proceedings of second international vegetable soybean conference, Washington State University, Tacoma, Washington, p 39

Gao ZH, Johansen E, Eyers S, Thomas CL, Noel Ellis TH, Maule AJ (2004) The potyvirus recessive resistance gene, *sbm1*, identifies a novel role for translation initiation factor *eIF4E* in cell-to-cell trafficking. Plant J 40:376–385

Garg AK, Kim J-K, Owens TG, Ranwala AP, Choi YD, Kochian LV, Wu RJ (2002) Trehalose accumulation in rice plants confers high tolerance levels to different abiotic stresses. Proc Natl Acad Sci U S A 99:15898–15903

Gegas VC, Nazari A, Griffiths S, Simmonds J, Fish L, Orford S, Sayers L, Doonan JH, Snapea JW (2010) A genetic framework for grain size and shape variation in wheat C W. Plant Cell 22:1046–1056

Gendall AR, Levy YY, Wilson A, Dean C (2001) The *VERNALIZATION 2* gene mediates the epigenetic regulation of vernalization in *Arabidopsis*. Cell 107:525–535

German-Retana S, Walter J, Doublet B, Roudet-Tavert G, Nicaise V, Lecampion C, Houvenaghel M-C, Robaglia C, Michon T, Le Gall O (2008) Mutational analysis of plant cap-binding protein eIF4E reveals key amino acids involved in biochemical functions and potyvirus infection. J Virol 82:7601–7612

Giraud E, Ho LH, Clifton R, Carroll A, Estavillo G, Tan YF, Howell KA, Ivanova A, Pogson BJ, Millar AH, Whelan J (2008) The Absence of ALTERNATIVE OXIDASE1a in *Arabidopsis* results in acute sensitivity to combined light and drought stress. Plant Physiol 147:595–610

Grandbastien M-A (1998) Activation of plant retrotransposons under stress conditions. Trends Plant Sci 3:181–187

Grandillo S, Ku HM, Tanksley SD (1999) Identifying loci responsible for natural variation in fruit size and shape in tomato. Theor Appl Genet 99:978–987

Gregory TR (2004) Insertion-deletion bases and the evolution of genome size. Gene 423:15–34

Gunstone FD (2001) Soybeans pace boost in oilseed production. Inform 11:1287–1289

Haake V, Cook D, Riechmann JL, Pineda O, Thomashow MF, Zhang JZ (2002) Transcription factor *CBF4* is a regulator of drought adaptation in *Arabidopsis*. Plant Physiol 130:639–648

Hamann A, Tafel J, Busing B, Munzberg H, Hinney A, Mayer H, Siegfield W, Ricquier D, Greten H, Matthaei JH (1998) Analysis of the uncoupling protein-1 (*UCP1*) gene obese and lean subjects: identification of four amino acid variant. Int J Obes Relat Metab Disord 22:939–941

Hansen LD, Church JN, Matheson S, McCarlie VW, Thygerson T, Criddle RS, Smith BN (2002) Kinetics of plant growth and metabolism. Thermochim Acta 388:415–425

Hauge BM, Wang ML, Parsons JD, Parnell LD (2001) Nucleic acid molecules and other molecules associated with soybean cyst nematode resistance. WO 01/51627 PCT/US01/00552 Patent # 20030005491

Hayashi K, Yoshida H (2009) Refunctionalization of the ancient rice blast disease resistance gene Pit by the recruitment of a retrotransposon as a promoter. Plant J 57:413–425

Hayashi K, Yasuda N, Fujita Y, Koizumi S, Yoshida H (2010) Identification of the blast resistance gene *Pit* in rice cultivars using functional markers. Theor Appl Genet 121:1357–1367

Heilbronn LK, Kind KL, Pancewicz E, Morris AM, Noakes M, Clifton PM (2000) Association of −3826G variant in uncoupling protein-1 with increased BMI in overweight Australian women. Diabetologia 43:242–244

Herrmann SM, Wang JG, Staessen JA, Kertmen E, Schmidt-Petersen K, Zidek W, Paul M, Brand E (2003) Uncouplin protein 1 and 3 polymorphisms are associated with waist-to-hratio. J Mol Med 81:327–332

Hilal M, Castagnaro AP, Moreno H, Massa EM (1997) Specific localization of the respiratory alternative oxidase in meristematic and xylematic tissues from developing soybean roots and hypocotyls. Plant Physiol 115:1499–1503

Ho LHM, Giraud E, Lister R, Thirkettle-Watts D, Low J, Clifton R, Howell KA, Carrie C, Donald T, Whelan J (2007) Characterization of the regulatory and expression context of an alternative oxidase gene provides insights into cyanide-insensitive respiration during growth and development. Plant Physiol 143:1519–1533

Holtzapffel RC, Castelli J, Finnegan PM, Millar AH, Whelan J, Day DA (2003) A tomato alternative oxidase protein with altered regulatory properties. Biochim Biophys Acta 1606:153–162

Hong Z, Lakkineni K, Zhang Z, Verma DPS (2000) Removal of feedback inhibition of pyrroline-5-carboxylase synthetase (*P5CS*) results in increased proline accumulation and protection of plants from osmotic stress. Plant Physiol 122:1129–1139

Horie T, Hauser F, Schroeder JI (2009) HKT transporter-mediated salinity resistance mechanisms in *Arabidopsis* and monocot crop plants. Trends Plant Sci 14:660–668

Hu C-AA, Delauney AJ, Verma DPS (1992) A bifunctional enzyme (Δ^1-pyrroline-5-carboxylate synthetase) catalyzes the first two steps in proline biosynthesis in plants. Proc Natl Acad Sci U S A 89:9354–9358

Hunt LT, Barker WC (1987) Von willebrand factor shares a distinctive cysteine-rich domain with thrombospondin and procollagen. Biochem Biophys Res Commun 144:876–882

Igarashi Y, Yoshiba Y, Sanada Y, Yamaguchi-Shinozaki K, Wada K, Shinozaki K (1997) Characterization of the gene for D1-pyrroline-5-carboxylate synthetase and correlation between the expression of the gene and salt tolerance in *Oryza sativa* L. Plant Mol Biol 33:857–865

Ingvardsen CR, Schejbel B, Lübberstedt T (2008) Functional markers for resistance breeding. In: Lüttge U, Beyschlag W, Murata J (eds) Progress in botany, vol 69, part 2. Springer, Berlin, pp 61–87

Jaillon O, Bouhouche K, Gout JF et al (2008) Translational control of intron splicing in eukaryotes. Nature 451:359–362

Jeddeloh JA, Stokes TL, Richards EJ (1999) Maintenance of genomic methylation requires a SWI2/SNF2-like protein. Nat Genet 22:94–97

Jia Y, Wang Z, Singh P (2002) Development of dominant rice blast resistance *Pi-ta* gene markers. Crop Sci 42:2145–2149

Jia Y, Bryan GT, Farrall L, Valent B (2003) Natural variation at the *Pi-ta* rice blast resistance locus. Phytopathology 93:1452–1459

Jia Y, Redus M, Wang Z, Rutger JN (2004) Development of a SNLP marker from the *Pi-ta* blast resistance gene by tri-primer PCR. Euphytica 138:97–105

Jin L, Lu Y, Shao Y, Zhang G, Xiao P, Shen S, Corke H, Bao J (2010) Molecular marker assisted selection for improvement of the eating, cooking and sensory quality of rice (*Oryza sativa* L.). J Cereal Sci 51:159–164

Jungk A (2001) Root hair and the acquisition of plant nutrients from soil. J Plant Nutr Soil Sci 164:121–129

Juwattanasomran R, Somta P, Chankae S, Shimizu T, Wongpornchai S, Kaga A, Srinives P (2011) A SNP in *GmBADH2* gene associates with fragrancein vegetable soybean variety "Kaori" and SNAP marker development for the fragrante. Theor Appl Genet 122:533–541

Juwattanasomran R, Somta P, Chankaew S, Shimizu T, Wongpornchai S, Kaga A, Srinives P (2012) Identification of a new fragrance allele in soybean and development of its functional marker. Mol Breed. doi:10.1007/s11032-010-9523-0

Kang BC, Yeam I, Frantz JD, Murphy JF, Jahn MM (2005a) The *pvr1* locus in *Capsicum* encodes a translation initiation factor *eIF4E* that interacts with Tobacco etch virus *VPg*. Plant J 42:392–405

Kang BC, Yeam I, Jahn MM (2005b) Genetics of plant virus resistance. Annu Rev Phytopathol 43:581–621

Kang BC, Yeam I, Li H, Perez KW, Jahn MM (2007) Ectopic expression of a recessive resistance gene generates dominant potyvirus resistance in plants. Plant Biotechnol J 5:526–536

Kanyuka K, Druka A, Caldwell DG, Tymon A, McCallum N, Waugh R, Adams MJ (2005) Evidence that the recessive bymoirus resistance locus *rym4* in barley corresponds to the eukaryotic translation initiation factor 4E gene. Mol Plant Pathol 6:449–458

Kiec-Wilk B, Wybranska I, Malczewska-Malec M, LeszczynnkaGolabek I, Partyka Q, Niedbal S, Jabrpcka A, Dembinska-Kiec A (2002) Correlation of the −3826 A → G polymorphism in the promoter of the uncoupling protein 1 gene with obesity and metabolic disorders in obese families from southern Poland. J Physiol Pharmacol 53:477–490

Kikuchi A, Sanuki N, Higashi K, Koshiba T, Kamada H (2006) Abscisic acid and stress treatment are essential for the acquisition of embryogenic competence by carrot somatic cells. Planta 223:637–645

Kim KS, Cho D, Kim YJ, Choi SM, Kim JY, Shin SU, Yoon YS (2005) The finding of new genetic polymorphism of UCP-1 A-1766G and its effects on body fat accumulation. Biochim Biophys Acta 1741:149–155

Kitashiba H, Ishizaka T, Isuzugawa K, Nishimura K, Suzuki T (2004) Expression of a sweet cherry *DREB1/CBF* ortholog in *Arabidopsis* confers salt and freezing tolerance. J Plant Physiol 161:1171–1176

Kotani K, Sakane N, Saiga K, Adachi S, Shimohiro H, Mu H, Kurozawa Y (2008) Relationship between A-3826G polymorphism in the promoter of the uncoupling protein-1 gene and high-density lipoprotein cholesterol in Japanese individuals: a cross-sectional study. Arch Med Res 39(1):142–146

Kovach MJ, Calingacion MN, Fitzgerald MA, McCouch SR (2009) The origin and evolution of fragrance in rice (*Oryza sativa* L.). Proc Natl Acad Sci U S A 106:14444–14449

Krattinger SG, Lagudah ES, Spielmeyer W, Singh RP, Huerta-Espino J, McFadden H, Bossolini E, Selter LL, Keller B (2009) A putative ABC transporter confers durable resistance to multiple fungal pathogens in wheat. Science 323:1360–1363

Lagudah ES, Krattinger SG, Herrera-Foessel S, Singh RP, Huerta-Espino J, Spielmeyer W, Brown-Guedira G, Selter LL, Keller B (2009) Gene-specific markers for the wheat gene *Lr34/Yr18/Pm38* which confers resistance to multiple fungal pathogens. Theor Appl Genet 119:889–898

Lata C, Bhutty S, Bahadur RP, Majee M, Prasad M (2011) Association of an SNP in a novel DREB2-like gene *SiDREB2* with stress tolerance in foxtail millet (*Setaria italica* L.). J Exp Bot 62:3387–3401

Leach JG (1933) A destructive Fusarium wilt of muskmelon. Phytopathology 23:554–556

Li SC, Shiau CK, Lin WC (2007) Vir-Mir db: prediction of viral microRNA candidate hairpins. Nucleic Acids Res 36:D184–D189

Li Y, Zheng L, Corke F, Smith C, Bevan MW (2008) Control of final seed and organ size by the *DA1* gene family in *Arabidopsis thaliana*. Genes Dev 22:1331–1336

Li YH, Zhang C, Gao ZS, Smulders MJM, Ma Z, Liu ZX, Nan HY, Chang RZ, Qiu LJ (2009) Development of SNP markers and haplotype analysis of the candidate gene for *rhg1*, which confers resistance to soybean cyst nematode in soybean. Mol Breed 24:63–76

Li Q, Li L, Yang X, Warburton ML, Bai G, Dai J, Li J, Yan J (2010a) Relationship, evolutionary fate and function of two maize co-orthologs of rice *GW2* associated with kernel size and weight. BMC Plant Biol 10:143

Li Q, Yang X, Bai G, Warburton ML, Mahuku G, Gore M, Dai J, Li J, Yan J (2010b) Cloning and characterization of a putative *GS3* ortholog involved in maize kernel development. Theor Appl Genet 120:753–763

Liang SS, Suenaga K, He ZH, Wang ZL, Liu HY, Wang DS, Singh RP, Sourdille P, Xia XC (2006) Quantitative trait loci mapping for adult-plant resistance to powdery mildew in bread wheat. Phytopathology 96:784–789

Ling KS, Harris KR, Meyer JDF, Levi A, Guner N, Wehner TC, Bendahmane A, Havey MJ (2009) Non-synonymous single nucleotide polymorphisms in the watermelon *eIF4E* gene are closely associated with resistance to *Zucchini yellow mosaic virus*. Theor Appl Genet 120: 191–200

Liu Q, Kasuga M, Sakuma Y, Abe H, Miura S, Yamaguchi-Shinozaki K, Shinozaki K (1998) Two transcription factors, *DREB1* and *DREB2*, with an *EREBP/AP2* DNA binding domain separate two cellular signal transduction pathways in drought- and low-temperature-responsive gene expression, respectively, in *Arabidopsis*. Plant Cell 10:1391–1406

Liu J, Van Eck J, Cong B, Tanksley SD (2002) A new class of regulatory genes underlying the cause of pear-shaped tomato fruit. Proc Natl Acad Sci U S A 99:13302–13306

Lorieux M, Petrov M, Huang N, Guiderdoni E, Ghesquiere A (1996) Aroma in rice: genetic analysis of a quantitative trait. Theor Appl Genet 93:1145–1151

Ma SQ, Xu Y, Gong GY, Zhang HY, Shen HL (2005) Analysis on the inheritance to PRSV-W and ZYMV-CH and their linkage in watermelon. J Fruit Sci 22:731–733

Macedo ES, Cardoso HCG, Hernandez A, Peixe AA, Polidoros A, Ferreira A, Cordeiro A, Arnholdt-Schmitt B (2009) Physiological responses and gene diversity indicate olive alternative oxidase as a potential source for markers involved in efficient adventitious root induction. Physiol Plant 137:532–552

Macedo ES, Sircar D, Cardoso HG, Peixe A, Arnholdt-Schmitt A (2012) Involvement of alternative oxidase (AOX) in adventitious rooting of *Olea europaea* L microshoots is linked to adaptive phenylpropanoid and lignin metabolism. Plant Cell Rep 31(9):1581–1590. doi:10.1007/s00299-012-1272-6

Mao H, Suna S, Yao J, Wang C, Yu S, Xu C, Li X, Zhanga Q (2010) Linking differential domain functions of the GS3 protein to natural variation of grain size in rice. Proc Natl Acad Sci U S A 107:19579–19584

Marcotrigiano J, Gingras A-C, Sonenberg N, Burley SK (1997) Cocrystal structure of the messenger RNA 5′ cap-binding protein (eIF4E) bound to 7-methyl-GDP. Cell 89:951–961

Martin GB, Bogdanove AJ, Sessa G (2003) Understanding the functions of plant disease resistance proteins. Annu Rev Plant Biol 54:23–61

Martin A, Lee J, Kichey T et al (2006) Two cytosolic glutamine synthetase isoforms of maize are specifically involved in the control of grain production. Plant Cell 18:3252–3274

Matson AL, Williams LF (1965) Evidence of a fourth gene for resistance to the soybean cyst nematode. Crop Sci 5:477

Maule A, Leh V, Lederer C (2002) The dialogue between viruses and hosts in compatible interactions. Curr Opin Plant Biol 5:279–284

Maule A, Caranta C, Boulton M (2007) Sources of natural resistance to plant viruses: status and prospects. Mol Plant Pathol 8:223–231

Maxwell JJ (2008) Genetic characterization and mapping of wheat powdery mildew resistance genes from different wheat germplasm sources. PhD thesis, North Carolina State University, pp 138

McDonald AE, Vanlerberghe GC (2006) The organization and control of plant mitochondrial metabolism. In: Plaxton WC, McManus MT (eds) Control of primary metabolism in plants, vol 22, Annual plant reviews. Blackwell Publishing, Oxford, pp 290–324

McMullen MD, Jones MW, Simcox KD, Louie R (1994) 3 genetic loci control resistance to *Wheat streak mosaic virus* in the maize inbred Pa405. Mol Plant Microbe Interact 7:708–712

Meyers BC, Dickerman AW, Michelmore RW, Sivaramakrishnan S, Sobral BW, Young ND (1999) Plant disease resistance genes encode members of an ancient and diverse protein family within the nucleotide-binding superfamily. Plant J 20:317–332

Millenaar FF, Gonzàlez-Meler MA, Fiorani F, Welschen R, Ribas-Carbo M, Siedow JN, Wagner AM, Lambers H (2001) Regulation of the alternative oxidase activity in six wild monocotyledonous species; an in vivo study at the whole root level. Plant Physiol 126:376–387

Morran S, Eini O, Pyvovarenko T et al (2011) Improvement of stress tolerance of wheat and barley by modulation of expression of *DREB/CBF* factors. Plant Biotechnol J 9:230–249

Munns R, Tester M (2008) Mechanisms of salinity tolerance. Annu Rev Plant Biol 59:651–681

Murphy JF, Blauth JR, Livingstone KD, Lackney VK, Jahn MM (1998) Genetic mapping of the *pvr1* locus in *Capsicum* spp. and evidence that distinct potyvirus resistance loci control responses that differ at the whole plant and cellular levels. Mol Plant Microbe Interact 11:943–951

Naderpour M, Lund OS, Larsen R, Johansen E (2010) Potyviral resistance derived from cultivars of *Phaseolus vulgaris* carrying *bc-3* is associated with the homozygotic presence of a mutated *eIF4E* allele. Mol Plant Pathol 11:255–263

Nasu S, Suzuki J, Ohta R, Hasegawa K, Yui R, Kitazawa N, Monna L, Minobe Y (2002) Search for and analysis of single nucleotide polymorphisms (SNPs) in rice (*Oryza sativa*, *Oryza rufipogon*) and establishment of SNP markers. DNA Res 9:163–171

Nesbitt TC, Tanksley SD (2002) Comparative sequencing in the genus *Lycopersicon*: implication for the evolution of fruit size in the domestication of cultivated tomatoes. Genetics 162:365–379

Nicaise V, German-Retana S, Sanjuan R, Dubrana MP, Mazier M, Maisonneuve B, Candresse T, Caranta C, LeGall O (2003) The eukaryotic translation initiation factor 4E controls lettuce susceptibility to the potyvirus *Lettuce mosaic virus*. Plant Physiol 132:1272–1282

Nieto C, Morales M, Orjeda G et al (2006) An *eIF4E* allele confers resistance to an uncapped and non-polyadenylated RNA virus in melon. Plant J 48:452–462

Niu X, Tang W, Huang W et al (2008) RNAi-directed down regulation of *OsBADH2* results in aroma (2-acetyl-1-pyrroline) production in rice (*Oryza sativa* L.). BMC Plant Biol 8:1–10

Noctor G, Paepe DR, Foyer CH (2007) Mitochondrial redox biology and homeostasis in plants. Trends Plant Sci 12:125–134

Nogales A, Muñoz L, Hansen L, Arnholdt-Schmitt (2012) Calorespirometry as a tool for the screening of carrot AOX polymorphic genotypes with different growth responses to temperature. Abstract book of XVII international Society for Biological Calorimetry (ISBC) Conference. Leipzig, Germany, p 48

Nogueira FTS, Borecký J, Vercesi AE, Arruda P (2005) Genomic structure and regulation of mitochondrial uncoupling protein genes in mammals and plants. Biosci Rep 35:209–226

Nogueira FTS, Sassaki FT, Maia IG (2011) *Arabidopsis thaliana* uncoupling proteins (AtUCPs): insights into gene expression during development and stress response and epigenetic regulation. J Bioenerg Biomembr 43:71–79

Oh SJ, Kwon CW, Choi DW, Song SI, Kim JK (2007) Expression of barley *HvCBF4* enhances tolerance to abiotic stress in transgenic rice. Plant Biotechnol J 5:646–656

Okamoto H, Hirochika H (2001) Silencing of transposable elements in plants. Trends Plant Sci 6:527–534

Okuley J, Lightner J, Feldmann K, Yadav N, Lark E (1994) *Arabidopsis FAD2* gene encodes the enzyme that is essential for polyunsaturated lipid synthesis. Plant Cell 6:147–158

O'Leary JM, Hamilton JM, Deane CM, Valeyev NV, Sandell LJ, Downing AK (2004) Solution structure and dynamics of a prototypical Chordin-like cysteine-rich repeat (von Willebrand factor type C module) from collagen IIA. J Biol Chem 279:53857–53866

Pasternak T, Prinsen E, Ayaydin F, Miskolczi P, Potters G, Asard H, van Onckelen H, Dudits D, Fehér A (2002) The role of auxin, pH and stress in the activation of embryogenic cell division in leaf protoplast-derived cells of alfalfa (*Medicago sativa* L.). Plant Physiol 129:1807–1819

Peixe A, Raposo A, Lourenço R, Cardoso H, Macedo E (2007) Coconut water and BAP successfully replaced zeatin in olive (*Olea europaea* L.) micropropagation. Sci Hortic 113:1–7

Plaxton WC, Podestá F (2006) The functional organization and control of plant respiration. Crit Rev Plant Sci 25:159–198

Polidoros AN, Mylona PV, Arnholdt-Schmitt B (2009) *AOX* gene structure, transcript variation and expression in plants. Physiol Plant 137:342–353

Popov VN, Simonian RA, Skulachev VP, Starkov AA (1997) Inhibition of the alternative oxidase stimulates H_2O_2 production in plant mitochondria. FEBS Lett 415:87–90

Potters G, Pasternak TP, Guisez Y, Palme KJ, Jansen MAK (2007) Stress-induced morphogenic responses: growing out of trouble? Trends Plant Sci 12:98–105

Qiu L, Wu D, Ali S, Cai S, Dai F, Jin X, Wu F, Zhang G (2011) Evaluation of salinity tolerance and analysis of allelic function of *HvHKT1* and *HvHKT2* in Tibetan wild barley. Theor Appl Genet 122:695–703

Quraishi UM, Abrouk M, Bolot S, Pont C, Throude M, Guilhot N, Confolent C, Bortolini PS, Murigneux A, Charmet G, Salse J (2009) Genomics in cereals: From genome-wide conserved orthologous set (COS) sequences to candidate genes for trait dissection. Funct Integr Genomics 9:473–484

Raghavendra AS, Padmasree K (2003) Beneficial interactions of mitochondrialmetabolism with photosynthetic carbon assimilation. Trends Plant Sci 8:1360–1385

Rains DW, Epstein E (1965) Transport of sodium in plant tissue. Science 148:1611

Ramensky V, Bork P, Sunyaev S (2002) Human non-synonymous SNPs: server and survey. Nucleic Acids Res 30:3894–3900

Ramkumar G, Sivaranjani AKP, Pandey MK et al (2010) Development of a PCR-based SNP marker system for effective selection of kernel length and kernel elongation in rice. Mol Breed 26:735–740

Ramkumar G, Srinivasarao K, Mohan K et al (2011) Development and validation of functional marker targeting an InDel in the major rice blast disease resistance gene *Pi54* (*Pik^h*). Mol Breed 27:129–135

Rasmusson AG, Fernie AR, van Dongen JT (2009) Alternative oxidase: a defence against metabolic fluctuations? Physiol Plant 137:371–382

Revers F, Lot H, Souche S, Candresse T, Dunez J, Gall LO (1997) Biological and molecular variability of Lettuce mosaic virus isolates. Phytopathology 87:397–403

Rhoads DM, Subbaiah CC (2007) Mitochondrial retrograde regulation in plants. Mitochondrion 7:177–194

Risch N (2000) Searching for genetic determinants in the new millennium. Nature 405:847–856

Risser G, Banihashemi Z, Davis DW (1976) A proposed nomenclature of *Fusarium oxysporum* f.sp. *melonis* races and resistance genes in *Cucumis melo*. Phytopathology 66:1105–1106

Rivandi J, Miyazaki J, Hrmova M, Pallotta M, Tester M, Collins NC (2011) A *SOS3* homologue maps to *HvNax4*, a barley locus controlling an environmentally sensitive Na + exclusion trait. J Exp Bot 62:1201–1216

Robaglia C, Caranta C (2006) Translation initiation factors: a weak link in plant RNA virus infection. Trends Plant Sci 11:40–45

Robbins J, Dilworth SM, Laskey RA, Dingwall C (1991) Two interdependent basic domains in nucleoplasmin nuclear targeting sequence: identification of a class of bipartite nuclear targeting sequence. Cell 64:615–623

Rokas A, Carroll SB (2008) Frequent an widespread parallel evolution of protein sequences. Mol Biol Evol 25:1943–1953

Ruben E, Jamai A, Afzal J et al (2006) Genomic analysis of the 'Peking' *rhg1* locus: candidate genes that underlie soybean resistance to the cyst nematode. Mol Genet Genome 276:320–330

Ruffel S, Dussault MH, Palloix A, Moury B, Bendahmane A, Robaglia C, Caranta C (2002) A natural recessive resistance gene against *Potato virus Y* in pepper corresponds to the eukaryotic initiation factor 4E (eIF4E). Plant J 32:1067–1075

Ruffel S, Gallois J, Lesage M, Caranta C (2005) The recessive potyvirus resistance gene *pot-1* is the tomato orthologue of the pepper *pvr2-eIF4E* gene. Mol Genet Genomics 274:346–353

Rundle HD, Nagel L, Boughman JW, Schluter D (2000) Natural selection and parallel speciation in sympatric sticklebacks. Science 287:306–308

Sakamoto T, Matsuoka M (2008) Identifying and exploiting grain yield genes in rice. Curr Opin Plant Biol 11:209–214

Sakthivel K, Shobha RN, Pandey MK et al (2009a) Development of a simple functional marker for fragrance in rice and its validation in Indian Basmati and non-Basmati fragrant rice varieties. Mol Breed 24:185–190

Sakthivel K, Sundaram RM, Rani N, Balachandran SM, Neeraja CN (2009b) Genetic and molecular basis of fragrance in rice. Biotechnol Adv 27:468–473

Sakuma Y, Liu Q, Dubouzet JG, Abe H, Shinozaki K, Yamaguchi-Shinozaki K (2002) DNA-binding specificity of the ERF/AP2 domain of *Arabidopsis DREBs*, transcription factors involved in dehydration- and cold-inducible gene expression. Biochem Biophys Res Commun 290:998–1009

Schnurbusch T, Bossolini E, Messmer M, Keller B (2004) Tagging and validation of a major quantitative trait locus for leaf rust resistance and leaf tip necrosis in winter wheat cultivar forno. Phytopathology 94:1036–1041

Shaposhnikov AS, Akopov SB, Chernov IP, Thomsen PD, Joergensen C, Collins AR, Frengen E, Nikolaev LG (2007) A map of nuclear matrix attachment regions within the breast cancer loss-of-heterozygosity region on human chromosome 16q22.1. Genomics 89:354–361

Sharma TR, Madhav MS, Singh BK et al (2005) High resolution mapping, cloning and molecular characterization of the *Pi-kh* gene of rice, which confers resistance to *M. grisea*. Mol Genet Genomics 274:569–578

Sheahan MB, McCurdy DW, Rose RJ (2005) Mitochondria as a connected population: ensuring continuity of the mitochondrial genome during plant cell dedifferentiation through massive mitochondrial fusion. Plant J 44:744–755

Shi H, Lee BH, Wu SJ, Zhu JK (2003) Overexpression of a plasma membrane Na+/H + antiporter gene improves salt tolerance in *Arabidopsis thaliana*. Nat Biotechnol 21:81–85

Shi WW, Yang Y, Chen SH, Xu ML (2008) Discovery of a new fragrance allele and the development of functional markers for the breeding of fragrant rice varieties. Mol Breed 22:185–192

Shomura A, Izawa T, Ebana K, Ebitani T, Kanegae H, Konishi S, Yano M (2008) Deletion in a gene associated with grain size increased yields during rice domestication. Nat Publ Group 40:1023–1028

Singh RP (1992) Genetic association of leaf rust resistance gene *Lr34* with adult plant resistance to stripe rust in bread wheat. Phytopathology 82:835–838

Singh A, Singh PK, Singh R et al (2010) SNP haplotypes of the BADH1 gene and their association with aroma in rice (*Oryza sativa* L.). Mol Breed 26:325–338

Skinner JS, von Zitzewitz J, Szucs P et al (2005) Structural, functional, and phylogenetic characterization of a large *CBF* gene family in barley. Plant Mol Biol 59:533–551

Smýkal P, Safářová D, Navrátil M, Dostalová R (2010) Marker assisted pea breeding: *eIF4E* allele specific markers to pea seed-borne mosaic virus (*PSbMV*) resistance. Mol Breed 26: 425–438

Song XJ, Huang W, Shi M, Zhu MZ, Lin HX (2007) A QTL for rice grain width and weight encodes a previously unknown RING-type E3 ubiquitin ligase. Nat Genet 39:623–630

Spielmeyer W, McIntosh RA, Kolmer J, Lagudah ES (2005) Powdery mildew resistance and *Lr34/Yr18* genes for durable resistance to leaf and stripe rust cosegregate at a locus on the short arm of chromosome 7D of wheat. Theor Appl Genet 111:731–735

Spielmeyer W, Singh RP, McFadden H, Wellings CR, Huerta-Espino J, Kong X, Appels R, Lagudah ES (2008) Fine scale genetic and physical mapping using interstitial deletion mutants of *Lr34/Yr18*: a disease resistance locus efeective against multiple pathogens in wheat. Theor Appl Genet 116:481–490

Sprang SR (1997) G proteins, effectors and GAPs: structure and mechanism. Curr Opin Struct Biol 7:849–856

Statistics Department, Ministry of Agriculture, Forestry and Fisheries (2009) Statistics on Production and Shipment of Vegetables Heisei 19. ISBN 4541036215

Stein N, Perovic D, Kumlehn J, Pellio B, Stracke S, Streng S, Ordon F, Graner A (2005) The eukaryotic translation initiation factor 4E confers multiallelic recessive Bymovirus resistance in *Hordeum vulgare* (L.). Plant J 42:912–922

Steward N, Ito M, Yamaguchi Y, Koizumi N, Sano H (2002) Periodic DNA methylation in maize nucleosomes and demethylation by environmental stress. J Biol Chem 277:37741–37746

Strizhov N, Abraham E, Okresz L, Blickling S, Zilberstein A, Schell J, Koncz C, Szabados L (1997) Differential expression of two *P5CS* genes controlling proline accumulation during saltstress requires ABA and is regulated by *ABA1*, *ABI1* and *AXR2* in *Arabidopsis*. Plant J 12:557–569

Strudwick S, Borden KL (2002) The emerging roles of translation factor eIF4E in the nucleus. Differentiation 70:10–22

Su Z, Hao C, Wang L, Dong Y, Zhang X (2011) Identification and development of a functional marker of TaGW2 associated with grain weight in bread wheat (*Triticum aestivum* L.). Theor Appl Genet 122:211–223

Sun SH, Gao FY, Lu XJ, Wu XJ, Wang XD, Ren GJ, Luo H (2008) Genetic analysis and gene fine mapping of aroma in rice (*Oryza sativa* L. Cyperrales, Poaceae). Genet Mol Biol 31: 532–538

Sun W, Xu X, Zhu H, Liu A, Liu L, Li J, Hua X (2010) Comparative transcriptomic profi ling of a salt-tolerant wild tomato species and a salt-sensitive tomato cultivar. Plant Cell Physiol 51:997–1006

Sweetlove LJ, Lytovchenko A, Morgan M, Nunes-Nesi A, Taylor NL, Baxter CJ, Eickmeier I, Fernie AR (2006) Mitochondrial uncoupling protein is required for efficient photosynthesis. Proc Natl Acad Sci U S A 103:19587–19592

Sweetlove LJ, Fait A, Nunes-Nesi A, Williams T, Fernie AR (2007) The mitochondrion: an integration point of cellular metabolism and signalling. Crit Rev Plant Sci 26:17–43

Takano-Kai N, Jiang H, Kubo T et al (2009) Evolutionary history of *GS3*, a gene conferring grain length in rice. Genetics 182:1323–1334

Takumi S, Shimamura C, Kobayashi F (2008) Increased freezing tolerance through up-regulation of downstream genes via the wheat CBF gene in transgenic tobacco. Plant Physiol Biochem 46:205–211

Tanhuanpää P, Vilkki J, Vihine M (1998) Mapping and cloning of *FAD2* gene to develop allele-specific PCR for oleic acid in spring turnip rape (*Brassica rapa* ssp. *oleifera*). Mol Breed 4:543–550

Tanksley SD (2004) The genetic, developmental, and molecular bases of fruit size and shape variation in tomato. Plant Cell 16:181–189

Tester M, Davenport R (2003) Na^+ tolerance and Na^+ transport in higher plants. Ann Bot 91:503–527

Thornsberry J, Goodman MM, Doebley JF, Kresovich S, Nielsen D, Buckler E (2001) Dwarf8 polymorphisms associate with variation in flowering time. Nat Genet 28:286–289

Tommasini L, Yahiaoui N, Srichumpa P, Keller B (2006) Development of functional markers specific for seven *Pm3* resistance alleles and their validation in the bread wheat gene pool. Theor Appl Genet 114:165–175

Tonsor SJ, Alonso-Blanco C, Koornneeff M (2005) Gene function beyond the single trait: natural variation, gene effects and evolutionary ecology in *Arabidopsis thaliana*. Plant Cell Environ 28:2–20

Tsaftaris A, Polidoros A, Koumproglou R, Tani E, Kovacevic NM, Abatzidou E (2005) Epigenetic mechanisms in plants and their implications in plant breeding. In: Tuberosa R, Philips R, Gale M (eds) In the wake of the double helix: From the green revolution to the gene revolution. Avenue Media, Bologna, pp 157–171

Umbach AL, Gonzalez-Meler MA, Sweet CR, Siedow JN (2002) Activation of the plant mitochondrial alternative oxidase: insights from site-directed mutagenesis. Biochim Biophys Acta 1554:118–128

Umbach AL, Lacey EP, Richter SJ (2009) Temperature-sensitive alternative oxidase protein content and its relationship to floral reflectance in natural *Plantago lanceolata* populations. New Phytol 181:662–671

Van Aken O, Zhang B, Carrie C, Uggalla V, Paynter E, Giraud E, Whelan J (2009) Defining the mitochondrial stress response in *Arabidopsis thaliana*. Mol Plant 2:1310–1324

Vanavichit A, Tragoonrung S, Toojinda T, Wanchana S, Kamolsukyunyong W (2008) Transgenic rice plants with reduced expression of *Os2AP* and elevated levels of 2-acetyl-1-pyrroline. US patent No. 7, 319, 181

Vanlerberghe GC, Cvetkovska M, Wang J (2009) Is the maintenance of homeostatic mitochondrial signaling during stress a physiological role for alternative oxidase? Physiol Plant 137:392–406

Vercesi AE, Borecký J, Maia Ide G, Arruda P, Cuccovia IM, Chaimovich H (2006) Plant uncoupling mitochondrial proteins. Annu Rev Plant Biol 57:383–404

Verslues PE, Agarwal M, Katiyar-Agarwal S, Zhu J, Zhu JK (2006) Methods and concepts in quantifying resistance to drought, salt and freezing, abiotic stresses that affect plant water status. Plant J 45:523–539

Wang X, Zhao X, Zhu J, Wu W (2005) Genome-wide investigation of intron length polymorphisms and their potential as molecular markers in rice (*Oryza sativa* L.). DNA Res 12:417–427

Wang Z, Jia Y, Rutger JN, Xia Y (2007) Rapid survey for presence of a blast resistance gene Pi-ta in rice cultivars using the dominant DNA markers derived from portions of the Pi-ta gene. Plant Breed 126:36–42

Wang C, Chen S, Yu S (2011a) Functional markers developed from multiple loci in *GS3* for fine marker-assisted selection of grain length in rice. Theor Appl Genet 122:905–913

Wang J, Rajakulendran N, Amirsadeghi S, Vanlerberghe GC (2011b) Impact of mitochondrial alternative oxidase expression on the response of *Nicotiana tabacum* to cold temperature. Physiol Plant 142:339–351

Wang S, Yang J, Zhang M (2011c) Developments of functional markers for *Fom-2*-mediated fusarium wilt resistance based on single nucleotide polymorphism in melon (*Cucumis melo* L.). Mol Breed 27:385–393

Watanabe A, Nakazono M, Tsutsumi N, Hirai A (1999) AtUCP2: a novel isoform of the mitochondrial uncoupling protein of *Arabidopsis thaliana*. Plant Cell Physiol 40:1160–1166

Watanabe CK, Hachiya T, Terashima I, Noguchi K (2008) The lack of alternative oxidase at low temperature leads to a disruption of the balance in carbon and nitrogen metabolism, and to an up-regulation of anti-oxidant defense systems in *Arabidopsis thaliana* leaves. Plant Cell Environ 31:1190–1202

Wei B, Jing R, Wang C, Chen J, Mao X, Chang X, Jia J (2009) *Dreb1* genes in wheat (*Triticum aestivum* L.): development of functional markers and gene mapping based on SNPs. Mol Breed 23:13–22

Weng J, Gu S, Wan X et al (2008) Isolation and initial characterization of *GW5*, a major QTL associated with rice grain width and weight. Cell Res 18:1199–1209

Wood TE, Burke JM, Rieseberg LH (2005) Parallel genotypic adaptation: when evolution repeats itself. Genetica 123:157–170

Wu D, Qiu L, Xu L et al (2011) Genetic variation of *HvCBF* genes and their association with salinity tolerance in Tibetan annual wild barley. PLoS One 6:e22938

Xiao H, Tattersall EAR, Siddiqua MK, Cramer G, Nassuth A (2008) *CBF4* is a unique member of the *CBF* transcription factor family of *Vitis vinifera* and *Vitis riparia*. Plant Cell Environ 31:1–10

Xie X, Lu J, Kullbokas EJ, Golub T, Mootha V, Lindblad-Toh K, Lander ES, Kellis M (2005) Systematic discovery of regulatory motifs in human promoters and 3'-UTRs by comparison of several mammals. Nature 434:338–345

Xing Y, Zhang Q (2010) Genetic and molecular bases of rice yield. Annu Rev Plant Biol 61:421–442

Xu ML, Melchinger AE, Xia XC, Lübberstedt T (1999) Highresolution mapping of loci conferring resistance to *Sugarcane mosaic virus* in maize using RFLP, SSR, and AFLP markers. Mol Gen Genet 261:574–581

Yahiaoui N, Srichumpa P, Dudler R, Keller B (2004) Genome analysis at different ploidy levels allows cloning of the powdery mildew resistance gene *Pm3b* from hexaploid wheat. Plant J 37:528–538

Yahiaoui N, Brunner S, Keller B (2006) Rapid generation of new powdery mildew resistance genes after wheat domestication. Plant J 47:85–98

Yahiaoui N, Kaur N, Keller B (2009) Independent evolution of functional *Pm3* resistance genes in wild tetraploid wheat and domesticated bread wheat. Plant J 57:846–856

Yaish MW, Colasanti J, Rothstein SJ (2011) The role of epigenetic processes in controlling flowering time in plants exposed to stress. J Exp Bot. doi:10.1093/jxb/err177

Yamaguchi-Shinozaki K, Shinozaki K (1994) A Novel cis-acting element in an *Arabidopsis* gene is involved in responsiveness to drought, low-temperature, or high-salt stress. Plant Cell 6:251–264

Yan CJ, Yan S, Yang YC, Zeng XH, Fang YW, Zeng SY, Tian CY, Sun YW, Tang SZ, Gu MH (2009) Development of gene-tagged markers for quantitative trait loci underlying rice yield components. Euphytica 169:215–226

Yang W, Liu XD, Chi XJ et al (2011) *Dwarf* apple *MbDREB1* enhances plant tolerance to low temperature, drought, and salt stress via both ABA-dependent and ABA-independent pathways. Planta 233:219–229

Yeam I, Kang B, Lindeman W, Frantz J, Faber N, Jahn M (2005) Allele-specific CAPS markers based on point mutations in resistance allele at the *pvr1* locus encoding eIF4E in *Capsicum*. Theor Appl Genet 112:178–186

Yeam I, Cavatorta JR, Ripoll DR, Kang BC, Jahn MM (2007) Functional dissection of naturally occurring amino acid substitutions in eIF4E that confers recessive *Potyvirus* resistance in plants. Plant Cell 19:2913–2928

Yoshiba Y, Kiyosue T, Katagiri T et al (1995) Correlation between the induction of a gene for D1-pyrroline 5-carboxylate synthetase and the accumulation of proline in *Arabidopsis thaliana* under osmotic stress. Plant J 7:751–760

Zavattieri MA, Frederico AM, Lima M, Arnholdt-Schmitt B (2010) Induction of somatic embryogenesis as an example of stress-related plant reactions. Elec J Biotechnol 13(1), ISSN: 0717-3458

Zeigler RS, Thome J, Nelson J, Levy M, Correa F (1994) Linking blast population to resistance breeding: a proposed strategy for durable resistance. In: Zeigler RS, Leong S, Teng PS (eds) Rice blast disease. CAB International, Wallingford, pp 267–292

Zhang Z, Gerstein M (2003) Patterns of nucleotide substitution, insertion and deletion in the human genome inferred from pseudogenes. Nucleic Acids Res 31:5338–5348

Zhang J, Kumar S (1997) Detection of convergent and parallel evolution at the amino acid sequence level. Mol Biol Evol 14:527–536

Zhang FZ, Wagstaff C, Rae AM et al (2007) QTLs for shelf life in lettuce co-locate with those for leaf biophysical properties but not with those for leaf developmental traits. J Exp 58:1433–1449

Zhu J-K (2001) Plant salt tolerance. Trends Plant Sci 6:66–71

Index

A
Affinity purification based methods, 255–256
AFLPs. *See* Amplified fragment length polymorphisms (AFLPs)
Allele effects, 340
Allogamous crops, 351
Amplified fragment length polymorphisms (AFLPs), 6, 24, 45, 46, 123, 232–234, 254, 260, 265, 269, 270, 314, 354, 355, 359, 360, 362, 363, 379, 380, 393
Association mapping, 52–55, 106, 119–137, 235, 237, 241, 330–332, 389, 406, 410, 422, 424, 425
Autogamous crops, 351, 382

B
Best linear unbiased prediction (BLUP), 361
Best-parent heterosis (BPH), 357
Biochemical and molecular techniques (BMT), 378, 379, 383
Bioinformatics, 56, 61, 109, 135, 218, 219, 224, 258, 282, 292, 302, 317, 432
Biotic stress, 25–26, 428–431, 443–445, 455, 486–498
Bisulfite conversion, 252, 255, 256, 258, 259
BMT. *See* UPOV Working Group on Biochemical and Molecular Techniques and DNA-profiling in particular
BMT Model 1, 2 or 3, 378, 383–397
Breeding lines, 241, 250, 274, 296, 488
Breeding programs, 4, 13–15, 20, 28, 29, 53, 55, 56, 91, 109, 110, 206, 234, 238, 242, 259, 265, 270, 272, 273, 275, 282, 293–297, 301, 302, 314, 341, 349–351, 356–358, 362, 363, 374, 405, 412, 416, 421, 424, 425, 451, 479, 483, 494

C
Cell-reprogramming, 498–500
Characteristics, 45, 72, 153, 188–189, 197, 200, 202–204, 265–269, 271, 272, 274, 288, 292, 297–299, 301, 302, 350, 358, 369–371, 374–378, 382, 383, 385–388, 391, 394, 397, 419, 456, 491, 497
Chromatography, 287, 291
Cis-and *trans*–regulation, 362
CNV. *See* Copy number variation (CNV)
Coefficient of parentage, 352, 354
Combining ability, 350, 358, 360
Common catalogue, 377
Community Plant Variety Office (CPVO), 372, 386, 389, 393, 398
Comparative genomics, 14, 136, 211, 212, 218–220
Conserved functional polymorphisms, 468
Copy number variation (CNV), 135, 206, 432
CPVO. *See* Community Plant Variety Office (CPVO)

D
Database, 20, 21, 42, 43, 48, 56, 71, 215, 216, 220, 284, 285, 287, 289, 301–303, 313–323, 338, 339, 377, 382, 395–397
Diagnostic marker, 7, 211, 222–224, 389, 390
Distinctness, Uniformity and Stability (DUS), 371, 372, 375–377, 379–385, 387–391, 393, 394, 396, 397
Diversity arrays technology (DArT), 132, 232, 236–237, 240–242, 380
DNA based marker systems, 379–381

DNA markers, 5–8, 12, 13, 15–17, 19–22, 24, 25, 27, 29, 127, 212, 221, 230, 260, 265, 268–270, 317–320, 353, 355, 356, 358, 359, 362, 363, 378
DNA markers for varietal identification, 381–383
DNA methylation, 13–14, 251–260, 265, 268, 275
Dominance hypothesis, 357
Doubled-haploid, 331, 358
Double strand break, 169–177
DUS. *See* Distinctness, Uniformity and Stability (DUS)

E
Early testing, 350
EcoTILLING, 145–160
Epistasis, 27, 266, 332–337, 357
Essentially derived varieties (EDV), 377–379
Euclidean phenotypic distances, 352, 354
Evolving marker technologies, 229–242
Expression profiling, 14–26, 269, 270, 360, 361, 363, 412, 424

F
Farmers' Rights, 373
FM. *See* Functional markers (FM)
FNP. *See* Functional nucleotide polymorphism (FNP)
Forward genetics, 41–56, 61, 72, 121, 125, 136, 145
Functional markers (FM), 5, 7, 55, 56, 121, 134, 135, 137, 222–223, 329–341, 383, 397, 406, 411, 412, 414, 415, 420, 424, 431, 444, 446, 456–459, 467–501
Functional nucleotide polymorphism (FNP), 211, 212, 223, 224

G
GAïA, 395
GBS. *See* Genotyping-by-sequencing (GBS)
GCA. *See* General combining ability (GCA)
Gene mapping, 44
General combining ability (GCA), 20, 270, 358–363
General Introduction to the Examination of Distinctness, Uniformity and Stability and the Development of Harmonized Descriptions of New Varieties of Plants, 375, 376

Gene-targeting, 61, 67, 79–80, 167, 169–171, 201
Genetic colinearity, 362
Genetic distance, 16, 17, 19, 24, 265, 268, 269, 273, 350, 352–361, 378
Genetic gain, 339–341, 349, 356, 357
Genetic potential, 281, 339
Genetic similarity, 15, 353–357
Genetic variability, 349, 350, 355
Genetic variance, 28, 333, 349, 351, 352, 354, 356, 359
Gene validation, 50–52
Genome wide association study, 387, 430
Genome wide DNA methylation, 253, 256–257
Genome-wide expression profiling, 14–23
Genomic selection (GS), 6–8, 28, 29, 91, 92, 95, 106, 109, 110, 135, 188, 201, 206, 237, 238, 240, 242, 337, 339–341, 350, 356, 357
Genotyping-by-sequencing (GBS), 49, 106, 201, 202, 204–207, 229–242, 322, 323
Germplasm diversity, 14–15, 24–25
GoldenGate assays, 195, 196, 235–236, 240, 241
Grain yield, 17, 19, 28, 44, 135, 212, 271, 353, 355, 357–361, 363, 405, 410, 416, 422–446, 449, 453, 455, 458
GS. *See* Genomic selection (GS)

H
Hemizygous complementation, 362, 363
Hemizygous genes, 362
Heritability, 21, 132, 136, 137, 260, 298, 349, 361, 446
Heteroduplex, 68, 152, 153, 155, 156, 201
Heterogeneous populations, 383
Heterosis, 7, 13, 14, 16, 17, 19, 20, 27, 54, 132, 135, 136, 265–276, 296, 297, 350, 357–359, 361, 363, 405
Heterotic groups, 16, 20, 350, 357–359
High-throughput SNPs, 49
Homozygous lines, 349
Hybrid
  breeding, 16, 17, 19, 20, 54, 265, 272, 276, 296, 349, 350, 358, 362, 363
  performance, 7, 13, 16–20, 27, 265–276, 349–363
  seed production, 19, 350, 358
  vigor, 260, 296, 357

Index

I

ILP. *See* Intron length polymorphism (ILP)
Immunoprecipitation based methods, 255, 256
Inbred lines, 4, 13, 43, 122, 214, 267, 296, 329, 350, 393, 405, 445, 475
Insertion/deletion (InDel), 5, 45, 123, 146, 170, 175, 188, 211–217, 221, 224, 236, 314, 329, 330, 332, 407, 412–415, 418, 420, 423, 426, 432, 447, 452, 457, 468, 469, 471, 474, 479, 480, 482, 484, 485, 490, 492, 497
Insertion-deletion (indel) polymorphism, 211, 221, 411
Intellectual property rights, 370–373
International Convention for the Protection of New Varieties of Plants, 370, 371
International Treaty on Plant Genetic Resources for Food and Agriculture (ITPGRFA), 373
International Union for the Protection of New Varieties of Plants (UPOV), 369–380, 383, 385, 386, 388–391, 393, 395
Intron, 171, 211–214, 218–221, 223, 224, 332, 386, 412–415, 427, 447, 448, 451, 468, 478–481, 484, 485, 488, 491, 497, 499
Intron-flanking primer, 214, 219, 220, 222
Intron length polymorphism (ILP), 212–215, 217, 220, 221, 224
Intron polymorphism (IP), 211–224
Isozymes, 6, 24, 45, 314, 359, 360, 395

L

Linkage disequilibrium (LD), 7, 19, 53, 107, 119–121, 125, 127, 128, 130, 133, 134, 223, 269, 330–337, 340, 355–357, 359, 363, 384, 413, 432
Linkage drag, 230, 341, 455–457
Low-throughput RFLPs, 230–231

M

Maize (*Zea mays* L.), 48, 118, 119, 212, 405, 410, 469
Marker aided selection, 444, 458
Marker-assisted selection (MAS), 6, 7, 13, 21, 28, 134, 135, 137, 188, 194, 236, 241, 242, 314, 317, 318, 356, 357
Mass spectrometry (MS), 24, 26, 152, 153, 191–193, 202, 286–289, 291, 322
Medium-throughput RAPDs, 231–233

Meganucleases (MNs), 171, 176
Metabolic networks, 26–27, 295, 303
Metabolomics, 11, 13, 26–28, 281–303, 313
Methylation-based markers, 251–260
Methylation-sensitive amplified polymorphism (MSAP), 254, 260
Microarray, 14, 19, 20, 22, 101, 102, 189–192, 194, 195, 221, 236, 258, 267, 270, 272, 313, 314, 361, 380, 427
Microcolinearity, 363
Mid-parent heterosis (MPH), 357–359
MNs. *See* Meganucleases (MNs)
Molecular markers, 5, 6, 12, 21, 41, 45, 46, 53, 55, 125–128, 134, 135, 187, 188, 212, 229–242, 265, 282, 293, 314, 317, 319, 330, 331, 351, 353, 360, 361, 381, 383, 384, 386, 388–390, 393–396, 431, 444, 455, 456, 458, 470, 476
MPH. *See* Mid-parent heterosis (MPH)
mRNA, 11, 14, 20, 51, 213, 266, 267, 274–276, 468, 491, 493
Mutagenesis, 41, 42, 44, 62, 64, 68–74, 79, 81, 146–150, 156, 157, 170–172, 175, 391, 410, 493, 496
Mutants, 41–44, 49, 55, 56, 61, 68, 73, 75, 78, 145–147, 149, 153, 156–158, 160, 169, 170, 255, 257, 288, 297, 316, 332, 339, 377, 411, 420, 421, 424, 443, 446, 448, 470
Mutation detection, 148, 152–158

N

NAM. *See* Nested association mapping (NAM)
National list, 377
Nested association mapping (NAM), 44, 45, 103, 127, 130, 132, 135, 331, 338, 339, 406, 409, 410, 430
Next generation sequencing (NGS), 14, 46, 49, 67–69, 78, 91, 94, 99, 100, 104, 107, 108, 120, 135, 176, 205, 213, 224, 257, 258, 357, 378, 381, 432
Non-DNA markers in plant breeding, 28–29
Non-homologous end joining (NHEJ), 167, 169–170

O

Orthologous gene functions, 476
Overdominance, 266, 357

P

Parental information, 350, 351, 353
Parental lines, 17, 159, 201, 254, 269, 271, 273, 294, 297, 351, 354, 355, 358, 361, 362, 393
PAV. *See* Presence/absence variation (PAV)
PBR. *See* Plant Breeders' Rights (PBR)
PCR. *See* Polymerase chain reaction (PCR)
Plant
  architecture, 406, 444, 447–449, 451, 455, 457, 459
  biomarkers, 281–303
  plasticity, 486, 498–500
Plant Breeders' Rights (PBR), 369–398
Plant Patent Act, 370
Plant variety protection (PVP), 133, 369–372, 374, 383, 384, 394, 396, 397
Polymerase chain reaction (PCR), 4, 12, 45, 46, 62, 64, 67–70, 73, 76, 78, 79, 92, 93, 95, 96, 100–103, 147, 151, 153, 155, 156, 172, 176, 191, 192, 194–197, 199, 200, 202, 205, 206, 211, 212, 214, 215, 218–221, 231, 233, 236, 240, 242, 254–256, 379, 414, 424, 446, 478, 489, 491
Population mean, 128, 349, 352
Prediction, 4, 5, 7, 17–20, 28, 121, 125, 265–276, 281, 282, 292–294, 296, 297, 299, 302, 303, 333, 341, 351–357, 359, 361, 363, 387, 394, 447
Prediction of hybrid performance, 17–20, 265–276, 349–363
Presence/absence variation (PAV), 136, 432
Proteome analyses, 24
PVP. *See* Plant variety protection (PVP)

Q

QTL mapping, 6, 21, 22, 26, 28, 44, 45, 49, 54, 120, 126, 135, 338, 356, 421, 432, 444
QTP. *See* Quantitative trait polymorphism (QTP)
Qualitative trait(s), 44, 123, 130, 134
Quantitative trait loci (QTL), 6, 7, 13, 19–23, 26, 28, 44, 45, 49, 50, 54, 107, 110, 119–126, 128, 134–137, 211, 221–223, 237, 241, 269, 270, 272, 274, 298, 314, 319, 320, 330, 331, 333, 334, 337–341, 355–358, 384, 390, 397, 406, 410, 411, 421, 422, 427–432, 443–449, 451–455, 458, 469, 476, 479, 480, 486, 501
Quantitative trait polymorphism (QTP), 8, 329–334, 337, 405–432, 457

R

Random amplified polymorphic DNA (RAPD), 6, 45, 46, 49, 123, 127, 231–233, 265, 355, 359, 360, 379, 380
Restriction fragment lengths polymorphisms (RFLPs), 6, 12, 45, 46, 229–242, 265, 314, 355, 359–361
Reverse genetics, 41, 50, 52, 55, 56, 61–81, 121, 145, 146, 157, 160
RFLPs. *See* Restriction fragment lengths polymorphisms (RFLPs)
RNAi, 55, 61, 66, 76–78, 80, 146
RNA marker, 269–272, 274, 314

S

Segregating populations, 21, 22, 49, 50, 52, 125, 132, 159, 234, 299, 331, 349–353, 430, 450, 475, 479, 489, 495, 498
Segregation variance, 355
Selection intensity, 349
Selection response, 14, 15, 20, 349
Selective sweep, 384, 423
Sequence-tagged site (STS) markers, 46, 234, 354
Sequencing, 6, 12, 46, 61, 91, 120, 147, 172, 191, 211, 234, 272, 302, 313, 339, 350, 378, 416, 446, 492
Serial analysis of gene expression (SAGE), 14, 361
Simple sequence repeats (SSR), 5, 12, 15, 45, 46, 120, 125, 212, 214, 223, 232–234, 240, 241, 314, 355, 359–361, 379, 380, 382, 393, 394, 443, 446–448, 457, 459
Single-cross hybrids, 19, 350, 359
Single nucleotide polymorphisms (SNP), 5, 12, 45, 78, 102, 123, 151, 187, 211, 234–236, 268, 294, 314, 329, 357, 379, 407, 445, 468
Specific combining ability (SCA), 265, 358–361
SSR. *See* Simple sequence repeats (SSR)
Stress tolerance, 13, 132, 133, 254, 260, 282, 374, 444, 458–500
Super D, 391, 393
Synteny, 50, 222, 469

T

TALE nucleases (TALENs), 80, 81, 177
Targeted mutagenesis, 81, 146, 169, 172
Teosinte, 319, 406–410, 419, 421
Test guidelines, 374–377, 386

Index

TILLING, 52, 61, 62, 68, 69, 78, 81, 125, 145–160
Total effects of associated markers' ('TEAM'), 361
Trade-Related Aspects of Intellectual Property Rights (TRIPs), 370, 373
Trait dissection, 133
Trait performance, 276
Transcriptome, 15–20, 23, 25, 27, 29, 104–105, 235, 265–276, 283, 295, 361–363
Transferability, 126, 218, 219, 234, 273, 333
Transfer-DNA (T-DNA), 41–43, 52, 61, 62, 70–73, 80, 146, 147, 169, 170, 176, 177, 255
Transgressive segregation, 349, 354
Transposon, 41–43, 64, 70–73, 80, 146, 147, 220, 274, 428, 491

U

Ultra high-throughput infinium assay, 232
UPOV. *See* International Union for the Protection of New Varieties of Plants (UPOV)
UPOV Convention, 370–378, 383
UPOV Working Group on Biochemical and Molecular Techniques and DNA-Profiling in Particular (UPOV BMT), 378, 379, 383
Usefulness, 4, 15, 158, 299, 349–363

V

Varietal ability, 349

Y

Yield, 11, 44, 123, 151, 212, 257, 265, 282, 339, 353, 369, 405, 443, 469–481

Z

Zinc finger nuclease (ZFN), 79–81, 146, 169, 171–177, 333, 341

Printed by Publishers' Graphics LLC
DBT130708.15.15.112